CW00447608

THE NAUTICAL INSTITUTE ON THE WORK OF THE NAUTICAL SURVEYOR

First published 1989 by The Nautical Institute,
202 Lambeth Road, London SE1 7LQ, UK.

©The Nautical Institute 1989.
All rights reserved. No part of this publication may be reproduced, stored in a retrieval system, or transmitted in any form or by any means, electronic, mechanical, photocopying, recording or otherwise, without the prior written permission of the publishers, except for the quotation of brief passages in reviews.

ISBN 1 870077 03 2

Although great care has been taken with the writing and production of this volume, neither The Nautical Institute nor the authors can accept any responsibility for errors, omissions or their consequences.

This book has been prepared by The Nautical Institute to address the subject of nautical surveying. This should not, however, be taken to mean that this document deals comprehensively with all of the concerns which will need to be addressed or even, where a particular matter is addressed, that this document sets out the only definitive view for all situations.

The opinions expressed are those of the authors only and are not necessarily to be taken as the policies or views of any organisation with which they have any connection.

This volume was edited by David J. Sanders, Ex.C, FNI, and compiled by C. J. Parker, B.Sc, FNI, under the general direction of The Surveyors' Working Group of The Nautical Institute.
Text set in Baskerville and printed in England by the Silverdale Press, Hayes, Middlesex UB3 3BH.

The Work of the Nautical Surveyor

THE SURVEYOR with nautical qualifications and background can find himself undertaking a wide variety of tasks. He may be employed by a Government, in which case he will be concerned with ship safety and certification, or he may be employed in the private sector, undertaking damage surveys, advising on collision cases, checking the loading of tankers or other duties covered in the text.

There is a difference between nautical and engineering disciplines. In the former case the deck officer is concerned with operating the ship at sea and in port. If this is done successfully there is no damage or other adverse consequences and no need to involve surveyors, so the average deck officer may not have much experience of survey practices. Engineers and naval architects operate in a much more tangible world of structures and materials which can be measured and costed. They are practising established disciplines linked particularly, so far as ships are concerned, to standards created by governments and classification societies that define strength, durability and maritime fitness. Because ship structures suffer from corrosion and engines are subject to wear there is an on going commitment to improved design, more economical construction methods and maintenance.

Nautical-related activities are more sporadic and create less predictable demands which are, by their nature, challenging. They do, however, require much experience for their solution. As opposed to engineer surveyors who are devoting their talents and skills to what is basically another diversification of their primary career, the nautical surveyor is in essence starting a second career and, coming from a shipboard environment, may find it somewhat difficult to find his way around shorebased institutions and practices.

This book on the work of the nautical surveyor has been designed to provide both insight and guidance and to set a professional standard for the nautical surveyor. Clearly, it is not possible to cover every aspect of what is a vast subject, but what we have decided to do is embrace a number of guide papers which demonstrate how a subject can be covered in depth and also other papers which generally illustrate the nature of the job.

In cases where the subject is too complex to be adequately covered here—for instance, cargo surveying—references are given for specialised reading. As a nautical surveyor must work closely with his colleagues in other disciplines a number of papers are devoted to wider aspects of surveying and embrace classification and other technical procedures.

I have much pleasure in paying tribute to the dedicated efforts of the Nautical Surveyors' Working Group and also those other individuals, both in the United Kingdom and other countries, all of whom contributed advice and practical experience to the compilation of what I am sure will be a thoroughly useful source of information. Appreciation is also expressed to the NI secretariat for its invaluable assistance in all the hard labour associated with a project of this nature.

I only wish that this book had been available to me when I left the comparatively sheltered life of a marine pilot to become a Government surveyor concerned with a jungle of regulations and practices relating to nautical standards on merchant ships. I am sure that future nautical surveyors, and indeed existing ones, will find this book of considerable value in providing guidance.

CAPTAIN A. CROMBIE, FNI
CHAIRMAN

Nautical Surveyors'
Working Group

THE NAUTICAL INSTITUTE
NAUTICAL SURVEYORS' WORKING GROUP

Captain J. J. Banister, MCMS, MRIN, MInstPet, ACIArb, FNI
Director, H. H. Bridges & Co. Ltd.

Captain A. Blackham, FNI
Noble Denton Weather Services Limited

Mr L. J. Crighton, C.Eng, FRINA,
Senior Principal Surveyor, Lloyd's Register

Captain A. Crombie, FNI
Former Head of Safety, The Liberian Bureau of Shipping; *Working Group Chairman*

Captain J. A. Cross, Ex.C, MCMS, MCIT, FNI
Director, Severnside Consultants

Mr N. C. I. de Spon, MNI
Marketing Manager, Caleb Brett Services

Captain W. V. Lusted, Ex.C, MBE, FNI
Shipmaster and Consultant

Captain A. Marshall, Ex.C, FNI
Principal Nautical Surveyor, The Salvage Association

Captain M. Macleod, MNI
Consultant, Macleod Marine

Captain J. M. M. Noble, B.Sc, FNI
Managing Director, Murray Fenton and Associates Limited

Mr T. M. Sims, Master Mariner,
Marine Co-ordinator, Coubro and Scrutton Limited

Captain B. Vale, Ex.C, MNI
Principal Nautical Surveyor, Department of Transport, UK

Mr R. I. Wallace, BA, FInstPet, MNI
Chairman, the Calibration Panel, PMAI

Mr C. J. Parker, B.Sc, FNI
Secretary

THE NAUTICAL INSTITUTE ON THE WORK OF THE NAUTICAL SURVEYOR

Introduction, scope of the book, areas of activity and definitions

The Nautical Institute Surveyors' Working Group

FOR ANY INDUSTRY to develop and prosper, it must attract capital resources and labour, adding value through manufacture or service, in line with demand. During this process, the community requires a measure of protection. Those who design, build and operate ships need an established system of limited liability, contractual protection and insurance. Those who entrust their goods to be carried by sea have a right to expect that they will be carried carefully, within the agreed price. The community has a right to expect that ships will not jeopardise life, property or the environment and crews have a right to expect reasonable employment conditions and operation of the ship by properly qualified personnel.

Ships are designed and given classification to withstand the rigours of going to sea—but exceptional conditions, Acts of God and perils of the sea can occur outside the bounds of reasonable experience. Shipping itself is international. It can be used, or misused, for piracy, fraud, embezzlement, pilferage and theft. Corners can be cut to reduce costs in ship operation which are illegal or deliberately exploit any system of permitted exemptions and certification. Not all acts of deceit or negligence are carried out on board ship, but cover-up attempts may be made to shift blame from shore to ship. Ships sometimes suffer damage as a result of internal cargo movement, collision or stranding. There can be general or particular average disputes about cargo condition and arguments about conditions of hire, fuel consumption, performance, delays, strikes, availability of berths, depth of water and so on.

Since all the items given above can give rise to claims or penalties, it is essential that they are properly reported and assessed. This, then, is the primary requirement of all marine surveyors, that they should be capable of assessing, either the condition of a ship, or her propulsive equipment or cargo and be sufficiently experienced to write a report on that condition and recommend remedies as appropriate. This principle involves the work of engineer surveyors, ship surveyors, chemists and other specialist consultants or advisers who have expertise in their own discipline. There is a common background to all surveying practices, which demands a basic knowledge in the appropriate discipline, an understanding of seafaring practices and the ability to record what has been seen accurately and concisely.

The **Nautical Surveyor** fits into this pattern, having direct involvement in ships' safety, navigation, ships' business and other deck department responsibilities, such as cargo stowage, lifesaving equipment, firefighting and general regulation of a ship's operation. At a higher level, those in authority must have the ability to appreciate the implications of survey reports and decide on the correct course of action. In many or most cases, a discussion will take place as most major incidents involve the skills and attributes of several disciplines. Should a dispute arise that cannot be settled on the basis of the reports, then the issues must be prepared for arbitration or settlement in a court of law.

The expertise of the surveyor will depend upon his training, professional development and subsequent experience. As with other disciplines, the nautical surveyor will be called upon to extend his 'core' knowledge so that he can deal competently with tasks that overlap other experts.

For instance, as well as giving advice on navigational matters, he may be called upon to survey a lifeboat (although he is not a boat builder), or verify the performance of a radar set (although he is not an electronics engineer), as well as giving expert advice on the stowage of cargo, he may be called upon to check the ship's stability (although he is not a naval architect); or verify the performance of a fire pump (although he is not an engineer). He should be however, above all, an authority on the practical operation of a ship. This is not to say that every nautical surveyor can be expert or competent in all fields; he must know, and let those who appoint him know, what his limitations are.

The nautical surveyor is usually a qualified master mariner, or equivalent, who has had practical experience in the day-to-day management of a ship; this 'core' knowledge is supplemented by further expertise in handling cargo and other maritime skills, according to his speciality. He will be able to balance safety considerations against commercial demands and provide survey and advisory services, within his competence and authority, in the fields listed below. He must often co-operate with other specialists in doing this.

Fields within which the nautical surveyor contributes

I Statutory

- Crew competence and officer certification.
- Safety equipment.
- Solas, Marpol and ILO requirements.
- Port State control.
- Navigational equipment verification.
- Passenger occupancy verification.
- Loading instruments verification.
- Casualty investigations.
- Safe loading of certain cargoes, such as grain, dangerous goods and deck cargo.
- Cargo gear surveys, annual, quadrennial and occasional.
- Survey of pilot ladders, gangways and navigation lights.

II Official and semi-official

- Advice on safe working practices, and contingency and emergency organisation.
- Compliance surveys of cargoes governed by rules, codes or advisory notices, such as grain, timber, bulk cargoes, dangerous goods, hazardous cargo, vehicles, steel products, livestock.
- Survey of hatch covers and tank lids.
- Practical interpretation of local laws and bylaws.

III Contractual

- On and off hire surveys.
- Draft surveys.
- Damage surveys.
- Surveys for cargo worthiness, including hold condition, tank conditions and cleanliness, ventilation, bilge arrangements, etc.
- Heavy-lift loading and stowage surveys.
- Hull and equipment condition surveys.
- Ship warranty surveys.
- Accident investigations.
- Cargo pumping and inert gas system efficiency surveys.
- Cargo handling and stowage advice, including advice on draft, stability, trim, safe carriage and discharge.
- Cargo condition surveys for insurance purposes.
- Cargo damage surveys for particular and general average and salvage awards.
- Meteorological and routeing advice; speed clause advice.
- Navigation and berthing advice.

IV Planning

- Advice on the value of navigational and operational equipment.
- Cargo-handling advice.
- Advice on the efficient use of the crew, taking into account their safety and welfare.
- Safety organisation advice.
- On the job training advice.
- Advice on safety in drydocks and other special situations.
- Advice on health, safety and hygiene on board.
- Advice on catering.
- Advice on hull and equipment maintenance.
- Advice on economic deck storing.

There may be other specialist categories of surveyors, not mentioned here. In practice, many surveyors, especially those in the private sector, have expertise in more than one discipline and can act accordingly, provided there is no conflict of interest.

Legal status of surveyors

Government nautical surveyors have authority and powers granted to them by the laws and statutes of their national government, by virtue of their status as State servants. They invariably have an absolute right to go on board their own-flag ships and foreign-flag ships within the surveyor's national jurisdiction, to the extent prescribed by national law. Government surveyors would, for instance, have statutory powers to detain a ship and prevent her from sailing, if that ship were suspected of being unseaworthy.

Private sector surveyors have no statutory rights of access to any ship and no powers of arrest or detention. Their activities are governed by their relationship, or the relationship of those who may employ them, to the circumstances of the case in hand. However, in all cases the surveyor is bound by the laws of the country in which he is operating. For instance, a cargo gear surveyor who issued a fitness certificate for cargo gear that subsequently failed and caused injury or death might find himself in a Court of law charged with anything up to manslaughter. A surveyor trying to conduct an inquiry might find himself, in some countries, in conflict with national law and end up in serious trouble.

Competent surveyors ensure that they are fully conversant with any laws that may apply to their particular activities and thus avoid problems. This is another argument in favour of experience. Private sector surveyors will normally obtain professional indemnity insurance. If employed on behalf of a ship or shipowner, the private sector surveyor may encounter no difficulties of access to a ship, but if employed or appointed by interests at variance to the best interest of the ship's interests, as might happen in the course of litigation, he might well be advised to exercise a great deal of care in his approach to the problem.

Definitions of surveying activities

It is implicit in these descriptions that all surveys will be conducted by competent persons—competent in the sense of being legally qualified, or alternatively being adjudged competent by an employer or the instructing principal.

Care should be taken when using terms, because customary usage can give rise to different interpretations. For example, an inspector in government has statutory powers and may detain a ship or require deficiencies to be made good. Traditionally, shipping inspectors are generally termed 'surveyors,' but this is not always so. In classification, the authority of an inspector is limited to verification and assessment. A classification surveyor, on the other hand, has discretionary management powers, can sign certificates and can instigate alternative courses of action.

Classes of survey

Marine or nautical surveys—Any survey concerned with the oceans or seas, ships, shipping, ships' cargoes, seagoing personnel, navigation, marine structures, and including oceanographic surveys.

Statutory surveys—Any survey conducted by authorised person(s) as required by Statute or by Statutory Instruments, or as required by the laws of the flag or host state.

Examples of these surveys include: tonnage measurement, loadline, port State control, safety equipment, cargo/passenger ship construction, radio installations, cargo gear/lifting appliances.

Commercial surveys—Any survey undertaken upon instructions received from a maritime commercial interest, and will thus include any survey other than statutory surveys.

Examples of these surveys include: cargo, marine engineering, hull/ship, ship condition, fixed and floating objects, container condition, cargo securing, cargo stowage, sale and purchase, on/off hire, bunker, warranty.

Ship/hull surveys—Any survey of the hull, structures, fittings, compartment layouts, weather decks of a vessel or other marine structure.

Marine engineering surveys—Any survey of the machinery, electrical installations, and performance trials of services and equipment, including air conditioning, ventilation, refrigeration arrangements, windlasses, winches, capstans, cranes, etc., of a vessel or other marine structure.

Dry-cargo surveys—Any survey of dry—i.e., as opposed to liquid—cargo to be carried, on board, having been discharged from, a vessel; to establish condition, value, suitability for shipment, damages which might have been incurred, and also cargo securing, cargo stowage.

Bulk liquid cargo surveys—Any survey to establish quantity and quality of mineral oils, petrochemicals, vegetable oils, fish oils, molasses, or any other bulk liquid loaded to, or discharged from, a vessel.

Draft surveys—Surveys to establish the weight of cargo loaded to, or discharged from, a vessel by comparing drafts.

Container condition surveys—Surveys to establish the structural condition of cargo containers, to provide a description of the damages which affect structural integrity together with an estimate of the repair costs.

On/off hire surveys—Surveys of vessels or marine structures undertaken either before the vessel or structure is delivered into a charter, or redelivered from a charter. The survey report should include a detailed description of a vessel's cargo spaces/deck areas structural condition, cargo space cleanliness, bunkers on board, listing of the vessel's statutory certificates, portable cargo securing equipment, etc.

Sale and purchase surveys—Any survey of a vessel or marine structure undertaken before a contract of sale is completed, and which should carefully detail condition of machinery and hull, and list an inventory of all equipment, bunkers and stores on board which are to be included in the sale.

Bunker surveys—Any survey to establish the quantity, and in some instances the quality, of the oil, fuel and fresh water remaining on board a vessel.

Condition survey—A survey of a vessel's machinery and/or hull in particular, or generally, and of the equipment on board to establish the condition, and perhaps fitness to trade of the vessel. Often required by Protecting and Indemnity (P&I) Clubs, prior to a vessel being entered with the Club, or at regular intervals during entry, or alternatively at the time any vessel reaches a predetermined age—sometimes 16 years.

Warranty surveys—Surveys which are usually a requirement of hull or cargo underwriters, and which are conducted prior to tug and tow, or specialised heavy-lift vessel/semi-submersible, being used to transport a large structure or vessel.

Fixed and floating object surveys—A term taken from the relevant rule of P&I Clubs to describe surveys to quays, wharves, jetties, dolphins or other fixed marine objects, buoys, pontoons, barges, lightvessels, or other floating marine objects—which may have sustained damages by collision or some other cause.

Recommissioning surveys—Surveys usually required under P&I Club rules when vessels have been laid up for a period of six months or more, and undertaken prior to the vessel leaving the place of lay-up, to ensure that cover under Club rules should be maintained.

Classification surveys—Any survey undertaken by an exclusive surveyor, who is directly employed by a classification society, or by a non-exclusive surveyor, whose company is contracted to a classification society, for the purpose of classifying a vessel or marine structure. The surveys cover the ship's hull and machinery including the electrical plant. Examples of the subject surveys are:

- Annual surveys of hull and machinery.
- Intermediate surveys, especially for tankers having attained an age of ten years.
- Drydocking surveys.
- Class renewal surveys.
- Hull and machinery periodical surveys.
- Special tests of machinery parts and equipment.
- Class extension surveys.
- Continuous class renewal surveys.
- Damage surveys; hull, machinery, including electrical equipment.

Other services—Marine surveyors and consultants will be asked to provide many other services, which may include investigations and opinions in respect of marine accidents, personal injury, ship collisions, charter-party disputes, cargo losses, ship losses, as well as advice in superintendency, ship's husbandry, ship management, ship economics, etc.

Survey companies

Historically, commercial survey companies associated with the maritime industry were categorised as follows:

Marine surveyors—Those companies who were traditionally, and only, employed/retained by ship owners, charterers, carriers of cargo, or their Protection and Indemnity (P&I) Club, or alternatively by marine lawyers representing the listed interests.

Underwriters surveyors—Those companies who were traditionally, and only employed/retained by the hull or cargo underwriters, or alternatively by lawyers acting on those interests' behalf.

This demarcation of the profession has, since the difficulties experienced during the late 1960s/early 1970s become eroded. Some would argue that it no longer exists, commercial surveyors now accepting instructions from whomever and wherever possible.

Some legal entities

Nautical assessor—One who assists the Admiralty Court or House of Lords at the Court of Appeal, in trying a nautical question, but who has no voice in the decision. Appointments to the panel of assessors are made by the Master of the Rolls, in practice, after consultation with the Admiralty Registrar. Nautical assessors may either be Elder Bretheren of the Corporation of Trinity House, or active or retired officers of Her Majesty's Navy, or Merchant Navy masters.

Experts—A person with special skills, technical knowledge or professional qualifications whose opinion on any matter within his cognisance is admitted in evidence. The decision as to the qualification of a witness to give evidence as an expert is made by the judge or arbitrators.

This introduction sets the scene for the content of this book which has been designed to provide both insight and guidance and to set a professional standard for the nautical surveyor. Clearly it is not possible to cover every aspect of what is a vast subject, but the working group decided to embrace a number of guide papers which demontrate how a subject can be covered in depth and a representative selection of exemplary contributions to illustrate the nature of the job.

The working group believe this book and training programme will be used widely by those seeking a second career in nautical surveying. Considerable effort has been made to encompass a broad and detailed examination of the subject which will prove valuable to the experienced surveyor also.☐

MANAGEMENT AND ADMINISTRATION FOR NAUTICAL SURVEYORS

C. J. Parker, B.Sc, FNI

Secretary, The Nautical Institute. Produced with the assistance of members of the Nautical Surveyors' Working Group

THE SEAGOING EXPERIENCE so necessary for nautical qualifications is not an infallible guide to working ashore. In this feature, I intend to touch on a variety of topics which are not so obvious and may trap the unwary seafarer, simply because he makes certain assumptions about office life based upon the way ships work at sea.

Organisation

Work on board ship often follows a pattern which does not require too much organisational effort; tradition, custom of the trade, the 24-hour routine, all ensure that little thought needs to be given to this aspect of sea life. Compare this with a medium surveying company employing 25 people, where new work has to be found, contacts made, surveyors allocated, the office covered, arbitrations slotted in, reports to be verified, fees to be collected, correspondence to be completed, cash in hand to be found, investment for the future to be available, surveys to be conducted all over the world in different locations and different times, holidays, and so on.

In a smaller company, staff will tend to cover all sorts of activities, whereas larger organisations tend to have specialist departments. The two administrations will be different and as with small ships and large ships, the larger will be much more formal. In a large organisation, it can take weeks to have a meeting with the managing director. The knack is to work with diaries, 'phone secretaries, get dates pencilled in and plan ahead.

On a ship, most people know what is going on and, because it is a close community, there is always somebody to ask. This is not so ashore, and a substantial amount of time and effort has to be given to keeping everybody else informed about what is happening. Most management books claim that working in an organisation is 85 per cent perspiration and 15 per cent inspiration, and real creativity may be as little as 5 per cent of the job.

There are many different styles of management. Some firms produce organisation manuals, but the majority do not and so it is necessary to enquire. The variety of work in a surveying company demands a flexible and willing approach to work.

Some managers hold on to their positions (real or imaginary) by controlling all the information flow to and from their departments. Many consultancy firms jealously guard their expertise and are very selective in what they release. 'Knowing the ropes' ashore is not always straightforward. For example, there are executive secretaries who are formidable, who think they know more (and sometimes do) about running the company than the boss. Get on the wrong side of them and life can be very difficult. At the other end of the scale, the secretary and filing clerk although the lowest paid can be potentially the most important people in your life. An occasional kind word of appreciation can go a long way.

Again, there is the government administration where people are employed to organise departments (Principals and Secretaries) and as a civil servant you comply with the rules, and the Official Secrets Act. This Act puts severe restrictions on the way information can be used.

Coming ashore is not only about learning a new job, it is also learning about the organisation. A good employee should contribute to both.

Communication

In any thriving business 'time' is the enemy and business communications must be conducted with this in mind. In the old Blue Funnel all the directors and heads of departments met every day at 12.00 for a ten-minute stand-up meeting, to review all the major points relating to a fleet of 70 ships. The stock market amplifies this system with computer data bases and telephones.

On a ship everybody knows who takes the decisions. Whilst there may be a few ditherers, most masters and chief mates confidently and competently take decisions as they come. This is quite different from justifying the employment of new members of staff, committing a client to litigation or deciding whether or not to take on a contract. In most surveying companies, where ex-sea staff abound, taking decisions is not usually a problem. Learning about the commercial implications of decisions, however, does take time.

Telephone—The advice may sound trite, but use the telephone properly. As the company's first point of contact with the outside world, it is essential to give a quick, courteous and helpful response. Most telephone systems have a hold button, so that an outside line can be held whilst an internal call is made. Go and *see* the switchboard and find out how to ensure that a call is properly transferred.

Telex—Most important commercial communications involve telex. If you are not familiar with it, see the operator and become familiar with the routine for sending and distributing these. Vital business has been lost on many occasions because the content of telexes were not acted upon by a manager away from the office!

Electronic mail and data processing—Increasingly computer-generated communication systems are becoming more widespread. The advantage is the use of high-speed data links which are more economical than telex. It costs about £2.50 to send international telex and £6 per minute to send high-speed data. Since the high-speed data has a density 40 times that of telex, it makes sense to use it if the company has enough traffic.

Fax—This is a popular instant form of communication, but it is relatively expensive for routine traffic. As with all forms of communication, the more concise the better.

Letters—These are still important whenever written confirmation bearing a signature is required. If you take office space, secretaries' salaries, stationery, stamps and mail room space into account, a letter costs between £3 and £5 to produce—not a resource to be wasted.

Reports—There are two types of report: those designed for internal consumption, and clients' reports. The first rather obvious point to make is that reports are not written for the benefit of the writer, but for the receiver. A report may be read by many people thereby taking up a lot of management time and it may be read unnecessarily by others. Put in this context, subject to company format, the first page inside the cover should be an abstract of the main points and who they are addressed to.

Next should follow the executive summary (written when the report is complete) and then the main body of the text. If it is a long report, readability is important with nicely headed and/or numbered sections. There should be short sentences, no frills, starting with an overview and working to conclusions.

Written like that, the key points can be quickly assimilated; those who do not need to read will not waste their time, and those who do get the picture and can understand the content.

Writing reports for clients is somewhat of an art form. Communication may not be top priority; instead, legal protection, getting more work, sheer eye-wash and a variety of other factors may play their part. The tricks of the trade have to be learnt. Nevertheless, most marine survey companies thrive on the standard of their reports and require them to be approved by a senior partner. Never be too proud to offer a report for comment before it is sent out.

As a general rule, properly quantify all statements and accurately specify details. It is always much better to spend time and effort getting all the relevant facts into the first document. Invariably the work has to be done, particularly if there is a claim involved.

All new surveyors should understand that the end result of most instructions will be a report. The speed at which this report is produced might well prove to be one of the criteria on which the consultant is judged. Time and effort spent learning to touch type is a priceless asset.

Office management—The object of good office management is to make the organisation as efficient as possible. The ability to access the right file which contains all relevant information in the shortest time demands organisation, discipline and adherence to common standards. The balance between communal and personal files can be tricky, but is easily resolved if the operational efficiency of the company is put first! Filing, like photocopying, is a soul-destroying job. Every office worker must follow the 'little and often' rule and share in the long and tedious jobs whenever possible.

Computers

One of the first applications of the microprocessor was for word processing. The value of having this ability has been recognised early in survey and consultancy firms, but less widely used are computers for general administration. Many firms have installed computers and do not use them. One firm known to the author installed a system for invoices and stock control and could not get it to work. After 18 months they went bust.

A computer simply holds and provides access to a data base. Depending upon the software, it can manipulate the data on command and make calculations. It can print out items quickly and be used for communications. After word processing, surveyors probably want a client data base and fee invoice system. Larger firms will want staff and payroll systems and, of course, technical staff will want their own specialised systems for evaluation and research.

Getting the right system is most important and demands a lot of time and discussion. It is amazing how consultants can advise clients on their needs and fail to evaluate properly their own requirements! Because computers influence the way data are handled throughout an organisation, everybody has to be involved when choosing a system.

The operation has to be planned like a military exercise: consult staff, draw up rough specification, hire consultant to complete the specification and put forward a number of options costed and maintained, confirm with staff, draw up detailed implementation timetable, include formal testing in the computer contract for both hardware and software. Work out the test routines to ensure as many bugs as possible are picked up before the system is operational. Keep a check on deadlines and ensure staff have ample time to compile the data base.

Make provision for training and ensure that the system is adaptable to future demand. Network systems are not cheap. The Nautical Institute's administrative system for 5,000 members cost £30,000 in 1986. It has not saved the Institute any money directly, but has enabled us to do 30 per cent more work without increasing the staff. Computer systems impose their own schedules and disciplines. If you are expected to use the system, then comply.

Sleuthing

All surveyors are employed to find out what is happening or what happened or what might happen. In a competitive environment, not surprisingly, information is seldom provided on a plate and techniques have to be developed to get behind the facade and find out the facts. Observation is of course the best form of data, supported by film and accurate measurements. However, following a collision or cargo damage, for example, the jig-saw has to be pieced together.

Direct questioning is unlikely to produce all the answers either and so techniques have to be developed which are more the stock in trade of intelligence officers rather than ships' officers.

Like navigation, it is important to be prepared and plan ahead (not always possible). Where scanty information is available, only intuition (experience based) and logic can help and it is essential that as much detail is assembled either formally or 'at the bar' to help piece the jig-saw together. The problem is made worse by witnesses who have come to believe a different story to what actually happened (see Dyer Smith on the human factor and De Coverly on investigation). Also on many occasions, official information is withheld, particularly when representing cargo interests, so a good knowledge of the trade and practices is esssential when examining certain cases.

Under cross-examination, one experienced consultant—who was asked by counsel how he undertook his work—replied 'by using my five senses.' Even better advice might be to 'keep one's eyes and ears open, and one's mouth firmly closed.' Never forget that a consultant may well have privileged information that his clients would not wish other parties to learn. A good consultant should be able to retain this confidentiality, and conversely gather all relevant information from other parties involved, for the benefit of his own clients.

Problem solving

The major part of a surveyor's job is in essence problem solving. Like all attributes, you either have it or you do not, and this particular facility probably more than any other determines the aptitude of those who choose surveying as a second career from those who elect to do something different. Having said that, it is always possible to improve performance by study and reflection.

Consistently good surveyors develop a professionalism towards problem-solving which is systematic and thorough. Not surprisingly Captain Jack Isbester, who is chairman of The Nautical Institute's Technical Committee, found his metier in surveying, and it is worth quoting from a paper on this subject that he wrote at sea when he was a master:

'I joined a reefer ship in the banana trade as master some years ago. When loading a full cargo of bananas we were required to reduce the cargo return temperatures as quickly as possible to 57 degrees F, and when we had achieved this we had to report our achievement by cable to the ship operators. I soon learnt that, whilst 14 of the compartments performed as required, the fifteenth normally took almost twice as long to reduce to the required temperature. No-one was able to offer an explanation for this fact, which I discussed with chief mate, chief engineer and reefer engineer, all of whom had served in the ship for some time. They told me that the symptom had been in evidence for many voyages past—what they actually said was that it had 'always' been like that—and they avoided the embarrassment of reporting the very long cooling down time for compartment 3D by simply ignoring it. Although it took 56 hours to reduce compartment 3D to temperature the routine was to report all compartments down to temperature after 36 hours or so. I fell in with this procedure on the first loading, but decided that the explanation had to be found.

'This was the last occasion that I served on a ship with cadets and, as an exercise, I drew up with their help a list of all possible reasons to explain the symptom. I can't remember them all now, but they included: air delivery trunking blocked; insulation perished; cargo stacked too high and preventing air circulation; defective thermometer; and so on. We had about a dozen alternatives in all.

'When we began to check them all we discovered that the starboard return air fan was rotating in the wrong direction, having been wired up wrongly. Since it was not designed to be reversible there was virtually no flow of air past it! It took the reefer engineer no more than five minutes to readjust the wiring correctly and in subsequent voyages we were able to report a 30-hour reduction period in all honesty—which is much the best way, and much nicer for the bananas.

'I also recall a large bulk carrier in which I served some years ago. Her hatches were opened and closed by endless chain whips located one on each side of each hatch. A motor drove a pair of gypsies which rotated to drag the chains along the hatch coamings. The chains were secured to the leading hatch cover and thereby pulled the covers open or shut. When I joined the ship as mate a programme of renewing the hatch chains was being carried out by ship's staff.

'As the renewals progressed, the operation of opening or closing a hatch became more and more hazardous. Something was clearly badly wrong. Hatch panels which weighed ten or 20 tons apiece—it may have been more—regularly became slewed in their tracks, and righting them presented enormous problems. The second engineer, a very experienced and conscientious officer who was in charge of the renewals, was in despair. We suspected him of measuring the new chains inaccurately when he fitted them. When he convinced us that his measurements were accurate the ship's company in general developed the theory that the ship had been distorted in bad weather!

'This problem was another which responded to a rigorous and methodical examination of all the facts. We discovered by these means that the new chains were not a matched pair, as they should have been, and their links were slightly different in length. Fifty links of chain passing over the port gypsy pulled the port side of the hatch lid 1.6 metres, say, as against the 1.5 metres travelled by the corresponding 50 links of chain passing over the starboard gypsy.

'Astonishingly, the makers had supplied us with eight dissimilar lengths of chain. Once we knew this we were able to sort out three reasonably well matched pairs, and to return the two remaining ones to the makers for replacement.

'My final example on this theme concerns an observant chief mate who noticed that the run of the sounding pipes within a double-bottom ballast tank was not vertical. The pipes ran down the ship's sides and followed the turn of the bilge until they eventually terminated at the deepest point in the tank, near midships, by which time they were only 30 degrees or so from the horizontal. Returning to his cabin he was able to confirm his suspicion that the ship's calibration tables were compiled on the false assumption that the sounding pipes passed vertically through the tanks. A 20 centimetre sounding in the sloping pipe corresponded only to an 8 cm depth of water, but the calibration tables assumed a 20 cm depth of water, and greatly exaggerated the tonnage of strippings being carried.

'The result of this was that voyage after voyage (the ship was eight years old at the time) the draft surveys carried out by chief mate or cargo surveyor would have overestimated the tonnage of strippings remaining unpumped, and correspondingly underestimated the tonnage of cargo carried.

'When these facts had been verified we concluded that our first task was to take measurements and compile a set of amendments to the calibration tables. We then had to set out the information in a form which would be easily understandable to our management and our sister ships, and which would carry

conviction when presented to surveyors.'

To summarise: (i) diagnose the problem; (ii) assess the most appropriate solution; (iii) implement the change; and (iv) monitor the results.

Problems seldom appear individually. It is usually more effective to tackle the small ones first and build up a response to larger ones later. A complicated case will demand teamwork.

Upholding the law

The function of the government nautical surveyor, though including a good deal common also to commercial surveyors, must be seen against the background of the law. Upholding the law can be an unpopular function, when, for example, it leads to punitive action, but it is by no means always so. Often the surveyor's insistence on proper standards, the rectification of deficiencies and so forth, is seen as a welcome support for hard-pressed masters, beset by many pressures on the one hand, and lack of resource on the other.

However, the surveyor will very often have to deal with cases where pure and simple application of legal requirements does not suffice. Despite the volume of maritime law, it by no means covers every circumstance, nor does it always keep pace with developments. There is much that is obsolete or at least obsolescent, and changing legislation is a ponderous process. To carry out his work fairly and thoroughly, the surveyor must use his professional knowledge and experience, and this must be backed by sound administration, supported throughout by clearly prescribed practices and procedures.

His decisions must not only be justified individually; they must also conform with what has been established by his colleagues and predecessors, probably over many years. This is not to say that precedents must be followed blindly, and indeed often there is no precedent, but there is a pattern. It takes time and experience to learn what is expected, and it is often wise—and sometimes a requirement—to seek a second opinion. Even surveyors with many years' service are not too proud to discuss difficult questions with others.

Administration and management

These words are used indiscriminately these days, but it is important to be able to differentiate between a number of different concepts. Not surprisingly it pays 'to know the ropes' ashore in equal measure to those at sea.

The British Empire spawned an administration, thus providing the wherewithal to control countries, trade, and enrich the exchequer. To do this the subjects, as in the Roman Empire, lose some personal freedom but are rewarded by a better standard of living than might otherwise be possible. Administrative models tend to be more static in format and are so large that there has to be a hierarchy.

In administrations, personnel are generally employed through personnel departments, are administered by an appraisal system and this is generally linked to salary administration with tied steps and grades.

Obviously, many books have been written on the subject, but these notes are designed as a pointer so that those who go into government will be aware that their performance is being measured by superiors and recorded at appraisal interviews. Appraisal is the time to tell superiors how good you are and that you want to advance, do extra courses, seek new opportunities, etc. If this is not fully recorded on the appraisal form, the civil servants who do not know you from Adam will never know of your hidden desires and wants.

Not surprisingly, the Americans broke the mould with the concept of modern management which puts self-determined economic results as the top priority. Management then becomes the art and science of getting results. The catch phrase 'management by objectives' is an apt description.

Some commercial companies may not be large enough to offer regular progress, but often find ways of providing perks for those who perform well. Remuneration and accounting are likely to be linked to optimising the company's tax position and may seem odd at first glance. Nevertheless, it is important in any organisation to negotiate and secure a contract of employment, written and signed by the organisation before starting. The exercise will not only provide a more realistic expectation of the job and its prospects, it will also provide the only security you have got if there is a dispute or the company goes into liquidation.

The nautical surveyor will ply his trade in many different types of organisations, embracing the variety and interest contained in this book. The following notes on background reading may come in handy at odd times throughout your career. □

References

Organisation in Business Management by Sir Walter Puckey is published by Hutchinson in paperback. It provides a good background in the subject and looks at the interplay between individuals, groups, organisations, structure, communication, cohesion and leadership. It does not cover the impact of electronic equipment in the office, but is good on human nature and principles.

The Interview at Work is a thin book by John Fletcher and published by Duckworth. It is designed for the interviewer and is written in plain language. It deals with most instances when a face to face interview has to be handled. Techniques can be developed and this book covers difficult situations well.

The Rational Manager—A Systematic Approach to Problem Solving and Decision Making is written by C. H. Kepner and B. B. Tregor and published by McGraw Hill. This is an excellent book for its practical approach and detailed observations. There is a very useful bibliography with it, too.

The Practice of Management. Management and Managing for Results by Peter Drucker (Heinemann) are worth reading. Although first published almost 30 years ago, the books have a unique freshness as they capture the essence of management purpose and by so doing brought a new awareness of the subject.

Leadership and Motivation by Douglas McGregor, MIT Press, is a classic. Once read it becomes possible to analyse one's own performance. Whilst most situations can be improved, there are cases where no amount of good management technique makes any difference: so be discriminating.

Management by Objectives by J. W. Humble, McGraw Hill, was first published in 1970 and has had a major impact on organisational practices. The book contains some excellent case studies both government and commercial and puts the appraisal interview into context.

Up to date office practices and the introduction of computers. The best advice is to consult periodicals and visit the local library.

Criminal Investigation (2nd Edition) by Charles E. Merrill Publishing Co 1980. No book ever tells you all you want to know about investigative methods, but this is a start.

Police Powers and Duties: A Practical Guide to the Police and Criminal Evidence Act 1984, Fourmat Publishing 1985. This is rather specialised but it does show how policing powers are constrained within the law and there is a valuable educational content for the government surveyor, trying to orientate himself to the role of law enforcement officer.

Making Crime: A Study of Detective Work by R. V. Ericson, Butterworth 1981. It is not the usual job of surveyors to solve crimes, but there are lessons to be learnt from this well-researched book.

A more specialised bibliography is contained at the back of this book.

THE NAUTICAL INSTITUTE

THE NAUTICAL INSTITUTE is an independent international professional body for qualified mariners, whose principal aim is to promote a high standard of knowledge, competence and qualification amongst those in control of sea going craft. To keep the membership up to date the Institute publishes its monthly journal called *Seaways*.

The Institute has an international membership of some 5,000, of whom two-thirds are actively employed at sea as deck officers, masters, pilots, and officers in the armed forces. The shore-based members include fleet managers, harbour masters, surveyors, educationalists, and those employed in ancillary services.

Membership of the Institute is based on proven qualifications and experience at sea—e.g. a Class 1 deck master mariner's certificate of competency or naval command examination from a recognised authority.

Associate Membership is awarded to those holding approved watchkeeping certificates. Further information on membership can be obtained from The Secretary.

THE NAUTICAL SURVEYOR'S CERTIFICATE SCHEME

REQUIREMENTS

Associated with the text book *The Work of The Nautical Surveyor,* the Institute has established a programme for sea staff and others who wish to follow a formal course of study, leading to the award of a certificate.

It must be emphasised that this study programme is intended only to introduce the subject and is designed for those who are thinking of changing course and coming ashore, who want a good foundation *before* applying for a nautical surveyor's job.

A Nautical Institute Certificate will be awarded to those who complete, to the satisfaction of the examiners (members of the Working Group), nine written essays.

All candidates must be at least 25 years old to be eligible for entry to the scheme which is open to deck officers and naval officers of all nationalities, who hold at least a watchkeeping certificate or the rank of Lieutenant. The work must be completed in English.

Readers requiring more information about this scheme should write to The Secretary, The Nautical Institute, 202 Lambeth Road, London SE1 7LQ, UK.

PART I
THE NAUTICAL SURVEYOR IN GOVERNMENT

ROLE OF THE GOVERNMENT NAUTICAL SURVEYOR

Ib Matthiesen
Deputy Chief Nautical Surveyor, Danish Government Ship Inspection Service

Examination of new tonnage

SHIPPING AND SHIPBUILDING are both industries which, due to their international character, to a great extent have gained an advantage from the endeavours which have been brought forward by the international organisations such as IMO and ILO in preparing accepted standards for the construction, equipment and operation of ships.

It is a general obligation of contracting governments to undertake that all laws and regulations, etc., necessary for the effective implementation of coventions are promulgated. The ultimate aim is that a ship, from a safety point of view, is fit for the service for which it is intended; from a social point of view, offers healthy working and living conditions; and does not present any threat of harm to the marine environment.

The basis for the implementation of internationally agreed conventions on safety of ships in Denmark is the Safety of Ships, etc., Act. In pursuance of this Act, the Ministry of Industry has issued an order which authorises the director of the Danish Government Ships Inspection Service to stipulate technical regulations on the construction and equipment of ships.

When the regulations contained in the annexes to IMO conventions have to be adapted to national legislation, it is found that even if the relevant subcommittee, when drafting, has endeavoured to be as detailed as possible, the phrase 'to the satisfaction of the Administration' appears from time to time. This is quite natural in a forum where so many countries have to come to a compromise. In the Danish administration it has, in these instances, been found necessary and appropriate to complement the convention text with the national interpretations printed in italics.

In the process of adapting the international standards to national legislation it is considered of great importance to have close contact with all relevant parts of the industry and organisations. Having finalised and implemented the subsidiary part of the legislation, the provisions of the Safety of Ships, etc., Act are operational.

Documentation

It should be noted that in addition to the preparation of national standards (adaptation of international standards), all the appropriate documentation required needs to be prepared and available to all concerned at the same time as the legislation enters into force. In this connection it is worth mentioning that a minor thing such as a list of material and equipment approved for use on board (Danish) ships very much facilitates the work of the industry as well as the administration.

The subsidiary legislation should contain provisions according to which the shipowner or yard at the earliest possible stage has to submit to the administration all information, drawings and plans, necessary for the administration to carry out plan approval.

In practice, this obligation for the owner/yard requires that the administration at the request of the owner/yard submits an 'information and drawing list' for the size and type of ship in question. Leaving it to the discretion of the yard/owner to submit the relevant drawings, the administration may either receive drawings which are insufficient to form the basis for preapproval or the administration may receive such an amount of 'paper' that vital details may be ignored.

Nautical surveys

During the monitoring of a newbuilding as well as during the process of plan approval those parts of the ship which naturally fall within the scope of work to be done by the Nautical Surveyor are mainly: lifesaving appliances; shipborne navigational equipment (including radar and ARPA); lights, shapes, means of making sound signals and distress signals; and fire-extinguishing appliances on deck.

As far as radar and ARPA are concerned, it may in some administrations be radio surveyors or employees from the tele-administrations who deal with that part of the equipment, since the radar formally 'belongs' to the Safety Radio Certificate. However, the conference on the revision of Solas and the Load Lines Convention will probably, with the revision of Chapter I of Solas, result in the radar being 'transferred' to the Safety Equipment Certificate.

Lifesaving appliances

Only a few parts of the traditional equipment which is required on board are as closely connected to the safety of the crew and passengers as are the lifesaving appliances. They are the last resort, when the requirements of all other parts of the ship have proved insufficient to support the view of the ship being the best lifeboat.

A systematic approach to safeguard human lives depends upon: technology (the rules), personnel (the crew); and procedures (periodical surveys). It is in the interaction of these elements the causes of accidents have to be found.

Safety is achieved by establishing some 'defence lines' against envisaged hazards. As the first line of defence, one may consider proper design and technology which requires adequate rules. As the second line of defence one may consider in-service training, muster and drills, carried out by the crew. And as the third line of defence one may consider the periodical surveys which are carried out, after it has been ascertained that the rules of construction, etc., are complied with.

Safety improvements may, according to the circumstances, be obtained by adequate change in the three elements. By changing the order, certain effects will occur in the other elements. However, they must all three be complied with to achieve the ultimate aim.

LSA rules

Following the tacit acceptance procedure which applies to the Solas Convention, except Chapter I, the revised requirements with respect to lifesaving appliances apply to ships the keels of which were laid or which were at a similar stage of construction on or after 1 July 1986.

Although the revised Chapter III has been in force only for a short period, it is possible to comment, at least in general, on major parts of the revised LSA requirements.

Basic philosophy

The basic philosophy and proposed layout for the revised Chapter III were described by the Maritime Safety Committee at its 30th session as follows:

1. The main objective of Chapter III is that of providing adequate lifesaving systems at sea. In order to achieve this it is desirable to define the types of casualties which should be considered important, both with regard to severity and the frequency with which they occur.
2. Consequently, as a background material for the priorities which have to be given with regard to requirements for a lifesaving system, casualty statistics qualifying ship and environmental conditions should be available. If not, at least a good knowledge of predominant ship casualties must be available.
3. On this basis general functional requirements and objectives for a lifesaving system should be established. These should be qualitative requirements regardless of ship type or type of equipment, and be present in Part B or Chapter III.
4. The functional requirements should then be considered for different ship types, sizes and trades, etc., whichever characteristics are of importance for a translation of requirements into functional criteria. Such criteria should form Part C of Chapter III, forming the qualitative basis toward which lifesaving equipment should be designed. These criteria thus define the standards for lifesaving equipment dependent on its utilisation.
5. These functional criteria can normally not be incorporated in one appliance only, without causing too many compromises. It may therefore be necessary to equip a ship with different types of lifesaving appliances which together form a lifesaving system in compliance with the criteria. Part D of Chapter III should therefore contain the detailed definition of various appliances and the specific requirements both with regard to the appliance itself and to possible special arrangements or design on board the ships. This means that the requirements for a lifesaving appliance may vary with ship type, etc.
6. Such a layout of Chapter III should accommodate new designs, ensuring that these are evaluated as pertinent to their actual use on board a vessel.

Improvements

The improvements in the revised requirements can mainly be divided into four parts: the new equipment which provides seafarers with an increased probability of survival; procedures for standardisation of requirements, provision for the acceptance of novel appliances; and crew training, muster and drills.

New equipment

Cargo ships of 85m in length and above shall be fitted with totally enclosed lifeboats. The administration may, however, permit cargo ships (except oil tankers chemical tankers and gas carriers) operating under favourable climatic conditions and in suitable areas, to carry self-righting partially enclosed lifeboats, provided the limits of the trade area are specified in the Cargo Ship Safety Equipment Certificate.

Passenger ships shall be fitted with either partially enclosed lifeboats, self-righting partially enclosed lifeboats or totally enclosed lifeboats.

Chemical tankers and gas carriers carrying cargoes emitting toxic vapours or gases shall be fitted with lifeboats with a self-contained air support system.

Tankers (oil, chemical and gas) carrying cargoes having a flashpoint not exceeding 60°C (closed cup test) shall be fitted with fire-protected lifeboats also having a self-contained air support system.

Passenger ships of 500 gt and over shall carry at least one rescue boat on each side of the ship. Cargo ships and passenger ships of less than 500 gt shall carry at least one rescue boat. Embarkation shall be with the lifeboat in the stowed position on cargo ships. Immersion suits and thermal protective aids shall be carried.

Enclosed lifeboats

Totally enclosed lifeboats offer far greater protection than the open ones. Many, if not the majority of, deaths at sea have been caused by exposure, and for this reason enclosed lifeboats are a great step forward. They do, however, require information to be given by surveyors to crew members about the importance of wearing safety belts and having all hatches closed, if the boat is to retain its self-righting capability. That seasickness is more likely to occur is just considered as an unpleasant experience compared to exposure.

Compared with open lifeboats, which for many years to come will still be found on board tankers, the requirements as to the new lifeboats, which have to be fitted aboard chemical tankers and gas carriers (self-contained air support system) and oil tankers, etc., carrying cargoes having a flashpoint not exceeding 60°C (fire-protected), constitute nothing less than a revolution.

That the revised requirements ended up with the possibility of cargo ships (except oil tankers, chemical tankers and gas carriers) operating under favourable conditions being fitted with partially enclosed lifeboats, seems to be one of the rare instances at IMO where the need for obtaining a compromise has led to a solution for which the practical justification seems hard to find.

Standardisation

The amount of detail in the revised Chapter III is much greater than in the previous version. Thus it should be possible to achieve a higher degree of standardisation between the national regulations of different administrations. Besides, the revised chapter contains a provision requiring administrations, before giving approval to lifesaving appliances and arrangements, to ensure that they are tested in accordance with the IMO recommendations. The Recommendation on Testing of Lifesaving Appliances (res.A521(13)) has thereby been made mandatory.

The probability exists that the reciprocal acceptance of appliances in the future will be facilitated between administrations. This is of importance not only to those manufacturing and selling lifesaving appliances, but also upon transfer of a ship from one flag to another.

Acceptance of novel appliances

Even if the revised chapter contains a provision as to the approval of novel lifesaving appliances or arrangements, it is doubtful whether any shipowner or manufacturer will find it tempting to embark on major projects of that kind. First, some of the already existing arrangements, which with regard to the 'old' chapter III would be considered as novel, have been included in the revised chapter—e.g., free fall lifeboats. Second, the costs involved in designing, manufacturing and testing a novel device, without any guarantee as to the economical prospects, will inevitably cause some reluctance on the part of manufacturers.

Drills

With the new and sophisticated equipment, proper musters and drills have become even more vital, besides which the drastic reduction in manning which has been seen worldwide in recent years leaves no room for any hands to be idle when an emergency threatens.

It should therefore be noted, that the requirements regarding muster list and emergency instructions, operating instructions, manning of survival craft and supervision, abandon ship training and drills, operational readiness, maintenance and inspections, and drills (passenger ships) apply also to existing ships, without a period of grace.

Surveys and safety certificates

With the entry into force of the Solas Protocol/78 on 1 May, 1981, the provisions concerning inspections and surveys were tightened up. It should be mentioned that additional inspections/surveys were required and nomination of surveyors or recognised organisations to conduct surveys and inspections can only be done if they also are empowered to require repairs to a ship, and carry out inspections and surveys if requested by the appropriate authorties of a port State.

The present Solas 74/78 requirements concerning survey and certificates have been in force for quite some years now, and it seems as if most of the problems arising from the complexity of the system have been overcome by both administrations and shipowners. However, the variation of certification and survey intervals in different IMO conventions is inconvenient for both administrations and shipowners. As a consequence a harmonised system of surveys and certification was planned in 1988, including Solas 74/78, the Load Line Convention, 1966, and Marpol 73/78, but will not enter into force before 1 February 1992 at the earliest.

Cargo ship safety equipment certificate

According to the provisions of Solas 74/78 the inspection and survey of ships in relation to the convention may be entrusted to surveyors nominated for the purpose or to recognised organisations. In the implementation of the Solas 74/78 requirements by national Danish legislation it has been decided that the initial survey as to safety equipment has to be carried out by surveyors from the administration. This applies also to ships built abroad. On completion of the survey a Cargo Ship Safety Certificate is issued.

Besides lifesaving appliances, those parts of the ship and its equipment which *inter alia* are dealt with by the nautical surveyor are shipborne navigational equipment and lights, shapes and means of making sound, and distress signals.

Shipborne navigational equipment

Upon the entry into force of the 1981 amendments to Solas 74 the requirements as to navigational equipment were increased significantly; they cover such items as:

- Gyro and magnetic compasses.
- Radar installations.
- Automatic radar plotting aids.
- Echo-sounders.
- Devices to indicate speed and distance (log).
- Rudder angle indicators.
- Propeller revolution indicators.
- Rate-of-turn indicators.

The requirement that the equipment has to be type approved by the administration requires resources which only a few administrations have available. Having overcome the problems in connection with the type approval, the field surveyor still has to make sure that the installation of the equipment has been carried out satisfactorily. This comprehensive job can hardly be carried out without the surveyor being present during a sea trial. As an example reference could be made to the performance standards for automatic pilots (resolution A.342(IX),) where it is prescribed:

2 General

2.1. *Within limits related to a ship's manoeuvrability the automatic pilot, in conjunction with its source of heading information, should enable a ship to keep a preset course with minimum operation of the ship's steering gear.*

2.2. *The automatic pilot equipment should be capable of adapting to different steering characteristics of the ship under various weather and loading conditions, and provide reliable operation under prevailing environmental and normal operational conditions.*

The ability of the automatic pilot to comply with para. 2 is dependent on the hull, rudder, propeller and the ship's horsepower, and can not be a part of the type approval, since these factors differ from ship to ship. A verification of compliance with the performance standards should consequently require a sea trial, even under different wind and sea conditions.

Lights, shapes and sound signals

The exemptions given during a nine year period in Reg. 38 of the International Convention for Preventing Collisions at Sea, 1972 expired on 14 July 1986.

From that date where applicable the vertical and horizontal positioning and spacing of lights have to be in accordance with Annex I and the technical details of sound signal appliances have to comply with Annex III of Colreg 72.

Certificates

Being satisfied that all applicable requirements are complied with, the surveyor can issue a Cargo Ship Safety Certificate. During the survey, close cooperation should exist with surveyors of other categories, mainly the engine surveyor in order to ensure that the requirements as to the ship's fire-fighting capability are met.

Periodic Surveys

Administrations have to a great extent delegated the statutory surveys (renewal), mandatory annual surveys and intermediate surveys of tankers ten years of age and over to classification societies. The concept of unscheduled inspections is not being practised.

It should be noted that following the 14th Assembly, revised guidelines on surveys required by the 1978 Solas protocol, the *International Bulk Chemical Code,* and the *International Gas Carrier Code* have been issued.

It should be emphasized that, regardless of the work being delegated, the administration in every case shall fully guarantee the completeness and efficiency of the inspection and survey, and shall undertake to ensure the necessary arrangements to satisfy this obligation.

Port State inspections

Looking back to search for the causes leading to international and regional action on substandard ships, it is plain that there were several roots from which it all began. It was in particular following the *Amoco Cadiz* casualty that the European countries realised that preventive action against substandard ships would have to be given priority. It had become a political matter.

The term 'substandard ship' has been used in many instances in a way giving rise to misunderstanding and misinterpretation. In this context, reference is made to section 3 of IMO Resolution A. 466 (XII), to Article 4 of ILO Convention No. 147 and section 3.7. of the Memorandum of Understanding on Port State Control, 1982.

Since the official definitions of a sub-standard ship are given in very general terms, a few examples to illustrate circumstances that will inevitably qualify a ship as 'substandard' should be mentioned:

- Launching of life boats not possible due to absence of greasing, accumulation of paint and failure to carry out regular drills.
- Emergency fire pump inoperative.
- A ship staying in a Scandinavian port during the winter season is found to have no accommodation heating system.

Reasons for substandard ships

No introduction into the complex problem of substandard ships would be complete unless it includes the reasons behind the phenomenon. Without ranking such reasons, and without pretending to draw up a complete list of them, it is safe to say that the underlying reasons include the following:

- (Old) age of the ship.
- Lacking (or no) operational control by the shipowning company.
- Lacking (or no) training of officers and/or crew.
- Lacking (or no) supervision on part of the flag State.

Responsibility

Varied are the reasons for substandard ships, and so are the responsibilities for combating the phenomenon. The responsibility rests mainly with:

(a) The international organisations such as IMO and ILO, whose task includes the preparation and adoption of relevant safety standards.
(b) The flag State which has the primary responsibility for the effective implementation of the standards embodied in the relevant conventions.
(c) The shipowner—he must ensure the safe and seaworthy condition of his ship as well as her safe manning (it is provided that the ship's flag State has transformed into national law the provisions of the relevant conventions).
(d) The master, who bears the full responsibility for the observance aboard his ship of operational and safety standards.
(e) Every crew member—last but not least.

The following instruments include the right and the obligation for the port State to verify that foreign ships and their crew comply with the relevant standards: ILO No. 147, Article 4; Solas 74/78, Chapter I, Reg. 19; STCW, 1978, Article X; Marpol 73/78, Article 5; and ILLC, 1966, Article 21.

The subject of port State inspections is covered in the chapter by Captain R. L. Newbury on page 51.

Incidents and investigations

The obligation for administrations to conduct an investigation after any serious marine casualty or incident is included in the following conventions: Solas 74/78, Chapter I, Regulation 21; Marpol 73/78, Article 6; LLC, 1966, Article 23; and ILO No 147, Article 2 (g). Further, the *Memorandum of Understanding on Port State Control* in Section 5 imposes the obligation for the participating administrations to cooperate in securing evidence relating to suspected violations of the requirements on operational matters of Rule 10 of Colreg and Marpol 73/78.

As far as Marpol 73/78 and the Memorandum are concerned the investigation may also be carried out with respect to foreign ships. However, the legal action towards the ship/master, according to the above conventions, has to be taken by the flag administration.

Apart from the interest of the flag State to prosecute suspected violations of national legislation, investigations are additionally conducted to supply the organisation with all pertinent information concerning the findings of such investigations in order that they are able to determine possible changes in the conventions. Due to the experience gained during many years at sea as mate and possibly as master, the contribution of the nautical surveyor in connection with investigations of marine casualties and operational violations is invaluable.

Health and safety

The idea of health and safety taken in general is a very comprehensive one and might in fact form the basis of a complete syllabus. The following brief reference to the topics is not thought to be complete and will with respect to safety be limited to operational safety and as to health to medical care and examination.

A precondition for obtaining safe working conditions aboard a ship, as well as ashore, is all groups and individual members of the crew taking an active part in the prevention of occupational hazards. The primary responsibility rests with the master, chief engineer and the senior officers. However, full support from all crew members can be expected only if a mandatory system is set up through which appointed members of the crew can inform the master, etc., about all minor and major hazards, to which they feel exposed, without corrective action has been taken.

Records of meetings in this committee for accident prevention should be kept in the ship's file. During inspections the surveyor should go through the records and consult crew members representing the various categories of the crew, in order to ensure that requests regarding hazards that may exist in their daily work aboard the ship have been taken appropriate care of. All complaints regarding conditions on board should be investigated thoroughly and action taken as deemed necessary by the circumstances.

It is impracticable to mention all the items to which attention should be paid. As examples the following items fall within the scope of the nautical surveyor:

- Safe means of access to the ship.
- Safety measures on and below deck.
- Loading and unloading equipment.
- Dangerous cargo and ballast.
- Personal protective equipment for seafarers.

If deficiencies or operational circumstances give rise to a serious hazard to the safety or health of persons on board, the surveyor shall take appropriate action to remove the hazard by requiring rectification of the deficiency or prohibiting continuation of the operation. In Annex 1 to the *Memorandum of Understanding on Port State Control,* comprehensive guidelines have been given to the surveyors on this issue when carrying out port State control. These guidelines are based on ILO Convention No. 134 on Prevention of Accidents.

Medical care and examination

Medical certificates are required for persons who are employed in any capacity on board a ship. In ILO Convention No. 73 on the Medical Examination of Seafarers, small ships as well as certain categories of seafarers are exempted from the provisions. In addition, being a statement as to the general health of the seafarer, the certificate shall attest that the person concerned is fit by reason of hearing and sight. For persons to carry out deck watchkeeping duties, it shall

MAJOR CATEGORIES OF DEFICIENCIES

Major categories of deficiencies	Number of deficiencies			Deficiencies in % of total deficiencies			Deficiencies in % of No. of inspections			Deficiencies in % of individual ships		
	1984	1985	1986	1984	1985	1986	1984	1985	1986	1984	1985	1986
Ships' certificates	1,852	1,637	1,568	12.50	12.27	9.98	18.11	15.71	13.36	24.10	20.78	17.98
Crew	688	632	671	4.65	4.74	4.27	6.73	6.07	5.72	8.95	8.02	7.69
Accommodation	430	252	230	2.90	1.89	1.46	4.20	2.42	1.96	5.59	3.20	2.64
Food and catering	87	83	102	0.59	0.62	0.65	0.85	0.80	0.87	1.13	1.05	1.17
Working spaces	67	70	67	0.45	0.52	0.43	0.66	0.67	0.57	0.87	0.89	0.77
Life saving appliances	3,591	3,288	3,783	24.25	24.64	24.08	35.11	31.56	32.22	46.72	41.73	43.38
Fire-fighting appliances	2,656	2,121	2,753	17.93	15.90	17.52	25.97	20.36	23.45	34.56	26.92	31.57
Accident prevention	98	106	84	0.66	0.79	0.53	0.96	1.02	0.72	1.28	1.35	0.96
Safety in general	1,447	1,617	2,205	9.77	12.12	14.04	14.15	15.52	18.78	18.83	20.52	25.28
Alarm signals	51	39	32	0.34	0.29	0.20	0.50	0.37	0.27	0.66	0.49	0.37
Cargo	192	198	238	1.30	1.48	1.52	1.88	1.90	2.03	2.50	2.51	2.73
Load lines	757	669	773	5.11	5.01	4.92	7.40	6.42	6.58	9.85	8.49	8.86
Mooring arrangements	65	70	81	0.44	0.52	0.52	0.64	0.67	0.69	0.85	0.89	0.93
Propulsion + auxiliary machinery	349	247	292	2.36	1.85	1.86	3.41	2.37	2.49	4.54	3.13	3.35
Navigation	1,854	1,584	1,973	12.52	11.87	12.56	18.13	15.21	16.81	24.12	20.10	22.62
Radio	149	184	172	1.01	1.38	1.09	1.46	1.77	1.47	1.94	2.34	1.97
Marine pollution—Annex I	313	427	578	2.11	3.20	3.68	3.06	4.10	4.92	4.07	5.42	6.63
Deficiencies specif. for tankers	23	85	72	0.16	0.64	0.46	0.22	0.82	0.61	0.30	1.08	0.83
All other deficiencies	60	25	19	0.41	0.19	0.12	0.59	0.24	0.16	0.78	0.32	0.22
Deficiencies not clearly hazardous	82	8	16	0.55	0.06	0.10	0.80	0.08	0.14	1.07	0.10	0.18
Totals (Deficiencies/Inspections/ Individual Ships)	**14,811**	**13,342**	**15,709**	**14,811**	**13,342**	**15,709**	**10,227**	**10,417**	**11,740**	**7,686**	**7,879**	**8,721**
	Deficiencies			Deficiencies			Inspections			Individual Ships		

attest that the colour vision is satisfactory. The certificate shall not be older than two years and as regards colour vision not older than six years from the date of issue.

Annex 1 to the Memorandum also includes detailed guidelines on this topic to be followed by the surveyor in connection with port State control. These guidelines could, as well as those previously mentioned on occupational safety, be adopted when carrying out flag State control.

ILO Conventions No. 73 and No. 134 are two of the conventions under the umbrella of ILO No. 147 on Minimum Standards for Merchant Shipping, which is one of the major—if not the most important—of the maritime conventions adopted by the International Labour organisation.

Considering the authority which lies in the hands of a surveyor, it should be evident that one of his first duties, when boarding a ship, is to report to the master or his representative. Without being requested he should present evidence that he is duly authorised by the administration to carry out the survey and investigation required by him. □

CONTROL OF SHIPS, SAFETY, ANTI-POLLUTION STANDARDS AND METHODS OF ENFORCEMENT

Captain A. Crombie, FNI

Formerly Head of Marine Safety, Liberian Bureau of Maritime Affairs

ANY DISCUSSION on methods for the enforcement of maritime standards must obviously begin with some reference to the volume and complexity of regulatory requirements generated by the various international conventions in force. The Load Line Convention of 1966 is a relatively simple instrument and one hopes it will remain so. Solas 74 was relatively simple until the need arose to amend it. Marpol was the first international convention to create, from its inception, serious difficulties for shipowners who have to comply with it and the maritime administrations who have to enforce it. In the past few years, due to the deluge of amendments to conventions, the vast number of resolutions and the proliferation of interpretations, Marpol has lost the dubious honour of being the biggest headache. They are all headaches on an equal basis. It becomes difficult to determine just why so many amendments to conventions have been necessary over the past few years. No doubt politicians, international orgaisations and the increasing influence of the environmentalist have had much impact, but even if allowance is made for improved technology and increasing problems which may have initiated amendments, there must be some doubt regarding the quality of the original conventions when they were initially drafted. The responsibility rests with all of us who were involved.

Use of authorised agencies

Most of the enforcement effort of the Liberian administration is accomplished by means of non-governmental agencies of one kind or another, including the six classification societies approved by Liberia. It is a condition of Liberian registration that a ship must comply with the classification rules of one of the approved societies. These authorised agencies conduct surveys and issue, on behalf of Liberia, all certificates required by international conventions. There are advantages and disadvantages associated with this system. So far as the actual survey of ships is concerned Liberia can, through its authorised agencies, call upon the services of skilled surveyors who have a great deal of expertise and experience. The services of these surveyors are paid for by the shipowner and not by the Liberian administration which, once again, is good economics, although the surveyors when carrying out any survey on a Liberian ship are always responsible to the administration in respect of quality of the survey and the integrity of any certificate subsequently issued.

There is also much merit in having an agency that may well have approved the design of a ship and supervised that ship's construction continue to monitor the ship's compliance with international requirements on behalf of the administration. Some, if not all, of these authorised agencies possess impressive research and development capabilities which are available to Liberia. This is a particular advantage when related to casualty investigation for which the skilled expertise of a specialised agency may be retained on an *ad hoc* basis, thus avoiding the necessity to maintain inhouse capabilities which may be under utilised for most of the time.

Liberia exercises a reasonable but firm control over its authorised agencies by means of regular discussion and consultative sessions at executive level and day-to-day contacts at operational level as may be necessary.

Unfortunately there are disadvantages. Some of these authorised agencies are in competition with one another, which of course breeds initiative. Initiative thrives on a lack of uniformity and this, uniformity, is precisely what the Liberian administration desires to see applied to the survey and inspection of Liberian ships. Another aspect appears to be that control by some of these agencies of their field staff seems in some instances to lack direction and what Liberia wants has not, upon some occasions in the past, penetrated to field surveyor level. Some field surveyors tend to be very independent and this sometimes contributes to that lack of uniformity.

The authorised agency surveyor is always wearing two hats and is trying to please two masters, his own agency and the Liberian administration, and thus conflicts of interest do occasionally arise. It is not perhaps surprising that some agency surveyors may feel that their loyalty lies towards the shipowner and their own agency rather than the Liberian administration for which, at the material time, the surveyor is acting. This is an undesirable situation but fortunately it does not happen very often and it is gratifying to note how most agency surveyors cope with their dual status in what can often be very sensitive circumstances.

One of the aspects of regulatory control of ships which has attracted much attention and criticism in past years is that of postponing surveys and issuing extensions to statutory certificates. In some circumstances it may be unreasonable to require a ship to carry out repairs or routine survey in a port where no suitable facilities exist and in the present economic conditions in the maritime world a shipowner may possibly request deferment of, for instance, drydocking so that he does not lose a charter.

Based upon his personal experience of the Liberian system of almost totally delegating international convention survey and certification functions to authorised agencies, the author would argue that the concept is efficient provided there is a firm control and oversight of those agencies by the administration. For any administration simply to hand over all its regulatory functions to authorised agencies without some form of oversight would be an abdication of its responsibilities. Although the economies of delegation must be attractive to any nation, especially a developing nation, there must be in the national administration a nucleus of experienced and trained personnel who can exercise supervisory and monitoring functions. The author believes that most authorised agencies would agree and would welcome a competent national administration from which guidance and a measure of control can be obtained.

One fact of the matter is that authorised agencies are unwilling to be seen to give instructions to shipowners in matters that lie within the responsibility of the government authority, particularly surveys, certification and other regulatory matters. Liberia has recognised this difficulty faced by authorised agencies, although not always agreeing with it, and has consequently evolved a method of conveying information to shipowners by means of marine notices, guidance documents on Solas and Marpol, and also other circulars on various matters where this has been found necessary. Activities of this nature place great strain on staff capabilities, but any maritime administration must be prepared to cope with the responsibility of giving advice and assistance.

Most traditional maritime countries have elected to have a maritime administration wherein a designated government department is held responsible for marine regulatory functions. These departments are normally staffed with all the necessary expertise and personnel under direct and strict control of a centralised authority. The author would agree that the Liberian system does not permit such a rigid control of, for instance, classification society surveyors who are administratively subject to their own organisation although responsible to the Liberian administration for their professional activities on board Liberian ships. If there is evidence or suspicion of less than satisfactory professional competence an agency surveyor can be, and one or two have been, prohibited from acting on behalf of Liberia. However, effective control can be maintained provided the individuals, within the administration, who are given authority to exercise control within their specific area of responsibility, have the expertise and will to do so.

Liberian administration inspectors

In addition to authorised agency surveyors who act on behalf of the Liberian administration, Liberia has its own inspection organisation, financed by levies made upon shipowners, composed mainly of master mariners, although there are a few engineers and possibly people from other disciplines. These inspectors are all retained on a contractural fee-earning basis and are not employees, although most of them are controlled directly. Under the terms of their contract they are required to provide self-insurance cover against accidents occurring in the course of their duties. Fees are negotiated in accordance with local conditions and are not uniform.

Most have other sources of income, although one or two are totally employed in carrying out inspections of Liberian ships. Some are also non-exclusive surveyors for other agencies, and this is another situation where one has to ensure that conflicts of interest do not arise. As an example, some Liberian nautical inspectors are also non-exclusive surveyors for classification societies, but they are not permitted to act in any dual capacity in the same ship. Some are also nautical inspectors for other national flags, a practice not regarded with favour.

These inspectors carry out the required annual inspection of Liberian ships which basically monitors officer/crew certification and operational practices on board. The Liberian inspection looks at safety and anti-pollution standards so far as these are operational and not equipment related. The inspectors are empowered to require boat and fire drills and to generally ensure that the ship's officers and crew are competent to deal with any routine or emergency situations that they may be faced with. In the event of serious operational deficiencies being detected, the nautical inspectors will normally report to their control office but they are empowered to prevent any unsafe ship proceeding to sea using whatever methods are appropriate, including a request for port State assistance.

As a back up to the routine inspections by contract inspectors, there are a number of marine safety officers, who are salaried staff and are stationed at certain selected ports. In addition to other duties they make 'spot' inspections of Liberian ships without notice, which is not a popular exercise but has produced some very satisfactory results. Unfortunately the 'spot' type inspection, if not carefully monitored, can attract accusations of discrimination.

Inevitably, there is some duplication of effort, especially in the areas of safety and firefighting equipment which are subject to statutory survey by other authorised agencies but Liberian inspectors are not there to spy on other people's efforts.

The personnel certification requirements of the STCW Convention have to be monitored by the contract inspectors as this task is inappropriate for delegation to other authorised agencies.

Enforcement of safety standards

In the event that problems are found with a Liberian ship that may affect safety of ship or crew, a number of correctional procedures are open. In most cases the operators of the ship will be co-operative in their own interests, but if not Liberia will issue an order of detention which prohibits the operators and master from taking the ship to sea. This is normally effective as it has legal implications and can impact upon insurance and charter-party provisions, in addition to having significant financial impact if the detention period is prolonged.

In some instances assistance is requested from a port State and some port States have been very helpful in this respect. In cases where there is a serious neglect of international or Liberian rules, the ship's Liberian registration can be suspended or revoked although action of this nature has significant legal and financial implications and is not implemented without serious consideration of the consequences. The governing factor must always be to prevent an unsafe or badly operated ship from proceeding to sea and creating conditions that cause difficulties for itself or other ships.

Procedures for dealing with delinquent ships must be formulated and implemented by the administration and cannot be delegated to outside agencies. In all cases where a Liberian ship is found to be deficient in any respects, whether by Liberia's own inspectors or by a port State, it is standing policy to have follow-up inspections by the inspection organisation to ensure continued compliance with safety standards. The cost of these additional inspections are charged directly to the shipowner and thus become a form of indirect punishment.

Human factors

Much has already been said in this chapter, and elsewhere, about regulatory enforcement of maritime rules, how it affects the administration, the authorised agency and the shipowner, but not much attention is given to the man at the sharp end of enforcement—the fellow at sea. The author does not know the extent to which surveyors or inspectors in other administrations are ex-seagoing personnel, but most Liberian nautical inspectors are in this category. Many can remember very well how they were harassed by various authorities upon arrival in port, probably after a night on the bridge under pilotage or in fog, and how they hated those people. That particular problem is compounded in this day of fast turnarounds which affect nearly all ships, especially the smaller ones.

The Liberian Marine Safety Department is very conscious of this problem and inspectors are encouraged to cause the minimum of disruption to ship's routine, consistent with carrying out an efficient inspection. However, for one reason or another, it is very easy for an inspector to arrive on board in an aggressive frame of mind, and equally the master or chief officer may have had all night out or even a bad egg for breakfast and they are just as aggressive. In those circumstances we have all the potential for an explosion of tempers which can easily happen during an inspection. Liberian inspectors are warned that this personality conflict situation can arise and must be guarded against, and whilst there is not the same control over agency surveyors one hopes that they, too, recognise the danger of lack of understanding and co-operation between inspectors and ship's personnel.

This problem shows to a very great extent during the quarterly inspections of passenger ships where, within 12 hours, a thousand passengers have to be disembarked, the ship provisioned, cleaned and another thousand passengers embarked. For these ships arrangements have been made to eliminate unnecessary paper work, thus permitting both inspector and crew to concentrate on matters directly related to safety such as boat and fire drills.

There has to be understanding by both marine inspectors and ships' crews of the problems of each other and in the opinion of the author it is the maritime administration that should identify any such difficulties and show an example by formulating reasonable inspection procedures which achieve satisfactory results without creating friction.

Control of ships

Finally, the author would like to make some comment on port State control in the context of Liberia's relationship with port States. Liberia supports this concept as it feels that it complements the country's efforts to promote safety at sea. Liberia does not object when a port State cites or even detains one of its ships because it prefers to hear about problems before they become potential disasters.

On the very important matter of manning which is fundamental to the promotion of safety standards, Liberia has some concern over the apparent disparity in manning levels for different flags. It follows a policy of equating reduced manning to increased technology and training; it does not permit purely economic reasons to override safety considerations when crew levels are being considered. Liberia believes that port State control of manning standards could be improved by increasing the powers of port State inspectors beyond the current position where any manning certificate has to be accepted on its face value. Inspectors cannot be blamed for this problem as they have nothing to guide them. Whilst they have the Solas, Loadline and Marpol conventions as guidelines in the areas of structure and equipment, in manning matters there is only an IMO resolution which is solely a recommendation and does not appear to have had much impact. The author argues that there should be an international minimum manning scale for the guidance of both flag and port States and that the maritime community should address this question. Creation of such an international manning scale is of course a problem, but no more so than some of the many problems that have been resolved in the past by IMO on the basis of goodwill and commonsense. A ship navigated by overworked, tired officers and crew is a menace to itself and other ships that may be encountered.

Scope of statutory surveying

Statutory surveying is the application of any country's national rules for the regulation of ships or, in the absence of national rules, the application of international conventions from which national rules are derived.

The principal international conventions in the maritime field are dealt with below. Conventions relating to safety at sea and the prevention of marine pollution are created by international agreement and are promulgated, reviewed and administered by the International Maritime Organization (IMO), the maritime forum of the United Nations, headquartered in London. Some other safety Conventions such as the cargo gear rules are sponsored by the International Labour Organisation (ILO), based in Geneva. ILO also sponsors and enacts rules for crew accommodation in ships and conditions of employment of seamen. All these conventions are enforced to a greater or lesser extent by surveyors.

Solas 74

The Safety of Life at Sea Convention, 1974 (Solas 74), is the successor to prior post-war conventions of 1948 and 1960. It has been subject to significant amendments enacted in 1981 and 1983, all of which are now in force. There are also many resolutions and interpretations issued by the International Maritime Organization that affect the provisions of the convention, all of which must be understood by nautical surveyors. The amendments, resolutions and interpretations are very complex, whilst most apply only to ships built after the particular rules enter into force some are retroactive and others have effective dates sometime in the future.

A surveyor must be absolutely certain of his decisions when applying the Solas Convention, especially in cases where his actions may result in the detention of a ship. The IMO has recently published a new edition of Solas 74 which incorporates the 1981 and 1983 amendments but not all resolutions and interpretations. The IMO also publishes guidelines and checklists for the conduct of surveys required by Solas 74. For relevant information reference must be made to IMO publications.

Marpol 73-78

The Internation Convention for Prevention of Pollution from Ships 1973 (as amended in 1978) (Marpol 73-78) was first enacted in 1973 but before it entered into force it was substantially amended by a conference held in 1978. The primary sections were Annex I (oil), Annex II (chemicals). Annex I entered into force on 2 October 1983 and Annex II entered into force in April 1987. Both these annexes deal with the structure of the ship, equipment and operational procedures to prevent pollution of the sea. Annex II requirements are the most rigorous and complicated.

In addition, the Marpol 73-78 Convention has three other annexes which were optional in the sense that countries ratifying the oil and chemical sections of the convention could opt out of legalising the other three sections (and many did). The optional annexes are Annex III (harmful substances carried in packages or other containers), Annex IV (Sewage) and Annex V (Garbage). Annexes III and IV are not yet in force (March 1989) as they have not yet received the necessary number of ratifications by Governments, but Annex V came into force in 31 December 1988.

The Marpol 73-78 Convention is extremely complex. Its application to ships is dictated by ship size, ship type and date of build and great care is necessary in interpreting the requirements of the various sections. There was also some retroactive application but with the passage of time and the effect of economic forces (scrapping) the number of ships so affected has been reduced. Much experience is required from surveyors as the consequences of a mistake can be very costly.

Annex II (chemicals) of the convention is affected by amendments adopted by IMO in December 1985 and by subsequent unified interpretations approved by IMO. The object of these unified interpretations is to achieve uniformity by flag States. Guidelines and check lists for the conduct of surveys required by the Marpol 73-78 Convention have been produced by ILO.

Load lines

The International Convention on Load Lines, 1966, is concerned with the watertight integrity of the ship. It covers hatch covers, ventilators, double-bottom tanks vents, all openings in the shell and many other items. Some surveyors consider that the load line surveys require much more exercise of experience than other conventions which may involve only the application of concise rules. Load line surveys require assessment of condition and good judgement as to whether equipment or material is satisfactory and will be serviceable until the next survey or whether it should be replaced or repaired without further delay.

It should be noted that, unlike some other convention certificates, a Load Line Certificate cannot be extended. When the expiry date is reached the certificate expires and becomes invalid. If it is not convenient or difficult to perform the full renewal survey, the survey must be commenced and progressed to the point where seaworthiness is verified at which point a conditional Load Line Certificate may be issued permitting the balance of the survey to be completed within a specified, and limited, period of grace.

Tonnage

The International Convention on Tonnage Measurement, 1969—Tonnage measurement is a complex process and is usually done by specialists just before a new ship enters into service, although many ships are remeasured for one reason or another during their lifetime. The average practical nautical surveyor would not be involved in tonnage measurement although, in the course of his visits to ships, he might note alterations or additions to the ship's structure that might alter or increase the measurement and affect such things as port/pilotage dues.

Manning

The International Convention on Training and Certification of Seafarers, 1978, specifies standards for examination, certification, qualifications and training of seamen. It also prescribes manning standards for certain categories of seamen, primarily on some types of tankers. The practical nautical surveyor would be involved only to the extent of verifying that the correct numbers of suitably qualified and experienced seamen are on board any ship. The nautical surveyor has some difficulties with this convention. He may find himself faced with a flag State manning certificate over which he may have some doubts, but which he is obliged to accept unless he has reason to believe that the manning of the ship is not in accordance with the levels stated in the manning certificate. A great deal of practical seagoing experience is necessary to enable decisions to be made on whether a ship is efficiently manned or not.

IMO Codes of safe practice

There are a number of IMO codes of safe practice which affect safety at sea and with which the nautical surveyor has to be familiar. Of these the principal ones are:

- Carriage of grain (chapter VI of the Solas 74 Convention).
- Carriage of bulk cargoes (other than liquid cargoes).
- Carriage of dangerous goods.
- Carriage of timber deck cargoes.
- Lashing and securing of Ro-Ro cargoes.

It should be noted that, whilst reference is made in this section to the international rules affecting ships and safety at sea, many countries write the substance of international rules into national legislation. The nautical surveyor would therefore be obliged to be familiar with his own national rules for his own-flag ships and international rules for foreign-flag ships. In many instances, national rules are applicable to foreign-flag ships within national jurisdiction and, in some instances, national rules are more severe that the international rules from which they derived.

This section does not provide a comprehensive list of all rules relating to ships and safety at sea. The reader should refer to the many IMO publications on the subject.□

MARINE INSPECTION DIVISION

MARINE SAFETY DEPARTMENT
OFFICE OF DEPUTY COMMISSIONER
THE REPUBLIC OF LIBERIA
MINISTRY OF FINANCE
BUREAU OF MARITIME AFFAIRS

INSPECTION PORT	
LAST INSPECTED	DATE THIS INSP.
TIME COMMENCED	TIME COMPLETED
☐ INITIAL ☐ ANNUAL	☐ SPECIAL
☐ QUARTERLY	☐ SPOT

REPORT OF SAFETY INSPECTION

NOTE: This form is to be completed in triplicate by the Nautical Inspector on the occasion of each inspection. The original and all copies are to be signed by both the Nautical Inspector and the vessel's Master (or his authorized representative in his absence). The green copy shall be delivered by the Nautical Inspector to the Master and retained on board by the latter as part of the vessel's official documents. All remaining copies must be returned to the office of the Deputy Commissioner. The Master shall produce the vessel's copy of the previous report on request to the Nautical Inspector on the occasion of any subsequent inspection.
REFER TO THE LIBERIAN MARITIME REGULATIONS, CHAPT. VII. REG. 7. 191

PART A. GENERAL

NAME OF VESSEL	2. OFF. NO.	GROSS TONS	KW PROPULSION PWR.	3. TYPE OF VESSEL	4. YR. BUILT

5. LAST PORT OF CALL	6. NEXT PORT OF CALL	BRIDGE ☐ AMIDSHIP ☐ AFT

7. VESSEL CLASSED BY				8. NAME, ADDRESS LOCAL AGENT	9. NAME, ADDRESS PRINCIPAL MANAGING OPERATOR
	ABS		GL		
	BV		LRS		
	DNV		NK	TLX OR TWX	TLX OR TWX
			RIN	TELEPHONE	TELEPHONE

PART B. SHIP DOCUMENTS

10. NAME OF CERTIFICATE	Expiry Date	Last Date of Endorsement	NAME OF CERTIFICATE	Expiry Date	Last Day of Endorsement
a. Cargo Ship Safety Equipment			g. Fitness for Carriage of Liquefied Gases in Bulk		
b. Cargo Ship Safety Radiotelegraphy/Phony			h. Passenger Ship Safety		
c. Cargo Ship Safety Construction			i. Mobile Offshore Drilling Unit (MODU)		
d. Load Line					
e. International Oil Pollution Prevention (IOPP)			j. National Cargo Ship Safety (Vessels under 500 GT or storage/special service)		
f. Fitness for Carriage of Chemicals in Bulk			k. *Liberian Financial Responsibility Civil Liability Convention 1969		

PART B-1

11b. Ship Radio Station License Number ______
Expiration Date ______

PART C. PUBLICATIONS (Indicate if on board)

12. NAME OF PUBLICATION	Y	N	Yr. of Issue		Y	N	Yr. of Issue
a. Combined Publications Folder RLM-300 Contents apparently intact. Latest transmittal sheet No.				l. MARPOL 73/78			
b. Liberian Articles of Agreement				m. IMO-SOLAS 1974-Amended to 1983			
c. Medical Guide				n. IMO-INT. Conv. on Load Lines, 1966			
d. Medical Log Book				o. IMO-INT. Maritime Dangerous Goods Code 1986			
e. Int. Health Regulations				p. IMO Code of Emergency Proc. for Dangerous Gds.			
f. ICS Tanker Safety Guide (Gas Chemicals)				q. IMO Code of Safe Practice for Sld. Blk. Cargoes			
g. ICS Int. Safety Guide for Oil Tankers and Terminals				r. IMO Code of Safe Practice for Tmbr. Deck Cargoes			
h. ICS Guide to Helicopter/Ship Operations 1982				s. Manual for Use by the Maritime Mobile and Maritime Mobile Satellite Services "ITU Blue Book" 1982			
i. OCIMF/ICS Peril at Sea & Salvage 1982				t. Medical Guide for Dangerous Goods			
j. Clean Seas Guide for Oil Tankers				u. IMO Regulations for Prev. Coll. at Sea			
k. Standard Marine Vocabulary				v. Accident Prevention Code			

* Must be issued by the Republic of Liberia
Applies only to Tankers carrying more than 2000 tons of oil in BULK as cargo.

FORM RLM-252A REPORT OF SAFETY INSPECTIONS (Rev 1/87)

PAGE 1

Name of Vessel: ______________________ Official Number: ______________________

PART D. LICENSING AND MANNING

13. DECK OFFICERS Note: O—Original license C—Copy of license P—Posted (if posted mark with a check sign)

a. Name	b. Date engaged	c. Capacity in which serving	d. Liberian License Grade					WS*		Serial Number	Date Issued	Sighted			e. Foreign License Grade							Country Of Issue
			M	1	2	3		Y	N			O	C	P	M	1	2	3			N	
		Master																				
		Ch. Mate																				
		2nd Mate*																				
		3rd Mate*																				

RESTRICTED DECK AND ENGINEERING LICENSE MUST BE VERIFIED AS BEING ADEQUATE FOR TONNAGE AND HORSEPOWER OF VESSEL.

14. ENGINEERING OFFICERS

a. Name	b. Date engaged	c. Capacity in which serving	d. Liberian License						WS*		Serial Number	Date Issued	Sighted			e. Foreign License Grade							Country Of Issue
			C	1	2	3	S	M	Y	N			O	C	P	C	1	2	3	S	M	N	
		Ch. Eng. (1E)																					
		*1st Asst. (2E)																					
		*2nd Asst. (3E)																					
		*3rd Asst. (4E)																					

15. VESSEL HOLDING REDUCED MANNING CERTIFICATE ISSUED BY BUREAU OF MARITIME AFFAIRS ☐ YES ☐ NO EXP. DATE: ______

Vessels with reduced manning certificate issued by BMA must produce same to inspector. Chief of Division must be notified if vessel does not carry certificate.

*Citation must be issued when these officers are not holding Liberian licenses but they are holding national licenses in appropriate grades. Citation will be issued even when application for Liberian license has been submitted to BMA.

16. RADIO OFFICERS

a. Name	b. Date	c. Liberian License Grade								Serial Number	Date Issued	Sighted			d. Foreign License Grade							Country Of Issue
		GEN	1	2	S	Rtf			N			O	C	P	1	2	S	Gen	Rtf		N	

*WATCH STANDERS

FORM RLM-252 B REPORT OF SAFETY INSPECTION (Rev. 1/87)

PAGE 2

Name of Vessel: ______________________ Official Number: ______________________

PART E VESSEL HOLDING MINIMUM SAFE MANNING CERTIFICATE ISSUED BY REPUBLIC OF LIBERIA ☐ YES ☐ NO

17. SHIP'S COMPLEMENT		
OFFICERS	Manning* Requirements	Actual Number
Deck		
Engine		
Others		
CREW		
Deck		
Engine		
Catering		
Others		

*As shown in Minimum Safe Manning certificate

PART E-I. 18 LIBERIAN SEAMAN'S IDENTIFICATION AND RECORD BOOK

18a. Are all crew members holding Liberian Seaman's Books ☐ YES ☐ NO

18b. Are there on board the minimum number of persons holding Liberian Special Qualifications as required by the Minimum Safe Manning Certificate

☐ YES ☐ NO CHECK BELOW

	ABLE SEAMAN		FIREMAN/ WATER TENDER		TANKERMAN (PETROLEUM) ASSISTANT		TANKERMAN (PETROLEUM) PERSON IN CHARGE		SURVIVAL CRAFT CREWMAN
	ORDINARY SEAMAN		OILER		TANKERMAN (LIQUID CHEM) ASSISTANT		TANKERMAN (LIQUID CHEM) PERSON IN CHARGE		OTHER
	JUNIOR ORDINARY SEAMAN		WIPER		TANKERMAN (LIQUIFIED GAS) ASSISTANT		TANKERMAN (LIQUIFIED GAS) PERSON IN CHARGE		

PART E-II. CREW ACCOMMODATIONS

19. VISUAL OBSERVATION OF CREW ACCOMMODATIONS

1. Medicine chest with instructions ______________________
2. Access and escape arrangements ______________________
3. Stowage of ship's stores or equipment in crew spaces (not permitted) ______________________
4. Lighting ______________________
5. Ventilation (Efficiency-obstructions) ______________________
6. Interior finishing or decoration ______________________
7. Sanitary (check toilets) ______________________
8. Drinking water (Check supply & arrangements) ______________________
9. Galley ______________________

FORM RLM-252C REPORT OF SAFETY INSPECTION (Rev. 1/87)

PAGE 3

Name of Vessel: ______ Official No.: ______

PART F. BRIDGE LOG BOOK

20. BRIDGE LOG BOOK

LANGUAGE—

Entries	Y	N		Y	N
Gear Tests			Entries		
Fire and Boat Drills			Shift from Auto to Hand Steering		
Arrival and Departure Drafts			Speed Changes		
Monthly Test of Emergency Steering Gear			Sea State and Weather Conditions		
Positions and Times			Helmsmen		
Courses and Changes			Lookouts		
Monthly Inspection of Lifesaving Appliances			Distance Off Navigational Aids		
Soundings-Bilge and Tanks			Line-throwing Gun Drills Instructions		

PART F-I. NAVIGATIONAL AIDS

21. EQUIPMENT	Fitted		Op.		Remarks:					
	Y	N	Y	N						
a. Radar No. 1.					True motion		Relative motion			
b. Radar No. 2					True motion		Relative motion			
c. Decca Navigator										
d. Loran										
e. Echo sounding device					Graph		Visual			
f. Radio direction Finder					Date last calibrated:			Date last check bearings taken:		
g. Radiotelephone — mf					Wheelhouse		Chart Rm.		Radio Rm.	
h. Radiotelephone — uhf					Wheelhouse		Chart Rm.		Radio Rm.	
i. Radiotelephone — vhf					Wheelhouse		Chart Rm.		Radio Rm.	
j. Gyro Compass Repeaters					Steering		Emercy. Steering			
k. Gyro Compass — Master										
l. Magnetic Compass					Steering		Emercy. Steering	Dev.		Last adjusted:
m. Rudder Indicator										
mm. Rate of Turn Indicator										
n. Speed and Distance Indicator										
o. Course Recorder										
p. Course Recorder paper					Spare rolls on board: ( ) Yes ( ) No					
q. Auto Pilot										
r. Radar plotting facilities/ARPA					Describe:					
s. Signal Light(s)					Fixed: ( ) Yes ( ) No Portable: ( )Yes ( ) No					
t. Signal Flags										
u. On-board communications system					Describe:					
v. Aldis Lamp										
w. Satellite										

FORM RLM-252D REPORT OF SAFETY INSPECTION (Rev. 1/87)

PAGE 4

Name of Vessel: ______________________ Official Number: ______________________

PART G. NAVIGATIONAL CHARTS, PUBLICATIONS AND RECORDS

Where applicable inspectors must enter below Year of Issue and Date of Last Entry or Correction.

22. TITLE	On board Y	On board N	Date of last entry, correction	Remarks:
a. B.A. Chart List				Year of Issue:
b. U.S. Oceanographic Off. Chart List				Year of Issue:
c. List of Charts on Board				
d. Navigational Charts				
e. Pilot Charts				
f. Pilot Books/sailing Directions				
g. Notices to Mariners				Correction System on board and used ( ) Yes ( ) No
h. Light List				Year of Issue:
i. Radio Aids To Navigation				Year of Issue:
j. Tide Tables				Year of Issue:
k. Tidal Stream Atlas				Year of Issue:
l. Navigation Tables				Issuing Authority:
m. Nautical Almanac				Year of Issue:
n. Bridge Bell Book				
o. Engine Rm. Bell Book				
p. Compass Error Book				
q. Chronometer Rate Book				
r. Radar Log Book				
s. Radar Maintenance Record				
t. Int. Code of Signals				Year of issue: Last Amendment:
u. RDF Calibration Chart				
v. Master's Night Order Book				
w. Master's Standing Orders				
x. AMVER Instructions				Vessel participates in AMVER: ( ) Yes ( ) No
y. IMO Merchant Ship Search/Rescue Manual				Year of issue:
z. RPM/Speed Table				Posted: ( ) Yes ( ) No
aa. Maneuvering Characteristics				Posted: ( ) Yes ( ) No
bb. Liberian Oil Record Book 1983 Edition in use.				
cc. Oil Record Book Properly Maintained				
dd. Liberian Cargo Record Book For Chemical Carriers				
ee.				

N.B. INSPECTORS MUST ENSURE THAT ALL REQUESTED DATA HAS BEEN ENTERED IN PART "G" ABOVE.

Note-Tankers must carry separate oil record books for machinery space and cargo and ballast operations.

FORM RLM-252 E REPORT OF SAFETY INSPECTION (Rev. 1/87)

PAGE 5

Name of Vessel: ______________________ Official Number: ______________________

PART I. GENERAL SAFETY

23. Pilot Ladder condition		SAT		UNSAT	25. Date lifeboats last lowered into water
23a. Mechancial Pilot hoist		SAT		UNSAT	26. Date last 3 boat drills
24. Muster list and emergency instructions (posted language)		YES		NO	Date last 3 fire drills.
24a. Fire Control Plan (language) Gangway		YES		NO	26a. Training manual on Life-Saving appliances and language
24b. Fire Control Plan (language) accommodations					YES ☐ No ☐
24c. Instructions for maintenance and operation of fire fighting equipment					26b. Audio-Visual Aids in lieu of the manual YES ☐ NO ☐

27. Crew instructed in use of line throwing apparatus ☐ Yes ☐ No Date Expiry date of rockets ______________

28. Fire drill held in conformity with Guide No. 5 ☐ Yes ☐ No If not, give reason

State conditions of fire fighting equipment (Hoses, nozzles, fire extinguishers, etc.)

Number of Fireman's outfits ______________

Locations______________

Date Equipment Recharged______________

29. State conditions of lifesaving equipment (Boats, rafts, lifejackets, lifebuoys etc.)

Expiry date of rockets______________

Boat drill held in conformity with Guide No. 5 ☐ Yes ☐ No If not, give reason

Number of lifeboats carried Color

Number of liferafts carried__________

Total number of persons carried by liferafts__________

inflatable Liferafts(s)—GIVE DATE OF LAST SURVEY__________

Lifebuoys equipped with lights and self-igniting smoke signals ☐ Describe

30. General Remarks:

IMPORTANT NOTICE

NEITHER THE INSPECTION NOR THE REPORT CONSTITUTES A CERTIFICATION, WARRANTY OR OTHER REPRESENTATION AS TO THE SEAWORTHINESS OF THE VESSEL DESCRIBED HEREIN NOR DO THEY RELIEVE ANY PERSON OR ORGANIZATION FROM THEIR RESPECTIVE RESPONSIBILITIES AND OBLIGATIONS TO ENSURE THAT THE VESSEL IS MAINTAINED IN A SEAWORTHY CONDITION.

INSPECTOR		MASTER (Or Representative)	
Signature	Date	Signature	Date
Name (Print)		Name and Title (Print)	

FORM RLM-252 E (a) REPORT OF SAFETY INSPECTION (Rev. 1/87)

NOTICE TO MASTERS: Masters are required to review the Report before signing it and discuss with the Nautical Inspector any findings which may not agree with the actual condition prevailing on board.

PAGE 6

NATIONAL TRANSPORTATION SAFETY BOARD USA

Captain L. A. Colucciello, USCG (Retired)
Chief Marine Accident Division

THE National Transportation Safety Board was created by the Transportation Act of 1966 as an agency of the US Department of Transportation. As a result of the Independent Safety Board Act of 1974 (United States Code, Title 49, Section 1901), the Safety Board was established in 1975 as a totally independent agency with broadened responsibilities. The Board consists of five members appointed by the president of the United States of America with the advice and consent of the US Senate to serve for terms of five years. The appointments are made so that one term expires each year. The Safety Board and its staff comprise a relatively small agency with fewer than 350 employees.

The Safety Board seeks to ensure that all modes of transportation (air, highway, marine, pipeline, and railway) in the USA are operated safely, by investigating accidents, conducting studies, and issuing safety improvement recommendations. All Safety Board investigations are factfinding investigations, and compliance with Safety Board recommendations is voluntary. The Safety Board has no authority to conduct administrative, civil, or criminal proceedings against the principals involved in the accident, but does possess a review role with respect to actions taken by the US Coast Guard or the Federal Aviation Administration to suspend or revoke the licences or certificates of seamen and airmen.

The Safety Board is authorised to investigate maritime accidents which occur in US waters, regardless of the nationality of the ships involved, or which involve a US-registered ship. The Safety Board has promulgated joint regulations (US Code of Federal Regulations, Title 49, Part 850) with the Coast Guard that state the Safety Board will investigate accidents which result in: the loss of six or more lives; the loss of a mechanically-propelled vessel of more than 100 gross tons; property damage initially estimated as $500,000 or more; or serious threat to life, property, or the environment by hazardous materials.

The Safety Board also investigates accidents involving US public vessels and nonpublic vessels which result in at least one fatality, more than $75,000 in property damage, or significant safety issues relating to Coast Guard safety functions.

Investigative procedures

Safety Board investigations are conducted using a technical party system where all participants are involved in the development of the facts relevant to the accident. Parties are selected to participate in Safety Board investigations based on their ability to provide technical expertise to assist in the investigation. Parties are designated by the Safety Board as needed for each individual accident and may include government agencies, companies, associations, and unions.

A party may be selected because of the party's expertise in the use of its products, operation of its transportation vehicle, its activities in training personnel, or its standard-setting responsibilities. Party representatives serve as assigned and as directed by the Safety Board investigator-in-charge. Parties must be represented by suitable qualified technical employees who do not occupy legal positions. Parties cannot be represented by any person who also represents claimants or insurers.

One of the more visible aspects of a Safety Board investigation is the use of a 'go-team.' After an accident, evidence perishes and witnesses disperse quickly, especially in the maritime mode, so 'go-team' personnel must be ready to proceed to the scene of an accident quickly. The 'go-team' is a group of Safety Board employees who possess a wide range of professional expertise. In the maritime mode, the group could include a licensed deck or engineering officer, naval architect, ship inspector, expert trained in witness interrogation, metallurgist, meteorologist, hazardous materials expert, survival factors specialist, or human performance specialist.

When possible, party representatives are also designated to participate in the initial on-scene investigation. During the on-scene investigation, wreckage parts or failed machinery components are identified and selected for further testing. Logbooks, course recorder tapes, charts, bell logger tapes, and other documentary evidence is acquired. When possible, witnesses are interviewed by investigators at a convenient place near the accident site shortly after the accident. If necessary, subpoenas are issued and sworn testimony may be taken at another location at a later time.

Public hearings

The Safety Board sometimes holds fact-finding public hearings so that persons involved in the accident can testify and be questioned. In such cases, the Safety Board establishes a Board of Inquiry, which is presided over by a Safety Board member and witnesses are called to testify and be questioned under oath. Parties associated with the investigation are generally invited to participate in the public hearing process. Such hearings are always public and are designed to help the Safety Board determine the cause of an accident and provide information on which the Safety Board may be able to develop recommendations to prevent other similar accidents.

Since laboratory tests, engineering studies, and taking of sworn testimony from initially unavailable witnesses are frequently necessary, the fact-finding stage of an accident investigation is seldom completed on scene. Meetings with party representatives are held throughout the fact-finding stage to ensure that all parties have an opportunity to review all evidence and testimony, and to be sure that no important aspects of the investigation are overlooked. A final 'technical review meeting' is held to ensure that all parties are in agreement as to the pertinent facts uncovered. Parties are invited to provide proposed analyses, conclusions, and recommendations for the Safety Board to consider in the formulation of its report.

Report and recommendations

At the conclusion of each major investigation, the Safety Board determines the probable cause of the accident, issues a public report describing the facts, conditions and circumstances of the accident, and makes safety improvement recommendations to prevent similar accidents in the future. The report includes the Safety Board's analysis of the facts of the accident and specific conclusions drawn from the analysis. The Safety Board determines the 'probable cause' from its analysis of the facts, conditions, and circumstances surrounding the accident and identifies those factors which contributed to the accident.

The probable cause of the accident as stated by the Safety Board is not a legal determination of fault, guilt, or liability. No part of any report of the Board relating to any accident or investigation may be used as evidence in any suit or action for damages growing out of any matter mentioned in such report. In the event that additional important information becomes available after the Safety Board has issued its report, a party may submit the information and request modification of the Safety Board's report, if necessary.

Reports and publications

The Safety Board publishes its accident reports and safety studies. Individual reports and studies may be obtained at cost from the National Technical Information Service (NTIS), 5285 Port Royal Road, Springfield, Virginia 22161, USA, or all marine reports may be obtained from the NTIS for an annual fee. The Safety Board also makes an annual report to the US Congress regarding its accident investigations and recommendations. □

MARINE ACCIDENT INVESTIGATION: GOVERNMENT SURVEYOR'S PERSPECTIVE

Captain J. de Coverly, FNI
Principal Nautical Surveyor, Casualty and Survey Policy Branch, UK Department of Transport, Marine Directorate

THE FIRST THING a surveyor or inspector charged with the investigation of any marine accident should bear in mind is why he is doing the job. Of course, the short answer is that, like most of his other work, it is in furtherance of marine safety, but it is worth considering how such investigations contribute to this aim. They do so broadly in two ways. First and foremost, they show whether there are lessons to be learned. Evidently to find out whether there are such lessons we must find the causes of the accident, but this is not the end in itself. Secondly, casualty investigations assist in keeping up professional standards, through their association with disciplinary measures.

One cannot avoid the fact that the causes of accidents often lie partly or wholly with individuals, nor the fact that human nature being what it is, the presence of sanctions on unsatisfactory or inadequate performance does provide a healthy reminder of the need not to fall below the proper standards. The punishment or admonition of individuals must never be looked at as the principal object of the exercise, but it must be recognised as a necessary part of it.

To meet its objectives, the investigation must seek to answer four questions:

- What happened?
- How did it happen?
- Why did it happen?
- What can be done to prevent it happening again?

These questions are, I suggest, in ascending order of importance; but you cannot answer the latter ones without dealing with the others first.

Established investigation system

Regulation 21 of the 1974 Convention for the Safety of Life at Sea (Chapter 1) says that:

'Each administration undertakes to conduct an investigation of any casualty occurring to any of its ships subject to the provisions of the present Convention when it judges that such an investigation may assist in determining what changes in the present Regulations might be desirable.'

Clearly, to fulfil this obligation every country which has accepted the convention ought to include in its maritime administration some form of marine investigation system. There are, of course, several ways of achieving this, and although in the following paragraphs I concentrate on how the United Kingdom has tackled the subject, I do so to provide an illustration and not to suggest that this is the only way, or even necessarily the best. (Indeed, we are proposing some quite considerable changes in the near future and I include later a brief note about these.) But I do suggest that our system, well tried over very many years, has some substantial virtues, certainly enough to make it a useful peg on which to hang a discussion.

To sketch in a little background, it was in 1850 that under the Merchant Shipping Act of that year a new marine department was set up as part of the Board of Trade. The accident rate for British ships had been causing concern for some time and the new Act, and another which followed it a few years later, introduced or strengthened a number of provisions to regulate shipping and it was hoped improve its safety record, governing such matters as surveys of ships, the qualifications required for officers, collision regulations and so forth. It placed on the Board of Trade responsibility for the general superintendence of merchant shipping, and to assist the new department in its duty to implement that responsibility, a staff of surveyors was appointed, forming the genesis of the Marine Survey Service which still remains the technical arm of the Marine Directorate today.

It is worth saying that although we have more than once changed our name and the government department to which we are attached, the present marine Directorate of the Department of Transport is a direct and clearly recognisable descendant of the original marine department, and in essence our responsibilities are the same. Among those responsibilities, from a very early stage it seems, work on casualties has been included.

To look at how we carry out our work in this field, I am breaking the total process down into stages: information, investigation review and follow-up action.

Information

Clearly the first element required for the system is a means of knowing that a casualty has occurred, with sufficient detail to allow an assessment of whether or not an inquiry needs to be taken further. There is a statutory requirement for British ships to report both casualties to the ship and accidents to her people (and also certain specified 'dangerous occurrences'—incidents which might well have led to an accident but did not in fact do so). To back this up we study the daily reports of casualties in *Lloyd's List* and we also receive reports from our survey offices and from HM Coastguard of incidents in UK waters. While our interest is mainly in accidents to British ships we also look into incidents involving foreign ships in British waters; something like 600 or 700 reports of casualties to ships are received in a typical year (and perhaps twice that number of accidents to men and dangerous occurrences).

The report at this initial stage is usually brief, but even so in a substantial number of cases it is sufficient to indicate that the incident is minor and tells us all that we want to know. For example, a misjudgement in berthing alongside may lead to damage to the ship, but the damage is not serious, nobody is hurt and it is most unlikely that any major lessons are to be learnt from further enquiry. In such cases the casualty officer will decide 'no further action' and the matter ends there.

On other occasions, of course, the first report is sufficient to tell us that we do need to carry out a full investigation. In between there are those instances where we need more information before deciding what is required. Most commonly we obtain this by going back to the ship's owner or master, perhaps simply for a fuller report on the incident as a whole or perhaps for more detail on some specific point. If a search and rescue operation in UK waters was involved we may well ask for a report from the Coastguard. If the incident occurred in harbour waters we can ask the port or pilotage authority for information. The further report often tells us all we need, and either we close the case, simply noting the details for our records, or go straight to the final stage of follow-up action. If, however, we still need to know more we must go on to the next step, which is a departmental investigation.

Investigation

As I have mentioned, a large part of the department's responsibility for maritime safety is executed through the Marine Survey Service, a body of some 240 professional officers posted in various ports around the UK or in Marine Directorate Headquarters in London. They consist of nautical surveyors, who are master mariners, engineer surveyors (marine engineers) and ship surveyors (naval architects). There is a small section at headquarters devoted principally (though not exclusively) to casualty work and this includes a principal surveyor of each discipline, who is the casualty officer I have mentioned already. The main function of this section is to consider reports and pursue follow-up action, and although the casualty officers carry out some investigations most are undertaken by other surveyors, from whichever port is most convenient.

The investigation may be informal, when the surveyor simply visits the ship, asks questions, looks at whatever documents are to hand and makes a brief report to headquarters summarising his conclusions, but if the case is more serious the surveyor will be given an appointment specifically to carry out an inquiry into the casualty. He will then have the very considerable powers of an inspector laid down in the Merchant Shipping Acts (see Annex 1) and under them the surveyor can take statements by way of formal declarations; he can require any person he considers may have helpful information to attend him; he can call for the production of documents and if he deems it necessary, remove them; and he can also remove any items of equipment which he wishes to have examined.

His questions to witnesses must be answered (and there is a penalty for giving false answers). The proceedings are not held at any fixed place or time but where and when it is convenient for the witnesses and the inspector (this will often be on board the ship herself) and while declarations are being taken no third parties may be present except a person nominated by the witness, unless it is with the mutual agreement of both the witness and the inspector. In practice it is quite common for an owner's representative to apply to attend and the inspector will usually allow this if he is satisfied that the witness will not be inhibited and has no objection but the representative may only listen and must take no active part in the proceedings.

The surveyor when carrying out an investigation is very much his own master. He can of course call on colleagues within the department as well as external bodies if he needs some particular expertise outside his own field, but it is up to him how he proceeds with his inquiry, and the eventual report which he makes, together with his findings and recommendations, is his alone.

The report, which is confidential in its nature, goes to the casualty section in headquarters and leads us to the next stages in the process. These may include yet further investigation by way of a public inquiry. However, bearing in mind that we are looking at things chiefly from the surveyor's angle, it is appropriate to this paper to regard such proceedings—held by a specially-appointed Court, and conducted in large measure by lawyers and under the adversarial system beloved of English law—as part of follow-up action, despite their very clear investigative function.

Review and follow-up action

I have bracketed these two together because, although the review stage is a necessary and distinct part of the total system, its object is essentially to see what should be done consequent upon the casualty and it therefore seems sensible to look at the two stages together.

The report by the surveyor will conclude with his findings as to cause and recommendations on what action should follow. In reviewing the report, the casualty officer will not often seriously dissent from the former, for the surveyor is clearly in the best position to interpret the evidence he has gathered. But the latter may well require careful consideration, for in this field the casualty officer will be able to bring to bear his knowledge of other cases and perhaps has a greater familiarity with the pros and cons of the various possibilities, of which there are many.

Much more often than not, further action will be informal. For example, weaknesses in operational practices may have been shown and a senior officer of the department may interview the ship's master in confidence and point out to him how they may be improved, or if it seems that inadequate procedures were being followed because of company practice rather than that of the particular ship involved we may take the matter up with the owners, as we will if the report shows that some improvement in the ship's equipment is desirable.

In incidents within harbour limits we may think that port authority procedures need some revision and make suggestions accordingly. In other cases we have suggested to the appropriate authorities that there should be changes or additions to navigational lights or marks, or to the charts or sailing directions. When a wider promulgation of the lessons learned seems desirable we may issue an M Notice (that is, a general notice to merchant shipping), while internally lessons from casualties may be passed on to other branches within the Marine Directorate and may perhaps lead to change in the regulations or to some particular emphasis on relevant matters in the examinations.

Sometimes, of course, there is no distinct follow-up action except for adding the case to our records, but this does not mean that no useful purpose is served. Our records form a valuable source not only for the statistical summary which we publish every year but also for research. This indeed would be a worthy subject in itself and I would be wrong not to mention the importance of a good recording system designed both to show up short-term trends and to assist longer-term study of particular factors in casualties. At the moment, for example, we have in hand a project on the human element in shipping casualties, quite an ambitious piece of research which may eventually yield results of significance beyond the purely marine field.

Public Inquiries

There is a small minority of cases where some form of public action in respect of the casualty is called for. Where there has been a breach of statutory regulations, this may take the form of a prosecution of the person responsible in the courts; but though this can be appropriate in relatively straightforward cases it is clearly unsatisfactory if aspects of the casualty other than the breach of law need to be canvassed or if questions of professional competence are at issue. It is then that a public inquiry is needed, either a formal investigation or an inquiry into the conduct and fitness of a certificated officer.

The difference between the two is indicated by their titles: the former is an investigation into all aspects of the casualty and the latter specifically a disciplinary hearing. I do not wish to say a great deal about public inquiries because although they are the aspect of casualty investigation which most catches the eye, they are very much the tip of the iceberg: the bulk of the matter lies in the day-to-day work carried out quietly, and unseen by all but those directly involved. However, it would be wrong not to give a brief outline, particularly of formal investigations which form as it were the ultimate stage in the investigatory process.

Formal investigations are ordered by the Secretary of State for Transport after considering the views of the inspector who carried out the department's inquiry and the advice of the casualty officers and other professional staff at headquarters. In deciding whether or not a formal investigation is required, the following questions need to be considered:

- Is it likely that such an investigation will throw additional light on to the cause of the casualty?
- Will a formal investigation assist in publicising and pursuing the lessons of the casualty so as to prevent a recurrence?
- Did the casualty involve heavy loss of life or in some other way attract extensive public attention?
- Was there default or negligence on the part of the ship's master or officers so that disciplinary action is desirable? (If this is the main consideration then an inquiry into conduct is likely to be more suitable than a formal investigation. By and large the procedures of both types of inquiry are similar, though the inquiry into conduct being more circumscribed in its aims is generally much the briefer of the two.)

The Court before which the formal investigation is held consists of a Wreck Commissioner, who is a senior practising lawyer, and assessors of nautical engineering, naval architecture or other specialist background. The Court is nominated by the Lord Chancellor, and is thus entirely independent of the Department of Transport and of the Secretary of State, who indeed appears before it as one of the parties to the investigation. Other parties are likely to be the owners and masters of the ships involved, any officers whose certificate we consider ought to be dealt with, and any other persons or bodies who it is expected may be subject to criticism. Any other body or person can apply to be made a party, for example representatives of next of kin.

The Secretary of State does not of course normally appear in person but is represented by a lawyer, usually a barrister who specialises in maritime cases. He is not an official, though he is assisted by government solicitors and advised on technical and policy matters by the department's casualty officers (who are also often required to give evidence, as is the surveyor who carried out the initial inquiry). It is common for the other parties also to brief counsel, though there is no requirement for this and sometimes parties are represented by solicitors or indeed appear on their own behalf.

The position of those representing the Secretary of State is interesting. They must manage the case in the sense of seeing that all the necessary preparations are made, witnesses are warned to attend, documents produced and so forth. They must present an account of the incident based on the departmental inquiry, and they must give an opinion on what happened and why. But I must emphasise that when the Court is in being these duties do not give special privileges. The Court is entirely and indeed robustly independent, and this is a most important principle.

As to the proceedings, I cannot do better for a short description than quote from the rules under which they are held:

'The formal investigation shall commence with an opening statement by the Secretary of State, followed at the discretion of the wreck commissioner with brief speeches on behalf of the other parties. The proceedings shall continue with the production and examination of witnesses on behalf of the Secretary of State; and the Secretary of State may adduce documentary evidence. These witnesses may be cross-examined by the parties in such order as the wreck commissioner may direct and then be re-examined on behalf of the Secretary of State. The Secretary of State shall then cause to be stated the questions relating to the shipping casualty or incident and to the conduct of persons connected with the shipping casualty or incident upon which the opinion of the wreck commissioner is desired . . .

'Any other party to the formal investigation shall be entitled to make a further opening statement, to give evidence to adduce documentary evidence, to call witnesses, to cross-examine any witnesses called by any other party and to address the wreck commissioner in such order as the wreck commissioner may direct. The Secretary of State may also produce and examine further witnesses who may be cross-examined and re-examined . . . A party who does not appear in person at a formal investigation and is not represented may make representations in writing to the wreck commissioner . . .

'Every formal investigation shall be conducted in such manner that if substantial criticism is made against any person that person shall have an opportunity of making his defence either in person or otherwise . . .'

The hearing ends with each party making its submission to the Court, which then retires probably for some weeks before producing its report.

This is a brief description of a long process. Witnesses are commonly called in considerable numbers, usually including not only witnesses of fact but also experts in appropriate fields, and the amount of documentary evidence produced is voluminous. Most recent inquiries have gone on for several weeks so it will be appreciated that they are only resorted to rarely (on average, about one a year, with perhaps twice as many inquiries into conduct). However, they do undoubtedly have value firstly because of their complete independence and secondly because, as I have been at pains to say, the fundamental purpose of casualty investigation is to learn lessons and the virtue in learning lessons is lost unless they are transmitted to the maritime community in general. There is no more effective way of achieving this than through a public inquiry.

Discipline

A Court of inquiry, whether an FI or an inquiry into conduct, has the power to cancel or suspend an officer's certificate, but it is only in a small minority of casualties that personal fault is so serious that this extreme step is called for. As a less drastic alternative, the Court may censure an officer, but much more often, when some disciplinary action is required, it is far more appropriate that it be taken by owners—ranging from a reprimand through loss of seniority or rank to dismissal—than by a Court. As to the department, we have no general power over certificates and of course since we are not his employer we have no direct power in respect of an officer's career. Thus, the disciplinary side of the system is, in the UK and I suppose other countries where the Merchant Service is composed principally of independent companies, more often practically exercised by shipowners than by authority.

This of course emphasises the need for owners to adopt a responsible attitude to the investigation of casualties to their ships, and by and large they do so. However, having said at the beginning that whether we like it or not discipline is an important part of the total system, I do not want to give the impression that the department neglects it. The object is to maintain a presence without usurping the proper functions of owners. Most often when action is taken by the department it is through the interview which, as I have mentioned, quite often takes place as part of the follow-up. If it is thought appropriate this will include a form of reprimand which, though confidential at the time, is placed on the officer's record and can be referred to should he be involved in a further incident. In practice we find that such further reference is rare, which suggests that these unostentatious admonishments are quite effective.

The practice of investigation

I hope all this gives some insight into one approach to the framework of casualty investigation, but I realise that it says little about the actual mechanics of carrying out the job. I am thinking now about the surveyor or inspector, well experienced in his own discipline, but perhaps without too much experience of casualty investigation, and now confronted with the need to carry one out. Let me say at once that I do not presume to be able to tell him how to do it, but perhaps I can make a few suggestions. In addition, Annex 2, which is extracted from guidance notes issued to the Department's Surveyors, is of some interest.

The first thing is to get as many *facts* as possible, and these should include background material, such as details of the ship and her complement, as well as matters directly relating to the casualty. Everything does not of course have to go into the eventual report, but at this early stage the aim is to build up a picture in the mind and therefore to get as much information as possible that will help towards this.

You have two sources: documents and witnesses. In a difficult or important case, the inspector will gather documents like a squirrel stocking up with nuts for the winter—not only the obvious such as charts, plans, log books and weather reports, but radio signals, correspondence, records of the ship's trading, repair and maintenance records, photographs and even such ephemera as press cuttings. I do not wish to encourage putting together a vast dossier of paper when what happened is clear and the documents add nothing of any substance; but in a minority of cases it is well worthwhile getting hold of everything you can.

As to witnesses, these are likely to include people ashore as well as the ships's crew—for example, owner's superintendents, stevedores (if a cargo shift is a possible factor in the incident), Coastguards on search and rescue aspects, the pilot who was last on board for an experienced and independent assessment of the ship and how she handled, and perhaps local residents who may have actually seen what happened if the casualty occurred near the shore.

The list of possibilities is very long and while again I do not suggest that it should be explored to the limit in all inquiries, for this would certainly be superfluous in straightforward cases, when you are confronted with a puzzle it is worth interviewing anybody who may be able to fill in a piece of the jigsaw. Often the interview can be very short, a few minutes sufficing to clear up the specific item which the particular witness can explain, or indeed to show that he cannot in fact help at all; but it must not be hurried and with major witnesses it is likely to be a lengthy affair.

This leads on to the important matter of how interviews should be carried out. The main object is to get the person to talk, freely and openly, but he or she is likely to be nervous (even if the person has nothing to be nervous about) and this may make him or her tongue-tied. It may also have the opposite effect and sometimes one is confronted with a witness who cannot stop talking. In any event, I think most casualty inspectors would agree that even with a major witness from whom a formal statement or declaration is obviously required, it is best to start informally and try to get a good outline of what can be contributed before attempting to put anything in writing.

This makes for a long interview with witnesses of importance, but it is worthwhile in the long run. Once you have a clear idea of what the witness can usefully say you can concentrate your questions so as to develop a statement which is reasonably concise and to the point, but first let him have his say. Of course, before you begin the interview you should have in mind the sort of questions you intend to ask, and indeed with some witnesses it is a good idea to set down a list in writing—but do be prepared to change and add to them. New and valuable information can sometimes come from unexpected quarters.

As you put together the documents and consider the witnesses' statements a picture of what happened will form, sometimes quickly and clearly, sometimes only bit by bit and perhaps with some aspects seeming to contradict others. It is with the latter type of incident particularly that it is worth going back in time to see how the ship in question was managed and operated in the period before the casualty, when quite often the inconsistencies will be explained and a likely solution for the missing parts of the pattern will appear.

At this stage of the investigation you must attempt something of a mental balancing act, for while it is important to develop an idea of what happened as soon as possible so as to point your inquiries in the right direction, yet at the same time you must keep your mind open to fresh evidence. You must also of course decide when to call a halt. It must be accepted that some investigations will never be complete in the sense of every possibility having been explored to the ultimate. When in your judgment you can answer with reasonable confidence the four questions I referred to at the beginning of this chapter your task is finished—except for one last, and vital, part of it.

For, no matter how thorough and competent the investigation has been, its value will be lost if the findings are not successfully transmitted to central authority, whoever that may be. A clear report is therefore absolutely essential to complete the job. The form of the report is, of course, less important than its content, but we have found in the UK that a standard framework is useful, and some suggestions made to our surveyors are contained in Annex 2.

I would particularly stress the importance of including your own appreciation of the incident, based on your interpretation of the facts and your observation—if this is possible—of the ship, her crew, and how she is run, and of the site of the casualty. You are an experienced person and your own views will not lightly be ignored; do not hesitate to express them. But at the same time make it clear what is fact and what is conjecture or opinion.

The report should conclude with your recommendations, and these need careful thought. Sometimes, inevitably, they are unpleasant to make, particularly when one has to advise action against an officer's certificate. Even in a system such as in Britain where the final decision on such action is the responsibility of an independent inquiry, such a recommendation will not be made lightly. But all is not gloom, and sometimes this part of the report provides an opportunity which will be welcome, to give praise for good work in the face of adversity and to make constructive suggestions for the future.

Brevity is a virtue and though I have said that when collecting information the rule should be 'if in doubt, gather it in,' when it comes to the report stage do not hesitate to do some pruning. But do not overdo it: the report must provide a full picture for the benefit of the reader and it must include sound arguments to support its conclusions and recommendations. Remember also the importance of the supporting documents, which if you follow the sort of procedures outlined in Annex 2 will be submitted as appendices —not necessarily everything you have collected, but all those items which will help to illustrate and substantiate what you have said. But it is wise to bear in mind that that is the only function of the appendices. The report by itself should be complete enough to tell the story and point the moral.

The future

On the technical side, the most obvious development is the 'black box.' Such a device, very broadly similar to that already fitted in aircraft, has already been produced and operated on extensive trials and if it comes into general use it will undoubtedly prove useful to investigations. It can maintain a record of the ship's movements and of various aspects of her condition, and of the conditions of weather and sea, and thus it will answer many questions without depending on the vagaries of human recollection. Moreover, it should be self-contained in a 'float-free' container so that in at least some cases it will be recovered even when the ship is lost.

But I give a word of warning. It will often answer my first question— what happened—and sometimes my second—how it happened. But it will frequently not tell us *why* it happened, and it can at best do no more than hint at what should be done to stop it happening again. It will certainly not relieve the investigator of the need to make a balanced assessment based on judgment and experience, though it should remove some—not all—of the need for inspired guess work.

As to procedures, as I have mentioned briefly, changes are afoot in the UK system and it is worth saying a little about them, not only for the sake of completeness—since I have largely used the British example in this chapter—but also because they at least to some extent reflect trends which can be discerned elsewhere and the problems which they seek to address are by no means peculiar to the UK. I must stress, though, that at the time of writing the detail of our new system has yet to be worked out. As this chapter goes to press, the head of the new accident investigation branch had been appointed.

New procedure in the UK

Firstly, it has been decided to set up a new body specifically charged with the duty of investigating marine accidents. This Marine Accidents Investigation Branch will be part of the Department of Transport but quite separate from the Marine Directorate. There should thus be no question of conflict of interest if an investigation raises doubts as to the actions of the directorate on its staff—for example, as to the adequacy of surveys prior to the accident or of SAR procedures after it.

Some countries have already taken a similar step, and indeed one or two have gone further and set up an organisation to investigate accidents with any type of transport. The National Transportation Safety Board in the USA for example, which is an independent Federal agency, even includes pipeline accidents within its remit. (The US method is an interesting attempt to gain the benefits of investigation by an independent body without losing those of investigation by the body charged with marine safety generally, for serious casualties are aften inquired into jointly by NTSB and the US Coast Guard). Logically, there seems no reason in principle why one should not go further still and make *all* accidents, whether involving transport or not, subject to inquiry by a single statutory body; but so far as I know this is not done at present.

Secondly, it is intended that the new branch will publish its reports on the more significant incidents, unless they are remitted to public inquiry, and will produce summary information on other investigations. We have been concerned for some time that less information than is desirable has been released on our inquiries, and the aim is to rectify this so that mariners generally can profit from the lessons to be learned. Some countries confronted with the same problem meet it by releasing the inspector's report, but our tentative intention is rather different: the published report will be that of the chief inspector, so that the inspector can retain his present independence and carry out the inquiry in his own individual fashion.

Thirdly, it is likely that a new type of public inquiry or review will be introduced, to give independent consideration to the branch's report particularly (though perhaps not exclusively) when the report's findings are critical of some person and he dissents from those findings. It will be much more limited in its scope than a formal investigation; it is not intended to do away with 'Formals' altogether, but as I have said they can only be held in a very small minority of cases, because of their complexity and expense, and it is felt that there is a place for a more simple and much briefer type of inquiry to augment them. It will be most interesting to see whether this proves a useful step, and indeed to see how the new system develops generally. I believe that the outline approach addresses real weaknesses in our present procedures, but it is vitally important that in pursuing it we do not destroy the sound foundations which have been laid over many years, but rather build upon them.

Finale

To conclude, I go back to the fundamental subject of this chapter, the appointed surveyor's duty to carry out an inquiry. Beyond doubt casualty investigation is one of the most interesting and at times challenging aspects of a surveyor's job. Moreover, it is one which is clearly worthwhile, for many inquiries have led eventually to real improvements in maritime safety. It must be said that it is also frequently rather a sad business and in a good many cases there is a considerable element of 'there but for the grace of God go I.'

Often, of course, the cause of the casualty is quite obvious, sometimes almost painfully so, but there are also many cases when the causes are complex and varied, and the investigation is particularly difficult in those tragic incidents where a ship is lost with all hands and there is therefore no first-hand evidence to help the surveyor in his enquiries. In the last resort, whether or not you have a 'black box' to help you, you must fall back on your own experience, to put yourself in the mind of the ship's master and others on board, so as to make the best possible assessment of what happened. If you do this I think you will find that more often than not the pieces fall into place.

In preparing this chapter, I have drawn heavily upon others, particularly colleagues both past and present, and I would like to thank them but the views I express are my own. In particular I must stress that they are not necessarily those of the United Kingdom Department of Transport.□

ANNEX 1: EXTRACT FROM THE (UK) MERCHANT SHIPPING ACT 1979

*Powers of Department of Trade Inspectors**

27.—(1) An inspector appointed in pursuance of section 728 of the Merchant Shipping Act 1894

(a) may at any reasonable time (or, in a situation which in his opinion is or may be dangerous, at any time)—
   (i) enter any premises in the United Kingdom, or
   (ii) board any ship which is registered in the United Kingdom wherever it may be and any other ship which is present in the United Kingdom or the territorial waters of the United Kingdom,

   if he has reason to believe that it is necessary for him to enter the premises or board the ship for the purpose of performing his functions as such an inspector;

(b) may, on entering any premises by virtue of paragraph (a) above or on boarding a ship by virtue of that paragraph, take with him any other person authorised in that behalf by the Secretary of State and any equipment or materials required to assist him in performing the said functions;

(c) may make such examination and investigation as he considers necessary for the purpose of performing the said functions;

(d) may, as regards any premises or ship which he has power to enter or board, give a direction requiring that the premises or ship or any part of the premises or ship or any thing in the premises or ship or such a part shall be left undisturbed (whether generally or in particular respects) for so long as is reasonably necessary for the purposes of any examination or investigation under paragraph (c) above;

(e) may take such measurements and photographs and make such recordings as he considers necessary for the purpose of any examination or investigation under paragraph (c) above;

(f) may take samples of any articles or substances found in any premises or ship which he has power to enter or board and of the atmosphere in or in the vicinity of any such premises or ship;

(g) may, in the case of any article or substance which he finds in any such premises or ship and which appears to him to have caused or to be likely to cause danger to health or safety, cause it to be dismantled or subjected to any process or test (but not so as to damage or destroy it unless that is in the circumstances necessary for the purpose of performing the said functions);

(h) may, in the case of any such article or substance as is mentioned in paragraph (g) above, take possession of it and detain it for so long as is necessary for all or any of the following purposes, namely—
   (i) to examine it and do to it anything which he has power to do under that paragraph,
   (ii) to ensure that it is not tampered with before his examination of it is completed,
   (iii) to ensure that it is available for use as evidence in any proceedings for an offence under the Merchant Shipping Acts or under regulations made by virtue of any provision of those Acts;

(i) may require any person who he has reasonable cause to believe is able to give any information relevant to any examination or investigation under paragraph (c) above—
   (i) to attend at a place and time specified by the inspector, and
   (ii) to answer (in the absence of persons other than any persons whom the inspector may allow to be present and a person nominated to be present by the person on whom the requirement is imposed) such questions as the inspector thinks fit to ask, and
   (iii) to sign a declaration of the truth of his answers;

(j) may require the production of, and inspect and take copies of or of any entry in,—
   (i) any books or documents which by virtue of any provision of the Merchant Shipping Acts are required to be kept; and
   (ii) any other books or documents which he considers it necessary for him to see for the purposes of any examination or investigation under paragraph (c) above;

(k) may require any person to afford him such facilities and assistance with respect to any matters or things within that person's control or in relation to which that person has responsibilities as the inspector considers are necessary to enable him to exercise any of the powers conferred on him by this subsection.

(2) It is hereby declared that nothing in the preceding provisions of this section authorises a person unnecessarily to prevent a ship from proceeding on a voyage.

(3) The Secretary of State may by regulations make provision as to the procedure to be followed in connection with the taking of samples under subsection (1) (f) above and subsection (6) below and provision as to the way in which samples that have been so taken are to be dealt with.

(4) Where an inspector proposes to exercise the power conferred by subsection (1) (g) sbove in the case of an article or substance found in any premises or ship, he shall, if so requested by a person who at the time is present in and has responsibilities in relation to the premises or ship, cause anything which is to be done by virtue of that power to be done in the presence of that person unless the inspector considers that its being done in that person's presence would be prejudicial to the safety of that person.

(5) Before exercising the power conferred by subsection (1) (g) above, an inspector shall consult such persons as appear to him appropriate for the purpose of ascertaining what dangers, if any, there may be in doing anything which he proposes to do under that power.

(6) Where under the power conferred by subsection (1) (h) above an inspector takes possession of any article or substance found in any premises or ship, he shall leave there, either with a responsible person or, if that is impracticable, fixed in a conspicuous position, a notice giving particulars of that article or substance sufficient to identify it and stating that he has taken possession of it under that power; and before taking possession of any such substance under that power an inspector shall, if it is practicable for him to do so, take a sample of the substance and give to a responsible person at the premises or on board the ship a portion of the sample marked in a manner sufficient to identify it.

(7) No answer given by a person in pursuance of a requirement imposed under subsection (1) (i) above shall be admissible in evidence against that person or the husband or wife of that person in any proceedings except proceedings in pursuance of subsection (1) (c) of the following section in respect of a statement in or a declaration relating to the answer; and a person nominated as mentioned in the said subsection (1) (i) shall be entitled,

---

*Note: It is intended that these powers will continue unchanged under the new system described previously in this chapter.

on the occasion on which the questions there mentioned are asked, to make representations to the inspector on behalf of the person who nominated him.

Provisions supplementary to s.27

28.—(1) A person who—

(a) wilfully obstructs a Department of Trade inspector in the exercise of any power conferred on him by the preceding section; or

(b) without reasonable excuse, does not comply with a requirement imposed in pursuance of the preceding section or prevents another person from complying with such a requirement; or

(c) without prejudice to the generality of the preceding paragraph, makes a statement or signs a declaration which he knows is false, or recklessly makes a statement or signs a declaration which is false, in purported compliance with a requirement made in pursuance of subsection (1) (i) of the preceding section,

shall be guilty of an offence and liable on summary conviction to a fine not exceeding £1,000 or, on conviction on indictment, to imprisonment for a term not exceeding two years or a fine or both.

(2) In relation to a person, other than a Department of Transport inspector, who has the powers conferred on such an inspector by the preceding section—

(a) that section and the preceding subsection shall have effect as if for references to such an inspector there were substituted references to the person; and

(b) that section shall have effect as if for references to the functions of such an inspector there were substituted references to the functions in connection with which those powers are conferred on the person.

(3) Nothing in the preceding section shall be taken to compel the production by any person of a document of which he would on grounds of legal professional privilege be entitled to withhold production on an order for discovery in an action in the High Court or, as the case may be, on an order for the production of documents in an action in the Court of Session.

(4) A person who complies with a requirement imposed on him in pursuance of paragraph (i) (i) or (k) of subsection (1) of the preceding section shall be entitled to recover from the person who imposed the requirement such sums in respect of the expenses incurred in complying with the requirement as are prescribed by regulations made by the Secretary of State, and the regulations may make different provision for different circumstances; and any payments in pursuance of this subsection shall be made out of money provided by Parliament.

(5) References in the Merchant Shipping Acts to a Department of Transport inspector are to an inspector appointed in pursuance of section 728 of the Merchant Shipping Act 1894.□

ANNEX 2: EXTRACT FROM GUIDANCE NOTES ISSUED TO UNITED KINGDOM SURVEYORS*

General procedures. Owners or masters have a statutory obligation to report shipping casualties to the department. However, in the majority of cases the department first learns of a casualty from some other source, most commonly either from HM Coastguard or from the daily casualty report in *Lloyd's List* which is scanned at headquarters each morning. The bulk of reports are received at headquarters but some go direct to the local marine office; in the latter case the district chief or senior surveyor may decide to take action but will also speak to the appropriate casualty officer at HQ so as to avoid duplication of effort (as a report may also have been received there), to agree the correct level of inquiry, and so that administrative procedures—the raising of a file and the despatch of casualty record forms and any other relevant papers—may be put in train.

In some minor cases the initial report is noted and recorded and no further action is necessary. In slightly more serious cases a request may be sent for an amplifying report. If, however, more than that is required then a surveyor will be called to investigate the incident and report his findings. The surveyor's investigation may be at one of three levels: an informal inquiry; a report with declarations; or a preliminary inquiry.

Informal inquiries. For these, the surveyor has and requires no special appointment and his powers are those attaching to his everyday duties as a surveyor of ships. Thus, he has no power to *require* witnesses to answer questions, give statements, or produce documents. This does not usually cause problems as in the very large majority of cases ship's personnel co-operate willingly; often they will volunteer a statement (possibly a copy of one already made to owners) and will generally if requested produce relevant documents such as charts and logbooks, if these are available. It should be noted that where a voluntary statement is taken it may be signed but must not take the form of a declaration, for declarations are only appropriate when an appointment has been issued.

When exceptionally the surveyor's lack of powers does give rise to difficulty, he should consult the district chief surveyor and if necessary an appointment as inspector can be issued. HQ should be advised of any such change in the status of an investigation. Surveyors should also report to the chief surveyor if their initial inquiries suggest that the casualty has more serious aspects than was first thought so that, again, consideration may be given to changing the level of the inquiry.

Surveyor's report with declarations. For rather more serious cases or where some particular consideration is thought to make it desirable a surveyor will be issued with an appointment giving him the powers of an inspector. Most importantly, he can *require* a witness to attend for interview, to answer his questions, to sign a declaration of the truth of his answers, and to produce documents. A person being questioned is obliged to answer and commits an offence by refusing to do so.

The examination of a witness is solely a matter between the witness and the inspector. But a witness has a legal right to nominate a person to accompany him. This person may advise the witness on any point but he may not prevent the witness giving a direct answer to any questions which the inspector may pose.

Owners sometimes ask if they or their bona-fide representative may be present when crew members from their ship are examined. This is entirely at the discretion of the inspector. If he suspects that a witness might be inhibited by the presence of the owner, the request should be refused. If the inspector is disposed to agree to the request, he should first seek, in private, the agreement of the witness to the owner's presence. In no circumstances should the owner or his representative be allowed to interfere with the course of the interview. Although less common, a person other than the owner may ask to be present during an interview. The same considerations apply in such circumstances.

Declarations should generally be obtained from the master (or senior surviving officer), the officer on watch, the chief (or senior surviving) engineer and the engineer on watch and from any other important witnesses such as the helmsman and the lookout according to the circumstances of the casualty. Members of the crew who are on watch below at the time of the casualty (e.g., engineroom personnel) can often furnish useful information. In certain cases, it might prove necessary to take declarations from others such as the owner, shipbuilder, shiprepairer, shipper or ship's agent, and any other witnesses who could make a useful contribution.

The text of a declaration should begin with the witness giving brief details of his experience and qualifications, and the surveyor's accompanying report should include comments on the way in which witnesses gave their evidence and the surveyor's views on any discrepancies between individual statements. A copy of the declaration should be given or sent to the witness but to no other person; though the witness may of course if he wishes hand his copy to, for example, his or the owners' solicitor, any requests by such persons to the surveyor for copies should be referred to headquarters.

Preliminary inquiries. In the most serious cases, a surveyor may be appointed to conduct a preliminary inquiry. His powers are still those of an inspector as outlined in the previous paragraph and, apart from signifying the serious view taken of the casualty, the difference between a PI and a surveyor's report with declarations is to a large extent an administrative one. [For this reason, no distinction has been drawn between the two types of inquiry in the body of this paper.] However, a point of some importance is that, although it is not an absolute requirement, it is normal practice only to hold a formal investigation after a PI. Therefore, if a surveyor carrying out a lower level of inquiry considers as the circumstances unfold that a formal investigation may be called for, he should report accordingly as soon as possible so that the inquiry, and his appointment, may be up-graded.

Reports. The depth of the surveyor's inquiries and the extent of his report should be related to the nature of the casualty. In many cases, particularly those covered by informal inquiries, the incident will be straightforward, its causes will rapidly become apparent, there will be no new lessons to be learnt, and the surveyor will not wish to make many recommendations. In such instances, both the investigation itself and the surveyor's report should be kept brief. On the other hand some casualties, even among those which are quite minor in themselves, will require searching and lengthy investigation. Surveyors should remember that the object of casualty investigation is to prevent, or at least reduce the likelihood of, a repetition, and if points emerge from which valuable lessons may be learnt these should be fully explored irrespective of the gravity of the original incident.

*With minor alterations for presentation.

Suggestions for the presentation of a report on a major casualty are given in the next section, and in general reports on lesser incidents should follow the same lines. They should thus contain three sections: a brief introductory summary; a factual account; and the surveyor's comments, conclusions and recommendations. The surveyor should not hesitate to express opinions or draw inferences but it must be quite clear what is fact and what is conjecture; to facilitate this an additional section covering a supposed course of events may sometimes conveniently be interposed between the factual account and the concluding section, where the former alone is inadequate to describe the incident. The division into distinct sections is important and should be followed even in simple cases where each section may well consist of no more than a single paragraph.

The report should always include the surveyor's recommendations for further action, even if these are of an interim nature as for example when it is considered that more information should be sought from other sources, or indeed if the only recommendation is that no further action is required. Recommendations may include, but should not be confined to, disciplinary action; they may also cover any point (even if not directly related to the casualty) which the investigation has thrown up and which would, in the surveyor's opinion, help to prevent a recurrence. Acts of gallantry or merit should also be mentioned. Reference should be made to any action in respect of the casualty which has already been taken; this should include disciplinary action taken by owners, if details are known, for in moderately serious cases this action may well be sufficient and indeed more appropriate than any open to the department.

Reports of preliminary inquiries (major cases)

In general, the inspector should ensure that the following aspects of a casualty are adequately investigated and reported on as is appropriate to the particular incident:

(a) If loss of life occurred, how it was caused.
(b) Any defects in the hull, machinery or equipment of the ship which may have led to or contributed to the casualty.
(c) The ship's standard of stability. In cases where the standard may be held to be in question, then a full examination should be undertaken using the Department's computer facilities.
(d) If pollution occurred, its extent and nature and how it was caused.
(e) The adequacy and functioning of the safety appliances with which the ship was provided and the effectiveness of the precautionary or remedial measures which shipmasters are instructed or advised to take, and in particular whether any injuries or deaths may have been due to causes which may have been prevented if other appliances had been available or better arrangements had been made or if other advice or instructions had been issued.
(f) The operation and efficacy of the ship's radio equipment and navigational aids.
(g) The nature of the damage to the ship resulting from the casualty.
(h) The prevailing weather conditions, including the degree of visibility.
(i) The state of the cargo, how it was loaded and whether the ship was overloaded.
(j) Whether relevant statutory requirements, M Notices, codes of practice, etc., had been complied with.
(k) The cause or probable cause of the casualty.
(l) Details of any acts of gallantry.
(m) Rescue services rendered by other ships or any other help given to the ship or survivors.

In a complex case it may take some time to assemble every item of evidence, e.g., a full examination of the stability or calculation for alternative stability conditions, results of research, declarations from witnesses who are out of the country, etc. Unless such evidence is central to the inspector's conclusions, submission of the report should not be delayed on this account. A supplementary report can always follow the original submission.

The report should be submitted in standardised form; the paragraphs should be numbered in the 'decimal' system. It should be divided into summary; factual report; inspector's comments; and appendices.

Summary

The summary is a necessary aid to busy senior officers. It should form the first page of every PI report, and should normally be contained in not more than 300 words. It should contain a short summary of the sequence of events, the inspector's conclusions as to the reasons for the casualty, and his recommendation as to any action to be taken as a result.

Factual report

This section should describe the events leading up to the casualty by reference to the declarations of witnesses and other direct evidence. It should be confined to matters of fact, and should offer no interpretation of the evidence, nor draw conclusions. The sources of all the facts from which the inspector has determined the causes and/or possible causes of a casualty should be identified. Where an inspector has inspected a ship, its equipment, cargo, etc., following a casualty, he should make this clear and describe what he found. Cross-references should be made in the margin to the appropriate appendix document from which any other facts are drawn. In other words, it should always be made clear whether facts reflect the declarations of witnesses, what the inspector saw for himself or have been drawn from some other source.

The factual report should contain information on the following:

(a) Background (e.g., a description and factual information about the ship, the crew, the ship's equipment, cargo and voyage, weather, relevant operational arrangements and procedures, etc.).
(b) Events leading to the casualty.
(c) Sequence of events following the casualty, including search and rescue.
(d) Other relevant circumstances and events after the casualty.

Inspector's comments, conclusions and recommendations

This section should contain the inspector's comments on the reliability of witnesses or other evidence, comments on the attitude and behaviour of individuals, conclusions as to the reasons for the casualty, recommendations on any action necessary to prevent a recurrence, and a recommendation on whether or not to seek a formal investigation. The following headings may form useful guidelines but are not exhaustive:

(a) Reliability of witnesses and other evidence.
(b) Breaches of the Merchant Shipping Acts.
(c) Discussion on the sequence of events and other related matters.
(d) Execution of emergency procedures.
(e) Search and rescue operation.
(f) Cause of the casualty.
(g) Measures which might have prevented the casualty.
(h) Recommendations as to FI and any other action.

Appendices

Supporting documents should accompany the Report, and should include as appropriate: declarations, plans of ship, plans, diagrams or records of equipment, relevant statutory certificates, etc., charts, crew and passenger lists, log extracts, cargo details, SAR reports, weather reports, transcripts of distress traffic, press cuttings and photographs, and other appendices at the inspector's discretion.

Each appendix should be given a number and where it consists of several items, e.g., in the case of declarations, the separate items within each appendix should in addition be lettered, and numbered e.g., A2.1, A2.2 etc, and a contents list attached to the front of each appendix. Appendix 1 should always be a complete list of all remaining appendices, giving the number and title of each. The remaining appendices should be arranged in the order shown above if this is logical, given the circumstances of the casualty.□

THE HUMAN FACTOR IN ACCIDENT INVESTIGATION

Martyn B. A. Dyer-Smith, BA(Honours), AMNI
Tavistock Institute of Human Relations, London

IT IS SOMETIMES claimed that 80 to 90 per cent of accidents are caused by 'human error'. This term can be very misleading, however. With reference to marine casualties it is more accurate to say that 'human involvement' is found in a very high proportion of those cases *where there is sufficient evidence to make a judgement.* (This by the way does not necessarily imply that the human element was *the* causal factor.)

The evidence of human beings is rarely, if ever, unequivocal. If witnesses disagree, how do we decide whose testimony we should accept? And if they do agree on every single detail might we not have reason to suspect collusion? No task is more difficult than that of assessing the human factor.

What is an accident?

We usually think of an accident as an unexpected event, certainly unintended and with distinctly unpleasant outcomes. A 'true' accident is, almost by definition, neither predictable nor controllable. By way of illustrating this point, we would not normally consider the failure of a light-bulb to be an accident. Although we cannot predict the precise life of an individual bulb, the range of our expectations about its performance is fairly limited.

The same cannot be said for the behaviour of human beings. It is because the nature of the human element can not always be known in advance with absolute certainty, that events with a distinct human element involvement are unpredictable, and more difficult to manage or contain in consequence.

Of course, this is not to say that we are ignorant of the mental and physical capacities of human beings, and from the earliest times such knowledge—initially derived from trial and error—has been incorporated into technical design. For example, the leadline knotted and labelled at intervals with leather or bunting enabled a man by sight or by texture to sense the depth of water under his ship by day or by night. But the leadsman who economises his effort by 'swinging the lead' rather than letting it run to the bottom can easily defeat the accuracy of the measurement. Technology may change 'for the better', but uncertainties regarding the actual effect of the intended improvements will always endure as long as human operators are involved.

Uncertainty characterises many systems. In the accident domain, similar techniques to those found so useful for predicting the performance of light-bulbs have been applied to the behaviour of people, with some success. The evolution of the design and use of motorways provides a comparative example. Statistical analysis may reveal that accidents occur predominantly at bends, and that oncoming traffic is the main danger. By eliminating bends and separating trafffic flows we have produced, in theory, 'the safe road'. Unfortunately, driver behaviour changes with the new environment. The sensation of speed may be warped and the lack of activity may induce torpor. Similarly, we may propose that since so many marine collisions occur in fog then radar, by 're-sighting' the navigator, should eliminate the problem. But here also we have seen how changes in behaviour, particularly in terms of increasing ship speed, may offset the expected gains in safety.

One of the most worrying current features of marine accidents has been their apparent resistance to technological solutions. The human element is probably the reason.

Why investigate accidents at all?

In the field of marine casualties investigations are undertaken for three main purposes:

(i) To explain and understand the event, with the aim of preventing recurrence;

(ii) To establish blame and liability (and if necessary to punish or provide damage compensation);

(iii) To comply with legal requirements of the relevant maritime authorities.

Here we are concerned with the first issue only, although we cannot ignore the others, which can have critical influences on the access to and reliability of data necessary to explain and understand the part played by the human element in marine casualties.

What needs explaining?

Of the many problems confronting the investigator in explaining and understanding the event, the most fundamental is to decide what should constitute the facts which require explanation.

It is easy to forget how subjective this choice can be. As one senior surveyor remarked: 'I collect all the information I think is relevant.' This kind of selective data gathering may well be a function of the particular discipline of the casualty inspector appointed. For instance, a nautical surveyor may be expected to search for 'navigational facts' such as violation of the Collision Regulations. Indeed, his appointment indicates that the tentative decision has already been made that the particular casualty concerned is a 'nautical problem.'

The way we choose our facts for investigation will determine the method itself. This point can be neatly expressed by the well-known 'figure-ground' problem, where one may choose to see either a vase or two faces, depending on which part of the picture is regarded as being the figure and which part the background.

In this particular case we are able to decide consciously, but in many cases the decision is determined subconsciously by the context.

What about this one?

Once we know that the figure above is the logo of the National Express coaches we can entertain the notion that it could be the letter 'N' and its mirror or shadow image. Knowledge can thus be seen to be a critical element in deciding which facts to pay attention to and which to ignore.

Furthermore, much of what we will perceive, and indeed accept, as causal may well be dependent on the possibilities we believe exist for intervening with prevention in mind. For instance, we can say that 100 per cent of falls are 'caused by gravity', but we do not usually regard gravity as a cause since it is not feasible to remedy gravity.

Human factor implicit

The investigator must keep firmly in his mind that all information available to him is permeated by the human factors, including his own, involved in the very process of data collection itself. For example, the investigator may find that he has to rely heavily on the reasons that one particular witness gives for his actions, but for these reasons to constitute a 'valid' explanation in the investigator's estimation, some overlap will be required in the value systems of both the witness and the investigator.

It may be helpful for the investigator to approach the accident from a variety of different perspectives in order to gain as complete a picture as possible of what happened. A useful analogy to draw upon might be the inferences we can make about a tree from the contours of its shadow (where the tree represents the incident, and its shadow the information available on the incident). Given that the information about any particular casualty can never be 100 per cent 'total', the proposition is that we may gain deeper insight into an incident by 'illuminating' it with questions from different perspectives. From the shadow's contours, although each is individually limited in dimension, we may hopefully put together a composite picture of that tree.

Thus, the best approach may well be that of a questioning perspective which generates alternative hypotheses, rather than seeking to determine the most logical cause or 'single best explanation'. This approach also aims to avoid a prejudged stance or 'jumping' to conclusions. What questions should we then be asking to find out about the human element?

What is the human factor?

In developing a list of questions we may find it helpful to consider a number of alternative viewpoints about what consitutes an accident. Thus:

(i) The accident may be a pure chance event;

(ii) The accident may be due to some characteristic(s) of the actors;

(iii) The accident may be a result of a particular interaction between the actors and the situation they find themselves in; or

(iv) The accident may be caused by social or cultural factors.

Logically, we can never reject the possibility that the accident is a pure chance penomenon. However, it is very difficult to make further progress from this point of view, and it is psychologically a rather unsatisfactory approach. We all seem to have a fundamental desire for order, and not infrequently our all too human reaction to uncertainty is to lean heavily upon prejudice to counter it. Thus, in a collision where the only thing we know about the other vessel is that it is a foreign trawler we may assert that this is 'all we need to know', drawing on some stereotype we hold of the incompetent foreigner.

There exist personality stereotypes, too. For example, it is often maintained that there are certain types of people who are particularly prone to accidents. Certainly we may entertain the notion that facets of personality, particularly in some people whom we 'know' in our experience to be clumsy or persistently unlucky, might be expressed in a vulnerability to accidents. This commonsense notion is however not supported by psychological research. It may be more useful to the investigator to think more in terms of *attitude* of a particular actor towards his task, or safety in general. There are a whole range of motivational issues which we might consider. We have mentioned attitude and values, but we might for instance equally consider 'risk perception' or 'expectations'.

The interaction between an actor and the situation he finds himself in is a particularly useful perspective. The concept here is comparable with a 'disease model' in which some sort of host-agent interaction brings about the malady. Implicit in the 'disease model' is the assumption that, in principle at least, it is curable or preventable by removing the pathogen or by inoculation. This approach is particularly fruitful from the point of view of the number of answerable questions that may be generated.

If in a particular casualty we were to suspect that fatigue may have played some part in creating vulnerability to the 'disease,' we should then want to know the details of the watch system aboard ship, the hours worked in the preceding period, and whether or not those involved had eaten and slept properly. It is also important to know whether the accident happened by day or by night (since the diurnal rhythms of the body are disturbed by shift work, and this could be a factor in generating accidents). These are questions where it is possible to obtain concrete answers, yet they are not always asked by investigators.

Other answerable questions arising from the interactive viewpoint relate to the exact timing of the accident. This is important to know not only in order to ascertain the sequence of events, but also to establish the time scale within which the events unfolded. The instantaneous accident leaves no opportunity for premeditated action, and the man must react as best he may in the circumstances. Depending on the extent to which we see him driven forward by pressure of events, however, we may have to take a very different view of the human element and the effective impact it can have on the prevention or mitigation of the accident.

Other helpful questions may be generated from considerations relating to information provision. First we may ask questions relating to availability of information (e.g., was the information adequate? was it confusing? was it not communicated?). Then we may consider what use the men involved made of this information—i.e., how did they judge it (e.g., did they fail to notice it or did they make wrong assessments?) The provision of more information is often proposed as a key solution, but there are reasons to question this conventional wisdom. For example, we know from a number of studies that navigators only use a fraction of the information that is available to them at any one time. It is important therefore to discover what part information plays in the overall accident scenario.

The interactive approach may equally well be applied to groups and organisations as well as to individuals. We might think of interactions between actors and situations in terms of team ambiguity (e.g., master-pilot) or 'group-think' (e.g., where the consensus view overwhelms initial individual reservations). It has also been claimed that commercial pressure causes accidents, but this is a difficult area to investigate. A master might be ill-advised to report commercial pressure openly. In the light of the sensitivity of this particular issue the investigator must frame his questions indirectly (e.g., was an ETA sent to your next port? what speed would have been necessary to make the ETA? would any cost have been incurred if the deadline was not met?). There is nothing in principle to prevent the answers to questions put in this way from being known.

Finally, it is as well to bear in mind the possibility that cultural and social features may have contributed to the accident. Seafarers have been said to present a 'macho' image of themselves, which leads them to seek out, and to remain in, a 'devil-may-care' risk-taking culture. However far-fetched this notion may seem, there may well be organisational expectations that 'real men' accept the dangers associated with the seafaring profession. These issues are particularly tricky since some organisations might well expect their employees to 'use their initiative' (i.e., bend the rules most of the time), but will hit them over the head with the company 'bible' of rules and regulations when an accident occurs.

A 'check-list' approach

Depending on our particular education, background and set of individual prejudices, we would all no doubt formulate a different set of questions suitable for investigating the human factor in accidents. There is no substitute for the initiative and intuition of the individual surveyor, but some uniformity of approach, perhaps along the lines of a check-list, may help the inferential process. The surveyor will have his own theories of human element involvement in respect of seafarers. Where a check-list may help him is in relating his thoughts to the general body of human factor knowledge. Furthermore, it forces him to consider certain questions which are so easily forgotten in the heat of an investigation (e.g., was it day or night?). Equally, it may work as a safety awareness-raising device for those being questioned. This may be particularly attractive with respect to newly arising concerns of seafarers, such as stress and fatigue, or in order to check out the possibility of dangerous behaviour as a result of, rather than in spite of, training ('negative transfer of training').

We have already given some examples of the kind of questions which could be usefully asked. A check-list can spell these out in a systematic manner, by addressing the different variables (both human and other) which have been shown to be relevant to the explanation of accidents. Thus, the following accident 'dimensions' may be listed:

- Geographical (e.g., position, area, distance from land).
- Weather (e.g., wind direction and force, sea state, swell, visibility, tide and rate.
- Time (e.g., date, day of week, time of day, light or dark, season, first point in time at which someone could have reasonably intervened to reverse the trail of events, length of time during which the casualty itself occurs, and the last opportunities to avoid it).
- Ship data (e.g., name, flag, owner, age, type, gt, load state, cargo, speed at start/end/critical period of incident, closing speed, give-way vessel, time into voyage, port calls, deadlines relating to ETA).
- Socio-technical (e.g., equipment, crew certificates, information provision/communication/judgement, knowledge, practical skill, attitudes, values, risk perception/tolerance, personal problems, interpersonal relations, management, organisation).

Of course, no check-list can be exhaustive; this is theoretically impossible, as well as unnecessary in practice. Whole classes of questions may be eliminated because they are unanswerable within a particular casualty investigation. For example, 'disorientation of the watch-officer' might be suggested from the available evidence as a reason for a grounding, yet a casualty investigator can never know for sure what went on in the watch-officer's mind.

But for a casualty investigation to be both as meaningful and as 'objective' as possible, it is crucial to ensure that not just more questions are asked (and, in consequence, further information is gathered), but that the right amount of the right questions are asked in the right order and at the right point in time. Timing indeed may be the critical factor.

When should the questions be asked

Ideally, the best time to investigate a casualty may be while it is actually happening. However, this theoretical ideal requires the investigator to be omnipresent; the nearest alternative would be to bring him out to the scene the same day in order to begin asking questions immediately, but even this would be impossibly expensive in most cases. In practice, some delay is inevitable.

Delays of three months or longer between the casualty and the first taking of statements are not at all unusual. We might expect that the longer the time lapses between the event and the actual questioning by the investigating surveyor, the more detail of the incident will be forgotten. However, this decrement does not appear nearly as important as the amount of secondary elaboration which will occur as an actor mulls over the event and its possible causes in his own mind or in discussion with others.

Bearing in mind both the complexity of events and the speed with which they can develop in a casualty situation, we should not be too surprised to find some inconsistencies in the ordering of events and their interpretation by different key witnesses. But one of the many bizarre facts encountered in historic casualty records is the seemingly almost 'miraculous' consistency of witnesses' accounts overall, which if they differ at all do so often in only minor details. From all we know of human nature and, in particular, from the experimental work on the unreliability of eye-witness testimonies we may speculate that there are forces at work moulding the collective memory towards a unitary account of the casualty.

The most obvious of these forces may well be self-interest driven by a perceived need for self-protection (e.g., against possible legal action). Here the investigator can help by explaining to the witness exactly what the range and limitations of his powers are. Some investigators do take great pains with this aspect of their work. It certainly would help if the assurances could be written as well as verbal.

The crew may indeed be inclined to 'get their act together' with the specific intention to externalize the blame (e.g., it was the other ship's fault, or a docker left a strop on the deck). But this process is of course not necessarily dishonest; it occurs quite naturally at an unconscious level. After an accident the mind is busy, below the level of consciousness, trying to make sense of what happened. This process may require the real events to be bent and manipulated, with necessary inferences added or subtracted so that this unique event may be understood, along with all previous experiences, in an organised chunk or 'schema'. After a period of time, a person cannot distinguish the reality of the event from his own subconscious inferences. It is perhaps not surprising, therefore, that men of similar training and experience so readily tune their collective memory towards one unitary report.

It may also happen that the investigator himself may easily, at an unconscious level, be leading the witness towards his own theory. A number of studies have shown how remarkably easy it is to warp a witness's memories by altering the shape of the questions. The investigator must be continuously aware of this possibility.

Some obvious measures may be taken to counteract these effects by first requiring a witness to answer a series of questions, such as those on a check-list, before an attempt is made to solicit a 'full' or 'comprehensive' verbal account. Since this check-list of questions would be in a standard format, the various items would also be independent of individual investigator bias. Secondly, we could arrange for the time lapse between the event itself and the first recording by the inspecting surveyor of a witness's statement to be in some sense 'optimal.' We have mentioned the attendant practical difficulties, but it might be possible to overcome these to some extent with improved shipboard reporting procedures. Anyway, the optimum delay may not be simply as small as possible after the event, since the actor may need some time to recover from the experience before giving evidence. A lapse of two days would seem sensible, but much more needs to be learned about 'optimal' delay.

In considering the human factor in accident investigation, the surveyor has to accept that he will rarely be able to explain and understand everything. He may turn to the key actors to learn the reasons for their actions, but their interpretations can never be regarded as conclusive. We have tried to show that despite there being no absolute test for the correctness of knowledge gained, there appears to be ample opportunity for at least narrowing the margin of uncertainty.□

Ship's Copy

Report on Inspection

In Accordance With The Memorandum Of Understanding On Port State Control

1. Issuin˛ Country – United Kingdom (GBI)

2. Name of s.'o

3. Type of Ship (code)

4. Flag of Ship

5. Call Sign

6. Gross Tonnage

7. Year of Build 19

8. Date of Inspection

Place of Inspection (MO Dist. ˋt)

9. Nature of Deficiencies **ˈferences** **10. Action Taken**

Note: Insert the appropriate code for deficienc. s, references and corresponding action taken (in accordance with the Aide – Memoire)

SPECIMEN

Any Other Remarks

Name and signature of the Surveyor Authorised by the Maritime Authority

Name (in BLOCK letters)

Signature

SUR/PS˙C

PORT STATE CONTROL AND THE EUROPEAN MEMORANDUM OF UNDERSTANDING *

Captain R. L. Newbury, Ex.C, ISO, FNI, FRGS, FRIN
Chief Surveyor, DTp, Bristol Channel and South Wales District

THE ORIGINS OF port State control lie in the memorandum of understanding between certain eight North Sea States that was signed at The Hague during 1978. This collective agreement was for the inspection of shipping of all flags that entered the ports of the member States, and ensuring that their standards were up to the international conventions.

Although this agreement was perhaps a little slow to get off the ground, it did at once show itself to be both desirable and effective in dealing with the sub-standard ship. The result was a European conference of ministers on maritime safety, giving birth to the memorandum of understanding (MOU) signed in Paris between 14 member States, which superseded the Hague Agreement and became effective on 1 July 1982.

Countries involved in this Paris Agreement comprised: Belgium, Denmark, Finland, France, Germany (Federal Republic), Greece, Ireland, Italy, The Netherlands, Norway, Portugal, Spain, Sweden, and the United Kingdom. The agreement is based upon compliance with the standards set by the following international conventions: Load Lines 1966; Safety of Life At Sea 1974; Protocol of 1978 relating to Solas 1974; Prevention of Pollution from Ships 1973 as modified by the Protocol of 1978 relating thereto; Standards of Training, Certification and Watchkeeping for Seafarers 1978; Regulations For Preventing Collisions At Sea 1972; and Merchant Shipping (Minimum Standards) 1976 (ILO Convention 147).

It is clear to all that when 14 European maritime nations unite together for the enforcement of standards set by such all-embracing conventions, each State is concerned about the sub-standard ships that visit its ports. Through the memorandum of understanding might be seen a course to prevent the operation of sub-standard ships within those agreement countries. The MOU effectively establishes a block of maritime nations that virtually covers the coast of Europe continuously from Norway southward to the eastern shores of the Mediterranean Sea, which means that every vessel trading to Europe is almost certain to be considered for inspection, if not at her first port of call then at a second or subsequent port.

Each of the MOU signatories have agreed that the national surveyors of each State will endeavour to make visits to 25 per cent of the foreign vessels visiting that State, and embraced in the agreement is a 'no more favourable treatment clause,' which, as its name implies, ensures that vessels flying the flag of a State not party to one or more of the conventions listed, are not permitted to adopt lower standards than those applied to a vessel flying the flag of a State that is party to all the latest conventions.

The Secretariat for the MOU is provided by the Netherlands and has been established in The Hague, and the information bank has been provided by France and established in St Malo.

Inspection and reporting system

A great deal depends upon the method of selection of vessels for port State control (PSC) inspection. Systems vary not only from one member country to another, but also from one geographical area to another within the countries. The underlying themes, however, are that no vessel is normally inspected more frequently than six months after the last inspection or certificate renewal. The latest condition for every vessel inspected or surveyed within the MOU States is recorded in the information bank to which reference may be made by any national surveyor from member countries at any time of day or night. Every vessel, can, therefore, normally expect no further attention during the six months that follow a survey or PSC inspection. There are a few exceptions, however, the principal one being a vessel that has been inspected and allowed to carry forward to a subsequent port some lesser deficiencies at a surveyor's discretion.

A number of factors generally govern the final selection of ships that might be inspected at any given time, and these might be seen to include age, track record and a vessel which may involve a special hazard, such as an oil, chemical or gas tanker. When the surveyor boards a vessel for PSC he is not biased in any way, but will be acting purely as his own professional judgement dictates, and one of three situations develop:

(a) Certificates all in order and the surveyor has no misgivings concerning the condition of the vessel.:
(b) Certificates invalid or missing.
(c) Certificates all in order but, the surveyor decides there are clear grounds for believing that the condition of the vessel and/or her equipment does not substantially correspond with the particulars on the certificates.

The first case of everything in order clearly requires no further attention. The other two, however, require some consideration which will depend entirely upon the severity of the deficiencies and the intended voyage, but will, sooner or later, culminate in the surveyor being satisfied with the vessel's condition. If the surveyor's satisfaction is only partial when the vessel is ready to sail, he may, at his discretion, communicate with the maritime authority of the next port of call if it is within a Paris agreement State, and

Crown Copyright Reserved. Opinions are those of the author only and do not necessarily represent Crown opinion or policy.

the outstanding deficiencies will be carried forward for attention at the next port.

The inspection now has to be recorded at the St Malo information bank. This is done daily, and for UK inspections a single report is sent by telex from London covering every UK inspection during the previous 24 hours. In order that all communications are smooth, simple and not unduly time consuming, everything is handled in a set format with the use of a numerical code, thereby achieving both clarity and brevity and breaking down language barriers at a stroke. A copy of the information to be passed to St Malo is also left with the master of the vessel.

These messages contain the following information:

- Issuing country.
- Name of ship.
- Type of ship.
- Flag of ship.
- Call sign.
- Gross tonnage.
- Year of build.
- Place of inspection.
- Date of inspection.
- Nature of deficiencies.
- Action taken.

A copy of a specimen report form is shown. The signatory States of the MOU do not inspect and report their own flag vessels, only those flying the flag of any other State in the world.

Funding of MOU

The overall funding of the secretariat is met by the member States on an equal contribution basis. The information bank at St Mao and the computer serving it are charged to member States on the basis of the individual time used.

All other costs lie with the State that incurs them. It will be clear, therefore, that there is no charge upon the shipowner, ship's agent or the port authority for the operation of the memorandum of understanding.

No charge is made for the first inspection of a vessel, including the recording of her particulars and condition in the St Malo information bank. Should the ship have some deficiencies, still no charge is made. If however, the deficiencies necessitate a second visit by a surveyor to see they have been rectified, then a charge is made for the second and subsequent visits where deficiencies continue to be involved. In the United Kingdom the charge is on an hourly basis.

Aims of the Paris MOU

It will be appreciated that the sub-standard ship is not by any means a phenomenon of this decade or even of this twentieth century. It is simply that the means of ridding the seas of such vessels has changed over the years. During the 19th century the loss of British seamen and British ships produced some very unwholesome statistics indeed, the average loss of life amongst seamen being about 800 seafarers per year, whilst in one bad year 2,350 seamen died, and 1,313 British ships were wrecked or lost. The situation then was largely rectified by legislation and we saw the establishment of the Board of Trade Marine Department in 1850, followed by the various Merchant Shipping Acts, Load Line Acts, and a Royal Commission appointed to enquire into unseaworthiness of ships.

Since the turn of the century through to the 1970s there have been five international conventions on safety of life at sea, and several international conventions relating to such matters as the pollution of the sea, loadlines of ships, standards of certification of seamen and minimum standards on ships. This period might perhaps be described as one in which the standards were kept up by a combination of legislation and education.

Moving on into the 1980s it is clear to all that, although legislation and education are plentiful, the sub-standard ship is still to be found. They are not worthy of definition. Surveyors are all familiar with sub-standard ships, recognise them instantly, and not one amongst them has any desire to be associated with such ships, even remotely. Its demise would seem perhaps to lie in the close enforcement of existing legislation, not on a unilateral basis which enables such sub-standard ships to avoid those countries that interfere with them, but on an international basis that enables a block of countries to turn the cold shoulder to that category of ship.

The Paris memorandum of understanding, known to all as port State control, gives the cold shoulder to sub-standard ships visiting those 14 countries. The attractive feature of this system is that the ordinary properly-built, properly-equipped and properly-maintained vessel does not have to pay a penny piece towards the cost of uplifting the sub-standard ship, for this port State control is a simple scheme in which the transgressor is inconvenienced, the transgressor pays, and the greater the transgression the greater the end costs and inconvenience to that same transgressor.

Port State control effects

To consider how successful or otherwise PSC has been, we must bear in mind that it really started on 1 July 1982, and, like everything new, was brought in progressively. As surveyors gained experience on the system so inspections became more numerous. The exchanges of information between the different maritime authorities of the 14 signatories gradually became commonplace. The information flow on a direct exchange basis has now reached a quite high level. Whereas initially exchanges between the maritime administrations of the 14 States were conducted by telex, many have now moved into the era of the 'electronic mailbox' enabling direct question/answer communication between ports, either in plain language or with the use of the MOU numerical codes.

During the first four and a half years of operation the principal coverage and deficiencies found in ships in the 14 MOU countries can be expressed in the following small table:

Principal Deficiencies

	1-7-82 to 30-6-83 %	1-7-83 to 30-6-84 %	1-7-84 to 30-6-85 %	1-1-86 to 31-12-86 %
Ships inspected	8,839	7,350	7,665	8,721
Different flags	108	106	112	116
Ships with defects	3,175	1,495	1,875	1,356
Ships detained and/or delayed	271	436	328	307

The cutting edge of the port State control shows us, however, that in the MOU countries during its first four-and-a-half years of operation, 1,442 sub-standard ships have been detected and delayed or detained, and there is no reason to suppose that this would have happened but for the memorandum of understanding. When viewed statistically, it appears that standards generally have improved, because the number of ships upon which defects are detected has dropped markedly over the four-and-a-half years, as shown in the table.

An additional side effect of PSC is that privately-owned tidal jetties are now using properly administered and operated ships since they want the ships off the berth once cargo is completed in order to work the next ship. Any ship delayed or detained affects the jetty owner's fortunes, since these tidal berths have no lay-by berths or anchorages, and a ship is either on the berth or at sea. It would seem, therefore, that PSC is having the desired effects in two ways: (a) improving standards in ships not equal to Convention requirements; and (b) indirectly creating difficulties of employment for ships if known to be of doubtful standards.

An interesting feature to note is that moving away from sub-standard ship to the ordinary vessel with ordinary defects, it is very simple to see which defects were associated with inadequate repair or maintenance organised by the shipowner's shore staff, and which defects were a direct result of lack of proper between-survey maintenance by those on board the ship.

Defects and subsequent action

The whole concept of MOU is for the surveyor to look first at the ship's certificates, and it is only when these are defective in some way, or the surveyor has clear grounds for further investigation, does the inspection progress beyond the point of a documentation check. A ship that is found to be unsatisfactory in some way is involved in further attention, in this connection the options being along these lines:

(i) Surveyor assesses the defects as being of fairly minor nature. He will probably require the necessary repairs to be carried out at once, but, if the vessel is bound for another European port, may permit the vessel to proceed there at his discretion, in which case the authorities at that port will be notified in advance of the ship and the defects to be remedied.

(ii) Surveyor assesses the defects as being in need of replacement items but of a minor nature. If the items are not immediately obtainable he may permit a short voyage to another European port, depending upon the nature of the defects.

(iii) Serious defects are found. Surveyor requires defects to be made good before the vessel leaves port. This is usually achieved during the vessel's scheduled stay in port without need to delay her, although from time to time delays are involved.

(iv) Surveyor finds the vessel is dangerously unsafe and has to resort to detention.

When the surveyor resorts to delays or detention the harbour master is customarily contacted. It is not to place any responsibility upon him to keep the ship in port, but simply to advise him of the action taken. This enables the harbour master to plan his shipping movements and berthings in the knowledge that one ship will not be sailing as originally scheduled. A detained ship may be moved about the harbour to the requirements of the harbour master. Detention does not tie a ship to any particular berth and the berth she occupies is purely at the discretion and direction of the harbour master. It is only to sea that the ship may not proceed, and when a vessel has been detained whilst occupying a non-tidal berth harbour masters customarily advise the surveyor concerned if in the direction of shipping within his harbour he finds it necessary to move the detained vessel to a tidal berth.

To resort to detention is really the final sanction held by a surveyor in the interests of safety of life at sea and the implementation of port State control. It is not lightly that they tread this path and when doing so have to proceed with meticulous care.

Firstly, one must not confuse arrest with detention. Arrest is associated with legal proceedings against a ship, often for non-payment of bills, and the actual arrest follows a Court order. Surveyors are not associated with arrest, this work being discharged by the Admiralty Marshal, but from time to time at tidal berths surveyors will, upon the written request of the Admiralty Marshal, arrange for a vessel to be immobilised. Detention, on the other hand, is associated with contravention of the Merchant Shipping Acts, and has not become associated with legal proceedings, but the actual detention follows a Government surveyor's inspection or survey. This in turn can be broken down again into two sections:

(i) Unsafe ships, as covered by Sections 459 and 462 of the Merchant Shipping Act 1894 as amended. These sections provide for detention, when, having regard to the nature of the service for which the ship is intended, she is considered unfit to proceed to sea without serious danger to human life by virtue of defective condition of hull, machinery, or equipment, undermanning, overloading, improper loading and certain loadline contraventions.

(ii) Ships detained under specific provisions of the Merchant Shipping Acts, generally for a direct breach of the Merchant Shipping Rules and Regulations (the Statutory Instruments) or failure to produce safety certificates.

Detention and sanctions

A ship detained is of considerable concern. The instrument of detention is a piece of paper. For detention a ship is not physically restrained, but the penalty for breaking detention is substantial, quite apart from penalties associated with taking or sending an unsafe vessel to sea. Physical restriction of a vessel is usually associated with arrest, and achieved by immobilising the vessel concerned. Similarly, British authorities do not detain British ships whilst in foreign ports or, naturally, whilst they are at sea, since that is the very environment for which they are temporarily unsuited.

It is important to mention that ship detention is not the only sanction available, but is the one often associated with MOU matters and unsafe ships. It might be regarded as an instant penalty, and of course there are procedures for those who feel detention has been improperly applied, including the recovery from the marine administration of sums of money to defray the losses involved with wrongful detention. Apart from detention, the Merchant Shipping Acts embrace various other penalties which include fines of up to £50,000, terms of imprisonment of up to two years, cancellation or suspension of certificates of competency and withdrawal of seamen's documents for either a specified term or permanently. (These ceilings on penalties applied on 31 December 1987.)

On the subject of detention, it is important to stress that it is never undertaken lightly, it being a serious matter and only even considered for dangerously defective vessels. It should not be thought that marine administrations shrink from detaining a vessel when all else fails, but it occurs only when all else fails.

In fact the way it must be viewed, so far as detention is concerned, can perhaps be clarified by the advice gratuitously given, which is to the effect that one is not interested in having ships lying around in port in an unsafe condition. The aim is to keep ships moving and at sea in a safe and satisfactory condition. One must assume that to be the wish of the shipowner, too, so if detention has to be considered, make certain that every possible and reasonable avenue has been considered collectively and objectively before taking one's fountain pen to sign detention documentation.

By the same token it should not be thought that surveyors shrink from detaining a vessel when all else fails, but it occurs only when all else fails. One ship detained was eventually cut up into manageable pieces in the dock and passed the office window on the backs of lorries, bound for the scrap yard. Perhaps that might be defined as having been finally detained, but the expression 'Finally Detained' is one in fact used in detention procedures, and relates to a ship that reaches the condition of detention when all appeal procedures, if made, have been unsuccessful.

A further sanction applied from time to time is the serving of improvement and prohibition notices under the Merchant Shipping Act 1984. This in effect translates the Health & Safety At Work Act into the marine sector and might properly be said to provide the overlap of interests between Health & Safety inspectors and the marine administration surveyors. It is gangways and other means of access to ships that create the principal overlap, since one end is shipborne and the other land-borne. It is proper to have an overlap, since without an overlap of interests there might be a gap where no concern is shown.

Improvement notices are associated with safe equipment and procedures, and commonly gives the person upon whom it is served 21 days in which to rectify a fault. Prohibition notices are rather more severe in that they prohibit instantly the further use of certain defective equipment or instant termination of unsafe procedures associated with any process being operated by the vessel. They may also prohibit the vessel from going to sea.

Paperwork

The paperwork involved with MOU really is short and simple, almost to the extent of being negligible. On completion of an inspection the surveyor hands the master a report which shows any defects, in a numerical code, thereby overcoming any language barriers at once. The master is recommended to keep this report for six months, but he may, if he wishes, destroy it instantly. A copy goes to the St Malo information bank and a copy is retained by the surveyor.

No other paperwork is involved, for the implementation of port State control is not concerned with paper mountains, but simply with the observation of maritime conventions, which benefit not only seafarers, but also those ashore and the environment. Further, in avoiding paper mountains, it ensures that ships inspected are real ships which are well placed to withstand the real hazards of the seas.

British ships in UK ports

United Kingdom ships are given their MOU inspection in foreign ports. However, when they are under survey in the UK, the matters outside the survey in question but within the MOU requirements are looked at and commented upon by the surveyor in order that the owner might ensure his vessel complies fully with domestic legislation. This in turn means compliance with the port State control instruments. Non-UK-registered British ships are treated in the same manner as any other non-UK-flag vessel so far as port State control is concerned.

In conclusion one must consider the vessel and crew that manages by design and intent to avoid all contact with port State control. For a United Kingdom ship to achieve it by wilful avoidance is almost a certain indication that she has something to hide. Different people have different ideas concerning her progress and some will say she ends in Davy Jones' locker whilst others say she trades to that Great Haven in the sky. There is a chance that if carrying defects and exercising wilful avoidance of detection, she will one day form an entry in an annual publication of Her Majesty's Stationery Office entitled *Casualties to Vessels and Accidents to Men,* which is very valuable reading to all concerned with the operation of ships. □

The author acknowledges and appreciates the assistance given by Mr H. Huibers (Secretariat Memorandum of Understanding on Port State Control, Rijswijk) for permission to reproduce statistical information from MOU Secretariat annual reports.

(A specimen copy of the MOU Report on Inspection is reproduced on page 50)

MAINTENANCE OF OPERATIONAL EFFICIENCY IN THE ROYAL NAVY

Commander P. J. Melson, MNI, RN
Commander, Sea Training, HM Naval Base, Portland, UK

A RECENT ARTICLE in the *Sunday Times* on the UK National Health Service contained a remarkable paragraph that discussed that paper's view of the state of the service and ended with the observation that its inefficiency was rivalled only by that of the Royal Navy. It occurred to me as I read it that whatever Royal Navy the author had seen was not the Navy that I knew, nor was it the Navy known and admired not just by NATO allies but by other navies the world over. The Royal Navy does have an international reputation for safety and for operational efficiency that is virtually unrivalled, but that reputation has its costs and it is the relationship between efficiency and costs that I will attempt to spell out in this chapter as I develop the theme of the maintenance of operational efficiency. (Come to think of it, I've never really been badly let down by the NHS either!).

Even to begin to climb the slope towards operational efficiency you must presuppose trained manpower, and here the Royal Navy is lucky in that it operates within a fixed budget which is not at present subject to commercial pressures. The balance between training, hardware, salaries and research and development is thus to some extent our decision. We could, for instance, decide to have a fleet of 200 elderly ships, filled with untrained manpower, to keep an impressive presence in the oceans of the world. Some navies do follow this philosophy. On the other hand, we could have half-a-dozen highly capable units of, say, the US *Ticonderoga* class, filled with very highly trained personnel but able to cover only a very small area of sea.

The balance has to be struck and in the Royal Navy we have settled at a fleet of three aircraft carriers, two assault ships, 50 destroyers and frigates, 45 mine counter measures vessels, four ballistic missile submarines, 17 nuclear attack submarines and 10 conventional submarines, all supported by a fleet train of some 40 auxiliaries and manned by about 65,000 men and women. It is a small fleet in superpower terms, but is highly trained at all levels and, above all, it keeps the sea. *Invincible* during the Falklands conflict was at sea continuously for 150 days, and her escorts were at sea for over 120. This obviously does not sound all that impressive to a tanker man who regularly transits from Europe to Japan via the Gulf and back, but to 250 men inside a 2,500-ton ship (even the carriers are only 20,000 tons) it is a long time and only continuous training makes it possible.

Initial training

All naval personnel receive an initial training which varies from 16 weeks for a seaman missileman, through four years for an engineering apprentice, to six years for an engineering officer or a Harrier pilot. Apart from the degree course of some seaman officers, undertaken as a mind-broadening exercise for those destined to be future policy makers, all training is objective and is orientated strictly around the person's future employment.

At rating level, little of it is academic because it is designed to fit a man to fill a particular billet in a modern warship and, much as we would like to stretch people's minds, neither the time nor the money exist to do it, so we train to the job. Fortunately the unemployment of recent years has enabled the Navy to be selective in its recruiting to the rating structure, and both 'O' and 'A' levels are common amongst recruits; for artificer apprentices they are mandatory. The officer scene is different, because we are fishing for people in the top 5 per cent of the UK's academic spectrum, and demographic trends are diminishing this pool. Put simply, we haven't got enough officers and badly need more.

Initial training, then, fits a person to do a particular job in a particular ship. For example, the Royal Navy has three types of computerised command systems, all designed to assimilate the vast amount of information available today and present it to the commanding officer for decision. These systems are manned by seamen who are trained to operate just one of the three and are thus only draftable to a ship fitted with that system. This obviously introduces inflexibility into the drafting cycle, but again a balance has to be struck between the cost of overtraining and the cost of career inflexibility. Seaman officers, too, are trained in one of the three systems, although engineer officers receive a broader training and are theoretically capable of maintaining any equipment.

A ship's company is, therefore, composed of men all of whom have been trained in the particular skills necessary to operate and to maintain that ship at sea. The problem now is the welding of those men into a fighting team capable of dealing with threats both expected and unexpected and of repairing their ship if damaged. It is a demanding problem, not faced by the Merchant Navy except in the case of fire on board, and it is one that occupies a very large percentage of a ship's life.

Shore courses

The task begins with various forms of team training ashore while the ship is in the last stages of build or refit. The command team of officers and ratings spend two weeks in the operations room simulators at HMS *Dryad,* near Portsmouth. Here various tactical scenarios are played out on equipment that exactly replicates that fitted in their ship, with the only difference being that ashore a whole arsenal can be simulated in attack, an event that at sea would surely signify Armageddon. At the same time that the command teams are training at *Dryad,* the damage control teams are training at HMS *Phoenix* in Portsmouth or HMS *Raleigh* at Plymouth.

Damage repair units at both these establishments represent a section of a ship that can be flooded and set on fire. If the teams get it wrong, free-surface effect will eventually almost capsize the unit, to the intense discomfort of all inside it. The aim here is to develop expertise and to build up resistance to stress; it is the only place that we can subject our people to real fire and flood, for obvious reasons. Eventually, with shore team training over, the ship sails, usually for a period of trials which last from around six weeks for a ship out of refit to up to two years for a first-of-class ship fitted with a new generation of equipment. However long the trials period though, eventually the ship, if it be a surface ship and in this chapter I am concentrating mainly on surface ships, arrives at Portland for her basic operational sea training, or BOST.

A precondition of arrival at Portland is that all of the ship's equipment is operational, and she is ready to fight in all environments. This is what the trials period has been about, and tuning sophisticated missile systems is a lengthy business. Once at Portland, though, this part of the ship's life is over and the equipment is handed to the users for operation. The BOST starts with a staff sea check when, in the case of a frigate or destroyer, some 60 staff officers and senior rates thoroughly check out the ship's organisation and material state. All equipment is functioned and checked to operational standards and every piece of documentation is mustered and verified for accuracy.

Date of survey

It is a baseline for training, and not all ships pass it. Some have to go away and spend more time in preparation, some even have to be returned to the dockyard that refitted them. An important part of the staff sea check is the establishment of the date of survey of every piece of equipment subject to survey on board. It is, of course, the responsibility of the ship's commanding officer to ensure that all equipment is in date for survey and the staff sea check of these dates is a very good indication of how well the ship is managed, as well as being essential to the safety of the ensuing training.

The staff sea check behind them, though, ships go into six weeks of intensive training in all disciplines from counter-insurgency through to fighting World War III after a nuclear exchange. Pressure and stress play a large part in this training, particularly in the latter stages, as the staff incapacitate primary methods of equipment control, and fill the ship with smoke to simulate damage, thus forcing the ship's company to fall back on little-known secondary and tertiary control systems to fight an intensive air, surface and sub-surface battle whilst wearing breathing apparatus and attempting to repair damage.

Veterans of the Falklands war testify to the realism of the Portland training and state that, as ships were hit, they fell instinctively into the routine learnt at such cost at Portland, while NATO allies, particularly the Dutch and Germans, confirm that judgment by sending their ships to be trained under the aegis of the Flag Officer Sea Training. Some training is more mundane, though, and revolves round safety and the establishment of safe practice on board. An example is the rule of the road test that all bridge watchkeepers have to undergo in the first week of training. Another is the training of the ship's company in abandon ship and liferaft boarding drills.

Maintenance of standards

Apart from operations, damage control and safety, the maintenance of engineering standards plays a large part in the Portland syllabus. It is perhaps a fault of human nature that left to ourselves many of us will accept second best and it is in the careful monitoring and maintenance of machinery that this is best observed. We have all walked past a piece of equipment that we know ought to be better fitted, or should be working, but familiarity breeds contempt and too often we leave it for another day.

It was in recognition of this very human failing that the Portland engineering staff were established. Their aim is to demand of ships' companies only the very highest of engineering standards. Some would say that the standards expected are impossible to achieve, but some ships do achieve them, and the point is that the best standards should be the aim of every ship. If Portland sets this as a target, then ships know that at least once every 18 months, when they return to Portland, they will be judged against the highest datum.

Indeed, the credo of 'absolute standards' is essential to the Portland doctrine. A ship is assessed by the staff as they see it. No mitigating circumstances are taken into account and reasons or excuses are not considered. Thus, a ship that is 20 years old and has just completed a commercial refit in the open on the Tyne is judged against exactly the same set of criteria as a brand new ship which has been built under cover at Birkenhead. This may seem unfair, but it is impartial and much of Portland's reputation is built on this impartiality. In fact, very often the older ship does better than the newer one, simply because engineering standards and conscientious adherence to maintenance schedules are better than those found in the newer ship.

In common with the nautical surveyor, the need for Portland would disappear if man as a species was capable of self-regulation and discipline. Regrettably this does not appear to be so and external pressure has to be applied to ensure that acceptable standards are set and maintained. From that point of view the Portland staff have much in common with the nautical surveyor and aim to ensure that ships keep the sea in safety and can fight efficiently should the time come.

The BOST finishes with a final inspection by the Admiral and provided all is well the ship is cleared to join the operational fleet. Her employment for the next four years or so, until her next refit, will be an amalgam of directed operational tasks, such as the Falkland Islands or the Gulf; NATO exercises such as Northern Wedding or Ocean Safari; overseas deployments to the Pacific or Western Atlantic;

maintenance periods in home ports and leave. She will spend something over 200 days at sea each year, with the ship's company being entitled to 48 days' leave in that year. The rest of the year is spent in maintenance and the occasional foreign visit.

Standards at sea

The maintenance of operational efficiency once away from the supervision of Portland falls entirely on the individual commanding officer, although he will be assisted in his task by the small, mobile, staffs of the flotilla flag officers and also by the even smaller staffs of the squadron commanders, the Captains' 'D' and 'F' of the destroyer and frigate squadrons. Since these staffs have their own jobs to do, in their own ships, though, the help they can offer is limited. To assist the commanding officer in planning his programme of continuation training, the Commander-in-Chief Fleet publishes a list of target achievements for all the various disciplines that go to make up a ship's fighting capability.

Some of these exercises can be done internally. First aid training, damage control exercises and machinery breakdown drills, for instance, require no external assets and are done on a continuing basis, planned internally by the ship. Other exercises, such as air defence training and anti-submarine warfare, depend on the availability of aircraft and submarines and these assets are both scarce and expensive. As a result, the ship's operational efficiency tends to tail off the longer it spends away from Portland, where such assets are concentrated for reasons of economy. If a ship is lucky she may be programmed to take part in a major NATO exercise, with a plethora of aircraft and submarines.

Another, more formal, training opportunity is the joint maritime course, organised jointly by the RN and the RAF, and geared around the needs of worked-up ships of all NATO countries. Up to six submarines are provided as 'Orange' forces, plus a large number of attack aircraft from various nations. The exercise takes place off the North of Scotland and is carefully analysed in order to gain maximum benefit from the training provided. It is really quite surprising to what extent the attacker's tale of what occurred differs from the attacked!

Measuring success

Operational analysis is a very important element in the job of maintaining operational efficiency. In a profession where true efficiency can only be measured in terms of enemy units sunk or shot down, it is important to establish some means of measuring the likely success of weapon systems and tactics. To do this, a rigid methodology of analysis has been formulated and careful records are taken of every weapons firing and tactical exercise. Records take the form of computer dumps and manual observation and recording and are analysed initially on board the ship by a specially-designed computer suite, and later ashore by permanent teams of officers, ratings and Wrens whose sole job is to establish what really happened, compare it with what the ship claimed happened and report accordingly.

Specific analysis of a particular subject of interest is also undertaken by the Defence Scientific Service, an example being the establishment of the number of air targets detected by an air defence destroyer in an intensive air environment, compared with the number of targets actually available. The difference between the two is then further analysed to determine how much is due to failure of computer auto-detection, which can be corrected by software changes, and how much is due to human failure to which psychological study has to be applied. These latter failings are then addressed by another unit of the Defence Scientific Service who deal in the area of man/machine interfaces and seek to establish how best to get the most out of an easily-bored human being.

However intensive a ship's programme, though, there will be gaps in its training opportunities as the months roll by. A ship on Armilla Patrol in the Gulf won't see a submarine the whole time she is on station, while a ship on independent passage in the pacific will be lucky (or unlucky) if she sees anything. Eventually, then, a ship's operational efficiency tails off to the extent that she has to return to Portland for refresher training. The period between this training is generally between a year and 18 months and depends critically on the percentage of the ship's company that has changed, and the ship's programme. But once back, the whole Portland process starts again, and, although shorter, begins with the staff sea check and ends with the inspection, just as the basic operational sea training did all those months before.

Accurate assessment

Over the years various methods of measuring operational efficiency have been tried and have varied from an annual inspection by the Captain 'D' or 'F', to a questionnaire, completed by the Captain, which requires him to give an honest assessment of his ship's effectiveness. None of these methods has been wholly successful and in the end the impartial and absolute judgment of Portland has been found to offer the best indication to the Commander-in-Chief of the state of his ships.

A fully-operational unit is obviously deployable worldwide into any threat scenario; a ship that has not been quite so successful at Portland either needs more training, or must have its programme planned with care, although, in the final analysis, the fleet is so thinly stretched that all frontline units must be considered as deployable. The real threat, of course, is the one hanging over the commanding officer's career. It is quite amazing how often a ship's overall performance is wholly attributable to the strengths and weaknesses of her Captain and, in this, command at sea still remains one of the most accurate assessments yet devised of an officer's suitability for the higher ranks. What is encouraging is how few fail this most strenuous of tests.

In summary, in this chapter I have tried to show the means by which the Royal Navy attains, and maintains, operational efficiency. It is a process of balancing the cost of sophisticated weapon systems against the pay necessary to persuade men and women of the right calibre to join to design and operate these systems, and against the cost of training to fight when called upon. The three legs of equipment efficiency then have to be set against governmental requirements for directed tasks, during which efficiency tails off due to lack of practice, and an optimum method of retraining devised.

The result, of individual shore training, followed by team shore training, and then whole ship sea training, achieves generally high standards and good results, results that are generally accepted to be the best in the world. The Navy does not always get it right though and, although the experience of the Falklands War showed that we were generally on the right track, it was also very clear how many of the lessons of World War II had been forgotten. It is, perhaps, in an honest recognition of weaknesses, and in a determination to put them right, that the Royal Navy's greatest strength lies.□

PORT HEALTH AND RESPONSIBILITIES FOR PEOPLE AND CARGOES

Lieutenant-Commander J. C. Strachan, MNI, FIEH, RNR
Chief Port Health Inspector, London Port Health Authority

PORT HEALTH AUTHORITIES in the UK (PHAs) were first established by Act of Parliament in 1872 at a time when the country was threatened with an invasion of cholera. They were charged with taking proper steps, under Orders in Council, to prevent the introduction of cholera and other quarantinable diseases and to enforce related legislation aimed at improving sanitary conditions aboard vessels in port.

As Dr Harry Leach, the first medical officer of health for the Port of London, wrote in 1873: 'it is acknowledged that as a natural result of the insular position of the kingdom and the vast extent of our commerce, the sanitary conditions of shipping and of the floating population must exercise a considerable influence on the health of the country as regards the importation and transmission of epidemic diseases.'

Whilst infectious disease control remains an important objective of PHAs, their responsibilities for protecting the community at large have widened to include controls over cargoes of imported food intended for human consumption, meat intended for the pet food trade, food hygiene, rodent control, the safe processing of shellfish harvested from their district, water supply, living conditions aboard houseboats, controls over noise and smoke emissions and general sanitary conditions aboard ship.

The following account is intended to describe in general terms the role of port health authorities in the United Kingdom, with particular emphasis on the work of the London Port Health Authority.

Administration

Each PHA is administered by a council or board comprising members of one, or more, 'riparian' local authorities bordering the port and the extent of each district and its responsibilities are defined by an Order in Council. Financial provision is drawn from local rates and government rate support grants.

Since its inception in 1872 the London PHA has been administered by the Corporation of the City of London and now encompasses a district extending from Teddington Lock in the west for a distance of 94 miles to the Tongue lightvessel. The area includes a large part of the Thames Estuary, Tilbury Dock and the lower course of the River Medway, with the ports of Sheerness, Queenborough and Ridham Dock. Health controls at London City Airport are also administered by the authority.

In its capacity as the London Port Health authority, the Corporation employs a medical officer of health—with responsibilities for the port and City of London—15 port health inspectors, eight technical assistants and six launch crew, together with administrative support staff. In addition, there are two groups of part-time medical officers who provide cover both within the port and at London City Airport. The authority also uses the services of a public analyst and of four public health laboratories situated in Kent, Essex and London, including that of the food reference laboratory at Colindale.

The London PHA district is divided into three divisions—Tilbury, Sheerness and River—with offices located throughout the area. Two launches operated by the LPHA are based at Gravesend and Charlton. All port health inspectors employed by PHAs are required to be qualified as environmental health officers—a four-year degree or diploma course—while at London and certain other ports a number of them have, in addition, considerable experience as Merchant Navy officers.

London and all other principal UK port health authorities, together with those of major airports, are members of the Association of Port Health Authorities, committees of which meet regularly to consider general policy and, in particular, draft EEC and UK legislation. There is close liaison with the Ministry of Agriculture, Fisheries and Food (MAFF), the Department of Health and Social Security (DHSS), other relevant government departments and the UK representatives of a number of nations exporting to the UK.

The primary objective of PHAs is to undertake a range of responsibilities aimed at safeguarding public health, by applying United Kingdom Government and EEC legislation, which may be categorised into the following fields of work.

Imported food

Britain continues to import large quantities of food through its seaports from a variety of countries whose standards may vary considerably from those of the UK in terms of hygienic production, handling, storage and the quality of the raw material used.

It is the responsibility of PHAs to enforce the imported food regulations and other related legislation aimed, *inter alia,* at preventing the entry of food which is unfit, unsound or unwholesome. In the case of meat, meat products, milk and certain other categories of food—including those originating in specified countries following the Chernobyl nuclear accident—it is necessary to ensure that each consignment is derived from an acceptable source and is accompanied by valid public health certificates.

The process of imported food control begins by an inspector selecting from cargo manifests those consignments of food which he wishes to be produced to him for examination and those which he intends to detain pending receipt of the correct health documentation. Where this documentation is not forthcoming or where it has been shown to be incorrect then the consignment is usually detained pending its re-exportation.

Foods selected for physical or 'organoleptical' examination are inspected to ascertain the country of origin, any necessary certification, physical defects, composition, with particular reference to the possible presence of non-permitted additives—e.g., preservatives, colours, etc.—and labelling irregularities. Physical defects may take the form of transit damage, including taint, carriage at incorrect temperature, oil or water damage, rodent or insect damage, fire damage, mould growth, 'freezer burn', 'sweating', or goods crushed in stow. The inspector will also check for evidence of canning defects, including 'blown', rusty or leaking cans. Fresh meat, including poultry meat, is examined at one of the meat inspection buildings specially constructed for that purpose, to check for evidence of disease, and to ensure that it is imported in full compliance with the regulations.

The inspector will then decide whether to draw samples for chemical analysis—e.g., for the presence of non-permitted preservatives, colours, anti-oxidants, heavy metals (lead, cadmium, etc.), pesticide residues and other non-permitted or otherwise harmful substances—or for bacteriological examination—e.g., the presence of pathogenic bacteria, including salmonella, styaphylococcus aureus, bacillus cereus, etc.

Foods selected for chemical analysis may include wines (some brands of which from certain EEC countries have recently been found to contain added methanol and diethylene glycol, both harmful substances), groundnuts (for the presence of aflatoxin, a mould known to be highly carcinogenic when consumed by animals), foods originating in the area of fall-out from Chernobyl (as a check on radiation levels), sauces and other compositional foods of uncertain origins (to test for the presence of non-permitted additives), fruits (for the presence of excess or non-permitted pesticide residues) and many other additives in a wide range of commodities. Foods regularly sampled for bacteriological examination to check for the presence of pathogens may include imported, cooked prawns, shrimps, oysters and other shellfish, egg albumen, milk products, including cheese, pâtés and other meat preparations, including corned beef and canned salmon.

The inspector may detain consignments for up to six 'working' days until the completion of any special examination or until the results of tests have been received. When, as a result of physical examination or unsatisfactory chemical or bacteriological tests, the food is considered to be unfit, unsound, unwholesome or otherwise unacceptable, he may 'seize' the consignment and seek its voluntary surrender from the importer for destruction. The inspector may, alternatively, agree to the use of the food for some other purpose, including reconditioning, fumigation, animal feed, etc., which he considers acceptable subject to the receipt of any necessary guarantees. Where an importer does not voluntarily surrender the food then a representative sample is taken before a Justice of the Peace, together with the public analyst's certificate or other supportive evidence, for an opinion as to whether or not the food is 'unfit' and should be condemned; this action is rarely necessary as most foods are voluntarily surrendered.

In the London PHA area, where in 1986 some 4 million tonnes of food were imported, including over a quarter of a million tons of meat and meat products, around 15,000 inspections were carried out of foods in containers, Ro-Ro vehicles or upon discharge from break-bulk vessels. In addition, 13,000 public health certificates were received and checked for consignments of meat and meat products, 551 samples of food were submitted to the public analyst for chemical analysis and 831 samples for bacteriological examination to public health laboratories. Several hundred tonnes of food were considered to be unfit or otherwise unacceptable and were disposed of by destruction, re-exportation, or under guarantee, for some other purpose acceptable to the authority.

Attention to satisfactory food hygiene practices is essential in preventing contamination in the transport, handling and storage of imported food. PHAs administer the Food Hygiene (Docks, Carriers, etc.). Regulations which are aimed at safeguarding the hygienic integrity of such food.

Imported meat for pet food

Many thousands of tonnes of meat are annually imported into the UK for use in the pet food trade from a variety of countries, including Australia, New Zealand, the Americas, China and the Continent of Europe. This meat may not have been subjected to public health inspection procedures and could therefore constitute a great risk to the public were such meat to be fraudulently switched, after importation, for sale for human consumption.

To avoid this occurrence, PHAs are required to issue a movement permit for each consignment/vehicle load before it may leave the port. One copy of this permit is later countersigned by the local authority at destination and returned to the PHA as proof of safe arrival. Particular vigilance is required by PHAs as profits from the illegal sale of such meat on the 'human food' market would be very substantial. In addition, checks are made at the port to ensure that the packaging is correctly marked 'not intended for human consumption' and, in the case of carcase meat, as opposed to 'offals' it is adequately stained. The transporting vehicle is also required to be properly marked. In London, during 1986, a total of around 26,000 tonnes of 'pets meat' was imported and 1,600 movement permits issued.

Infectious disease

Early shipborne diseases—The prevention of the spread of infectious disease between nations has been a preoccupation of government since the earliest times, particularly with island communities such as the UK. This problem is further illustrated by an extract from the Registrar General's report for the quarter ending December 1865 which reads: 'The importance of attention to the hygienic condition both of our merchant vessels and our seaports is clearly seen; for a foul ship, instead of merchandise, carries from land to land the seed of depopulatory diseases, and a foul seaport supplies the soil in which they readily germinate.'

One of the earliest recorded incidents of the spread of disease by sea transport occurred during the winter of 1346/47 when a small band of Genoese traders took shelter at Caffa (now Theodosia) on the Black Sea. The town was besieged by attacking Tartars, many of whom were infected by the plague or 'Black Death' as it later became known. On returning to Genoa, these seamen brought with them the deadly disease which rapidly spread throughout Northern Europe, to arrive in England at the port of Melcombe (now part of Weymouth) on the Dorset coast in June 1348, from where it quickly spread to other parts of Britain.

It is understood that one of the earliest forms of 'quarantine' occurred in 1374, when the Venetian republic ordered the inspection of incoming vessels and the exclusion of infected ships. This successful operation was extended to other ports, and three years later the town of Ragusa insisted upon infected persons being detained for 30 days before entering the gates. The port authorities at Marseilles introduced similar measures and in 1383 extended the period to 40 days, while at Venice in 1403 travellers from the Levant were ordered to be 'isolated' for the same period, 'quaranta giorni', from which the word quarantine is derived.

In England, the first of several orders in Council was enacted in 1709 establishing quarantine procedures for ships and persons from infected areas.

The voyages of discovery of Christopher Columbus in 1492, Vasco de Gama in 1497 and Ferdinand Magellan in 1520 played a substantial role in destroying the geographic barriers that had hitherto isolated populations around the world and resulted in the spread of infectious disease from countries where they had become endemic and thereby relatively harmless, to those where their population had no immunity to attack. There followed devastating outbreaks in various nations of smallpox, syphilis, measles, and many other illnesses.

Present position—A few of the diseases which may be considered of particular concern to the international traveller include cholera, which continues to reach epidemic proportions in a number of countries within Africa and Asia. There were 40,500 cases reported to the World Health Organisation (WHO) from 36 countries in 1985, although the real figure is thought to be higher. When visiting countries where the disease is prevalent it is advisable to take a number of basic precautions, including the chlorination or boiling of water supplies, the use of heat-treated milk and dairy products, the protection of food and drink from contamination by flies and strict attention to personal hygiene. Vaccination also gives a measure of personal protection.

AIDS (acquired immunodeficiency syndrome), which results from HIV infection, is a growing international concern, with 75,392 cases so far reported from 130 countries (December 1987), a six-fold increase in three years. At the World Aids Conference held at London in January 1988 the director of the WHO AIDS programme reported that, assuming a conservative figure of 5 million people now infected, the total number of AIDS cases (i.e., showing symptoms of the disease) could rise to 1 million by 1991. Every precaution must be taken by individuals to prevent infection, the principal means of transmission being through sexual intercourse with an infected partner. Advisory leaflets are distributed by the London PHA and at a number of other ports in the UK. Advice may also be obtained by contacting the AIDS Information Service (see references at the end of this chapter).

Malaria continues to be a major public health problem in endemic countries. The number of cases reported to the WHO in 1985 was 4.8 million (provisional figure). This figure excludes those of the WHO African Region from where reports are few but where the disease is particularly prevalent. Of a total world population of 4,818 million it is estimated that some 2,318 million (48 per cent) live in areas where antimalarial measures are carried out. It is understood, however, that in many of these areas controls are minimal and 405 million people inhabit areas where no specific measures are undertaken to control the disease. Resistance by the anopheles mosquito to some insecticides, often caused by their earlier indiscriminate use, is of particular concern in certain malarial areas. There is also an increasing problem arising from the resistance of some malarial parasites to antimalarial drugs. It is therefore essential to seek medical advice before proceeding to malarial zones in order to acquire the correct prophylactic treatment.

Cases of **yellow fever** continue to be reported from some countries in Africa and South America. Travellers to affected areas are advised to be vaccinated against the disease and an international vaccination certificate is a requirement of entry to some countries if travelling from an infected area.

Plague is still present in parts of Africa, Asia and the Americas and in 1985 the number of cases reported to the WHO was 483 (including 51 deaths). However, the disease is not generally considered to pose a problem for the international traveller, although seafarers must remain cognisant of the precautions needed to prevent shipborne infestations of rodents as it is fleas from rats that are the principal vectors of the disease to mankind. Containers and LASH barges loaded in plague areas can pose a threat to public health when becoming rodent-infested.

Smallpox was confirmed by the WHO as being globally eradicated in 1980. Post eradication policies continue whch include the investigation of suspected cases and the holding of emergency stocks of vaccine.

Naturally, there are many other diseases which may affect the international traveller and, as previously mentioned, it is advisable to seek medical advice on the precautions to be taken before leaving the UK. Certain basic information is contained in a DHSS leaflet entitled *Protect Your Health Abroad.*

Health controls at UK ports

Port health authorities each have an administration system which is aimed at providing early diagnosis and the prevention of transmission of infectious disease as well as arranging for any necessary treatment. In London a group of port health inspectors and medical officers operate a 24-hour stand-by system to respond to reports of illness from vessels bound for the port.

Under provisions contained in the Public Health (Ships) Regulations it is the responsibility of the master to report to the port health authority the presence aboard his vessel of a person suffering from infectious disease or having symptoms which suggest disease of an infectious nature. It is also a requirement to report any condition which might lead to the spread of infection—e.g., contaminated food or water, etc., and the presence of animals or captive birds. This information is to be reported either directly to the PHA via a coast radio station or by the vessel's agents, to arrive between 4 and 12 hours of her arrival. Full details of these reporting procedures for London appear in the Admiralty Channel Pilot; these are similar to the requirements for other ports.

The regulations provide for the inspection of ships, the examination of persons suffering from infectious disease, or suspected of being so infected or having been exposed to a disease of an infectious nature. There are a number of measures that may be taken by the PHA to prevent danger to public health including the removal of infected persons, surveillance procedures, restriction upon boarding and leaving and deratting.

Rabies control—Rabies is a disease to which most warm blooded animals are susceptible and is present in all but a few countries in the world. It is usually transmitted to man through virus-laden saliva when bitten by a rabid animal.

The United Kingdom exercises strict control over animals arriving from abroad in an endeavour to remain rabies-free. Imported animals are subjected to a period of six months' quarantine whilst those remaining aboard vessels in transit are required to be kept securely confined in an enclosed part of the ship throughout the stay in port. It is essential to ensure that these animals are prevented from coming into contact with those ashore.

In addition to receiving reports from the masters of inward bound vessels concerning the presence of animals under provisions of the Public Health (Ships) Regulations, a number of PHAs, like London, enforce anti-rabies controls under the Animal Health Acts.

Rodent control—Rats and mice have followed man to every corner of the world carrying with them disease and causing great economic loss through the consumption and contamination of food.

The ship rat, *Rattus rattus,* and other rodents breed prolifically with seven or eight young being born per litter and a breeding pair producing up to eight litters per year. The young become sexually mature at about three months with the result that, under ideal conditions, one breeding pair can be responsible for producing up to 800 offspring per annum.

In addition to plague *(Pasteurella pestis*—usually spread by the rat flea Xenopsylla cheopsis), rats are known to carry murine typhus fever *(Rickettsia mooser*—also spread by the rat flea as opposed to the generally more severe epidemic form of typhus fever transmitted by lice), bacterial infections such as *Salmonellosis (Salmonella typhimurium,* etc., causing food poisoning through the consumption of food contaminated by rat urine or faeces), *Leptospirosis* (a group of illnesses including Weil's disease characterised by haemorrhagic jaundice generally associated with water contaminated by infected urine), *trichinosis (Trichinella spiralis,* a round worm which may be acquired by man through the consumption of pork meat contaminated by rats or other animals), and rat-bite fever (*Streptobacillus moniliformis* and *Spirilum minor,* both transmitted by a rat bite).

Infestations of rats and mice aboard ship must be prevented not only because of the risk of disease, but also in view of the serious loss to cargo that can arise from the consumption and contamination of food. Damage to structures may occur by rats and mice not only in their search for food and water, but also because their incisor teeth continue to grow throughout their lives and gnawing is part of their regular behaviour. It is not uncommon to find severe damage to lead pipes and metal-sheathed cables resulting from these gnawing activities.

The international health regulations call upon PHAs to carry out regular inspections of foreign-going vessels for evidence of rodents, to require treatment where necessary and, where authorised, to issue deratting and deratting exemption certificates. In addition, they are required to take steps to ensure that ports are kept free from rodents in order to reduce the risk of ship-borne rats and mice from making contact with those of the indigenous population.

Shipmasters should report evidence of a rodent infestation to the PHA as soon as possible to enable advice to be given on the most effective treatment. Such evidence may take the form of faeces, footprints, gnawing, smears, urine and nesting material.

Food hygiene

Good hygiene practice is essential in preventing outbreaks of food poisoning. Such illness when occurring aboard ship may result in many of the crew becoming seriously affected, thus preventing them from undertaking the essential tasks necessary to maintain the safety of the ship.

Food poisoning may be caused by a variety of bacteria and viruses, some of the more common of which include the following. *Salmonellosis* is commonly associated with poultry meat, eggs, egg products, milk and dried milk which have been inadequately heat treated or, frequently, from the cross-contamination of cooked foods by bacteria in the raw produce when using the same preparation surface without thorough cleansing or by the use of contaminated knives, etc. Infections may also be passed from uncooked food on the hands of staff. *Clostridium perfringens* is usually associated with meat and meat dishes which have been inadequately heated or re-heated. *Staphylococcus aureus* is caused by the contamination of food—often pastries, sandwiches, salads, and sliced meats—by discharges from skin infections, including boils and septic cuts. *Bacillus cereus* is generally associated with rice and vegetable dishes which have been kept warm (below 63°C) after cooking. *Campylobacter* is found in cattle, poultry and dogs and contamination of food is thought to be from the faeces of these animals; unpasteurised milk has also been implicated in several outbreaks. Other food poisoning organisms include *Escherichia coli, Bacillus subtilis, Yersiniosis, Vibrio para-haemolyticus* and *Scombotoxin*.

Illness may also be caused by the consumption of meat containing the viable cysts of parasites which may develop into tapeworms in the intestine—*Taenia saginata* from infected beef and *Taenia solium* from pork. Infection may also arise from eating pork meat containing the larvae of the parasite *Trichinella spirallis*. These parasitic infections, which are more prevalent in countries where meat is not subjected to adequate inspection procedures at abattoirs, are destroyed by thorough cooking.

Attention should be given to maintaining a high standard of repair and maintenance in galleys and other food preparation and storage areas, regular and thorough cleaning routines, hand washing by the catering staff after use of the toilet, the wearing of clean, washable overclothing, the withdrawal from work of food handlers suffering from a gastrointestinal infection, the covering of cuts and abrasions with waterproof dressings and insistance of no smoking in food preparation areas. In addition, the effective control of cockroaches and other insect pests is essential as they can cause the contamination of food with harmful bacteria. Foods must be handled as little as possible in the course of preparation and care taken to ensure that meat and poultry meat is thoroughly defrosted before cooking. Joints of meat should be sufficiently small to ensure thorough heat penetration (say, 6 lb maximum) and all cooked foods must, unless for immediate consumption, be either kept at a temperature of above 63°C or rapidly cooled and stored at below 10°C to reduce bacterial growth. Only potable (drinking) water must be used in galleys and other food preparation areas.

Surveys have shown that the principal causes of foodborne illnesss are as follows: (in descending order of importance) inadequate refrigeration (highest incidence); foods prepared well in advance of serving; infected persons and poor hygiene; inadequate cooking or heating; food kept 'warm' at the wrong temperature; contaminated raw materials in uncooked foods; inadequate reheating; cross-contamination; and inadequate cleaning of equipment. This list serves to illustrate the need to observe good hygiene practices in order to minimise the risk of illness.

Port health authorities are responsible for investigating the cause of foodborne illnesses. They also carry out inspections of catering spaces in order to detect unsatisfactory conditions and unsafe working practices that may give rise to infection. Although very occasionally resorting to legal action to secure improvements, most remedial work is obtained with the co-operation of the master.

Water supply

As polluted water supplies may be responsible for outbreaks of illness, including typhoid and cholera, it is essential that, unless known to be of high quality, supplies are chlorinated on receipt or are subjected to some other satisfactory treatment—e.g., ultra violet (UV) or boiling. Hoses must be clean and with an impervious lining, never permitted to 'trail' into dock water and, when not in use, stored in a dry, protected area of the deck. It is advisable to flush the supply through the hose for five minutes or so before filling the tanks.

Tanks containing potable (drinking) water must be regularly inspected, cleaned and, where necessary, cement washed or treated with a proprietary coating. After work has been carried out in such tanks they should be filled with super-chlorinated water, left to stand for 12 hours, pumped out, and refilled with clean water. Equipment, pumps, etc., used for potable water must never be permitted to come into contact with non-potable supplies—e.g., ballast water drawn from the sea, rivers or docks. Non-potable supplies must never be supplied to galleys and other food preparation or storage areas for risk of contamination of food. Water filters must be regularly cleaned.

Legionnaire's disease and the milder condition known as Pontiac Fever, are transmitted by the inhalation of contaminated water droplets, generally from shower-heads or air-conditioning systems. Water installations should be designed to minimise the colonisation and multiplication of bacteria. Regular overhaul and cleansing of holding tanks, shower-heads and air-conditioning systems should minimise the risk of illness from this source and, ideally, water should be circulated at temperatures of below 20°C or above 50°C. However, at temperatures of above 50°C there is a serious risk of scalding so that, unless thermostatic valves can be incorporated to provide cooler, 'blended' hot and cold water, notices warning of the danger of scalding must be posted—see the relevant 'M' Notices.

Shellfish

Each PHA is responsible under the Public Health (Shellfish) Regulations for ensuring that molluscan shellfish—e.g., cockles, mussels, oysters and clams—harvested from polluted waters within their district are subjected to a satisfactory form of treatment before being sold for human consumption. These shellfish are notorious for causing illness as they filter large volumes of water through their gut which causes them to retain bacteria and viruses at up to six times the concentration found in the surrounding water.

In the London PHA district, where many thousands of tonnes of cockles are annually harvested from sewage-contaminated waters, the local traditional, cooking arrangements have recently ceased to be approved. This action has resulted following a number of reports of viral gastroenteritis associated with the consumption of cockles from the area. New approvals will be granted only provided equipment is installed that conforms to a MAFF specification designed to ensure that the cockle meat reaches a minimum temperature of 90°C for $1\frac{1}{2}$ minutes. This action should safeguard the public from further illness attributed to these shellfish.

Other enforcement activities include enforcement of the Food Hygiene (General) Regulations aboard vessels plying on rivers and on short coastal voyages; a particularly time-consuming operation for the

London PHA in ensuring the maintenance of satisfactory standards on the many passenger launches and permanently moored catering vessels serving food within the district. The monitoring of living conditions aboard houseboats and of the waterborne carriage of refuse (over half-a-million tonnes is annually transported to land-fill sites by barges on the Thames) to prevent 'nuisance' from spillage, etc., is enforced by the London PHA under local bylaws.

Controls over noise 'nuisance' and of dark smoke emissions are exercised under appropriate legislation; noise from disco launches operating in the early hours of the morning can be particularly troublesome. The lecturing of overseas visitors and of trainee environmental health officers in port health work continues to form a time-consuming though important part of the duties of the Port Health Inspectorate.

This account is intended to provide the nautical surveyor, the harbour master and other port officials with an insight into the work of port health authorities and to provide information that may prove helpful in their understanding of some of the basic concepts that contribute to maintaining good health aboard ship. □

Further reading

Cartwright F. F. *A Social History of Medicine.* Longman.

Coleman V. *The Story of Medicine.* Longman.

Glasschieb H. S. *The March of Medicine.* MacDonald.

Bassett W. H., Davies F. G. *Clay's Handbook of Environmental Health.* H. K. Lewis.

Benenson A. S. ed. *Control of Communicable Diseases in Man.* The American Public Health Association.

Lamoureux V. B. *Guide to Ship Sanitation.* The World Health Organisation.

Smith J. V., Pal R. eds. *Vector Control in International Health.* The World Health Organisation.

AIDS What Everyone Needs to Know—Health Education Authority (booklet).

AIDS Don't Die of Ignorance—Department of Health & Social Security (leaflet).

The Ship Captain's Medical Guide. HMSO.

PART II
THE NAUTICAL SURVEYOR IN COMMERCE

MANAGEMENT OF A MARINE SURVEY COMPANY AND THE EMPLOYMENT OF MARINE SURVEYORS

J. M. M. Noble, B.Sc, FNI
Managing Director, Murray Fenton and Associates Ltd

THE TITLE SUGGESTS there are two different elements in the running of a marine surveying company; the second half of the title is, in a sense, an expansion of the first. Why should management of a marine survey company be any different from managing any other business? The answer is quite straightforward; every company is different and the style of management depends very much upon the nature of the business and personalities of those involved. It is not the purpose of this article to explore the fundamentals of management, but rather to examine where the marine field is different.

One of the first points to mention is that a marine surveying company is a specialist company. It can be seen from other chapters in this book just how specialist. The manager of a company must recognise the skills of his surveyors as well as their sense of independence. Marine surveyors are professionally qualified and often hold additional academic qualifications. A manager must be able to deal with qualified staff. However, qualifications are not sufficient alone and it is essential to ensure that survey staff are properly trained.

When operating in the 'field,' the marine surveyor has considerable autonomy; even with modern communications he may find himself in a position where considerable sums of money are at stake—on his word!

Starting up

There is a common myth that a mariner is automatically qualified to become a marine surveyor because of extensive sea service. Nothing could be further from the truth and this misconception has resulted in many failures. No surveyor can be 'all things to all people' and before setting off in the survey world, an individual must have an idea of what direction he wishes to go. The surveying market is finite and many customers do not necessarily want the attention of highly qualified and experienced mariners. A good example of this is seen in tanker outturn surveys, where many of the inspectors employed have never been to sea in a responsible capacity.

An aspiring surveyor, setting up in business, must research his market, appreciate his own limitations and know what he wants to do!

Types of company

There are a number of different types of marine survey company operation. Firstly, there is the 'one man band'; such companies usually consist of a single operator, very often working from home with little back up. Management of such companies is highly individualised.

At the other end of the scale there are the large 'group' type companies, which offer surveying services either nationally or internationally; management of these firms is often left to specialist managers rather than surveyors. However, in this chapter, I am dealing with management of small survey firms employing a small number of staff where management techniques cannot be stylised in the business school sense.

The manager in a small company has the responsibility of ensuring a sufficient work load to keep his staff busy. This means identifying the market and approaching potential customers. The first role of the manager is to 'sell' the company to potential clients. This means establishing contact, following up with personal calls—and then giving the service. Follow-up includes monitoring both field work and report writing.

Supervision and training

Being a marine surveyor is not only a question of holding a sea qualification, having appropriate experience and possibly an academic qualification as well, but also having the right approach and mentality, with a willingness to learn. There is no formal training, in the sense that there is an ultimate qualification in marine surveying, as least not yet.

Individual organisations do operate their own training courses which are designed specifically to meet the requirements of the survey services they offer. Small and medium-sized surveying companies usually offer little or no formal training, but it is a fundamental management responsibility to ensure that a surveyor is not asked to undertake a task beyond his capability or experience. The manager has a particular role to watch, or even personally supervise, new surveyors when they start.

Experienced surveyors will meet problems not encountered before and an 'open' type of management is very often the most productive. By encouraging open inter-office discussion, problems can be aired and the clients get the benefit of combined experience.

Survey report

The customer will require a report on a surveyor's activities and results of the survey commissioned. The final 'product' is the report itself; therefore, it is extremely important that it meets the customer's requirements and is presented in a pleasing and logical way. This is an area where the manager's responsibility is paramount.

The most observant, best qualified and conscientious surveyor is quite useless if the report he writes does not make sense. The ability to write clear, concise, accurate English is as important as the surveying ability itself. Careful scrutiny of reports by the manager is exceedingly important and constructive criticism is often necessary. Any surveyor is only as good as his last report and, because the report is the product, it is important that it is accurate.

Financial

The life blood of any company is its cash flow. Sloppy financial management has led to the downfall of many a company. The manager must have a thorough knowledge of his clients' requirements and local commercial practices. In some locations it is necessary to seek payment, at least in part, before undertaking survey work. Ideally, the invoice for a survey should be submitted with the report, although this is not always possible, particularly when initial urgent verbal or telex reports are required.

The manager should ensure that each file maintains a cost record, to cover, for example, telephone calls, telex, photographs, transportation and subsistence. If this is not done, cost control becomes impossible and invoicing inaccurate.

The most difficult task is often collecting the payment.

Apart from routine surveys, most surveys are commissioned as a result of some incident or casualty, be it major or minor, and the client is usually expecting an immediate response to his problem. Once the survey has been conducted and the problem solved, then the client's priorities change. How individual companies react to non, or slow, payment of bills is a matter of judgement; an even-handed fair approach is the one which is more likely to get results. The option of resorting to the Courts, in the UK at least, is always there, but such an option should only be exercised as a last resort.

Employment of surveyors

As mentioned above, professional qualifications are a prerequisite if a surveyor is to operate in marine or nautical areas. The level and extent of qualification will vary from a minimum seagoing certificate to a final seagoing certificate supplemented by a degree or senior sea or shore experience. The level of qualification required really depends on the style of customer being served and the level of business at which the company is aiming to operate.

A main reason for requiring adequate qualification is the credibility of the surveyor. Whilst relatively few survey reports end up forming part of a formal dispute, the thought that it might must always be there. It is important that a surveyor, who may be making judgements in particular circumstances, is of sufficient stature that his opinion will stand up to cross-examination in Court or arbitration. As a matter of routine, anyone employing surveyors should check their qualifications.

Relationships between surveying companies and their staff do vary. Some surveyors are employed by companies as staff surveyors, whereas other companies are prepared to act more as 'agencies,' in that they operate as 'clearing houses.' Most surveying companies rely on *ad-hoc* staff to supplement their own permanent staff when times get busy. Companies have to make certain that their support staff are also properly qualified and that the relationship between them is clearly defined. This is particularly important when dealing with the Inland Revenue!

Freelance surveyors tend to operate more commonly in the offshore world, where there is a marked seasonal factor in employment. The acceptance by the customer of an *ad-hoc* surveyor depends very much upon the relationship between the surveying company and the client. As in any occupation, a new surveyor will undergo a probationary period when much of his work will be closely supervised, allowing an opportunity for on-the-job training.

Most established companies receive several letters a week from those wishing to take up surveying as a career. Whereas in the inspectorate type of companies there can be a relatively high turnover, in the consultancy firms the turnover is very low indeed and new job opportunities usually arise from expansion of a business rather than staff turnover. A well-presented summary CV, on one side of A4 paper only, will attract more attention than pages of career details; these can be asked for later if interest is being shown. Where a potential surveyor has a particular expertise (for example, gas carrying), there is more likely to be a positive response from a would-be employer.□

ROLE OF THE NAUTICAL EXPERT WITNESS

J. J. Banister, Master Mariner, FNI, FCMS, MRIN, MInstPet, ACIArb, ARINA
Marine Surveyor and Consultant, H.H. Bridger & Co Ltd, London

BEFORE DISCUSSING the role of the nautical expert witness in particular, it is desirable to consider the 'expert witness' in general. An expert has been defined as 'a person with special skills, technical knowledge, or professional qualifications whose opinion on any matter within his cognizance is admitted in evidence.'

Originally an expert was regarded as an assistant of the Court; it evolved, however, when oral evidence became usual procedure for an 'expert' to take on the role of a witness. This role is the most usual method by which experts provide evidence to the Courts or tribunals in England.

However, expert evidence is not relevant and thus not admissible when the Court or tribunal is competent to decide the matter in dispute without assistance. Under the rules of the Supreme Court, there is provision for a judge to sit with expert assessors—i.e., in the Admiralty Court. The role of the nautical assessors is to advise the judge during the Court proceedings on matters of seamanship, nautical knowledge, skill and experience. However, it is within the discretion of the Court, in cases involving unusual or specialised equipment or vessels, to permit expert witnesses to appear.

The rules of the Supreme Court also provide for the appointment of 'Court Experts,' but this rule has apparently been little used.

The number of expert witnesses who become involved in a dispute will of course depend upon the number of specialised issues raised by the dispute. The rules of the Supreme Court permit a limitation on the number of expert witnesses which each side may call, the numbers being equal for both sides. Arbitrations usually follow a similar procedure.

The practice of calling experts on both sides has led, in some cases, to the unfortunate fact that their respective testimonies sometimes appear to be advocacy, and as such will be discounted by the Courts or arbitration tribunals. The true role of the expert witness is to assist the Court, the tribunal, or the official inquiry in reaching the truth—clearly a position of privilege and trust.

Procedures

Once a serious dispute has arisen, conflict is probably inevitable. Civilised resolution of conflicts, however, will be attempted either by recourse to the Courts, or by appointing a third person to settle the dispute—arbitration.

The disputes in which a nautical expert may become involved will usually concern large sums of money, and thus lawyers will have already been instructed whether the route for resolution of the dispute is the Courts or an arbitration tribunal. The lawyers will draw up 'pleadings,' the formal documents delivered alternately by the parties to one another, until the facts and questions of law to be decided in the action (Courts) or hearing (arbitration) have been ascertained. The pleadings delivered are as follows:

Plaintiff/Claimant	Defendant/Respondent
Points of claim.	Points of defence and perhaps counterclaim.
Reply.	

During these exchanges, and interposed between the stages outlined above, there may be 'requests for further and better particulars.' These particulars are the details of either the claim or the defence which are necessary in order to enable the other side to know what case they are required to meet, thus enabling the full formal pleadings to be submitted.

When the stage of the 'pleadings' has been completed, 'discovery of documents' (usually abreviated to 'discovery') may take place. Discovery is a statement, which may be under oath, by one party for the information of the other, which enumerates all the relevant documents bearing on the points at issue, and on which that party intends to rely.

The experts' reports, as far as Court proceedings are concerned, are exchanged in advance of the hearing. This is usually also the case in arbitrations, but is at the discretion of the tribunal. The date and time of the hearing will have been fixed well in advance. Pleadings must have been served and discovery completed before the due date.

During the pre-hearing procedures, it is usual for each side to consider the merits of a settlement of the dispute before commencement of the Court/arbitration hearings. Should no amicable resolution of the dispute be reached, the hearing will proceed.

The procedures at the hearing at Court or arbitration ordinarily follow similar lines.

Opening address—The advocate for the claimant (or the claimant himself) opens his case and, if there is a counterclaim, he will at the same time open his defence to such counterclaim.

Claimants witnesses—The advocate for the claimant calls each of his witnesses in turn, and after taking the oath, each may then bear three oral examinations:

(a) By claimant's advocate. This is called 'examination-in-chief' (or direct examination).

(b) By respondent's advocate. This is called 'Cross-examination.'

(c) By claimant's advocate. This is called 're-examination.' The witness may only be re-examined on any matter raised in the cross-examination.

Respondent's address—The advocate for the respondent (or the respondent himself) will then address the Court/tribunal, and open his case.

Respondent's witnesses—The advocate for the respondent calls each of his witnesses in turn, and after taking the oath, each may then bear three oral examinations:

(a) By respondent's advocate—examination-in-chief.'
(b) By claimant's advocate—'cross-examination.'
(c) By respondent's advocate—'re-examination.'

At any stage in the proceedings, the judge or arbitrator may put any question to a witness.

Summing up—The respondent's advocate addresses the Court/arbitration tribunal. The claimant's advocate replies to the Court/arbitration tribunal.

In Court actions, the judge will normally deliver his judgment immediately, although he may 'reserve' judgment for a later date. Arbitrations are closed, as a general rule, by the arbitrator(s) informing the parties that an award will be made, in writing, in due course.

Duty of skill and care

Persons who engage in any exacting profession requiring special skills or knowledge have, from the Middle Ages, been recognised as owing a duty of care to those who would be affected by their work. Persons professing special skills must exercise such skills and care of the ordinary competent member of the same profession.

When acting as an expert witness, whether in Court or at an arbitration, all questions must be answered truthfully and honestly. False evidence obviously must not be given. On the other hand, an expert witness need not volunteer information which is not specifically requested of him.

An expert witness's prime duty is to the Court, and in as much must avoid misleading the Court. Thus, very careful consideration should be given before providing any answer to ensure that it does not misrepresent the facts as he sees them. Any subsequent elaboration of an answer should ensure that the intended true meaning has been made clear—but no more than that.

Nautical expert witness

The nautical expert should thus have a special nautical skill, nautical knowledge, or a nautical qualification. It should be noted, however, that professional qualifications are not an absolute prerequisite for a person to be accepted as an expert. Experience has shown that the evidence of a person who is regularly advising—i.e., a consultant—is usually of far more value than, say, an unqualified person, no matter how competent, skilled and experienced that person may be.

Nonetheless, the decision as to the qualification of a witness to provide evidence of opinion as an expert is made by the judge, or the arbitrators. Despite this acceptance of a witness as a nautical expert, the scope of his knowledge will then almost certainly be the subject of cross-examination, and the extent of his skills or experience will affect the strength of his evidence.

The expert is not permitted to express opinions above and beyond his own expertise, and a further special knowledge may also be essential as a qualification for an expert witness to provide opinions on 'other' matters. The Judicial Committee held in *United States Shipping Board* v *Ship St. Albans (1931) A.C.632, P.C* that marine surveyors were not competent as experts to ascertain from photographs the position of a collision between ships in a harbour.

Involvement

A nautical expert may become involved in a dispute in one of two ways. It may well be that through some pre-involvement during the course of his business or profession the expert finds himself involved in a dispute, which then requires to be resolved by the Courts or by an arbitration tribunal. The expert then has knowledge of the facts of the case and may be required to give evidence of these facts within his own knowledge, but also to give opinions as an expert.

A danger does exist in these circumstances when, because of their prior involvement, such witnesses might not be sufficiently impartial. It is certainly far easier to be objective when one has no involvement in the matters which led up to the dispute.

The second means of involvement is for a person who is considered to have the special skills required, and, although not pre-involved, at the request of one of the parties to the dispute is prepared to give evidence of opinion.

At this stage the expert should request enough information from those seeking his assistance in order to satisfy himself that he has no conflict of interest, that the evidence he will be able to provide lies within his own experience and expertise, and will be of use to his clients—and most importantly that he will be able to devote sufficient time and energy to the case. Then and only then should the appointment be accepted.

Function

The nautical expert may become involved in a dispute at any stage prior to the hearing. Obviously, as far as the expert is concerned, the sooner instructions are received the more time and energy can be devoted to the issues.

However, from experience, this is not always the position, and it has even been known for an expert to be appointed, submit a report, meet counsel for a pre-hearing conference, and appear in Court as a witness all in the period of three days.

Nonetheless, ideally the opinions of the expert will be sought before the 'pleadings' have been finally settled. At this stage the expert may be providing all relevant information counsel needs to settle accurate points of claim or defence, advising the instructing solicitor on which documents might be sought at discovery, and generally providing any technical assistance which the lawyers may require in order to prepare the case, or alternatively, which may be required to reach an amicable commercial settlement before the dispute reaches the Court or arbitral tribunal.

After the 'pleadings' have been settled, the expert should prepare his report, sometimes known as 'proof of evidence,' which he should write with the relevant pleadings before him.

This report is an account of what the expert is prepared to say to the Court/tribunal on oath. The report should be based on the expert's experience and personal knowledge, using the evidence provided in the documents supplied by the instructing solicitor. The expert may refer to text or specialist books on the subject in dispute, as well as supporting opinions from previous law cases and judgments in which he has given evidence, and which had been accepted by the Court.

Preparation, style and construction of the expert's report are matters for the individual. Clearly the report is of very great importance, and much care, consideration and attention to detail should be taken in its composition.

Drafting a report

Some of the essential factors to be borne in mind when drafting a report are:

(a) Simplicity of language. Even though the report necessarily will be technical in content the language should be reduced as far as possible to non-technical terms, and where not possible a glossary or explanation of technical terms may be appended.

(b) Clarity is essential. The report must be written in lucid, straightforward language.

(c) Positive statements are, wherever possible, preferable to negative ones, the latter being less straightforward and more liable to confusion.

(d) The report should be 'readable' and capable of being understood by the lawyers, and must serve the purpose for which it was intended.

(e) The general appearance of a report ought to be pleasing, sections should be numbered, and sub-headings used to advantage, unless the report is very short. This facilitates reference, and to this end an index should be included.

(f) Where the report makes reference to published data, sources should be identified and copies appended to the report. In such cases, care must be taken to number the appendices and to ensure that they are cross-referenced to the report.

The report should contain: the name and address of the witness; a brief description of his qualifications and experience—although these latter points may be contained in a separate curriculum vitae; factual observations; and conclusions, preferably in the order in which they are set out in the points of claim or defence. Skills in the arrangement and presentation of the facts and testimony are essential in order to provide lawyers with worthwhile material which they can use to best advantage.

The report is then submitted to the instructing solicitor, who will in turn pass it on to counsel.

Exchange of reports

The exchange of experts' reports should now take place—and the opposing expert's report will be passed to the expert. Careful study and consideration of the views, opinions and perhaps evidence not previously known to the expert contained within this document will almost certainly demand further advices or replies in writing to the instructing solicitor.

If meetings have not already taken place by this stage of the preparations for the hearing, the expert will almost certainly be asked to attend a conference with the plaintiffs or claimants, their lawyers and other experts who may have been appointed to give evidence on other disciplines. Should the dispute not then be resolved before the duly appointed date the hearing will proceed.

Court and arbitration hearings usually last about five hours per day, 10.00 to 16.00, or 10.30 to 16.30, with an hour's adjournment for lunch. This may seem a short working day, but it is a very long time if spent in the witness box.

Giving evidence

If the expert has not had prior experience of giving evidence, it is prudent and advisable that he should attend a case in the High Court in order to acquaint himself of the atmosphere and surroundings of a Court. It can be, to say the least, a daunting and intimidating experience for any witness under cross-examination, even those with some previous experience. It is not quite so easy to arrange an attendance at arbitration tribunals as these are private hearings.

Witnesses of fact may be barred from attending the proceedings until their evidence has been heard, thereby ensuring they do not become influenced by other witnesses of fact. In most civil cases the expert is permitted to attend the Court throughout.

The expert may, in complicated cases, be asked to attend the hearing from day one, or alternatively, only to attend whilst the opposing expert is giving evidence, and to provide his own evidence. However, whilst in Court the expert is working and would be well advised to take notes of any evidence given which touches on his report, expertise and knowledge. For this purpose he should sit near his instructing solicitor, and provide advices in written brief note form, who may in turn if the advices are considered valid pass them on to counsel.

In some important disputes, daily transcripts can be made available. The expert may be required to study his opposing expert's evidence, advising the lawyers of any points which should be pursued in examination.

If the expert is acting for the claimant he will be giving his evidence before that of the respondent's expert. On the days of the hearing, the expert should arrive in court in good time and suitably dressed. This may seem an unnecessary comment but appearances do mean a great deal; a neat and well-groomed appearance helps to inspire more confidence in the witness.

Upon entering the witness box, the expert will be required to take the oath or make a solemn affirmation. The evidence then becomes 'sworn.' In arbitrations, the evidence may be heard 'unsworn' at the discretion of the tribunal. Once a witness has taken the oath (affirmation) he commits perjury if he makes a statement which he knows to be false or which he does not believe to be true.

Also it is essential to understand that a witness remains under oath until all three stages of his examination have been completed. During adjournments in his examination for meals, overnight, over a weekend, the witness must not discuss the case with any other person. Only when his examination is completed should he enter into discussion. If recalled to give further evidence he will remain 'under oath' for the remaining period of the examination.

Examination

When the examination of the expert commences, counsel usually takes the witness through his curriculum vitae in order to show to the Court that he is duly qualified to give evidence as an expert. Sometimes there is discussion on this point, but invariably the expert is accepted by the Court/ tribunal.

The counsel will then question the witness about the matters in issue, but in doing so should not use leading questions. Although counsel puts the questions, the witness should address his answers directly to the judge or the arbitrators. Courts and arbitral chambers are usually laid out so that the answers directed to the judge/arbitrators can be heard by the lawyers and the Court officials.

The witness should obviously listen carefully and confine answers to the specific questions. The witness should speak slowly and clearly bearing in mind that the judge/arbitrators will be taking notes.

The witness is usually provided by the Court/arbitration with a copy of his report. He will not be allowed to provide his own copy in case it is marked in any way. The witness may call for, or be provided with copies of any other documents or evidence which is relevant and in the Court/ arbitration chamber.

Upon completion of the examination-in-chief, cross-examination will begin. Cross-examination usually takes much longer than examination-in-chief. It is more often than not a trying experience. Obviously counsel for the opposing party will use his best endeavours to discredit the evidence, but provided the witness has given strictly honest opinions based on the facts known to him he should have no fears. He should always endeavour to maintain a cool, calm, dignified manner.

The witness should confine his answers to the precise question and resist qualifying his answers, prevaracation, offering additional information. He should confine his answers to his true opinions. Do not fear saying that a question cannot be answered. Reasonable time will always be given to consider a matter. If mistakes have been made in the report or during the examinations these should be unequivocally admitted.

Upon completion of the cross-examination, counsel may wish to re-examine for clarification of some points which arose during the cross-examination. The counsel can only re-examine on matters which arose out of the cross-examination, after which the witness may be subject to further cross-examination.

Once the expert has completed his evidence he should obviously remain in the Court if he has still to hear the opposing experts' evidence-in-chief. If not, he should remain to provide assistance to the lawyers, and only leave on the express instructions of his solicitor.

Further reading

It has not been possible within the space of this paper to cover completely the role of a nautical expert witness. Further reading and study is essential and commended, both in respect of attaining some legal knowledge, as well as learning from the more experienced expert witnesses.

The Chartered Institute of Arbitrators hold annual seminars to give those attending some knowledge of judicial proceedings and practice and the role of the expert witness.□

Bibliography

The Expert Witness, by H. J. Millar.

An Introduction to Evidence, by G. D. Nokes.

The Giving of Expert Evidence, Guidelines of Good Practice, The Chartered Institute of Arbitrators.

A Concise Law Dictionary, by P. G. Osborn.

Professional Negligence, by G. E. Stringer.

Duties of the Expert Witness at the Hearing, by R. W. Bishop.

Russel on the Law of Arbitration, by A. Walton.

The Law of Arbitration, by W. H. Gill.

The Law and Practice of Arbitration, by John Parris.

Nothing But the Truth, by John Watson.

The Discipline of Law, by the Rt Hon Lord Denning.

Law of Evidence, by L. B. Curzon.

ROLE OF THE NAUTICAL ASSESSOR

Captain David T. Smith, FNI, RN

An Active Elder Brother of the Trinity House, London

IT IS UNLIKELY that anyone, on first choosing to follow the sea as a career, might ever contemplate becoming a nautical assessor. However, with advancing years there can have been few who, on attaining the highest level of professional competence, have not been proud to sit on the bench as assistant or adviser to a judge or magistrate considering marine causes. As a breed they are in a unique position to give advice that stems from their professional knowledge and many years of practical marine experience, in particular that which is related to their technical qualifications, sea time and command. In short, for a given set of circumstances, the nautical assessor will be called upon to advise the judge on how a prudent mariner and/or a ship might behave or should be expected to behave.

New personnel joining the ranks of assessors will of necessity already be fully qualified profesionally. They will, however, almost certainly be totally inexperienced in the background and general environment of the exclusive circle which they are about to enter. How then should they be indoctrinated? This chapter will but provide a key to the door of a very fascinating subject.

In the writer's experience no book of reference is, so far, readily available in the nautical library or elsewhere. This chapter makes no apology for making full use of such sources as do exist in an effort to present the reader with as concise an introduction to the subject as possible. An article on the province and function of assessors in English Courts[1] is a most authoritative source document and will be referred to extensively.

In the Courts

The first reported case published of assessors assisting the High Court of Admiralty was in 1541, since when the practice of calling upon nautical assessors to advise the Courts on matters of nautical skill and seamanship has continued. To this day, for example, the Admiralty Court is regularly assisted in its proceedings by two Elder Brethren of Trinity House sitting as assessors.

Assessors are especially valuable in proceedings where specialist knowledge or experience is frequently required. Without their assistance, complex and often conflicting evidence would have to be provided by experts. Alternatively, further specialist tribunals would have to be created to deal with such matters. There has been a trend since the second half of the last century for statutes to provide for Courts and judicial inquiries to be assisted by assessors in both nautical and non-nautical proceedings. Similar provisions apply today to many statutory committees, tribunals and inquiries[2].

The Assessor[3] is literally a person who sits by the side of another: a person called to assist a Court in trying a question requiring technical or scientific knowledge. In Admiralty business, it is the practice of the Admiralty Court and the Court of Appeal to call in assessors in cases involving questions of navigation or seamanship; they are commonly called 'nautical assessors.' Technical matters are covered by assessors who are specialists in engineering or naval architecture.

Royal Courts of Justice

Within the Royal Courts of Justice marine cases are initially dealt with by judges of the High Court of Justice, attached to the Queen's Bench Division, sitting in the Admiralty Court, which in time of war sits as a Prize Court. An appeal lies, as of right, to the Court of Appeal. But a dissatisfied litigant can only appeal to the House of Lords with leave, which is rarely granted. The application for leave to appeal is made first to the Court of Appeal and, if refused, by petition to the House of Lords.

Admiralty Court

In the Admiralty Court[1] one or more assessors are summoned to assist the judge almost as a matter of course in collision and similar 'damage' actions, in salvage actions, and in actions in respect of personal injuries or death on board ships and fishing vessels. However, the Court may order the trial of such actions without assessors. It is likely to do so if the parties agree that no questions of seamanship arise.

In all Admiralty proceedings, assessors will be summoned at the request of the parties or by order of the Court. An order at the summons for directions will determine whether the trial is to be with assessors. Although these are strictly matters for the Court, the parties may agree among themselves whether assessors are required, and what kind of assessors, and then seek the approval of the judge at the hearing of the summons for directions.

The assessors in the Admiralty Court are usually two Elder Brethren of Trinity House, otherwise known as Trinity Masters. The Masters for any case are selected by Trinity House and the choice notified to the Admiralty Registry before each hearing. However, the assessors need not be Trinity Masters, and must not be if Trinity House is a party to the action (e.g., if a Trinity House vessel or buoy is involved). Likewise a Trinity Master cannot sit if he has been in the service of any of the parties to the action, although a Trinity Master who is a retired Royal Naval officer may still sit if one of HM vessels is involved.

In cases where Trinity Masters do not or cannot sit, assessors, will generally be drawn from either the Court of Appeal panel of assessors or from the Home Office list. Assessors from the Home Office list are usual in cases where specialist nautical assessors, such as fishery experts or engineering experts are required. The Trinity Masters maintain their charts and nautical publications in the Trinity Masters' Room at the Royal Courts of Justice. Publications are retained over a span of several years to ensure that the ephemeral data appropriate to the case being heard are available.

Court of Appeal

The Appeal Court selects its nautical assessors to attend the hearing of Admiralty Appeals[1,4] either from the list of Trinity Masters (ex officio) or from the Master of the Rolls' list of four assessors appointed by him under the Act; normally consisting of two senior post Captains (usually retired) from the Royal Navy, nominated by the Ministry of Defence, and two mercantile marine assessors nominated by the Honourable Company of Master Mariners (though nominees need not be members of the company). Collectively these assessors comprise the Court of Appeal panel and it is customary to select one assessor from each list.

The Master of the Rolls, who is the senior judge in the Court of Appeal, has laid down a practice direction as to when Admiralty Appeals shall be heard with assessors and who shall be summoned.

Both High Court[3] or the Court of Appeal[3] may call in the aid of one or more assessors in any action or matter. Under the Prize Court Rules (1939) one or more Trinity Masters or other assessors may be called to advise on matters requiring nautical or other professional knowledge.

House of Lords

The House of Lords may, in appeals in Admiralty actions, call in the aid of two assessors and hear such appeals wholly or partially with the assistance of such assessors. The two assessors who attend the hearing of Admiralty Appeals are drawn one from the list of Trinity Masters and the other from a list of Royal Navy officers, either active or retired, nominated by the Ministry of Defence[5].

County Courts and judicial inquiries

Provision is made for assessors in County Courts as well as in various inquiries of a judicial character. Assessors appointed under the latter include those called to attend formal investigations into shipping casualties[6] and these may be appointed by selection from one of the five classes in the Home Office list as follows:—

Class I	Mercantile Marine Masters.
Class II	Mercantile Marine engineers.
Class III	Officers of the Royal Navy.
Class IV(a)	Fishery assessors.
Class IV(b)	Special panel for naval architects.

In the 45 County Courts that exercise Admiralty jurisdiction a nautical assessor will only be called at the request of one of the parties and with the permission of the judge. He will be chosen from the list of assessors held by the Court or, if no suitable assessors are listed, by agreement between the parties and with approval of the judge. If a County Court is to hear a pilotage appeal the judge is required to sit with an assessor who has both nautical and pilotage experience. The Elder Brethren of Trinity House customarily assist the Mayor's and City of London Court.

Similarly, formal investigations of shipping casualties and inquiries into loss of life from a fishing vessel's boat are held by a Wreck Commissioner, or a Court of summary jurisdiction, with the assistance of one or more assessors with nautical, engineering or other special skills as required, appointed from the lists of persons approved by the Home Secretary and selected from the Home Office list. At least two assessors having experience of the merchant service are necessary if the formal investigation may involve cancellation or suspension of the certificate of a certificated officer.

The Home Office list of assessors is important because it provides the sources of assessors, in addition to the Trinity Masters, to assist the Supreme Court in Admiralty actions. The list remains in force for three years, though assessors may be reappointed.

It must be remembered that an assessor differs from a referee or arbitrator in having no voice or power in deciding questions, his duties being confined to assisting the deliberations of the court[3].

Court of Session, Scotland

In Scotland the list of nautical assessors in the Court of Session is approved and published by the Lord President under the appropriate Act[7] and lasts for three years' duration. The current list expires on 1st November 1990 and consists of nominees from: the Trinity House of Leith; the Elder Brethren of Trinity House of London; the Northern Lighthouse Board of Edinburgh.

References

For those who wish to consult books of reference in order to ascertain the current composition of the Admiralty Court of the United Kingdom and also legal advisers to the admiralty board, a useful guide will be found in the most recent edition of the *Navy List.* This publication lists the Judges of the Admiralty and Prize Court, together with the Registrar and his officers. It also provides, in the case of Royal Naval officers, rules for the appointment of nautical assessors. Candidates for the Mercantile Marine lists are invited to apply directly to the Home Office. Further references may be found by consulting works such as *Whitaker's Almanack* which may be found in any public library.

Function and practice of assessors

After several years' practical experience as a Trinity Master in the Admiralty Court, it is possible to say that the following advice, already published in the *Modern Law Review* in 1970, is still relevant today.

Assessors are sources of information on matters within their own special skill or knowledge. Assessors, however, are not called by the parties, are not sworn, and cannot be cross-examined. Indeed their advice is both sought by and given to the Court in private and is only disclosed to the parties at the Court's discretion and then usually at the end of the case in the judgment. It is the practice of the present judge of the Admiralty Court to set out in his judgment the questions which he has put to the assessors and their answers. This has long been the practice of the Court of Appeal.

Viscount Simon, LC, in one of the very few House of Lords cases concerning non-nautical assessors (although equally relevant to nautical cases), said:

'To treat . . . any assessor, as though he were an unsworn witness in the special confidence of the judge, whose testimony cannot be challenged by cross-examination and perhaps cannot even be fully appreciated by the parties until judgment is given, is to misunderstand what the true functions of an assessor are. He is an expert available for the judge to consult if the judge requires assistance in understanding the effect and meaning of technical evidence.'

Although assessors are not sworn, for them to deliberately tender false advice would probably constitute a criminal contempt of Court.

From the above it is apparent that the assessor's function is not to supply evidence but to help the judge understand the evidence. From this it follows that assessors may properly be required to answer any questions of fact within their special skill or knowledge that is relevant to the case. Assessors may be used to the full for information; they are not only technical advisers; they are sources of evidence as to facts, within their own expertise[8].

This wider view that assessors may assist the Court by advising on any matter within their competence is supported by the rule that expert evidence is inadmissible on matters within the special skill or experience of the assessors assisting the Court.

Different role

The role of an assessor is thus quite different both from that of a judge and that of a jury. In Admiralty actions juries are never called, whether assessors are summoned to assist the Court or not. The judge must decide all the issues both of law and fact. As a matter of historical interest, in the 19th century the assessors were 'the jury,' who decided the facts. The judge summed up the evidence to the assessors, who then reached their decision as to the true facts.

When considering the role of the nautical assessors in the High Court of Admiralty, Sir Joseph Napier stated:

'[The judge] is advised and assisted by persons experienced in nautical matters, but that is only for the purpose of giving him the information he desires upon questions of professional skill, and having got that information from those who advise him, he is bound in duty to exercise his own judgment, and it would be an abandonment of his duty if he delegated that duty to the persons who assisted him. The assessors merely furnish the materials for the Court to act upon, and, for convenience sake, they are allowed to hear all the evidence.'

Assessors, therefore, have nothing to do with the credibility of witnesses. Assessors should not be asked questions which are tantamount to asking them to decide the issue.

Naturally, great weight should be given by the Courts to the opinions of their assessors, and Brett, MR, even went so far as to say that it would be impertinent in a judge not to consider their opinion as 'almost binding,' but nevertheless each Court has a duty to make up its own mind both on questions of nautical skill and on the value of the advice given on these matters. It is not at all unknown for judges to disagree with and refuse to follow the advice tendered by their assessors, as may happen, for example, when they feel that the assessors' standards are too high, though it has been recommended in the House of Lords that if a Court decided not to follow its assessors' advice it should give reasons for so doing. Assessors may, of course, disagree among themselves as to what advice to give the judge on any point, and then the judge must make up his own mind which advice to follow.

Solely to advise

However, the judge may not request or permit an assessor to exceed the bounds of his function, which is solely to advise. Nor may assessors put questions directly to witnesses, although Viscount Simon, LC, said:

'[An assessor] may, in proper cases, suggest to the judge questions which the judge himself might put to an expert witness with a view to testing the witness's view or to making plain his meaning.'

It has been recommended in judgments from the House of Lords that, whenever possible, questions should be put to assessors, and their answers received, in writing, as the questions and answers would then be available to an appellate Court in cases of an appeal.

This is now the general practice in the Court of Appeal, where the issues are clearly defined. In a Court of first instance, however, where the consultation between judge and assessors is informal and frequent, advice is usually sought and given orally. However, it is not infrequent for the judge to confirm such advice by the asking of questions in writing to which a written reply is given by the assessors.

Vicount Simon has said:

'It would seem desirable in cases where the assessor's advice, within its proper limits, is likely to affect the judge's conclusion, for the latter to inform the parties before him what is the advice he has received. The judge will usually refer to the questions he has put to his assessors and the answers he has received in the course of his judgment, and if these were put in writing they may be set out verbatim. Yet, even when the advice of assessors has been submitted in writing, the parties have no right to see this advice or have copies for the purpose of an appeal.'

An appellate Court may make full use of the advice given by assessors in the Court below, and will obtain this either from the judgment or, if it were given in writing, from the original statements. Thus, if an appellate Court is assisted by its own assessors it may consider both their advice and that of the assessors of the Court below. The resulting situation has been pertinently described by Scrutton LJ concerning an appeal from the Admiralty Court to the Court of Appeal.

'It is necessary to point out that the four assessors are a very peculiar sort of witness. The judge in the Admiralty Court talks to them, and gets information from them. The parties do not know what the witnesses are telling the judge; they have no opportunity of cross-examining the so-called witnesses. Indeed, in the Admiralty Court, the practice is not followed which we—in obedience to the direction of the House of Lords—follow, the practice of asking questions in writing, and obtaining answers in writing, and sending them up to the Superior Court. We do not know the terms of the questions

except from what the learned judge says in his judgment. One starts, therefore, with two witnesses whose evidence the parties do not hear, and whom the parties have no opportunity of cross-examining, and the case then comes to this Court, and we have to decide the case with two witnesses whom the judge below did not hear.'

Similarly, when a case which has been heard with assessors reaches the House of Lords, the Law Lords may use not only the advice of their own assessors, but also the advice given by the assessors in the Court of Appeal and in the Court of first instance.

As there is no hierarchy of assessors, the appellate Court does not have to pay more attention to its own assessors than it does to those of the Court below. The assessors in an appellate Court are simply in addition to those previously consulted. It is therefore open to the Court of Appeal, in considering the totality of advice available, to reject the advice tendered by its own assessors in favour of that given by the assessors in the Court below. The equal status and the advisory function of assessors also precludes an appeal from the advice of the assessors in one Court to the assessors of an appellate Court. Lord Birkenhead, LC, said: 'It would, I think, be intolerable if . . . appeals were treated as being not from one judge to another but from one assessor to another.'

Lord Summer has said:

'There is no such thing as an appeal either from or to assessors. It is really a criticism of the conduct of one Court or of both Courts below. It specially describes the effect on an appeal, when an appellate tribunal adopts a certain attitude towards its assessors, an attitude which is equally open to criticism in a Court of first instances. That attitude is the Court's surrender to the assessors of the judicial function of itself deciding the issue, however technical it may be.'

Although the function of assessors is solely to give advice, the value of their role should not be underestimated. Indeed, it is unfortunate that the advantages of obtaining expert information from assessors rather than by the examination of experts in the witness-box are not more widely recognised.

This introduction to the role of the nautical assessor would not be complete without a reference to the historical background of the Admiralty Court and the growing relevance of the Arbitration Court which, in marine causes, is conducted with the same dignity as the Admiralty Court although in less formal surroundings.

Historical background

The Silver Oar of the Admiralty—The silver oar now preserved at the Royal Courts of Justice in London is probably the third to have been carried by a long list of Admiralty Marshals. This symbol, used since the 13th century, today symbolises the authority of the Admiralty Court, part of the Queen's Bench Division of the High Court of Justice. The possession of such a mace from the time of Richard I was deemed sufficient authority for a sergeant-at-arms to make an arrest. From the Admiralty Records of 1586 it is known that the Lord High Admiral directed Jasper Swift to arrest certain persons 'for piracies and felonies by them committed on the high seas and on the River Thames below bridge towards sea.'

This oar-shaped mace was formerly the visible sign of the authority possessed by the Admiralty Court to arrest persons and vessels in respect of certain occurrences such as collisions on the high seas: today the authority extends to vessels only, but since 1840 all creeks and rivers of Great Britain have also been under the jurisdiction of the Admiralty Court[9].

As the quality of Admiralty jurisdiction spread throughout the world, wherever British maritime interest developed, so, too, Admiralty maces are to be found in Sydney, Bermuda, New York, Boston, Cape Town and in the Cinque Ports. Of these, the Silver Oar of Admiralty in London enjoys pride of place and is still put to good use. With its embossed Royal Arms of the Tudors, garters, coronets and foul anchor, it routinely rests before the judge in the Admiralty Court whenever a trial is in session.

Tribute paid to the Admiralty Court[10] over 100 years ago is still relevant today. Sir Travers Twiss, QC, said then that 'the Silver Oar of Admiralty still exists in England, and may serve to remind other maritime states that it was not found difficult in former days to institute and to maintain international tribunals for the administration of justice in matters touching the use of the high seas between merchants and mariners, and between mariners and mariners of different nationalities, according to common maritime law. It is idle for nations to agree to supplement the ancient customs of the sea by written rules adapted to the altered circumstances of sea navigation unless they agree in like manner to adopt a common system of judicature by which those rules may be enforced, and the disregard of them visited with penalties.'

Thus the Silver Oar still symbolises the contribution made by the English High Court of Admiralty to achieving good order and discipline upon the high seas, and to the enforcement of the highest possible standards. Amongst maritime Courts throughout the world it enjoys a reputation without equal.

The Arbitration Court will be fully discussed in a later chapter but a note of caution at this stage is appropriate since an assessor may well find, in the nature of his other professional duties, that he could be consulted on an issue of substance that might become the subject of litigation at a later stage. Once having expressed an opinion in writing the evidence of an expert will be used by the recipient to best advantage and what may have begun as a friendly exchange of opinion may lead to the expert concerned being called as a witness and cross-examined at some later date should the matter be tried in Court.

In the event that such a case be programmed to go before one of the Admiralty Courts, at whatever level, the expert concerned will be disqualified from sitting in that instance as an assessor. However, should the matter be referred to arbitration the assessor concerned could, without embarrassment, fulfil the role of an expert witness and subsequently undergo the experience of being cross-examined in the Arbitration Court by leading and junior counsel, those whom he would otherwise view from the shelter of the judge's bench. There is much to be gained from such an experience and, with hindsight, the individual concerned will realise he has undergone a demanding process of law which even counsel

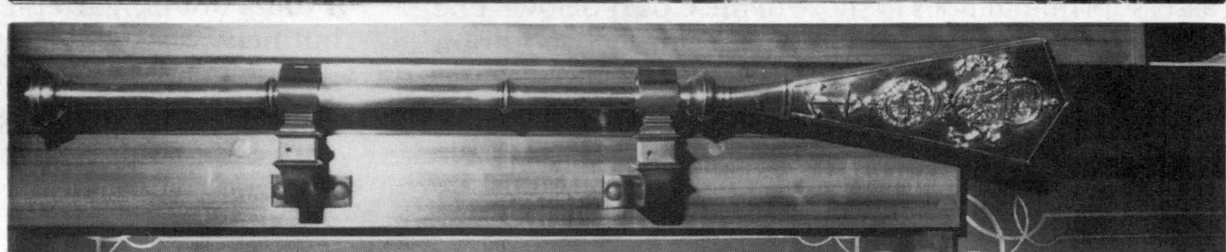

Photo by G. Rafter. *By courtesy of the Hon. Mr Justice Sheen,*

The High Court of Admiralty. The Admiralty Judge assisted by two Elder Brethren of the Trinity House sitting as Nautical Assessors (Trinity Masters).

Left to right: Captain D. T. Smith, RN, Captain T. Woodfield, OBE, and (centre) the Honourable Mr Justice Sheen, sitting above and behind the Silver Oar of Admiralty. The Admiralty fouled anchor is shown (on the wall of the Court) to the left of the picture.

themselves are unlikely to have been subjected to; and if their own feelings can be imagined neither would they wish to be!

Before getting involved at any stage, whether as an assessor or expert witness, the golden rule is 'know your subject,' check the facts and, when giving advice, never to avoid declaring doubt as and when it may exist. The judge in Court has every reason to expect the highest possible standard of professional integrity from his assessors and the confidence he places in their advice will be broken should they fail to achieve the necessary high standards. Likewise, an assessor appearing before the chairman of an arbitration Court in the role of expert witness will be expected to provide the same high standard.

The foregoing is written to explain and to help those seagoers who become involved with placing their knowledge and experience at the disposal of bench and bar; in no way is it intended to deter such individuals from competing for these quite unique positions, whether as assessor or expert witness, which will ensure that experienced mariners continue to play their part with distinction in the administration of their country's maritime law.

Finally, the author is deeply indebted to the Hon Mr Justice Sheen, of the Admiralty Court in the Royal Courts of Justice, for his kindness and advice in proving the content of this chapter. The professional mariner will be interested to know that the Hon Mr Justice Sheen was himself in command of a Royal Navy corvette in World War II. □

References

(1) Written by Anthony Dickey and published in *The Modern Law Review* September 1970.
(2) *The Modern Law Review* September 1970.
(3) *Jowitt's Dictionary of English Law* 2nd Edition 1977.
(4) Section 98 of the Supreme Court of Judicature (Consolidation) Act 1925.
(5) Appointed under Section 3 of the Supreme Court of Judicature Act 1981.
(6) Under Sections 467 and 471 of the Merchant Shipping Act 1894 (57 and 58 Vict c 60).
(7) Under 4th Section of the Nautical Assessors (Scotland) Act 1894.
(8) Vicount Dunedin.
(9) G. Bernard Hughes, *Country Life,* April 1958.
(10) *Nautical Magazine,* 1877, Vol 46, p572.

ROLE OF THE CARGO SURVEYOR

Captain J. A. Cross, Ex.C, FNI, MCMS, MCIT
Severnside Consultants

IT MUST BE STRESSED that the subject matter is of an ever-widening nature and too complex for detailed examination. It is, therefore, intended that this chapter should present a general view of the broad principles, and it is a view based on the personal experience of the writer. Clearly, other marine cargo surveyors may have differing views of the subject and these, of course, are respected.

Experience and training

It is essential that the marine cargo surveyor is a person of adequate experience, having been suitably trained, hopefully by persons of long experience in the field. It has been suggested that a minimum of three years working with an experienced surveyor is the requirement before anyone can be considered ready to enter the field unaccompanied. The writer considers that extensive sea experience is an asset when acting as a cargo surveyor, particularly in an 'on board' situation.

The established surveying firms work on a jealously guarded reputation, and after many years of experience and successful practice enjoy the absolute confidence of their principals.

The field is not an easy one, yet it is a well-known legend amongst established surveyors that every ex-seafarer considers himself to be an expert surveyor immediately he leaves seagoing employment. This could not be further from the truth, as the profession is greatly dependent on educated experience and procedure.

There have been many instances, due to misunderstanding and possibly ignorance, where inexperienced surveyors have been employed on specific potential claims and have only succeeded in further complicating what could have been a straightforward case. It may be that the preceeding remarks are considered rather harsh, but too much stress cannot be laid on the fact that the marine cargo surveyor must be the experienced link between his principals and those resisting or submitting a claim.

Identification of principals

It is customary for a marine cargo surveyor to be appointed by one of the following: shipowner, charterer or cargo underwriter/receiver. The approach of the surveyor to a casualty, howsoever caused, will clearly be influenced to a great degree by the interest of his particular principal.

The experienced marine cargo surveyor will usually be recognised as regularly acting for owners or cargo interests, although there are of course exceptions to the rule. It is of paramount importance that the surveyor represents the interest of his principal, bearing in mind that the eventual settlement of the claim in whatever direction will be greatly dependent on his report and opinion.

Under no circumstance should any marine cargo surveyor consider acting for more than one principal unless it is with their complete permission and understanding. Should there be a conflict of interest, then all credibility will be lost, and to a large degree the professional reputation of the surveyor involved. The experienced and well-established marine cargo surveyor will undoubtedly enjoy the complete confidence of a wide range of principals who have become used to his method of working and reporting. Such confidence is an absolute necessity in all spheres of the procedure of casualty and eventual claims.

Relationships and professional outlook

The marine cargo surveyor acting for various interests must be completely aware of the requirement of his principals and of the other parties to a possible claim. If acting for shipowners' interests the surveyor must formally establish the facility to be granted, particularly in a shipboard situation, to those surveyors acting in the interests of charterers or cargo.

It is quite usual for surveyors acting in various interests to be personally known to each other and to be fully aware of those facilities and courtesies which they can customarily expect. A satisfactory relationship and degree of co-operation between surveyors acting in opposing interests is essential for the progressing of a particular casualty and satisfactory settlement of a subsequent claim.

The particular principals will look to their appointed surveyor to provide them with the true facts of the case, coupled with an expert opinion as to the nature and extent of damage and the various related circumstances leading to the casualty. Principals will also be looking for sound advice as to how any potential loss may be reduced. It must always be remembered that the marine cargo surveyor is in fact the 'eyes' of his principals and in presenting a detailed and knowledgeable report must be to a degree responsible for any action his principals may decide to take.

Presentation of the report

There can be little doubt that the ability to present a clear factual report and opinion couched in language and terminology such that the principals can clearly understand is one of the greatest assets of the marine cargo surveyor. In many cases he may well be reporting on a casualty and subsequent damage several thousand miles from his principals' office and any action taken may be entirely dependent on the tone of opinion expressed in the report. It may well be that in particularly onerous cases the report is passed to other parties and is used as evidence in legal proceedings.

Practical surveying

The reasons for the appointment of a marine cargo surveyor can be many and varied. It is quite usual for surveyors generally to work in their own geographical area, but they may be requested to travel greater distances if they are considered expert in dealing with a particular commodity or procedure, or are requested to carry out follow-up surveys to maintain continuity.

Some examples of the attendance brief given to the surveyor are as follows:

Precautionary—If owners or cargo interests have reason for concern that a vessel has encountered difficulties on passage such as exceptional heavy weather with associated heavy rolling and pitching and other unusual circumstances, it may well be that a surveyor is appointed to be present when the vessel's hatches are first opened. First sight and associated impressions before the cargo is disturbed or handled for discharge is absolutely vital to the prudent surveyor and can greatly influence his opinion and report. If damage is apparent it is also important that the surveyor is present to monitor the development, or otherwise, of any evidence. If a surveyor is appointed by cargo interests and makes an application to board the vessel, it is customary for the shipowner to appoint his own surveyor to accompany him.

Discovery of damage—All too often shipowner or cargo interests will only be alerted to the existence of a problem when damage is exposed during the course of discharging operations. In complete contrast to the circumstances described in the previous paragraph, the surveyor in this instance can find his work hampered due to vital evidence having been disturbed or destroyed by stevedores. Slight damage patterns which are vital to the surveyor may appear unimportant to the stevedores, whose only concern is to clear the cargo.

Routine cargo survey—Particularly in the case of the old-established liner companies, it was considered a routine procedure to appoint a cargo surveyor to be in attendance at intervals throughout the discharge programme. The surveyor would report on the general stowage, ventilation arrangements, dunnaging and handling, etc. This presented a shipowner with a clear account of the discharge and on many occasions a sound defence if subsequent claims were entered. The survey is of particular value if stevedore damage is prevalent.

Cleanliness—When loading bulk or particularly sensitive cargoes it is extremely important that the vessel is presented clean and in all respects ready to load. In this case it is customary for shippers to appoint the cargo surveyor to inspect the holds and prepare a report and certificate. Such a report is also of value to a shipowner as in the case of damage it is clearly established that the holds were in suitable condition prior to the commencement of loading.

Transhipment and preshipment—In recent years it has become more the custom to ship cargo in large vessels and tranship in coasters from a central Port to a variety of outports. This again frequently requires the presence of the cargo surveyor to record the condition of the material on opening hatches, to note the preloading condition of the coaster, and to finally record the condition following the transhipment operation. The final report presented can be of great assistance to shipowners and cargo interests if damage is subsequently claimed at final destination.

Regarding preshipment condition, certain commodities are very sensitive to handling and require particular supervision during loading. The cargo surveyor may often be appointed to record the condition of certain cargoes at the time of loading. Such a requirement is frequently found on the shipment of steel coil in its various forms, particularly by shipowners. Continuous attendance during loading enables the surveyor to note any damage and so avoid a subsequent claim on the vessel.

Cargo shift—It may be that a master will contact his owners whilst on passage in adverse weather conditions to report a shift of cargo resulting in damage to the commodity. In such circumstances, surveyors for the owner and cargo interests will be alerted prior to the vessel's arrival. Again, it is extremely important that the surveyors have a full opportunity to examine the stow before stevedores commence work. In this way firm opinions can be drawn as to the adequacy of securing arrangements, etc., and a full report presented accordingly. If a vessel diverts to a port following a cargo shift to restow and resecure, the surveyor may be appointed to supervise the operation and approve final securing.

Containers—With the rapid development of container carriage, the cargo surveyor has become increasingly involved, on occasion many miles from the transit port. Damage and movement only discovered when the unit reaches its final destination frequently result in requests for survey, both in the interests of the receiver and the carrying vessel. Insulated units, with their own diesel drive cooling unit, and carrying frozen or chilled products, are very susceptible to breakdown resulting in excessive temperature variation and requiring survey to agree any deterioration. In the early days of container traffic, the cargo surveyor was frequently consulted to advise on the securing of goods within the unit to ensure safe carriage.

The above instances cover a number of the situations which may require the attendance of a marine cargo surveyor, but it must be said that the ever-widening range of commodities and types of vessel present the surveyor with an ever-widening brief.

Documentation

One important aspect of the surveyor's work is precise and clear interpretation of available documents, such as bills of lading, mate's receipts, log books, sounding books, etc. In the case of cargo damage, the cause can quite frequently be dramatically narrowed down by thorough examination of documentation.

Records of temperatures, humidity and soundings can be vital clues of the voyage history, and the experienced surveyor will spend much of his time sifting through records to ensure that any irregularities are detected and considered.

Expert consultation

Prudent use of commodity experts and analysts can assist in some situations and the experienced surveyor will not hesitate to consult such parties when it is felt to be of benefit to his principals. On occasion the surveyor's opinion as to cause of damage can be confirmed with the assistance of an analyst's report covering moisture content, degree of chlorine content, etc.

In attempting to define broadly the role of the marine cargo surveyor, it is hoped that an indication has been given of the many aspects of cargo carriage with which he may be faced. The prudent surveyor is always prepared to learn and widen his experience to assist further those who retain his services. Finally, it must be stressed that, in the marine cargo surveyor's field, there is absolutely no substitute for long experience.□

THE AVERAGE ADJUSTER

Contributed by the Association of Average Adjusters, London

THE FUNCTIONS of the average adjuster are principally the following: the adjustment of general average, the adjustment of claims on policies of insurance on any interest, e.g., all types of vessels and cargoes, directly or indirectly exposed to maritime perils, the preparation of statements of claim against third parties, the division of recoveries from third parties, or of proceeds of sale, and the arbitration of disputes arising in relation to the above or associated matters.

In the discharge of these functions the average adjuster may be appointed by any member of the maritime or marine insurance communities having an interest in the matter concerned, and, irrespective of the identity of the party appointing him, the average adjuster shall act in an impartial and independent manner.

The average adjuster may advise any party seeking his opinion on any matter within the area of his expertise; the average adjuster may assist in the collection of general average, salvage or other security; and the average adjuster may assist in effecting settlements under an average adjustment, or otherwise as required.

His main function is the preparation of adjustments and whilst the preparation of adjustments for loss of/or damage to ships, cargoes and freight and the application of such claims to the respective policies of insurance provides most of his work by the very nature of the subject, he is better known internationally for his work in connection with general average. The professional average adjuster appeared on the scene at the beginning of the 19th century and offices sprang up in all the major ports in Great Britain. At that time, Britain was a thriving maritime nation and as there were numerous major ports so too were there numerous average adjusters.

In the middle of the 19th Century, when average adjusting as a separate profession was in its infancy, there was very little in the way of established law to guide the practising adjuster, and consequently many points of practice had to be decided in accordance with custom. Some of these customs were subsequently ratified by legal decisions, but others were disapproved, and it became evident that unless steps were taken to establish a reasonable measure of uniformity among average adjusters, the profession would fall into disrepute. It was to remedy this situation that the Association of Average Adjusters was founded in 1869, with the object, among others, of 'the promotion of correct principles in the adjustment of averages and uniformity of practice amongst average adjusters.'

After the formation of the association, one of its first tasks was to consider the areas of divergency in practice, and decide how the various so-called 'customs' could be brought together into a uniform, if not universal, practice. This aim was largely achieved by the association in the first 15 years of its existence, by a two-fold approach:

- By the collection and refinement of the Customs of Lloyd's. This task was undertaken by a special committee which reported to the association in 1876. In the preamble to the Customs it was stated: 'Nothing can be called a Custom of Lloyd's which is determined by a decision of the superior Courts; for whatever is thus sanctioned rests on a ground surer than custom. A Custom of Lloyd's then must relate to a point on which the law is doubtful, or not yet defined, but as to which, for practical convenience, it is necessary that there should be some uniform rule.'
- By the adoption of rules of practice, relating to the adjustment of averages and the duties of adjusters in connection therewith. In the early days of the association it was hotly debated whether these rules of practice should bind members or not, and in the event it was decided that they would not be binding, although, naturally, they would carry considerable authority. Even now, if an average adjuster draws up a statement which is at variance with a rule of practice, he must place a note in his adjustment referring to the rule of practice and stating why he differs from it.

Since 1890, when the Customs of Lloyd's were reviewed and assimilated into the rules of practice, various new rules and amendments to existing rules have been adopted from time to time in order to regulate the practice of average adjusters in areas where the law is silent. Since the rules of practice were last printed in 1981, new rules or amendments have been introduced in order to bring English practice on the subjects concerned into line with York-Antwerp Rules 1974 (the current code dealing with the adjustments of general average).

In order to perform his task of preparing adjustments, the adjusters will have the reports prepared by the attending underwriters' surveyors, owners' superintendents, etc. Invariably there are further questions to be answered by the attending surveyors whether it concerned a survey of damage to a vessel's hull or machinery, for example, the time required to effect repairs, cause of damage, etc., or damage to cargo, for example, separation of damaged cargo between damage caused by fire and/or smoke (particular average) or damage caused by water (general average). Where more technical questions may be required, particularly with regard to the cause of damage, the matter may be referred to the Salvage Associaton's technical department, Lloyd's Register of Shipping or on occasions naval architects.

Claims on policies of insurance on ship

There may be a claim upon a marine insurance policy on ship when, by the operation of insured perils, a total or partial loss may occur.

A total loss may occur in two circumstances:

- *Actual total loss*—This arises when a ship is destroyed or is so seriously damaged as to cease to be a ship, or when the shipowner is irretrievably deprived of her (e.g., when she has sunk in deep water and cannot be salved).

- *Constructive total loss*—This arises should all or any of the following occur:
  - (i) When the ship's actual total loss appears to be unavoidable.
  - (ii) When the shipowner is deprived of his ship and her recovery is unlikely.
  - (iii) When the cost of recovery and repair of damage would exceed the ship's insured value.

Notice of abandonment must be tendered to the insurers as soon as it is apparent that the ship is likely to become a constructive total loss. In the absence of such notice a claim for constructive total loss cannot be made.

A partial loss includes claims for the following:

- *Particular average*—This comprises damage to the ship caused fortuitously. It does not include damage brought about by the ordinary action of the wind and waves, nor gradual deterioration on account of ordinary use (e.g., the chafing of a mooring rope), nor, unless the policy otherwise provides, a defect in the hull or machinery in existence at the time the insurance attaches. Furthermore, it does not include damage brought about by a voluntary act (which, if done in time of peril for the common safety, will form a general average sacrifice: see below). Examples of causes of particular average damage are collision, contact (including stranding and grounding), heavy weather and fire.
- *General average*—General average exists independently of insurance. It arises whenever a sacrifice of property or an extraordinary expenditure is reasonably and voluntarily made or incurred for the common safety of the interests concerned in a maritime adventure, and it may also arise in some instances of such expenditure voluntarily incurred for the common benefit. Since general average embraces the losses suffered or expenditure incurred by other interests as well as ship, a shipowner's claim upon his insurances on ship may concern:
  - (i) General average sacrifice, being damage to the ship voluntarily sustained for the common safety.
  - (ii) General average expenditure, incurred by the shipowner.
  - (iii) General average contribution, being what the shipowner has to pay towards the sacrifices suffered and expenditures incurred by other parties.
- *Salvage charges*—Salvage charges comprise the sum or sums paid in settlement of a claim by salvors for remuneration for salving the ship, or both ship and cargo, from a position of danger, together with legal costs and other charges which may be incurred in this connection. In nearly all instances, the amounts so paid will be treated as general average expenditure.
- *Charges incurred to avert or minimise a loss*—It is the duty of the assured under all policies of marine insurance under English law to take such measures as may be reasonable in order to avert or minimise a loss which could found a claim upon the policy. In traditional forms of policy, these are called *sue and labour charges,* and although not so described in modern forms of policy, charges incurred by the shipowner or his agents in fulfilment of his duty to preserve or attempt to preserve the ship from the consequences of an insured peril, or to minimize its effect, will be reimbursed by the insurers in addition to any loss recoverable under the insurance, whether total or partial.
- *Third-party liability arising from collision with another vessel*—This arises when the shipowner has paid damages in respect of his tortious liability to the owner of another ship or any property on it, arising out of a collision between the insured ship and the other vessel. Although some marine insurance policies provide full cover, such liability is normally insured only to the extent of three-fourths, the balance of one-fourth remaining the responsibility of the shipowner (unless he insures it by entering his ship in a Protection and Indemnity Association).

Insured perils

The standard policies on ship in use in the London market covers the assured for loss of or damage to his ship caused by perils of the seas rivers lakes or other navigable waters; fire, explosion; violent theft by persons from outside the vessel; jettison; piracy; breakdown of or accident to nuclear installations or reactors; contact with aircraft or similar objects, or objects falling therefrom, land conveyance, dock or harbour equipment or installation; earthquake volcanic eruption or lightning; accidents in loading discharging or shifting cargo or fuel; bursting of boilers breakage of shafts or any latent defect in the machinery or hull; negligence of master, officers, crew or pilots; negligence of repairers or charterers provided such repairers or charterers are not an assured thereunder; barratry of master, officers or crew; provided such loss or damage has not resulted from want of due diligence by the assured, owners or managers.

The cover under the last five perils is subject to the proviso that the loss or damage has not resulted from want of due diligence by the assured, owners or managers. Also, only the consequential damage is recoverable, the part which breaks is not recoverable.

Amount recoverable

Where a claim is admitted the amount recoverable for a total loss is the value insured by the policy. For a partial loss the amount payable by the policy, subject to any deductible provided therein, is as follows:

(a) *For particular average*
  - (i) In respect of such damage as has been repaired, the reasonable cost of the repairs effected.
  - (ii) In respect of such damage as has not been repaired, the reasonable depreciation in the value of the ship by reason of the unrepaired damage, not exceeding the estimated reasonable cost of repairs.

(b) *For general average*
  - (i) *For sacrifice of ship* allowed in general average, the amount payable by the policy is computed in the same way as for particular average, less the contributions received from other parties.
  - (ii) *For general average expenditure* incurred by the shipowner, the amount payable by the policy is the proportion which falls upon him; similarly *for general average contribution* the policy will pay the proportion attaching to the ship; but in both instances the claim is subject to the contributory value of the ship being fully insured, and if it is not, the claim will be reduced in proportion to the under-insurance.

(c) *For salvage charges*—The amount payable by the policy is computed in the same way as for general average expenditure.

(d) *For expenses incurred to avert or minimise a loss (sue and labour charges)*—The full sum expended, subject to the ship being fully insured.

(e) *For third-party liability arising from collision with another vessel*—Either the full amount or three-fourths, according to the policy conditions, of—
  - (i) the amount paid in respect of such liability, but not the exceeding the sum insured, and
  - (ii) legal costs incurred in respect thereof.

Deductible

Most forms of hull insurance provide that claims for partial loss will be subject to a deductible, that is to say, a fixed sum which the assured has to bear in respect of each claim to which the deductible applies. Institute Time Clauses, Hulls, for example, provide that the deductible shall be applied to the aggregate of all partial loss claims arising out of each separate accident or occurrence.

Recoveries from third parties

Under nearly every form of marine insurance policy, there are circumstances in which the claim payable by the insurer is reduced by an amount which may be recovered from some source, either at the time when the claim is presented or subsequently. From the point of view of the insurer, such recoveries can arise in two different ways:

- *Under the doctrine of abandonment*—This applies only when the insurer has paid for a total loss and has exercised his right to the proprietary interest in the subject matter of the insurance. For example, when a ship has been wrecked and the insurer has paid a constructive total loss, he is entitled to take over the wreck and if it can be sold, he may retain the proceeds of sale. However, if the insurer decides to exercise his proprietary rights, then he is likewise responsible to pay all charges attaching to the property as from the time of the casualty causing the loss.
- *Under the doctrine of subrogation*—For all practical purposes, recoveries under this heading comprise those sums of money which can be recovered from third parties on account of their liability for the accident giving rise to the loss or damage which is the subject of the claim on the policy. Examples are:
  - (i) A recovery from the owner of a ship which is in fault for a collision.
  - (ii) A recovery from a charterer who is responsible for having ordered the ship to an unsafe berth where she sustains damage.
  - (iii) A recovery from a repairer or dry-dock owner for negligent work.
  - (iv) General average contributions paid by other parties in respect of a sacrifice of ship or goods, for which the assured has a direct claim on his policy.

From the point of view of the assured, it is necessary in practice to give the insurer due notice whenever there is a possibility of a recovery from a third party. The reason for this is two-fold:

- To give the insurer the opportunity of saying, even if he has not yet responded for the claim about which he has been notified, whether or not he approves of proceedings being taken against the third party, and
- In the event that the insurer wishes to exercise his right of recourse, to enable the assured to prosecute his claim against the third party in the sure knowledge that the insurer will respond in due course for the proportion of the costs and other charges incurred in the prosecution of that claim, in so far as it relates to losses for which the insurer would have been liable.

How do recoveries affect the claim upon the policy of insurance?

- *Under the doctrine of abandonment*—This only applies when the assured has substantiated a claim for a total loss. In such a case, any recovery in respect of the value of the subject matter insured, or by reason of the insurer's proprietary interest in the vessel, will not invalidate the assured's claim for a total loss. The insurer is, however, entitled to the benefit of any proceeds in respect of the subject matter insured, irrespective of their relation to the insured value, or to the amount which the insurer has paid.
- *Under the doctrine of subrogation*—In principle, the insurer is entitled to the benefit of any recovery from a third party, but only up to the amount of the claim which he has paid or is liable to pay.

When under the policy of insurance claims for partial loss are subject to a deductible, the policy may specify how the recovery is to be treated, or it may be silent, in which event the treatment of the recovery will be made in accordance with the applicable law of subrogation.

Under the same principle, when there has been a recovery from a third party whose act or fault brought about the general average act, the amount recovered in respect of the sums admitted in general average will be credited proportionately to the general average contributions paid or payable under the general average adjustment.

Steps from casualty to collection of claim

When an accident has occurred, it is essential for notice of the accident, giving such details as are available, to be given promptly to the insurers through the insurance brokers. In addition, when the vessel is abroad, the master should notify the nearest Lloyd's Agent, particularly if there is likely to be any difficulty in communication between the ship and the shipowner's office.

The object of giving notice is to enable the insurers or their agents to appoint a surveyor to attend the vessel and survey the damage. Most policies of insurance contain an express provision regarding notice, and the Institute Time Clauses, Hulls, for example, provides that in the event of non-compliance a penalty amounting to 15 per cent is to be deducted from the total of the ultimately ascertained claim. Notice should also be given to the Protection and Indemnity Association in any case involving loss or damage to cargo, and/or when there is likely to be a claim for general average contribution from cargo interests.

Appointment of the average adjuster

If the casualty takes place during the course of a current engagement, and the ship has to put into a port of refuge or is likely to lose time in order to effect repairs, there is likely to be a case of general average. Consequently it is prudent at this stage for the shipowner to appoint his average adjuster and consult him regarding any possible general average claim.

If the casualty is serious the shipowner will wish to send a marine superintendent and/or engineer superintendent to the casualty, or to the port to which the ship is proceeding, in order to obtain their reports upon the situation: from the marine superintendent—as to the navigability of the ship and as to the necessity for cargo operations, such as the discharge of cargo from a stranded ship to lighters or other craft, in order to refloat, or the discharge, storing and reloading of cargo in order to effect repairs to the ship; and from the engineer superintendent—as to the repairs to the ship which may have to be effected for the safe prosecution of the remainder of the voyage. Either or both of the superintendents should remain in attendance for as long as necessary to supervise the cargo operations and repairs.

It is desirable for the damage sustained by the ship to be surveyed jointly and concurrently by the owner's superintendent and the surveyor appointed by the insurers. Likewise, so far as possible, they should agree upon the recommendations for repair and the instructions to be given to the shipyard or other repairers. At the final stage, when the repair accounts are submitted by the shipyard, they should be examined critically by both the superintendent and the underwriters' surveyor to check the level of pricing and negotiate any reduction that may appear necessary. If London market insurers are involved, they will probably appoint a surveyor from the Salvage Association, who have issued notes for guidance to assist shipowners in this connection.

The shipowner will almost certainly be concerned at this stage at the extent of the extra expenses falling upon him by way of repair costs, port charges and ordinary ship's expenses during the delay for repairs. He may therefore wish to ask his average adjuster to consider the preparation of an interim report or certificate recommending a payment on account by his insurers.

In this event the average adjuster will require as preliminary documents: log book extracts, or at the very least, an extended note of protest or ship's declaration giving details of the casualty; interim reports of the damage as seen by the shipowner's superintendent and the underwriters' surveyor; firm evidence of the agreed cost of repairs. If the approval of any of the repairs, or of the costs involved, has not been communicated by the underwriters' surveyor to the shipowners via the owners' superintendent or ship's agents, then the average adjuster can, subject to leading underwriters' agreement, obtain the necessary approvals.

Prosecution of the voyage

If repairs to the ship are necessarily effected during the course of a current engagement, questions may arise relating to the continuation of the voyage. *For example:* Will the ship still be able to make her port of destination, or other scheduled ports of call? If the ship is in ballast under charter, will she meet her cancelling date? If considerable time has been lost, will the charterers want to exercise any option available under the charter-party to change the voyage? If the above circumstances should arise, the shipowner may wish to consult his average adjuster or his Protection and Indemnity Association about them.

At this stage the shipowner should begin to assemble the documents which will be required in order to substantiate his claim.

The average adjuster has a two-fold duty: to his client—to see that the claim presented is fully supported by the evidence, and that it is as complete as possible, i.e., that nothing is missing; to the underwriters—not to submit, without making an appropriate note or reservation, any item of claim which cannot be supported either in law or in practice.

The average adjuster will obtain the agreement of his client to the figures which he has prepared in his statement, and this provides the final opportunity of ensuring that nothing has been overlooked.

The average adjuster will then issue his statement to the parties concerned in it. In the case of a claim upon a marine policy, the adjuster's statement will be presented to the leading insurers by the insurance brokers. In the case of a claim in general average involving collection of contributions from the concerned in cargo, the practice varies from country to country, but when the adjustment has been prepared in the United Kingdom, it is usual for the average adjuster to be instructed by his client to send out copies of the adjustment (or extracts therefrom) to the various concerned in cargo.

As indicated above, the collection of the claim from the insurers on ship will be handled by the claims department of the insurance brokers. If the insurers have any questions to raise, they may be addressed to the assured or referred back to the average adjuster for answering. Collection of the amounts due from cargo interests will be handled either by the shipowner or by the average adjuster on his behalf.

General average and salvage

General average arises when property involved in a common maritime adventure is voluntarily sacrificed in time of peril for the purpose of preserving the adventure, or where any extraordinary expenditure is incurred for a like purpose. Three examples are:

When the vessel is aground in a position of peril, and in order to refloat her: tugs are engaged; the ship's machinery is used; part of the cargo is discharged into lighters, and/or is jettisoned; after the ship has been refloated, any discharged cargo is reloaded; all expenditure and damage arising out of one or more of these operations is general average.

When fire breaks out on board the ship, and in order to extinguish it: water is poured on or chemicals are applied; holes are cut in the ship's structure for this purpose; cargo is discharged and/or jettisoned to get at the seat of the fire; various articles of ship's equipment are lost or damaged in these operations; all expenditure and damage arising out of one or more of these operations, including loss of and damage to cargo on board by the water poured on, is general average.

When in consequence of a casualty: the vessel puts into a port of refuge or is detained in port for repairs; the cargo is discharged to enable repairs to be effected to the vessel; the cargo is reloaded on completion of those repairs (provided that the repairs are necessary for the common safety, or to enable the voyage to be prosecuted with safety); all expenditure incurred as a direct result of one or more of these operations is general average; also the wages and maintenance of crew incurred and fuel and stores consumed during prolongation of the voyage and/or extra detention of the ship brought about as a result.

As distinct from general average, loss or damage caused by the accident or peril is particular average, which is borne by the actual interest damaged. Thus, in the above examples, the grounding damage in the first example is not general average, but particular average; in the second example damage to the ship and cargo caused by fire and smoke is particular average; and in the third example damage to the ship and cargo directly resulting from the casualty is particular average.

The total of the general average, both sacrifice and expenditure, is shared by all the interests (e.g., vessel, freight and cargo) at risk at the time of the general average act, in proportion to the actual arrived values of the property at the termination of the adventure, i.e., on completion of discharge of cargo, with the addition of any amounts made good in general average for loss or damage. The value of the cargo is to be ascertained from the commercial invoice rendered to the receiver, and is inclusive of the cost of insurance and freight, except in so far as freight was at the risk of other parties. If the cargo has sustained loss or damage on the voyage, this is taken into account in arriving at the value for contribution.

General average arises by way of maritime law, and in the absence of contractual agreement, the adjustment would have to be prepared in accordance with the law in the country of destination. Under modern conditions, the method of adjustment will almost certainly be regulated by a clause in the bill of lading or charter-party. Except in the most unusual circumstances such bill of lading or charter-party clause will provide for the application of the York-Antwerp Rules, 1974.

Although he may not be obliged to do so by law, a shipowner will in practice appoint an average adjuster to prepare the adjustment of general average. Normally the shipowner has a free hand in the appointment of the average adjuster, but sometimes the bill of lading or charter-party may stipulate the place where the adjustment is to be prepared. If the shipowner's regular average adjusters do not have an office in the place named in the contract of affreightment, it would be customary for the shipowner's regular adjusters to seek the co-operation of one of their correspondents in the place named for adjustment.

York-Antwerp rules, 1974

These rules govern the adjustment in practically every case of general average. As between shipowner and cargo owner, they take effect through being imported by reference into virtually all bills of lading and charter-parties, and the obligation of ship and cargo interests to pay their respective proportions of general average adjusted in accordance with those rules is covered in all standard forms of policies of insurance on ships and cargoes.

The shipowner is responsible for the adjustment and collection of general average and, for this purpose, is invested with a right of lien upon the cargo for general average. The shipowner's duty in this respect is not limited to recovering his own general average losses from the other interests, but includes the obligation to protect the right of the concerned in such cargo as has sustained a general average loss to recover from the other interests at risk.

In practice the shipowner will probably not need to make use of his right of lien, although he is entitled to refuse delivery of the cargo until adequate security has been provided by the receivers or cargo owners. What is 'adequate' is for the shipowner to decide, but it must be reasonable. The form which such security may take will be considered below. Instructions respecting the provision of security for general average will normally be communicated to the ship's agents at the port of discharge, either by the shipowner direct, or by the average adjuster on his behalf.

The chief duties of the ship's agents in the event of a general average are: if time permits, to notify the receivers of cargo and assist the master of the ship to make such declaration of the general average as is required by the law and custom of the port; to obtain from the receivers of cargo the appropriate security for their proportion of general average, and to obtain from the receivers of cargo particulars of the value of their goods.

The shipowner or the average adjuster will advise the form of security to be obtained from the receivers of cargo. It will usually take the form of *an average bond* signed by each receiver of cargo and, in addition, *a cash deposit,* calculated at so much percentage upon the value of the goods, or *an underwriters' guarantee,* or *a bank guarantee.* In some instances, if the shipowner so decides, the additional security may be dispensed with, and cargo delivered upon signature to the average bond only. In all cases the form of security to be obtained is to be decided by the shipowner, and agents must abide by the instructions they receive.

Lloyd's form of Average Bond was completely revised in 1977. It may be referred to by the code title LAB 77 printed in the top right hand corner. It consists of two pages: the top page is the Average Bond part of the form, while the bottom page is for recording the value of the goods received.

The particulars common to both pages (name of vessel, voyage, port of loading, port of discharge and particulars of cargo) coincide, so that by the insertion of a sheet of carpon paper, only one typing is necessary.

The receiver of cargo should return the top page (average bond) duly signed, as soon as possible in order that his goods may be promptly released to him. Subsequently, after he has received his goods and ascertained their condition, he should complete the information asked for on the bottom page (the valuation form) and send it to the ship's agents or the average adjuster, together with a copy of the commercial invoice rendered to him.

Where a cash deposit is required, the shipowner or the average adjuster will advise the percentage of the deposit to be collected. The deposits collected must be placed in a separate account (earning interest where possible) generally in the joint names of the shipowners or their agents on the one hand, and Lloyd's Agents on the other—the former to protect their own interests, and the latter the interests of the concerned in cargo. Where deposits are collected, receipts must be issued only on Lloyd's form and a separate receipt should be given for cargo covered by each bill of lading if practicable.

If any original receipts are lost by depositors, duplicate receipts may be issued but only against a letter of indemnity against the subsequent production of the original receipt. No interim refunds of deposits in respect of which receipts have been issued should be made without reference to the average adjuster for approval.

On completion of the general average adjustment, the amount deposited in excess of the general average or other contribution required is refunded to the depositor, or other party holding the deposit receipt, and this work is sometimes carried out by ship's agents at ports of discharge under the guidance of the average adjuster.

In cases where a casualty has occurred and general average expenses and losses have been sustained, it is recommended that claims should be declined pending the consideration of the average adjuster. Where a short-landed certificate is issued, it is generally found satisfactory to endorse it as follows: 'In view of general average being declared a copy of your claim will be forwarded to the average adjusters for consideration, but in the meantime we recommend that you lodge a claim with your underwriters.'

If it is necessary in a case of general average to appoint a cargo surveyor at a port of refuge or at destination, the decision to do so will normally be made by the shipowner in consultation with the average adjuster. But if for some reason ship's agents have not received instructions to do so at ports of discharge, they should instruct Lloyd's Agents to appoint a cargo surveyor whenever there is a likelihood of general average damage to cargo, so as to enable the factual information to be passed to the average adjuster.

Salvage

Like general average, the obligation to pay remuneration to salvors arises by way of maritime law, and in the absence of agreement between the salvors and the master of the ship on behalf of the salved interests, this obligation is determined in accordance with the law applying at the place where the salvage services terminate. The basis of the law of salvage, which is common to all maritime countries, is to reward independent contractors for their efforts in saving the property imperilled at sea, and in order to secure the rights of the salvor, he is granted a maritime lien over the property salved at the place where his services end.

This is the main feature which distinguishes the obligation to pay salvage from the obligation to pay other items of general average expenditure; namely to provide appropriate security to the salvors on completion of their service. Two consequences follow: the liability of the salved property to pay their proportion of the salvage remuneration is individual, and not joint; and the obligation arises at the place where the salvors complete their service, and is based upon the value of the salved property at that time, whereas the obligation of the parties to contribute in general average does not arise until the completion of the adventure.

In practice, the parties to the salvage service usually sign a form of agreement which sets out the procedure relative to the provision of security and the determination of the salvage reward. The form of agreement most frequently used is Lloyd's Form of Salvage Contract—'No cure, no pay,' often referred to as 'Lloyd's Open Form.'

Under the latest version of this form (LOF 1980) the interests of the parties are protected by the following provisions, *inter alia:*

- *Security to the salvors*—This has to be provided by the salved interests in a form acceptable to the Committee of Lloyd's or to the salvors. The owner of the ship is required to use his best endeavours to ensure that the cargo interests (if providing separate security) furnish their security before being released from the ship.
- *Arbitration*—Unless the parties to the salvage can reach agreement on the amount due for the service, the question of the salvor's remuneration will be submitted to arbitration.
- *Prevention of oil spillage from a laden tanker*—To meet the case where a salvor provides services which prevent oil spillage and pollution, but which cannot be recompensed under the 'no cure, no pay' principle, because there is no effective salvage of the property, there is a provision, known as the 'safety net,' whereby the shipowner recompenses the salvor for his expenses (plus a modest increment).

When the amount of the salvage remuneration has been decided, either amicably or by arbitration, then the salved interests must pay the salvors the amount which is due from them. In some cases, each interest will be obliged to pay its own share, although, in other cases, the shipowner may give security for all interests and pay not only his share but also the shares which attach to the other interests at risk.

However the salvage reward may be paid, the total of the payments made, together with costs and other expenses incurred in connection therewith, will be admitted in general average by virtue of Rule VI of the York-Antwerp Rules, 1974, whenever the salvage operation was undertaken for the common safety of all the interests at risk.

Special charges on cargo

These comprise expenses reasonably incurred, not for the common safety or as a result of general average damage to cargo, but for the preservation or recovery of the cargo only, e.g., re-conditioning charges. Such expenses form a direct charge against the cargo in respect of which they were incurred, and the shipowner has a lien upon the goods in this respect. Thus, as in the case of general average, adequate security should be obtained from the receivers of cargo to cover the special charges before the goods are released. It will be noted that the forms of average bond and cargo underwriters' guarantee cover special charges in addition to general average, so that no separate security in respect of such charges need be taken where general average security has already been obtained.

When general average disbursements have been incurred at an intermediate port on the voyage, and particularly when bail or other security has been given to salvors, the amount at risk both for disbursements incurred and for bail and other charges still to be assessed, can and should be insured for the remainder of the voyage. Such insurance should be effected before the vessel sails from the intermediate port, and the shipowner should therefore consult his average adjuster at the earliest possible moment to ensure that this is done promptly. The cost of effecting this insurance will be allowed in general average, per Rule XX of the York-Antwerp Rules.

In certain circumstances when a ship is at a port of refuge, for example if the ship has sustained severe damage and repairs may take a long time, or if the cargo interests specially request it, the shipowner may have to consider transhipping the cargo, or part of it, on to another vessel and forwarding it to destination. By taking this course the shipowner may save both time and expense, but unless the parties to the adventure otherwise agree, the separation of ship and cargo (or part cargo) in these circumstances could bring the 'common adventure' to an end and thereby prejudice the shipowner's right to recover certain general average allowances.

When faced with this situation, the shipowner should consult his average adjuster, and if, after taking his advice, it should be decided to tranship the cargo, or part of it, and forward it to destination, the average adjuster will probably recommend that to protect the shipowner's position a form of 'Non-Separation Agreement' should be attached to the Average Bond for signature by the party providing general average security. The wording of such a Non-Separation Agreement, as recommended by the average adjuster, will depend upon the facts of the case, but to meet the majority of such situations a standard form of Non-Separation Agreement is available.

There are average adjusters who are in practice and who are not members of the Association of Average Adjusters merely because they have not thought it necessary to sit for the qualifying examination. To become a member of the association the applicant is required to pass a very stringent written examination. As regards the aims of the association, their rules are prefaced:

1. To promote professional standards and correct principles in the adjustment of marine claims by ensuring, through examination or otherwise, that those entering into membership possess a high level of expertise.
2. To achieve uniformity of practice amongst average adjusters by providing a forum for discussion and by establishing rules of practice where necessary.
3. To ensure the independence and impartiality of its members by imposing a strict code of professional conduct.
4. To provide a service to the maritime community by establishing procedures by which advice on all aspects of marine claims may be obtained so as to facilitate their settlement.

It must be stated that although the adjusters' work and knowledge is highly respected in the Courts of the UK, their statement does not have the force of law and an adjuster's opinion on any particular point expressed in his statement could be disputed and possibly overruled by a Court's decision. To adjusters' credit, very rarely is there a need to resort to litigation regarding an adjustment of average.

To end, it would be fitting to quote a passage from an address made by one of the past chairmen of the Association of Average Adjusters, Mr Ernest Robert Lindley, when he spoke of the adjusters' duty:

> *The use of the adjuster individually is to grease the wheels of commercial machinery, to do work which neither the assured nor the underwriter have either time, training or inclination for, in such a manner as to expedite settlements without resort to the expensive machinery of the law: His duty is to act fairly to both parties to the contract of insurance or the contract of carriage, to set down all material facts, withholding nothing of importance, to present the figures of the suggested settlement in such a manner as to be capable of being easily grasped, and above all, in all cases wherever definite law or practice is not clear, to place the matter before the parties interested in such a manner as to facilitate an agreement between them.*

This encapsulates the adjuster's role consisely. □

This chapter has been compiled by J. A. O'Shea, ACII, Partner, Ernest Robert Lindley and Sons, drawing largely upon the Marine Claims Handbook, *by N. G. Hudson, MA, and J. C. Allen, FCII, Partners of the same firm, and upon* From the Chair, *a compilation of some of the addresses of the chairmen of the Association of Average Adjusters. We are grateful to the publishers of these two books, Lloyd's of London Press Ltd, for their permission to use them in this way.*

CARGO SURVEYS—WITH PARTICULAR REFERENCE TO GENERAL AVERAGE AND HEAVY WEATHER

J. M. M. Noble, B.Sc, FNI
Managing Director, Murray Fenton and Associates Ltd

THERE ARE BASICALLY two types of cargo surveys, the first being where goods are damaged in transit, when a survey will be called for by cargo underwriters to establish the nature, cause and extent of damage. This type of survey is the 'routine' cargo survey and it is not the purpose of this chapter to go into any great detail except so far as the 'nature—cause—extent' survey forms part of the heavy weather or general average survey in any event. Secondly, there is the cargo survey which is usually associated with some specific form of maritime incident.

Heavy weather damage

Even the most well-found ship can sustain damage to cargo when in heavy weather. The first thing a surveyor will be looking out for is whether or not the damage sustained during heavy weather was the result of poor stowage, insufficient packing or improper stuffing of a container. Nowadays, it is not sufficient to say the ship encountered severe weather, therefore the damage was caused by it. Cargo must be loaded and stowed in a manner fit to withstand the reasonably anticipated weather which can be expected on a particular voyage.

For example, during a winter North Atlantic voyage it is reasonable to expect prolonged periods of severe weather and, if cargo damage results, it is very difficult to prove a 'heavy weather' defence within the meaning of The Hague or Hague Visby Rules.

Ship design itself plays a factor, particularly in container ships. At one time, when claiming heavy weather cargo damage it was customary to show that the ship itself had sustained damage as a direct result of the weather. In the days of cowl ventilators on smaller ships this was not difficult. However, today, ships are probably better able to withstand the rigors of severe weather without structural damage than was once the case.

When inspecting cargo which has sustained weather damage, it is still important to look at the ship to determine if there is any structural damage. It is equally important nowadays to examine the ship's logs to get as accurate a record as possible of the weather reported by the ship during the voyage. This information, together with sea protests and statements from the master and crew, will provide the basis of establishing 'heavy weather.' If the matter does proceed, then it is likely that steps will be taken to obtain independent meteorological data.

Nature, cause and extent

The survey itself will consider the nature, cause and extent of damage with a view to establishing the loss in financial terms. Whilst a bill of lading will give an indication, it is customary to obtain a copy of the commercial invoice to establish values. Once the cargo itself has been surveyed, then it is usually necessary to look at the means by which the cargo was secured. Lashing and securing cargo is very expensive and all too often securing is cut to a minimum, with disastrous results.

With so much cargo being transported in containers, it is very difficult for a ship's staff to check securing arrangements, other than for the containers themselves. Nonetheless, the master still has the overall responsibility to ensure the safety of the cargo carried aboard his ship and the surveyor will be taking a particular interest in the steps taken by the master to safeguard the cargo.

Even so, ships do encounter extraordinary unanticipated weather during voyages and no manner of attention will prevent damage. It is the surveyor's job to establish, as precisely as possible, the criteria and parameters which resulted in the damage.

Bulk cargo incidents

Ships carrying bulk cargoes experience incidents where the cargo has become damaged as the result of sea water ingress into the holds during heavy weather. When such damage occurs, a surveyor must examine closely the hatch closing and securing arrangements. Such a survey will incorporate a detailed examination of the hatch covers and other watertight openings.

At the same time, cargo damage can occur as a result of the inability to ventilate a cargo properly during severe weather, and this type of damage usually leads to the most contentious exchanges between the parties involved. Where this type of damage occurs, the surveyor will require careful ventilation records and must have knowledge of the nature of the cargo itself.

Too many surveyors place reliance on the silver nitrate tests. The silver nitrate test is merely a first resort and some will say it is only really useful when there is no reaction. It is not uncommon, indeed it is often prudent, for the surveyor to seek further advice from chemists who will be able to analyse the cargo to establish more precisely the actual cause of damage or the nature of the cargo itself.

General average surveys

A ship may suffer a casualty which results in the whole voyage being endangered and steps may then be required which protect the 'venture' as a whole. When this occurs, and the master has had to take steps to protect the interests of all those involved in the voyage, he may declare general average. For example, suppose a consignment of cargo catches fire;

the master may have to use his fire hoses to extinguish the fire. Whereas the cargo which was directly damaged by fire has suffered a particular average, any other cargo which has sustained damage, resulting in loss, will be included in the general average.

The general average surveyor's role is a supportive one. Once general average is declared, average adjusters will be appointed and it is their function to undertake the 'accounting' tasks resulting from the incident. This exercise is undertaken by experienced professionals, who rely to an extent on the survey report. One of the first tasks of the GA surveyor is to differentiate between 'particular' and 'general' average. From the example above, the surveyor will have to determine which cargo has been damaged by fire and which was damaged as a result of extinguishing the fire. The survey will then include an examination of all the cargo on board, bunkers and the ship itself. Those who have suffered loss will receive a contribution to that loss from the other voyage cargo interests.

There are other considerations which the surveyor must take into account. If the incident which gave rise to the general average was a result of the ship's unseaworthiness then those interests which are being asked to contribute to the general average may decline on those grounds. Whilst the general average survey is an extension of the ordinary cargo survey, it encompasses a much broader series of issues and it is important the surveyor is fully aware of the complexities of this type of survey and principles which will be applied by the adjuster. □

CONDUCTING WARRANTY SURVEYS

Captain A. Marshall, Ex.C, FNI
Principal Nautical Surveyor, The Salvage Association

FIRSTLY, WHAT IS A WARRANTY in terms of marine insurance? In English law this is defined in clear terms in the Marine Insurance Act 1906, but, for present purposes, it can be simply stated as an undertaking by the assured whereby he promises to comply with the terms of a warranty or, to put it another way, with specific conditions in the policy. Secondly, why are warranties inserted in policies? Underwriters impose warranties to control more precisely the nature and extent of the risk that they are assuming.

Warranties may be either express or implied. Under English law in a marine insurance contract there is an implied warranty that the adventure insured is a lawful one. In a voyage policy there is an implied warranty that at the commencement of the voyage the ship shall be seaworthy for the purpose of the particular adventure insured. Express warranties must be written into the policy or in some document incorporated into the policy. An express warranty may, for instance, limit the geographical area to which the insurance cover applies; it may impose a condition that the vessel shall not carry cargo or take vessels in tow.

Effect of warranty

What is the effect of a warranty? A warranty is a condition which must be exactly complied with and, if not adhered to exactly the insurer is discharged from liability as from the time of the breach of the warranty, whether or not the warranty is material to the risk or whether the breach contributed to a loss. Nor does it matter in law whether the assured is culpable of any breach. Warranties therefore form important conditions in an insurance contract.

Insertion of express warranties by underwriters has increased over the years and probably for a number of reasons. There has been an increasing amount of marine activity by non-marine industries; civil engineering and offshore oil exploration has led to the carriage and towage of expensive and massive structures, and in general marine transport has become more complex. Underwriters feel it is necessary to impose conditions by way of a warranty to safeguard their own position by attempting to ensure that the risks to which the property will be exposed are no greater than those they are content to bear for the premium charged. In particular underwriters who are accepting risks following discussions across a desk may consider it appropriate only to do so on the basis of the assurance of a competent marine surveyor physically inspecting the risk or the project.

So the underwriter might impose a warranty as a condition of the insurance, in the case of insurance of a tow:

'Warranted tug, tow, towage and stowage arrangements to be approved by . . . and all recommendations to be complied with'
in other words, the underwriters require a warranty survey to be carried out.

If the named surveyor or survey organisation does not approve the 'arrangements', as contained in the warranty wording or if their 'recommendations' are not complied with, then there is a breach of warranty and the underwriter is discharged from any liability in the event of a loss.

Imposed by underwriter

Although it is the underwriter who imposes the warranty, in practice it is usually the assured who has to engage and pay for the services of the warranty surveyor, in order to comply with the warranty. The surveyor, notwithstanding his engagement by the assured, conducts the survey for the protection of the underwriters concerned. The warranty surveyor is not there to act as adviser to the assured.

As in the above towage warranty, the underwriter requires that the surveyor's recommendations are complied with, and this in effect means that the surveyor's recommendations become warranties themselves. Because of this, the surveyor must take care to ensure that his recommendations are realistic and necessary. These recommendations might for instance cover bunkering instructions, speed limits on tows, weather limits, routing, keeping of log books and records.

The surveyor must remember that he has no contractual relationship with the assured's contractors, such as tug operators, stevedores, riggers, shipyards, etc., and can therefore issue no instructions whatsoever to them. Ensuring that a contractor carries out the surveyor's recommendations for the conduct of a voyage is the responsibility of the assured, but the surveyor should ensure that the assured appreciates this. It is for the assured to select and employ competent contractors, and the appointment of a warranty surveyor does not relieve the contractors or the assured of their obligation to provide safe equipment and seaworthy vessels.

The assured is also obliged to give adequate notice to the surveyor of the existence of a warranty, so that there is adequate time for him to make his recommendations, and for the assured to carry out those necessary prior to commencement of the project. If a tug or other craft is involved it is preferable if this can be surveyed prior to the assured entering into a contract.

Because of the strict nature of warranties, because the surveyor is appointed to safeguard the underwriter's interests, yet is usually paid by the assured, and because he has to deal with practical problems face to face, the task of the warranty surveyor is demanding. It requires not only technical skill and experience, but an ability to relate to people at the scene of operations and, above all, a capacity to be decisive but tactful.

Having set the scene for the conduct of warranty surveys, let us look at some of the practical problems of the most common types of survey and the warranties that dictate them.

Towage (tugs)

'Tug, tow, towage and stowage arrangements to be approved by . . . and all recommendations complied with'

The towage survey is the most common type of warranty survey. This reflects the view that towage presents an increased risk. These surveys involve almost every type of floating craft, both in the sound and the damaged condition, the majority carrying some form of cargo. The craft may not necessarily be designed for seagoing purposes; for instance, floating docks, oil field equipment, concrete caissons, lock gates, and even floating restaurants.

The first point with any tow is that the towing vessel should be suitably matched to the towed vessel. Up to the latter part of the 1960s, a 4,000-horsepower, 40-ton-bollard-pull, deepsea tug was considered large. Ten years later harbour tugs with that order of horsepower had become commonplace. Improvements in design have increased the thrusts of tugs even more, so that the tug of 90 tons bollard pull is no longer considered large. The result is that the industry has come to expect the use of powerful tugs. This in most instances is all to the good because, along with greater power, comes heavier towing gear. It must be borne in mind, however, that accidents can be caused by too much power as well as by too little. Though some well-found ships and barges may be towed by powerful tugs at speeds up to 10 knots, it is often preferable when moving delicate equipment or a damaged ship to have a tug which, rather than force the tow through bad weather, will seek shelter.

In calculating the amount of tug power required, either one can adopt the criterion of the tug holding the tow still in adverse conditions, or of achieving a certain speed in favourable conditions. Any result obtained by calculation must be tempered by experience and common sense.

The survey of a tug may usually cover: certification, crewing, navigational equipment, structure, towing equipment, towing gear, machinery, bunkers and stores. A tug, as in the case of any other vessel taking up a contract, must have its certification in order. In addition to the normal ship's certificates, a tug must have certificates for its wires and other towing gear. There may also be a bollard pull certificate, but unless that certificate describes how the trial was carried out it may be misleading.

In some areas of the world tugs are seldom entered with a classification society. The warranty surveyor then has no bench mark from which to assess the tug's condition and state of maintenance. In such cases it is incumbent upon the assured to disclose voluntarily all material information concerning the tug and its condition.

The crew should not only be adequate to work the tug, but there must be sufficient towing experiencce specifically amongst the officers, who should be appropriately qualified. It is the responsibility of the assured to provide crew in compliancce with the requirements of the flag State. The navigational publications should be up to date and adequate for the voyage, and the navigational equipment should be in working order. The structure of the tug should be in good condition and particular attention should be given to load line items.

Items such as the towing winch, gog rope and hook are crucial to the performance of the tug. Towing gear must be of adequate strength in relation to the power of the tug and the strength of the gear must be continuous from towing winch to towing point. It is essential that a tug should have at least sufficient equipment to re-rig the whole system in the event of a break adrift.

Assessing the condition of machinery from information provided by the assured and from a superficial inspection is problematic, especially if the tug is not entered with a classification society—however, the general state of the engineroom is an indication. If the survey of the engineroom leaves doubts as to the performance of the tug, then a full power trial attached to a suitable bollard may be the only convenient method of assessment.

When proceeding on long voyages, smaller tugs may have difficulty in carrying sufficient bunkers, water and food. An unscheduled deviation to obtain bunkers, etc., increases the risk attaching to the voyage. A reduction in power to save bunkers lengthens the voyage and may itself increase the risk.

Towage (towed objects)

The towed objects produce an almost infinite variety of problems owing to the varied nature of the craft involved, the most common being ships, warships, barges, oil rigs, oil field equipment, dredgers and caissons. These can vary from being sound to heavily damaged, from the well maintained to the utterly deplorable, or from new to the genuinely historical. Each presents unique problems calling for the utmost care by the assured.

Barges

The simplest objects that are towed are barges, the majority of which are flat-topped pontoons with swim ends or flat-topped with a spoon bow and perhaps a rudimentary forecastle. Some of these are of the submersible type and may have buoyancy chambers on deck. Their construction is fairly basic and they all have a high degree of subdivision, so that damage limitation is seldom a problem. The dimensions of most barges are such that they can carry massive deck loads without seriously impairing their stability.

Generally the larger barges carry their own towing bridles and emergency pendants and the better equipped ones have their own anchoring system and ballast pumps. Barges, of course, do not have crews, which means that maintenance is not under the supervision of anyone on board and hence minor casualties usually go unreported. Swim-ended barges are particularly vulnerable to contact and heavy weather damage to the bows.

When surveying barges it is well to remember all the cautionary advice that has been written about tank entry. The tanks are relatively shallow and there is a great temptation to 'just nip down' an unventilated tank.

Ships

Ships are only towed when for some reason they are immobilised or have no crew. Many ships are designed with little or no thought given to the possibility of towage. This means that bitts, fairleads and leads usually leave a lot to be desired. Occasionally it is necessary to fabricate and fit towage connections. Further, apart from the possibility of underwater damage, some ships are directionally unstable when under tow.

On older cargo ships, and particularly warships and passenger ships, ensuring the watertightness of the engineroom can be a problem. In the case of warships and passenger ships this can be a problem of some magnitude. It has become almost a practice of the trade for tows to proceed unmanned, and a recommendation for a riding crew can also cause conflict with the statutory surveyors.

Tows of passenger and warships are often for scrapping voyages. Most ships on scrapping voyages can be considered to present a higher risk for the underwriter. Their condition is often less than ideal and sometimes they have been severely cannibalised. In many instances the vessels have been laid up for years with no chance of a drydocking before leaving port. The worst feature of scrap tows is that the parties involved can be on slim profit margins, with the result that they are seeking to keep expenditure to the absolute minimum, and will be reluctant to undertake any measures which increase in expenditure. This situation is often exacerbated by all the contractual arrangements being entered into before a surveyor is appointed.

Sometimes the ship to be towed has hull damage. This places a particularly heavy responsibility upon the warranty surveyor; class is usually withdrawn and statutory surveyors are often unavailable in remote places. The loss of modulus must be assessed, often from limited information. The bending and shear stresses must be calculated for the anticipated weather conditions before the surveyor decides whether to issue any certificate.

Damage below the waterline often acts as a rudder, making the vessel difficult to tow and causing wear to the towing gear. For the same reason the rudder needs to be efficiently secured amidships. To prevent damage to the machinery the propeller must be properly secured. Ships under tow may roll more heavily than usual, so cargo securing must be efficient and, unless there is a riding crew, the securing must be maintenance free. If there is no crew then particular care must be paid to all closing appliances, hatches and airpipes.

Floating docks

Floating docks present problems in towage because they are rarely designed for sea voyages. The voyage itinerary must be carefully assessed for the expected wave spectra and the structural strength checked against these. It is often necessary to recommend strengthening or dividing the structure. Such a recommendation may not be welcomed by the assured as it will usually involve additional expenditure and probably delay, or even the project to be cancelled.

In strong winds the resistance of the wind and waves on a dock can overcome the power of the tug. When docks depart from some ports, restricted waters extend so far that the tow runs out of forecast time when still in the restricted waters, but by then they are committed to the voyage. It is normal for floating docks to proceed manned and regrettably normal that the docks often incur minor structural damage en route.

Caissons

Concrete caissons are often used in port construction and are not normally hydrodynamically designed. They are generally heavy to tow and are directionally unstable. Ensuring watertightness of openings in a concrete structure is often difficult, and, unless the quality control has been good, concrete is apt to be porous.

Oil field equipment

It is sometimes necessary to tow floating objects of non-shipshape. These are often associated with the oil industry. Typical are spar buoys. Longitudinal strength of such items is a particular problem that must be examined before such a tow can be accepted. Other problems associated with such tows are placement of navigation lights, placement of emergency towing pendants, stability and damage stability.

Dredgers

When dredgers are towed on their own bottoms they tend to be very vulnerable. Ensuring watertightness on some dredgers can be difficult and occasionally the structure is apt to be wasted. Once water enters the hull of a dredger the stability becomes critical. Normally, bucket dredgers for stability reasons cannot proceed on sea tows with the buckets in position. For these reasons and because speedy damage-free deployment is desirable, numerous specialised semi-submersible craft have evolved to carry such equipment.

Cargo warranty surveys

'Warranted loading and stowage arrangements to be approved by . . . and all recommendations complied with.'

These fall mainly into two categories, the large single item of high value or the large shipment where the value exceeds a pre-determined amount. The craft upon which the cargoes involved are loaded may be a general-cargo vessel, specialised heavy-lift vessel, submersible ship or barge. The survey may cover review of the loading method, preparation, stowage, securing and sometimes the discharge, depending on the precise wording of the warranty.

The loading survey may simply involve the inspection of normal stevedoring methods and confirming the condition of the cargo. On the other hand, it could involve examining the procedures for the skidding or rolling on of very large objects. In the case of submersible craft the arrangements for sinking of the craft and the floating on of the cargo may need to be examined.

Particular attention must be paid to the securing of the cargo. This may be the traditional timber and wire shipboard securing. Many securing surveys concern items where the forces have been calculated and the seafastenings designed to suit. The surveyor must examine any such special fastenings. Some cargo warranty surveys require involvement of the surveyor from the factory to destination. This can cover lifting, road transport, sea transport and final lifting into position.

Lifting surveys

'Warranted lifting arrangements to be approved by . . . and all recommendations complied with.'

Sometimes, quite apart from a cargo loading survey, when a large item is to be lifted, whether for loading or not, the operation is the subject of a warranty. It is therefore necessary to check that the crane and all the lifting apparatus is suitable for the job, that the design of the lifting points and surrounding structure is adequate. In harbour works it is not unusual to lift an object with floating sheer legs and transport it in the slings for several miles before lowering it again.

The most frequent problem that is encountered in these surveys is wrongly calculated or wrongly declared weights. Suitable allowance must be made for this. Allowance must also be made for dynamic factors if the lift is not in port.

Lay-up surveys

'Warranted lay up and mooring arrangements to be approved by . . . and all recommendations complied with.'

These surveys became an important feature when the freight market for VLCCs/ULCCs collapsed. It was obvious that virtually none of the normally accepted lay-up sites was in any way suitable for such large vessels. The result was that shipowners proposed sites in numerous remote locations. Surveyors attended these sites and reported on the nature of the sites and their suitability for laying up VLCCs/ULCCs and made their recommendations.

When a full lay-up survey is required many aspects have to be covered. The primary one is that of the moorings themselves to ensure that they are strong enough for the expected forces. A ship's condition after lay-up reflects the care taken at and during lay-up. Therefore items such as preservation of boilers and machinery, dehumidification, preservation of electronics, cathodic protection of ballast tanks, crewing, access and emergency procedures may all need to be considered.

Reactivation surveys

'Warranted reactivation to be approved by . . . and all recommendations for the voyage complied with.'

The obverse of the lay-up survey is the reactivation survey. If a ship has been laid-up for more than, say, six months, the underwriters may insert a warranty requiring survey upon reactivation. This is to ensure that all reasonable procedures have been carried out to ensure subsequent operational viability. The classification surveyor is normally closely involved with reactivation, but the warranty surveyor may be required to cover both the aspects required for class and the operational aspects not covered by class rules.

Limitation of warranty surveyor's duties

The surveyor is contracted solely for the purpose of carrying out the requirements of the underwriters as expressed in the warranty. He is not there instead of the classification or certification surveyor, nor to act as the assured or owner's superintendent. It is certainly not desirable for the warranty surveyor to direct the work of the crew, or the stevedores, or to supervise the securing or stowing operations. The surveyor is there to approve such arrangements, not to implement them. He may create potential problems for himself if he carries out duties which are properly the function of the assured's own employees, or of the contractor.

This chapter does not cover every type of warranty survey that a surveyor might be asked to perform, nor is it intended to cover every aspect which he needs to consider, but it is hoped that it gives an idea of the potential complexity of warranty surveys, and dispels any thought that such surveys are a mere formality. It is only after initial guidance and considerable experience that a marine surveyor will feel confident of his ability to cope effectively with whatever warranty survey comes his way.□

CONDUCTING HULL AND MACHINERY DAMAGE SURVEYS

Captain A. Marshall, Ex.C, FNI, Principal Nautical Surveyor
C. A. Sinclair, C.Eng, FIMarE, Formerly Chief Surveyor
The Salvage Association

THE VAST MAJORITY of these surveys are undertaken by marine engineer surveyors. So far as this publication is concerned, the subject is mainly of background interest and hopefully shows some of the interlocking surveys. If a nautical surveyor is involved in this type of work he is more likely to be as an owner's representative than an underwriter's surveyor, or if attending for underwriters will probably be most concerned with the salvage aspects.

Damage surveys are normally initiated because of an insurance involvement—i.e., a shipowner contemplates making a claim upon his insurance policy. Occasionally such a survey is for legal interests, though even then there is normally an insurance connection. Although these surveys are normally held because a party is intending to make an insurance claim, it is not the function of the surveyor to interpret the policy. That is the work of the average adjuster and the underwriters' claims department. On the other hand, it is important that the surveyor appreciates the basic insurance terms and practice, so that he presents his report in a useful manner and pursues his investigations into the cause efficiently.

The marine insurance industry requires the services of the average adjuster because there is almost inevitably more than one party affected by a marine casualty. The adjustment can be a complex business. This separation of the functions of surveyor and adjuster marks the difference between a surveyor and an assessor. The latter performs both functions, but there are usually only two parties involved and the policy terms are simpler.

Assured to call for a survey

It is the assured who should call for a survey when he intends to make a claim upon an insurance policy. The principal parties to a damage survey are the owner's representative and the underwriters' surveyor. Included may also be the classification surveyor and a repairer's representative. The assured, who is probably the shipowner, is the client, and it is the duty of his representative to draw to the attention of the underwriters' surveyor the extent of damage suffered owing to the alleged casualty. The underwriters' surveyor must assess whether such damage could reasonably be attributable to the casualty.

The owner's representative also has the duty to put forward the cause of the damage. Obviously, if he intends making an insurance claim the cause must cover a peril insured against. The underwriters' surveyor must investigate the cause to check if the owner's allegation is valid. Every effort must be made by the parties during and after the survey to agree on the cause and extent of the damage attributable. This is not always a simple matter, and failure to agree on these basic points results in years of argument and correspondence.

Machinery damages provide a fertile area for disagreement. The results of negligence and wear and tear can be similar. If negligence is accepted, the extent of the damage due to the same may not be at all clear. Structually a favourite cause frequently put forward is heavy weather, although the structure may be severely deteriorated due to corrosion. Further spheres of disagreement occur when more than one casualty covers the same area of damage.

Estimated cost of repairs

An aspect which is not the subject of mutual discussion is the estimated cost of repairs. Each party puts his own estimate forward to his principals. Thereafter the underwriters' surveyor may have the opportunity to comment upon the owner's estimate. In the event of a serious casualty it is of paramount importance to avoid spending more than the insured value. Should the estimated cost of repairs exceed the insured value, the underwriters' surveyor can only advise his principals of the fact. It is only the assured who can declare the ship a constructive total loss, and is quite within his rights to take a 100 per cent partial loss and fund the excess cost of repairs himself. In these circumstances the surveyor must avoid recommending expenditure under the sue and labour and salvage charges clauses.

Where damage is extensive, a prudent owner will call for tenders to be submitted by potential repairers. The underwriters' surveyor will assist in this and help to draw up a specification. Because of a special tender clause inserted in most policies, the surveyor should not insist on tenders being called for unless instructed to do so by his principals.

Once repairs have been completed, the owner's representative should pass the repair account to the underwriters' surveyor for his approval. Should there be any items on the account which the surveyor feels are unreasonably high, then, with the owner's agreement, he can take the matter up directly with the repairer. Once the surveyor has satisfied himself that the accounts are fair and reasonable, he will give written approval to that effect. The approval of accounts by the surveyor is claused in such a way that it does not bind his principals to payment. Once the surveyor has cleared the technical and financial aspects of the repairs, the insurance decisions can be taken by the average adjuster and the underwriters' claims department.

Stranding

This brief description of the general principles involved in a damage survey covers the simple situation where a ship has suffered hull or machinery damage, without involvement of third parties or salvors. The complexity of marine insurance is due to the number of parties that are affected by a casualty. For example, if there is a stranding requiring salvage services there are the following points to consider:

(1) The particular average damage to the ship.
(2) The particular average damage to the cargo. There may be one consignee or numerous who may or may not be insured.
(3) The salvage charges and sue and labour charges which may be admissible under general average.
(4) Damage due to efforts to refloat which may be admissible under general average.
(5) Loss of cargo by jettison which may be admissible under general average.

Third party involvement

The situation can become even more complex when the casualty involves a third party as in a collision. One surveyor cannot attend all the aspects because of conflicting interests. The same surveyor may represent more than one interest as long as they do not conflict. In the case of a collision, the following persons may be involved, excluding class and statutory surveyors:

- Two owners' representatives (one for each ship).
- Two underwriters' surveyors to report upon the damage to the respective ships.
- Two underwriters' surveyors representing the underwriters of one ship but to survey and report to their principals upon the damage to the other ship without prejudice to liability. They can be the same persons as surveyed the first ship, but location often prevents this.
- The lawyers representing each owner will appoint surveyors to survey the damage of each ship and make an estimate of the speed and angle of blow of each ship. It is possible without conflict of interest for the same surveyor to represent the underwriters of the ship A in the first instance, to survey ship B without prejudice and represent ship A for speed and angle of blow.
- Cargo surveyors would be appointed to cover the particular and general average aspects of the cargo.

It will be seen from the foregoing that what may be called the commercial aspects of a damage survey and report are at least as important as the technical, and the surveyor must appreciate who he is representing and why. Though most surveys concern money in one way or another, the damage survey has an immediate impact upon the transactions between the owner, underwriter, shiprepairer and salvor.

Later queries

A feature of damage surveys, or rather the reports thereon, is that months or years after the event queries are received from average adjusters requiring clarification or disputing items in the report. These are inevitable as the adjuster seeks to put forward his client's claim in a favourable light. However, a large number of queries often means that the report is not comprehensive or explicit enough.

Though we stated earlier that the surveyor is not concerned with interpreting the policy, there are a few basic terms with which the surveyor must be familiar in order to present a useful report and negotiate with the owner's representative. Normally a policy of insurance is not concerned with the time it takes to carry out repairs, or the loss of earnings that the owner suffers. It follows therefore that the underwriter is not liable for the excess cost of overtime, unless, this gives a saving to the underwriter, such as occasioned by a reduction of time in drydock.

It is sometimes the case that an owner having suffered a casualty will wish to take the opportunity to make modifications or improvements. The underwriter is only liable to return the ship to its condition immediately prior to the casualty. Therefore the surveyor must report upon the excess cost or otherwise of the modification or improvement. In addition, an owner may opt to carry out temporary repairs. These can fall into various categories as follows:

(a) To enable the ship to proceed safely to a repair port.
(b) To enable the ship to proceed to a more economical repair port.
(c) To suit the owner's trading convenience.

Details of the reasons and the costs incurred must be reported upon because some of the costs are claimable, while others may not be. A regular adjuster's enquiry requests the comparative costs of repair at port X in year A as opposed to port Y in year B. This is because the underwriter is only liable for the repair costs at the time of the casualty, unless a saving can be shown. Repairs to damages which do not affect the vessel's classification position are often deferred until the next survey or drydocking. Often part permanent repairs are carried out. For reasons given in the previous sentences in this paragraph, these aspects must be carefully reported upon.

Wear and tear not covered

An insurance policy does not cover loss due to wear and tear, although normally marine policies do not have new-for-old deductions. This is a contentious area, because it is not always clear whether the casualty was due to wear and tear or some other cause. For more than one reason an owner's representative will be reluctant to admit that a casualty was due to wear and tear, though the reasons for that are beyond the scope of this article. Where strong disagreements on this or other points occur, it is essential that the parties involved assemble their facts carefully, because in the event that they cannot eventually be reconciled the ultimate step is action in the High Court.

Owner's repairs or maintenance is regularly carried out concurrently with casualty repairs. The surveyor therefore must report upon the respective lengths of time that the drydock was used for owner's and casualty work, and state whether drydocking was necessary in each event. Repair accounts have to be carefully scrutinised to ensure that the owner's and the casualty accounts have not been confused.

In the case of major casualties, the underwriters' surveyor attends from the time of the casualty and it is normal for an owner's representative also to be in attendance. Unlike the surveys taking place in the repair yards, it is a quite normal feature for these early attendances to be by nautical surveyors. The most usual casualty where this occurs is a major stranding where salvage services are required. The report of this aspect of the operation may be used to assist in arriving at a fair and reasonable award to salvors. Such a report should contain information on the following points:

- Nature of the contract, when it was signed, by whom and with whom.
- Circumstances resulting in the contract being signed.
- The time and location that the services ceased and who accepted that the services had terminated.
- The nature of the services provided, the equipment, personnel and craft supplied.
- Damage caused by the salvage services, including damage to cargo and cargo jettisoned.
- Details of the expertise shown by the salvors.

Reinsurance market

The surveys discussed above have all been of the type where there is an owner with a simple hull and machinery policy. This is not always the case. There may be two surveyors representing underwriting interests, the second appointed by the reinsurance market. This is an aspect that has increased in recent years, owing to a tendency to insure 100 per cent in one country and then reinsure 95 per cent of that in another market.

Some owners who have valuable charters have a loss of earnings policy which is quite separate from the hull and machinery policy. This policy is sometimes represented by another surveyor who may recommend overtime and other payments to expedite the return of the ship to service and reduce the liability under the policy.

Where a ship is under time-charter, the charterer may have a liability policy to cover damage that may be his responsibility under the terms of the charter party. Occasionally there is a bank involvement, and there is a surveyor representing the policy covering the mortgage on the ship.

Owner's representative

We stated at the beginning of this article that it was more likely that a nautical surveyor would find himself dealing with underwriters' surveyors as an owner's representative. The owner's representative may also be the master or chief engineer. In that position he must assist the owner in putting forward his insurance claim. Points to consider are as follows:

(1) Keep clear, concise, log books.

(2) With the assistance of the P&I lawyer, obtain statements if necessary from all the relevant crew members as soon as possible after the casualty.

(3) If at all possible obtain a photographic record.

(4) Obtain receipts and give reasons for all sue and labour expenditures.

(5) Establish the reason for the casualty and gather all possible evidence to support the allegation—for example:

   (a) If the allegation is latent defect, then the offending part should be recovered and professional reports on it obtained.

   (b) If the allegation is crew negligence, then log book and other evidence should support this. It is counter-productive if investigation shows that the owners induced bad operating procedures.

   (c) Log book entries showing heavy weather should be credible. Entries indicating storms and mountainous seas with the engines turning 95 per cent of maximum appear ludicrous.

(6) Avoid hypothesis. Long hypothetical arguments may assist the well-being of average adjusters, but they do not expedite payment of claims.

(7) At the time of the survey the owner's representative, in order to be credible, should in most instances be in a position to put forward the cause of the damage. □

CONDUCTING CONDITION SURVEYS FOR INSURANCE

N. E. Wolff, Naval Architect
Surveyor, The Swedish Club

TO BE ENGAGED in shipping has always been looked upon as a dangerous business. Foundering, collision, grounding, stranding and fire are all well-known hazards to the shipping industry. The need to cover these risks by some kind of insurance has existed for a long time and therefore marine insurance is an old phenomenon. Marine insurers eventually realised that the risk for a vessel could be reduced under certain circumstances—that is, with a competent crew and with a strongly built and well-maintained vessel.

This led to the formation of the classification societies which from the beginning were set up by the marine insurers to issue rules for the design and building of vessels, and to survey the vessel throughout its lifetime in order to guarantee its standard. This state of things continued for many years to the satisfaction of all parties involved. However, during the last decade there have been quite a few cases where the insurers feel that the classification societies have not carried out their task in a proper way. It has, therefore, become necessary for the insurance companies to carry out their own surveys in order to guarantee the condition of the vessel.

Surveys for vessels entering a P&I Club

This problem was also recognised by the Swedish Club, especially when the Club began to take on international risks, and condition surveys for vessels entering the Club have been carried out regularly for the last eight-ten years. To give you an understanding of the extent of these surveys, it can be mentioned that in 1985 about 150 vessels were surveyed in connection with entry to the Club.

If possible we try to carry out the surveys with one of our staff surveyors from the Gothenburg office, but for practical reasons this often turns out to be impossible, and therefore the greater part of the surveys will be carried out by our representatives around the world. In order to secure the quality and the extent of the survey we have drawn up a form called 'Guidance for Surveyors in Inspecting Vessels on Behalf of The Swedish Club' which has been sent out to all our representatives and to be used by them when carrying out a condition survey on our behalf.

We would like—although we unfortunately often have to give up the thought—the condition survey to be carried out in connection with a drydocking. We are not in a position to demand that the owner should bear the costs for drydocking in connection with a condition survey before entry to the Club, and the Club could not afford to cover such costs by itself. Already the costs for the condition surveys carried out are high, even if we consider the money has been spent to good use.

Experience

Some vessels pass the survey without any comment. However, for the majority there are remarks of a more-or-less serious nature. These remarks are passed on to the owners, together with a request to rectify the defects within a given period of time. If the remarks are of a serious nature it may be necessary to resurvey the vessel after correction of the deficiencies. In a few cases we find that the condition of the vessel is so bad that we quite simply do not accept the vessel for insurance cover with the Club.

As can be seen from the Swedish Club form, parts 'A' and 'B' deal with vessels' particulars, class and certificates. In most cases this information is easily obtained on board, but sometimes there seems to be a problem with filing the valid certificates for easy access. (On the other hand, you have to admit that there is a considerable quantity of papers going round.)

Hull problems

Part 'C' deals with the hull and that is obviously a very important part for the survival of a vessel—not least items 1-5. The major problems for the hull are cracks and corrosion. Cracks will occur in areas where there are stress concentrations—i.e., hatch corners, ends of bilge keels, and where longitudinal scantlings change over to transverse scantlings. Cracks may also appear due to inferior welding. Corrosion is particularly aggressive in water ballast tanks and we have seen many top wing tanks with as good as no internals at all, even where the vessel has been retained in class. To cope with the situation, we have found it necessary to equip our surveyor with an ultrasonic thickness gauge.

Signs on air and sounding pipes are very often missing or have been painted over. This may seem a minor problem but it may easily lead to a cargo hold being filled with water instead of the intended double-bottom tank. Stuck ventilator dampers make it impossible to close the compartment in question in case of a fire. Fire equipment and fire protection are of great importance, so we put a lot of effort into item No. 12.

Items 17 and 18, dealing with hatches, have a high priority, not least when it comes to P and I cover. Old and damaged rubber packing, corroded compression bars and inadequate battening down equipment are the most common problems.

Engine inspection

Part 'D' covers the engine inspection. The importance of this may be understood when you know that about 25-30 per cent of all paid out damage costs on the hull and machinery insurance are related to damage to engine equipment. The major part of these costs is related to the main engines and again the major part of this is related to medium-speed engines. This means that all equipment related to fuel oil and lubricating oil treatment is of importance, as well as a good monitoring and maintenance system.

Items 12, 13, 14, 16 and 18 all have to do with fire protection and fire fighting, which again are of great importance. Regarding item No. 16, you could say that it has to do with a vessel's safety in general, because a clean engineroom is a safe engineroom and unfortunately we have noticed that the reduced number of crew (which has become very popular nowadays) has a tendency to affect this item in a negative way. If the vessel has sufficient quantity of adequate spare parts, this could very well mean that a salvage operation could be avoided in case of an engine breakdown—salvage operations are never a low-cost arrangement.

Galley and accommodation

Parts 'E' and 'F' are about the galley and accommodation. The standard varies considerably with the flag and with the nationalities of the crew. Again, the important thing for the insurer is fire safety and fire protection. The stove and the exhaust fan present a hazard if they are not kept clean.

When it comes to part 'G', navigation equipment, the important thing is to check that vessel is equipped according to rules and regulations and of course that all is in working order. For bigger vessels, the loading instrument is of great importance. Major damage to a ship's hull has been sustained due to improper loading or discharging.

Part 'H' is partly about the vessel's crew. It has become more and more obvious that the quality of the crew has a great influence on the behaviour of a vessel from an insurer's point of view. It is therefore important to check the number, the competence and the training of the crew. It is also of importance that all instructions, manuals, signs and drawings are available and written in a language that is understood by the present crew.

Sighting class records

As a complement to the condition survey, sighting the vessel's class records may provide valuable information. Within the Swedish Club this is standard procedure for vessels older than ten years. In the records one can see reoccurring problems on a specific item. The record sighting should if possible take place before the condition survey is carried out and the outcome of the sighting should be handed over to the surveyor.

As mentioned in the beginning of this chapter, it has not always been the case that insurers have carried out their own condition surveys and one may hope that the confidence between the insurers and the classification societies could be restored. However, it is not at all a bad thing that marine insurers have to see their insured vessels and that the people on board the vessels meet with insurance people. Something good could come from this situation. □

the swedish club

Guidance for surveyors in inspecting vessels on behalf of The Swedish Club

CONDITION SURVEY

A. Particulars

1. Vessel's name: ....

2. Surveyed at: .... ☐ Afloat

.... (place) .... (date) ☐ In dock

3. Owners: ....

4. Managing Operators: ....

5. Built: .... (yard) .... (Hull No.) .... (year)

6. Gross: .... 7. Net: .... 8. Deadweight: ....

9. Type: .... 10. Port of Registry: ....

11. Present at Survey: ....

B. Class, Certificates etc.

1. Classification (Society Symbols) ....

2. Loadline Certificate: ....

   Issued: .... (place) .... (date) Expires: .... (date)

3. Safety Construction Certificate: ....

   Issued: .... (place) .... (date) Expires: .... (date)

4. Safety Equipment Certificate: ....

   Issued: .... (place) .... (date) Expires: .... (date)

5. Safety Radio Certificate: ....

   Issued: .... (place) .... (date) Expires: .... (date)

6. Oil Record Book (Last Entry) ....

7. Survey Particulars: ....

   a) Hull: Next S.S. Date: .... CSH Yes/No

   b) Machinery: Next S.S. Date: .... CSM Yes/No

   c) Drydock: Last Date: .... Place: ....

   d) Propeller Shaft: Last Date: .... Clearance: ....

   e) Boilers: Next Date: ....

8. For Tankers Only: ....

   a) Inert Gas System ....

   b) Crude Oil Washing ....

   c) Segregated Ballast Tanks ....

C. Hull inspection

1. Shellplating, Internal/External: (Major Indents, last measurements of plate thicknesses etc.)

....

2. Frames etc.: ....

3. Decks: ....

4. Superstructure and deck-houses: ....

5. Tank Tops: ....

6. Bilges/Suction Boxes, Alarms: ....

7. Air Pipes: (Signs, Closing devices) ....

8. Sounding Pipes: (Signs, Covers) ....

9. Manhole Covers: ....

10. Ventilators/Dampers: (To be tested, Signs) ....

11. Deck Service Line: ....

12. Fire Hydrant/Fire Lines and Hoses: (Last fire drill, date) ....

....

13. Winches: ....

14. Windlass: (Spare Anchor) ....

15. Deck Cranes/Cargo Gear: ....

    Inspections: 4 years Date/Place: .... Annual Date: ....

16. Access Ladders and Handrails: ....

17. Hatches, Coamings, Packings, Cleats: ....

18. Hose Test of Hatches: .... (place) .... (date)

19. Hold Lighting: ....

20. Davits/Lifeboats/Rafts: (Last lifeboat drill, date): ....

21. Drydock survey: (Shell, rudder, propellers etc.) ....

....

....

....

....

D. Engine Inspection

1. Main Engine (Type): ....

2. Auxiliary Engines (Type): ....

3. Generators: (Emergency Generator, Last tested, date) ....

....

4. Switchboard: ....

5. Boilers: ....

6. Thrusters, Steering Gear: (Rudder clearance) ....

7. Oily Water Separator: ....

8. Purifiers and Filters: ....

9. Bunker Tank Gauges: ....

10. Bilge Pumping System Alarms: (To be tested) ....

....

11. Overboard Discharge: ....

12. Remote Stops and Shutoffs: ....

13. Fire Appliances: (Checked, date) ....

14. Closing Devices for Skylights/Bulkhead etc.: (To be tested) ....

....

15. Quantity and Condition of Spares: (Spares Shaft/Propeller) ....

....

16. General Condition of Engine Room: ....

17. Last Crankshaft Deflections Main/Aux. Engines: ....

....

18. Emergency Fire Pump: (Last Tested Date, Location): ....

....

E. Galley and store rooms

1. Generally: ....

F. Accommodation

1. Generally: ....

2. Fire Appliances: (Checked, date) ....

G. Navigation equipment (No., type)

1. Radars: ....

2. Echo Sounder: ....

3. Auto Pilot: ....

4. Direction Finder: ....

5. Gyro Compass: ....

6. Magnetic Compass: ....

7. Radio Station, VHF: ....

8. Other Navigational Aids: ....

9. Navigational and Signal Lights: ....

10. Navigational Reference Books: ....

11. Stability Book: ....

12. Trial Trip Records: (Crash stop, turning circle) ....

13. Loading instruments ....

H. General condition of Ship and Crew

1. Ship's Officers. Nationality: ....

2. Ship's Crew. Nationality: ....

3. Number of Crew: ....

4. Instructions, Manuals, Plans etc.: (Language) ....

5. State of Maintenance: ....

6. Training courses for crew etc. (STCW-Convention) ....

....

   a) Fire fighting and prevention ....

   b) Radar training ....

   c) Automatic radar plotting aid ....

   d) Dangerous cargo, Safety man. ....

7. Officers' knowledge of English ....

8. Any Other Relevant Information: ....

Note: Attending surveyor should try to obtain a copy of the ship's particulars, if available.

Place .... Date .... Signature ....

GUIDELINES FOR CONDUCTING INSURANCE WARRANTY SURVEYS FOR OFFSHORE INDUSTRY TOWAGE

S. K. Morgan
Noble Denton International Ltd

THESE GENERAL GUIDELINES (based on Noble Denton's past practice) are intended to cover some of the technical and marine requirements of the insurance warranty surveyor for the unrestricted ocean towage and transportation of items such as offshore platforms, modules and other industrial and marine equipment on barges which may be the subject of an insurance warranty. They are written to aid the selection and operation of tugs and barges, the suitability of seafastenings, and the suitability of the structural strength of the cargo, if applicable.

These guidelines are intended to assist a surveyor in issuing a certificate of approval as may be required by an insurance warranty clause. The surveyor should always ensure that he receives complete instructions from his client, including the warranty wording if at all possible. These guidelines are not intended to cover all towage methods. Surveyors must review and approve every towage on its own merits, having taken all factors into consideration, and these guidelines should never be used as a substitute for experienced surveying.

Warranty approvals should never imply that approval by designers, regulatory bodies, harbour authorities and/or any other party would be given. It is important to recognise that there are no universally accepted 'rules' which an assured *must* apply in order to satisfy a towage insurance warranty wording as described in the definitions section following. However, there may be other requirements specified within an insurance policy, such as maintenance in class.

Specific written recommendations should be prepared for each intended voyage. It is the responsibility of every surveyor to interpret correctly and apply these and any other guidelines to which he may make reference.

Definitions

Insurance warranty—For the purposes of these guidelines, an insurance warranty is defined as a clause within an insurance policy that requires the assured to seek the approval of an independent surveyor for the particular operation that is being insured. The underwriter will frequently name either a recognised firm of surveyors or a particular individual. A simple warranty wording for a conventional towage may be: *'Warranted (name of surveyor) to approve tug, tow and towage arrangement.'* This will be increased in scope for more complex items such as large offshore structures and will also frequently state that: *' . . . all surveyor's recommendations to be complied with.'*

Bollard pull—'BP' = certified continuous static bollard pull, tonnes.

Breaking load—'BL' = certified minimum breaking load of wire rope or shackles, tonnes.

Cargo—Where the item to be transported is carried on a barge or a ship, it is referred to throughout this report as the 'cargo.' If the item is towed on its own buoyancy, it is referred to as the 'tow.'

Towage—The 'towage' is defined as the operation of transporting a 'tow' or a 'cargo' on a barge by towing it with a 'tug.'

Tow—The 'tow' is defined as the item being towed. This may be a barge carrying a 'cargo' or an item floating on its own buoyancy. Approval of the 'tow' will normally include consideration of condition and classification of the barge; strength; securing and weather protection; draft; stability; documentation; emergency equipment; lights, shapes and signals; fuel and other consumable supplies; manning.

Towing arrangements—The 'towing (or towage) arrangements' are defined as the procedures for effecting the towage. Approval of the 'towing arrangements' will normally include consideration of towlines and towline connections; weather forecasting; pilotage; routeing arrangements; points of shelter; bunkering arrangements; assisting tugs; communication procedures.

Towline connection strength—'TC' = ultimate load capacity of towline connections, including connections to barge, bridle and bridle apex, in tonnes.

Tug—The 'tug' is defined as the vessel performing the 'towage.' Approval of the 'tug' will normally include consideration of the general design, classification, condition, towing equipment, bunkers and other consumable supplies, emergency and salvage equipment, communication equipment, and manning.

Towline pull required—'TPR' = towline pull computed to tow or hold the tow against a defined weather condition, in tonnes.

Ultimate load capacity—'Ultimate load capacity' of a wire rope, chain or shackle is the certified minimum breaking load, in tonnes. The load factors allow for good quality splices in wire rope. 'Ultimate load capacity' of a padeye, clench plate, delta plate or similar structure is defined as the load, in tonnes, which will cause general failure of the structure or its connection into the barge or other structure.

Safe working load—'SWL' is the load rating of the item of equipment as specified by the relevant competent organisation. Caution is recommended when relating SWL to ultimate load capacity, as the relevant 'factor of safety' may vary from application to application (typical values being from 3 to 5).

Barge documentation

Copies of the following documents should be carried on the barge or (lead) tug for any towage:

(a) Certificate of class issued by a recognised classification society.
(b) Certificate of registry.
(c) Tonnage certificate (if not incorporated in other certification).
(d) Certificate/approval of navigation lights and shapes issued by a recognised authority.
(e) International load line certificate.
(f) Customs clearance.
(g) Deratisation certificate, or exemption.
(h) Surveyor's certificate of approval for the particular towage (issued on sailing).

For manned tows the following should also be carried:

(i) Life saving apparatus (LSA) and fire fighting apparatus (FFA) certificates.
(j) Crew list (as supernumerary tug crew or barge crew).
(k) Radio licences (as required).
(l) Permission from certifying authority to man barge.

Barge towing gear

Towage should be from the forward end of the barge via a suitable bridle. The components of the system are: towline connections, including towline connection points, fairleads, bridle legs and bridle apex; intermediate pennant; bridle recovery system (with winch or with pennant); and emergency towing gear.

If two balanced tugs are to be used to tow the barge, then the bridle should be split and the tugs should tow off separate bridle legs, via intermediate pennants. A recovery system should be provided for each leg of the bridle.

The minimum strength of the towline connections should be not less than that shown in the table given later in this chapter. It should be noted that the above requirement represents the minimum value for towline connection strength. It may be prudent to design the main towline connections to allow for the use of tugs larger than the minimum required.

The breaking load of shackles forming part of the towline should be at least 10 per cent greater than the breaking load of the towline to be used. The breaking load of shackles forming part of the bridle should be at least 10 per cent greater than the required breaking load of the bridle. If the breaking load of a shackle is not known, then the safe working load (SWL) should be not less than the continuous static bollard pull (BP) of the largest tug proposed.

Towline connection points

Towline connections to the barge should be of an acceptable type. Preferably they should be able to be released quickly under adverse conditions, to allow a fouled bridle or towline to be cleared, but must also be secured against premature release. Sufficient underdeck strength must be provided for all towline connection points. They should normally be sited at the intersection of transverse and longitudinal bulkheads in order to transfer the load to the barge.

Where fitted, fairleads should be of a suitable type, located close to the deck edge. They should be fitted with capping bars and sited in line with the tow connections. Where the bridle can bear on the deck edge, the deck edge should be suitably faired to prevent chafe to the bridle.

Each bridle leg should be of stud link chain or composite chain and wire rope. If composite, the chain should of sufficient length to extend beyond the deck edge and prevent chafing of the wire rope. The length of each leg should be such that the angle at the apex is between 45 and 60 degrees. The end link of all chains should be a special enlarged link, not a normal link with the stud removed. All wire ropes should have hard eyes.

The bridle apex connection should be a towing ring or triangular-shaped plate. This is often called a delta flounder or monkey plate.

Intermediate pennant

An intermediate wire rope pennant should be fitted between the main towline and the bridle or chain pennant. Its main use is for ease of connection and reconnection. All wire rope pennants should have hard eyes. A synthetic spring, if used, should not normally replace the intermediate wire rope pennant. All synthetic springs should have hard eyes.

The length of the wire pennant for barge tows is normally 10-15 metres, since this can be handled on the stern of most tugs without the connecting shackle reaching the winch. Longer pennants may be needed in particular cases. The breaking strength of the wire rope pennant should not be less than that of the main towline with the possible exceptions in the paragraph which follows. It should have hard eyes and be of the same lay (i.e., left or right hand) as the main towline.

Surveyors could approve a 'fuse' or 'weak link' pennant provided that: the strength reduction is not more than 10 per cent of the strength of the main towline; the resulting strength of the pennant is at least 95 per cent of that required for the towline; and it forms part of a 'standard towing configuration' of the towing company involved.

Bridle recovery system

The bridle recovery system, type A with winch, consists of a winch and a recovery line connected to the bridle apex, via a suitable lead which may be an 'A' frame, roller or bolster plate. The recovery winch should have its own power source, or be manually operated, and should have an adequate barrel capacity for the required wire rope. If manually operated, it should be geared so that the tow bridle apex can be recovered by two men operating the equipment in bad weather and should be fitted with ratchet gear and brake. The winch should be capable of handling at least 75 per cent of the weight of the bridle plus attachments including intermediate pennant. It should be suitably secured to the barge structure.

This breaking load of the recovery wire, shackles, leads, etc., should be at least three times the weight of the bridle, apex and intermediate pennant. The wire should be at least 25 mm diameter. The recovery wire should be shackled on to the bridle apex, preferably to a padeye on the top or after side of the delta plate, or directly into the delta plate.

The bridle recovery system, type B with pennant, has an additional pennant, not less than 45 metres in length, as strong as the intermediate pennant. It is connected into the towing arrangement at the fore end of the bridle apex. This pennant is led back to the barge, being soft lashed to one of the bridle legs. It should be secured outside all obstructions along the deck edge with soft lashings every 3 metres, or metal clips opening outwards. The terminal eye should be located close to the barge side to enable it to be passed to the tug. A messenger line should be available to assist in this operation.

Emergency towing gear

Emergency towing gear should be provided in case of bridle failure or inability to recover the bridle. Preferably it should be fitted at the bow of the barge. It may consist of a separate bridle and pennant. Precautions should be taken to minimise chafe of all wire ropes.

The emergency system consists of: towing connection on or near the barge centreline (over a bulkhead or other suitable strong point); capped fairlead (if required); emergency pennant, minimum length 80 metres, with hard eyes preferably in one length—this length may be reduced for small barges and benign weather areas; extension wire (if required) long enough to prevent the float line chafing on the barge; float line, to extend 75-90 metres aft of stern; and a conspicuous easily handled plastic buoy trailing astern.

The strength of the first three items in the previous paragraph should be as for the main towline. The breaking load of the handling system should not be less than 25 tonnes, which must be sufficient to break the securing devices. If the emergency towing gear is attached forward, it should be led over the main tow bridle. It should be secured to the edge of the barge deck outside all obstructions, with soft lashings every 3 metres, or metal clips opening outwards.

If the emergency towing gear is attached aft, the wire rope should be coiled or flaked near the stern so that it can be pulled clear. The outboard eye should be led over the deck edge to prevent chafe of the float line. The type B recovery system should not be accepted as, or be connected to, the emergency tow gear as it relies on the main bridle connections.

Strength of barge, cargo and seafastenings

The strength of the barge, cargo and seafastenings should be of sufficient to withstand the loadings resulting from the design environmental conditions, including: loads resulting from vessel motions; loads resulting from the design wind; loads resulting from direct wave action such as slamming, water on deck and other effects of immersion; and the effects of barge flexibility.

Standard configurations

In general, for standard configurations and subject to satisfactory marine procedures, the requirements of the first two loadings mentioned above will be satisfied by the application of the loads resulting from the following table:

Type	Single amplitude (in 10 sec full cycle period)		
	Roll	Pitch	Heave
Small cargo barge (less than 76 metres loa or 23 metres beam)	25°	15°	0.2g
Larger barges	20°	12.5°	0.2g
Small vessels (see note 6)	30°	15°	0.2g

Notes:

(1) The roll and pitch centres are assumed to be at the waterline.

(2) Loadcases should normally combine: Roll +/− heave; Pitch +/− heave.

(3) Negative values may be as critical as positive values, and particular attention should be given to the generation of uplift forces.

(4) No allowance should be made for friction to reduce seafastening forces as this cannot easily be calculated nor can the value be guaranteed.

(6) 20 per cent should be added to the loadings resulting from pitch motions for small vessels to cover the effects of slamming.

In order to satisfy the requirements of the second two loadings, the following should be considered: the nature of the cargo geometry and position; the nature of seafastening philosophy and proposed ballasting sequences; and the design storm.

For inland towages where wave effects are expected to be negligible, the requirements will normally be satisfied by considering a 0.1g heave acceleration combined with the more severe of: a 5 degree static angle in either roll or pitch direction; or an angle of inclination resulting from the worst one-compartment damage condition, plus wind induced inclination and static wind loads resulting from the wind velocity specified in the damage stability section in this chapter.

Alternatively, the requirements may be satisfied by detailed motion response calculations to determine design loadings.

Design storm

The design storm for the towage should be the 10-year return period monthly extreme storm for the towage route, reduced as appropriate for exposures of less than 30 days. The design storm is *not* used for calculating bollard pull requirements, but is to be used in any calculations of motions which will be used in turn for strength calculations. Alternative methods of computing motions may be acceptable.

Free surface corrections to reduce metacentric height (GM) and hence to increase natural roll period will not generally be approved. The effect of any reduction in GM must, however, be considered in intact and damage stability calculations, and when computing wind induced heel or trim.

The overall strength of a classed barge, operating within its classification limitations, should normally be accepted. Calculations should be submitted where necessary to demonstrate that static and dynamic loadings from the cargo and seafastenings are suitably distributed into the barge structure and that the ballast distribution is satisfactory.

The stress levels should be in accordance with a recognised standard. Fatigue should be considered for long towages and for designs which are liable to suffer fatigue damage.

Non-destructive testing

Non-destructive testing (NDT) should be carried out on the primary structural members of the cargo and seafastenings, unless seafastening design stresses are very low. NDT may be by magnetic particle inspection (MPI), dye penetrant, ultrasonic or any other suitable method.

The following is a guide to the extent of NDT recommended: in all cases, up to 20 per cent of all welds, depending on design stress levels and welding quality; where higher allowable stresses have been allowed, up to 40 per cent of all welds; for critical areas, or where welding quality is poor, then 100 per cent inspection may be required.

Seafastenings should be designed to accept hog, sag and twisting of the barge in a seaway. The use of chains or wires should not be considered for unmanned tows. Even for manned tows their use should be carefully considered and avoided if possible except when they form part of a standard configuration for the vessel under survey.

All cargo should be protected from wave slam and wetting damage as appropriate. This may require provision of breakwaters or waterproofing of sensitive areas. Internal seafastenings may be needed to prevent items moving inside structures or modules.

Stability, draft and trim

The range of intact static stability about any axis should be not less than the angles shown in the table. The righting arm should be positive throughout this range.

Type	Intact Range
Small cargo barge (less than 76 metres loa or 23 metres beam)	40°
Larger barges	36°
Small vessels	44°

Alternatively, if maximum amplitudes of motion for any specific towage can be derived from model tests or motion response calculations, the minimum range of static stability should be not less than $(20 + 0.8\theta)$ degrees, where θ = the maximum amplitude of motion in degrees about the axis concerned caused by the design seastate, plus the static angle of inclination from the design wind.

Any opening giving an angle of down-flooding less than $(0 + 5)$ degrees, should be closed and watertight when at sea, where θ is the angle defined in the previous table or above, whichever is the greater. In general, maximum watertight compartmentation should be maintained. Attention must be paid to any openings in bulkheads or burnt off deck attachments.

The area under the righting moment curve to the second intercept of the righting and wind overturning moment curves or the downflooding angle, whichever is less, should be not less than 40 per cent in excess of the area under the wind overturning moment curve to the same limiting angle.

The wind velocity taken for overturning moment calculations should be the design wind as defined previously. In the absence of other data, 50 metres/second should be used.

Damage stability

The tow should have positive stability with any one compartment flooded or broached. Minimum penetration should be considered to be 1.5 metres. Two adjacent compartments on the periphery should be considered as one compartment if separated by a horizontal watertight flat. The emptying of a full compartment to the damaged waterline should be considered if it gives a more severe result than the flooding of an empty compartment.

The area under the righting moment curve from the angle of loll to the second intercept of the righting and wind overturning moment curves or the down-flooding angle, whichever is less, should be not less than 40 per cent in excess of the area under the wind overturning moment curve to the same limiting angle. Wind velocity taken for overturning moment calculations in the damage condition should be 25 metres/second or the wind used for the intact calculation, if less.

The towage draft should be small enough to give adequate freeboard and stability and large enough to reduce motions and slamming. Typically it will be 35-60 per cent of hull depth, which is usally significantly less than the load line (maximum) draft. Where water ballast is used it should, where possible, be kept in full tanks, pressed up. Other tanks should be clean and dry.

Trim should be selected to give good directional control. It will typically be about 1 per cent of the waterline length, by the stern, and should be obtained, where possible, by the position of cargo. Where barges with faired sterns are fitted with directional stabilising skegs, it may be preferable to have no trim. However, allowance should be made for trim caused by the towline force.

Pumping and sounding

Pumps may be required for the following:

- Ballasting during and after loadouts.
- Restoration of draft and trim after discharge (especially at sea).
- Damage control, including counterflooding.
- Deballasting to reduce draft to enter port, or after accidental grounding.
- Trimming to allow inspection and repair below normal waterline.

The use of a compressed-air system will not be practicable for all these cases, especially if the barge is holed above the waterline. A compressed-air system should have a compressor on board, working into the permanent lines.

It should be possible to sound and pump into or from critical (generally outer) compartments in severe weather. The following should be provided:

- Pumping system.
- Watertight manholes, with 'top hats' if necessary.
- Sounding plugs, extensions and tapes or rods. An additional remote sounding system may be needed for compressed air ballasting systems.
- Vents to all compartments.

Pumping system

It is recommended that all barges have one of the following systems, able to pump into and from all critical tanks, in order of preference:

- Two independent pumprooms or one protected pumproom, as described below.
- An unprotected pumproom with an independent emergency system that can pump out the pumproom.
- A system of portable pumps. These may alternatively be carried on the tug.

Independent pumprooms should have separate pumps, control and access and each be able to work into all tanks. To be considered protected, a pumproom, and any compartment required for access, should be separated from the bottom plating by watertight double bottom plating, not less than 65 cm high, and from the outer shell by other compartments or cofferdams not less than 1.5 metres wide.

If portable pumps are used, then *either* they should be portable enough to be moved around the barge (and cargo) by two men *or* enough pumping equipment should be carried, so that any critical compartment can be reached. Each portable pump should be able to pump out from the deepest tank (with top hat installed). This requires submersible pumps for barges over about 6 metres depth, due to suction head. Portable submersible pumps must be able to fit through tank manholes.

The total capacity of the fixed or portable pumps should be such that any one wing tank (or other critical tank or pumproom) can be emptied in 4 hours for an unmanned barge, or 12 hours for a manned tow. At least two pumps should be provided, except where there is a protected pumproom.

Watertight manholes

If manholes to critical compartments are covered up by cargo then *either* alternative manholes should be fitted *or* cutting gear should be installed and positions marked for making access, and welding gear and materials carried for remaking watertight.

Where the barge is classed, the owner should inform the classification society in good time of any holes to be cut or any structural alterations to be made. Access should always be available to pumprooms and other work areas. Ladders to the tank bottom are required from each manhole position. Suitable tools should be provided for removal and refastening of manhole covers and sounding plugs. All manhole covers should be properly secured with bolts and gaskets, renewed as necessary.

A sounding plug should be installed in each compartment (in manhole covers if necessary) to avoid removing manhole covers. For tanks that will be sounded regularly, a sounding tube and striker plate are recommended. On all tows where the deck may be covered with water, the following should be provided:

- A portable top hat of 60 cm minimum height, that can be bolted in place. The top should be constructed to avoid damage to hoses and cables.
- At least one sounding tube extension of 60 cm minimum height, threaded so that it can be screwed into all sounding plug holes, should be provided on all tows where the sounding plugs can be covered by water.

All compartments connected to a pumping system should have a 6 mm diameter breather hole fitted. This will give audible warning or reduce pressure differentials in event of mishap. This breather hole can be drilled into the gooseneck of the vent or through the wooden bung used to close the vent. For short unmanned tows, the vents should be closed with the wooden bungs or with steel blanks. Alternatively the vents may be of an approved self-closing type, provided they are fully operational.

Emergency anchor system

All barges should have an emergency anchor system, always capable of holding the barge and cargo in gale conditions. (Note that many anchors fitted to classification society rules are only suitable for holding the barge without cargo.)

The weight of one anchor must be at least 1/10 of the towline pull required (TPR) for the tow with a maximum requirement of a 10 tonnes anchor. Larger anchors will be accepted. A high holding power anchor with anti-roll stabilisation is preferred. For a barge that will be used for many tows the anchor should be chosen for the tow with the largest likely TPR.

The normal minimum effective length of anchor cable required is 180 metres, preferably mounted on a winch. If the cable runs through a spurling pipe, or other access, to storage below decks, then the pipe or access should be capable of being made watertight. If there is no winch and space is inadequate to flake out a cable properly, a minimum length of 90 metres may be acceptable.

For cable on a winch, or capstan, which can be paid out under control, the minimum breaking load of the cable should be 15 times the weight of the anchor, or 1.5 times the holding power of the anchor if greater. For cable flaked out, to allow for the extra shock load, the minimum breaking load of the cable should be 20 times the weight of the anchor, or twice the holding power if greater. The last few flakes of cable on deck should have lashings that will break and slow down the cable before it is fully paid out.

Attachment of cable

The inboard end of the cable should be led through a capped fairlead near the barge centre line and be securely fixed to the barge. Precautions should be taken to minimise chafe of the cable. The breaking load of the connections of the cable to padeye or winch, and padeye or winch to the barge structure should be greater than that of the cable.

If there is no suitable permanent anchor housing the anchor should be mounted on a billboard at about 60 degrees to the horizontal. The anchor should be held on the billboard in stops to prevent lateral and upwards movement. It should be secured by wire rope and/or chain strops that can be released manually without endangering the operator. The billboard should normally be mounted on the stern. It should be positioned such that on release the anchor will drop clear of the barge and the cable will pay out without fouling.

If the anchor is mounted and secured at the stern, this will become the bow when anchored. Anchor lights and shapes must be positioned accordingly. For any system, it should be possible to release the anchor safely, without the use of power to release pawls or dog securing devices. If the anchor is held only on a brake, an additional manual quick release fastening should be fitted.

Mooring arrangements

The barge should be provided with at least four mooring positions (bollards/staghorns, etc.) on each side of the barge. If fairleads to the bollards are not installed then the bollards should preferably be provided with capping bars, horns, or head plate to suitably retain the mooring lines at high angles of pull. Suitable chafe protection should be fitted to the deck edge for low angles of pull. At least four mooring ropes in good condition of adequate strength and length, typically about 50-75 mm diameter polyprop or nylon, and each 60-90 metres long, should be provided for a sea passage. These ropes may be carried on the tug if they cannot be stored or secured on the barge.

Damage control

When the length and area of the tow demand it, the following equipment should be carried on the barge in suitable packages or in a waterproof container secured to the deck: burning gear, welding equipment, steel plate, caulking material, sand, cement, nails, wooden plugs—various sizes, wooden wedges—various sizes, hammers and other tools.

Lights, shapes and sound signals

The barge should carry the lights and shapes required by the International Regulations for Preventing Collisions at Sea (1972) amended (1983), and any local regulations. These will include:

Normal lights and shapes (underway)—The following should be on board at all times, and lit or displayed when appropriate: 1 port side light; 1 starboard side light; 1 stern light; 1 black (towing) diamond shape (when the length of tow exceeds 200 metres).

Difficult tow and not under command (NUC)—The following should be carried at all times and lit or displayed vertically on the signal mast at the towmaster's discretion: 2 red lights (all round—NUC); 2 black ball shapes (NUC).

Anchor lights—2 white (all round) anchor lights (1 only if barge is under 100 metres in length).

Fog signals (if barge manned)—1 bell (for use forward when barge is at anchor); 1 Norwegian horn or other suitable sound signal; 1 gong (only for barges over 100 metres in length, for use aft when barge is at anchor).

Navigation lights should be independently operated (e.g., from gas containers or from independent electric power sources). Spare mantles/bulbs should be carried, and fuel and power sources should be adequate for the maximum anticipated duration of the voyage plus a reserve. Note: The British M notice 795 states that 'it is most desirable that a duplicate system of lights be provided.'

Manning

Manning may be recommended for high value and difficult tows, where feasible, for the following reasons: inspection of barge, cargo and seafastenings, with maintenance, damage control and repair as necessary; reconnection of towlines, or dropping of anchor when necessary; and maintenance of barge lights and shapes.

The riding crew should number at least four people, including an officer or bosun. Between them they should have a knowledge of seamanship and engineering, including electrics. They should be familiar with the use of the burning, welding and pumping equipment provided.

The crew should be organised into a watch keeping system to ensure a continuous monitoring of conditions on board. They should be familiar with the barge, equipment and cargo and make frequent inspections. All barge compartments should be sounded daily where possible.

The barge position should be marked periodically on a chart on board. A barge log should be kept giving:

- Position, speed and heading.
- Wind and sea conditions.
- Barge motion.
- Soundings of compartments and conditions of barge, cargo, seafastenings, equipment, lights and shapes.
- Anything unusual.

The results of all inspections and reports of barge motion or events should be radioed to the tug at least twice a day.

Requirements if manned

Suitable accommodation should be provided with facilities for watchkeeping, communications, cooking, sleeping and sanitary requirements. These may be subject to national and local regulations. At least two portable VHF radios should be on the barge. They should have spare batteries or means of recharging and have frequencies on which they can communicate with the tug.

Lifesaving and fire-fighting equipment should be carried in accordance with international and national regulations for the safety of the number of people carried, and the documents listed earlier under 'barge documentation' obtained. Safety lines should be rigged for the use of the riding crew at all times. Inclinometers should be fitted for the monitoring of barge motions. The decks and access ways should be adequately lit but such lighting should not interfere with or be liable to be confused with the navigation lights.

Whether the barge is manned or not, there must be suitable access. This may include at least one permanent steel ladder on each side or stern, from main deck to below waterline level. They should preferably be recessed, back painted for easy identification, be clear of overhanging cargo and well faired off to permit access by inflatable dinghies. A pilot ladder on each side of the barge or over the stern, secured to prevent the ladder being washed up on deck, may be accepted as a short term alternative. A clear space should be provided, with access ladders if necessary, so that men may be landed or recovered by helicopter.

Tug selection

The tug used for any towage to be approved should be inspected by the surveyor before the start of the towage. They survey should cover the vessel, its equipment, machinery and manning. It is preferable to use oceangoing tugs with raised fo'c'sles, or well-found tug/supply or tug/anchor-handling/supply vessels.

The following in-date documents should be held on the tug:

(a) Certificate of registry
(b) Load line certificate
(c) Tonnage certificate
(d) Certificate of class (hull and machinery)
(e) Cargo ship safety certificate
(f) Certificates for life saving appliances
(g) Radio certificate
(h) Deratisation certificate
(i) Certificates for all tow wires, pennants, nylons, shackles, etc.
(j) Stability booklet
(k) Bollard pull certificate.

For ocean towages the recommended manning is 10 men, comprising:

1 Master (certificated and experienced in towing)
2 Watchkeeping mates (certificated)
1 Bosun
3 Seamen (including an experienced leading seaman)
2 Engineers (certificated)
1 Catering
—
10 Total
—

This crew may be reduced in benign weather areas, when the tow is manned and for short towages. For some old or large tugs a larger crew may be required to handle the gear and for maintenance. Government regulations for some countries may also require a larger crew.

Bollard pull requirements

In general it is recommended that the minimum towline pull required (TPR) should be computed for zero forward speed against a 20 metres/second wind, 5.0 metre significant seastate and 0.5 metres/second current acting simultaneously. This approximates to gale force conditions. For benign weather areas, these environmental conditions may be reduced.

Continuous static bollard pull of a tug should be in excess of the minimum towline pull requirement to take account of tug efficiency in heavy weather. Note that tug efficiency can lie in the region 0.3 to 0.8, depending on the size and configuration of the tug and the seastate considered. Where the tow is expected to pass through restricted areas or an area of continuous adverse currents or weather, a greater BP may be required.

In general, only tugs with towing winches should be approved. Towing hooks may be accepted in some special cases where towline recovery and shortening up can be achieved by hand hauling, capstan or winch, in a time comparable to that with a towing winch, and where there are adequate anti chafe precautions.

If only one drum is used for towing wire, the spare wire should be stowed on a reel or other spooling device, and be readily accessible even in heavy weather and in such a position that transfer to the main towing drum can be easily and quickly effected. The winch should be adequately secured to the tug. The end of the wire must be adequately secured to the winch drum, though a quick method of releasing the wire from the drum may be provided.

The tug should have a spare towline. It is recommended that the minimum breaking loads of the main and spare towlines and the ultimate load capacity of the towline connections to the tow including each bridle leg, should be related to the continuous static bollard pull (BP) of the tug as follows:

Towline breaking load (BL)

Bollard pull (BP)	Benign areas	Other areas
Less than 40 tonnes	BL = 2.0xBP	3.0xBP
40 to 90 tonnes	BL = 2.0xBP	(3.8—BP/50)xBP
Over 90 tonnes	BL = 2.0xBP	2.0xBP

Ultimate load capacity of towline connections (TC)

Bollard pull (BP)	Benign areas	Other areas
Less than 40 tonnes	TC = 2.5xBP	3.0xBP
40 to 80 tonnes	TC = 2.5xBP	(2xBP) + 40
Over 80 tonnes	TC = (2xBP) + 40	(2xBP) + 40

The minimum length of each of the main and spare towlines (L) should be determined from the European formula: L = (BP/BL)x1,800 metres. One full strength wire rope pennant which is to be permanently included in the towing configuration may be considered when determining the minimum length. For benign areas the minimum length may be reduced to: L = (BP/BL)x1,200 metres.

Synthetic springs

Where a synthetic spring is used, its breaking load should be at least 1.5 times that required for the main wire rope. As synthetic springs have a limited life due to embrittlement and ageing, it must be in good condition and protected from wear, solvents and sunlight. All synthetic springs should have hard eyes. A synthetic spring should not be connected directly to the bridle apex if twisting problems could occur.

If used, the synthetic spring should normally be between the main towing wire and the intermediate pennant. A synthetic spring made up as a continuous loop with a hard eye at each end is generally preferable to a single line with an eye splice at each end.

A spare pennant should be carried on the tug or barge, depending on the method of reconnection.

The breaking load (BL) of any shackle in the tow should be at least 10 per cent greater than the BL of the towline (or *required BL* of the bridle if the shackle is part of the bridle). All tugs should carry adequate spare shackles of sufficient size, as well as a number of smaller ones. Split pins, seizing wire and lead for securing shackle pins should be carried. All tow lines and pennants should have hard eyes or sockets.

Where a towing tailgate or stern rail is fitted, the radius of the upper rail should be at least 10 times the diameter of the tug's main towline, and adequately faired to prevent snagging.

Towline control

Where a towing pod is fitted, its strength should be shown to be adequate. It should be well faired and the inside and ends should have a minimum radius of 10 times the towline diameter. Where no pod is fitted, the after deck should be fitted with a gog rope, mechanically operated and capable of being adjusted from a remote station. Where the anchor-handling wire is used as a gog rope, a spare must be carried. On square-sterned towing vessels, it is preferred that mechanically or hydraulically operated stops, capable of being withdrawn or removed, be fitted near the aft end of the bulwarks.

A powered workboat must be provided, for emergency communication with the barge while under tow, and must have adequate means for launching in a seaway. An inflatable powered by an outboard motor may be acceptable, provided it has flooring suitable for carriage of emergency equipment to the barge.

In addition to normal authorities' requirements, the tug should carry portable transmitter/receivers—i.e., walkie-talkies—for communication with the barge when tug personnel are placed on board during an emergency. Suitable spare batteries or means of recharging them should be provided.

It is suggested that tugs be fitted with a searchlight to aid night operations and for use in illuminating the tow during periods of emergency, or malfunction of the prescribed navigation lights. On any tow outside coastal limits, the tug should carry portable pumps, suitable for the requirements outlined earlier, equipped with means of suction and delivery and having a self-contained power unit with sufficient fuel for 12 hours' usage at the pump's maximum rating.

Anti-chafe gear should be fitted as necessary. In particular the towline should be protected at towing pods, tow bars and stern rail. It is advisable that all tugs be equipped with burning and welding gear suitable for use by crew members in emergency.

The tug should have a reserve of fuel and other consumables of at least five days' supply for any proposed towage. If refuelling en route is proposed, then suitable arrangements must be made before the tow starts.

Tandem double tows

This section covers the following cases:

Double tow—two barges each connected to the same tug with separate towlines. One towline is usually longer than the other so that it hangs below the first barge.

Tandem tow—two (or more) barges in series behind one tug—i.e., the second and following barges connected to the previous one.

Two tugs (in series) towing one barge—where there is only one towline connected to the barge and the leading tug is connected to the bow of the second one.

More than one tug (in parallel) towing one barge—each tug connected by its own towline to the barge.

Compared with single tows, tandem and double tows have additional associated problems including those of manoeuvring in close quarter situations, and reconnecting the towlines after a possible breakage.

It must be recognised, however, that these practices are followed by a number of companies and have been successful on many occasions. A surveyor should always ascertain from the assured that the intention of such a tow has been declared to underwriters through his broker. It is strongly recommended that special care is taken if a surveyor is to consider the approval of such towages.

Double tows should usually only be considered as acceptable in benign areas, and for coastal tows with suitable shelter and covered by good weather forecasts. The tug should be connected to each barge with a separate towline on a separate winch drum. It should also carry a spare towline, stowed on a winch, or able to be reeled on to a winch at sea.

Tandem tows are normally only acceptable in very benign areas or in ice conditions where the barges will follow each other. In ice conditions the towlines between tug and lead barge and between barges will normally be short enough for the line to be clear of the water. Care must be taken to avoid barges over running each other or the tug.

The tugs (in series) towing one barge are usually only feasible when a small tug is connected to the bow of a larger, less manoeuvrable tug to improve steering.

This configuration is generally only acceptable if (a) all the towing gear (towline/pennants/bridles/connections etc) between the second tug and the barge is strong enough for the total combined bollard pull; and (b) the second tug is significantly heavier than the leading tug (to avoid girding the second tug).

Multiple tugs to one barge are generally considered good practice, provided that each tug has a separate towline to the barge (via bridles or pennants as required). Care must be taken that the tugs do not foul each other or their towing gear. Consideration should be given to matching the size and power of the tugs. The use of eccentric bridles may be advantageous but care must be taken to avoid chafe. Normally there will be not more than three tugs.□

Wet patches on a dry hull, indicative of fastening problems.

Rust weeps from seams on a carvel hull, indicative of corrosion to fasteners and keel bolts.

Carvel hull in poor condition—planks split, seams open, and poorly shored up in boatyard.

SURVEYING YACHTS AND PLEASURE CRAFT

Captain K. Millett, MNI

Bureau Maritime Ltd, Marine Consultants and Surveyors

ON ANY FINE SUMMER WEEKEND or public holiday the rush to the sea is irresistible and UK coastal waters, estuaries and rivers become a teaming mass of colour and spectacle as many thousands of keen boat owners and their crews rush to get afloat. It is probably true to say that at such times the number of keen amateurs afloat outnumber the true professionals whose whole life and income is obtained from the sea in many diverse ways. The experience of amateur sailors spans the complete spectrum of ability from the totally incompetent, to the dedicated professional type, who is highly capable.

The craft they sail in cover an equally wide range, from dinghies costing a few hundred pounds to ocean-going luxury yachts costing tens or even hundreds of thousands of pounds. The condition of these craft also varies greatly, some having been home built, and others built by reputable yards. The majority will suffer from a lack of maintenance to some degree, as maintaining even a small boat is a time consuming job, not to mention expensive, and the urge to spend the season sailing and leave the maintenance until next year is very real.

Is it any wonder that the insurance underwriter, who will be asked to cover the risk of loss or damage to this diverse assortment of craft, operated and maintained by an equally diverse group of people, has trouble sleeping at nights? In the midst of all this, the one person the underwriter or public can turn to for advice and guidance is the surveyor of yachts and small craft.

Professional experience

The surveyor, like his big ship counterpart, is a professional in his field, with many years of experience in all kinds of construction and materials to draw on. Many surveyors are also designers of small craft and who better to have knowledge of all the detail that goes into the construction of what can be a very complicated sailing machine or high powered motor cruiser.

It is not just the nuts and bolts of the construction that has to be taken into account when surveying, but also those design aspects which will affect stability and performance which will of course in turn have an effect on the shape of the hull, its construction and its rig or power units. In recent years, the quest for even greater performance has led the marine industry to question whether some trade-off in the field of safety—i.e., stability or constructional strength—has taken place in the quest to obtain maximum performance under the various racing rules. The surveyor who also has experience as a naval architect or designer is best able to appreciate such problems.

This is not to say that all surveyors are designers, far from it. Many surveyors have spent years working in boatyards gaining an untold wealth of practical knowledge which is invaluable when carrying out condition surveys. After all, before anyone can pass an opinion on the structural condition of a craft they must know in detail how it was put together in the first place and equally the method of repair, should any damage or degradation to the structure occur. There is no substitute for experience in the surveying of small craft.

Relatively few surveyors in this field come from the big ship field of surveying and the reasons for this are obvious. Dealing with survey work on yachts and pleasure vessels is a completely different field requiring different expertise, knowledge and standards to that required of the ship surveyor. Small craft owners generally consider, quite rightly, that the man who surveys huge bulk carriers and tankers will not have the detailed experience and knowledge required to carry out a full survey on a 30-ft yacht.

The big ship surveyor is concerned basically with only one material, steel, and one method of propulsion, an engine driving the propellor. Whilst there are many variations in this field, the yacht surveyor has even more. He is concerned with varying substrates such as wood, FRP, steel, aluminium or ferro concrete. Each of these items can again be subdivided into different groups. The numerous types of timber available and used in boat construction vary greatly and it is the individual type of timber's properties that generally govern its specific use. Equally, boats built in different areas of the world will be constructed with different kinds of timber, some of which may be unsuitable if used outside their intended area of operation.

Fibre reinforced plastic (FRP) can also be subdivided into a number of categories depending on the properties required. Glass reinforced plastic (GRP) is the most common and well known, but other fibres such as kevlar and carbon fibre are used in the construction of such composite materials and may be referred to as carbon fibre reinforced plastic. Again, all these fibres can be subdivided into various types, each with individual properties and it is the job of the designer to decide how best to use these materials in order to meet the requirements of the craft's design. In order to understand the laminating of these fibres into a composite material, a basic knowledge of chemistry is required and though there is not space to go into the subject here—indeed, many complete books and papers have been written on the subject—the small craft surveyor must have a knowledge of the subject as inevitably the problem will arise at some time during the course of a survey.

Having mentioned the variation in the groups of materials used in small craft construction, again, there are wide and very different methods of

Cold moulded hull on which moisture penetration has caused timber seams to open. Owner has attempted to arrest situation by caulking seams but this is unsatisfactory and further deterioration will follow. Such a hull should contain no caulking, the hull laminates being held and sealed purely by the adhesive resin.

A carvel hull which has been sheathed with glassfibre. The GRP has come away from the planking due to moisture in the timber and the sheathing has been holed and torn. Note caulking cotton hanging out and the absence of stopping in the seams. Not a satisfactory method of making an old hull watertight.

construction using the same material. Yet again, it will be the designer who decides which method of construction best suits his requirements, but the surveyor must have a detailed knowledge of each of these methods in order to advise on the condition of a craft or on the method of repair of any damage.

Propulsion

Propulsion is not merely a question of sail or power but a knowledge of the various rigs and power units available. Account also has to be taken of their construction, installation and, most importantly, condition at time of survey, particularly as it may be up to 100 years or more since some of them were originally installed. Not everyone's pleasure craft was launched at the last boat show; some people spend years renovating with loving care wrecks that have lain in the mud forgotten for decades, restoring them to pristine condition. Usually the assistance or advice of a professional surveyor will be involved somewhere along the line in the reconstruction. It is evident therefore that the work of the small craft surveyor is fundamentally different from that of his commercial shipping counterpart. Whilst both are experts in their chosen field the likelihood of them exchanging roles and being able to carry out the other's work with complete confidence and knowledge without considerable retraining is extremely unlikely. Small craft surveyors will always consider themselves the ultimate specialists in their field, a fact of which they take great pride.

Having given an insight into some of the main differences between ship and small craft surveyors and their position in the marine world it's time to look at some of the more practical aspects of the subject and the materials they work with. With such a large subject only very basic information can be given here as numerous books have been written on boat-building, repair and surveying over the years. Many of these books are updated as new methods and systems are devised. As with any major industry, that of the yacht and pleasure craft is continually improving its product as new technology becomes available. It is not just the builders who are changing, but the suppliers of electronics, paint coatings, engines, and lifesaving appliances who are always striving to update their equipment and the modern surveyor has to keep abreast with all these changes and advances in technology.

Timber

As a boatbulding material timber has been used for centuries and even today with modern methods of construction it is still a very competitive material. Its main drawback is that it requires a skilled labour force to construct such craft. It is a time-consuming operation, so that labour costs are high when measured against those of other materials more suited to mass production methods. There is, however, nothing quite like a wooden boat, though as any owner of one will tell you, they require a great deal of time and effort to keep them in first class condition.

Traditional methods of timber hull construction are the clinker and carvel types which use mechanical fastenings such as copper nails, screws or iron spikes to hold the structure together. The advent of reliable adhesives has enabled modern-day craft to be built by cold moulded construction, using no mechanical fastenings at all and, using the double diagonal method, a strong lightweight hull can be built on a mould without any internal framing.

With the clinker-built boat, each plank overlaps, giving a very visible effect to the run of the planking. The upper outside edge of each plank is bevelled so that the next plank can be laid flat against it. This overlap is called the land and remains constant along the length of the plank except at the bow and stern. The planks, which are normally of mahogany, spruce, cedar or pitch pine, are secured together along their length by copper nails or rivets. The overlapping of the planks gives added longitudinal strength so that stringers are not required internally. Frames are usually steam bent, overlapping and fastening at the hog, planking being fastened to the frames with copper rivets. No caulking is used, the planks swelling as they absorb moisture to give a tight seam.

When surveying a clinker-built boat, special attention must be given to the frames, particularly at the turn of the bilge and the hog, where fractures and localised rot are likely to be found. If the planking is ageing, it can become brittle and fractures along the edge of the plank running down a line of fastenings can occur. Such damage invariably requires a new plank to be let in. Check the ends of the boat for caulking, as this may indicate that the plank ends are beginning to spring out. It may also be evidence of a crack developing between the stem and the apron which the presence of caulking will surely not make watertight.

Galvanic action

All hulls containing dissimilar metals in their construction will be the subject of galvanic action. Whilst the fitting of sacrificial anodes can control this problem, it can still lead to further problems. Fastenings such as copper are often more noble in the galvanic series than other metals used in the hull, such as a cast ballast keel. This can cause a galvanic couple to be set up between the two with the fastening as the cathode and the ballast keel as the anode. Whilst it is the anode which suffers corrosion the timber in contact with the cathode fastening can become very alkaline leading to decay and loss of strength over a period of time. Such deterioration is termed 'electro-chemical decay,' and is better known as nail sickness. As the timber softens, the fastenings become slack, leading to movement and leakage. The usual remedy for this is to place another fastening in the vicinity in sound timber to take over the job of the weakened one, though in serious cases larger fastenings or even replacement of timbers may be necessary.

Carvel hull

The carvel-built hull consists of flush laid planks fastened to internal frames. The planks are caulked with cotton or oakum and stopped with a sealant to make the joint watertight. On the more upmarket carvel hulls, the plank seams are splined by coating a tapered section of batten with glue and then driving it into the gap between the planks. Once the glue has set

Damaged keel stub—note cracking in GRP laminate at bolt holes, a common problem with GRP hulls fitted with bilge keels.

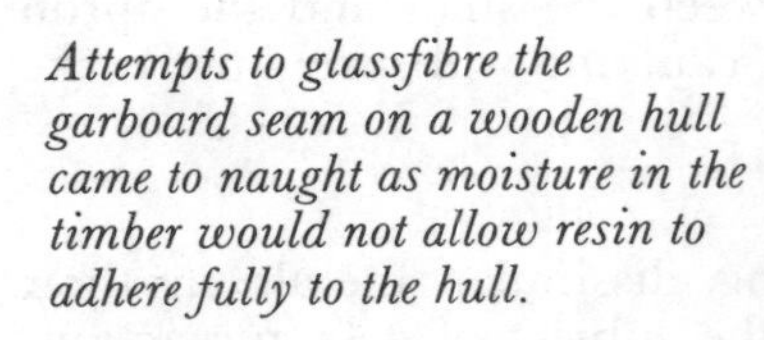

Attempts to glassfibre the garboard seam on a wooden hull came to naught as moisture in the timber would not allow resin to adhere fully to the hull.

Surface cracking in a ferro cement hull.

the spline can be planed down flush with the planking to give a smooth finish.

A carvel hull is a composite structure and depends on the keel, hog, frames, planking and fastenings all being maintained in good order. Deterioration of any one part will lead to a loss of strength in the structure and additional stress being placed on other components if not promptly attended to. The keel or hog forms the central backbone of the craft, with the frames being attached to it. Keels and frames in the UK have been traditionally made of oak, a timber that has always been renowned for its strength and resistance to rot in the marine environment. Frames can be grown, sawn or laminated depending on choice. Grown frames are not readily available these days and sawn frames are therefore the usual choice.

A sawn frame is constructed of several pieces of timber known as futtocks. These may be paired with staggered joints to make a very strong frame of double futtock thickness. Equally the futtock ends can be butted and supported by a butt block type cleat over each joint. Laminated frames are generally found in the more modern craft when light weight and cosmetic appearance are required. The laminates are steamed and bent into shape with an adhesive resin between each layer. This structure is then held tight by clamping in a jig until the adhesive has cured, when the laminate can be finished by planing to final shape. It is not unusual to find a mixture of all three types of frames in a carvel hull.

The frames are secured at the base to the hog and at the upper end are all connected to the beam shelf which runs along the top inboard edge of the frames. As its name implies the beam shelf supports the ends of the transverse deck beams. When surveying a hull, great care should be taken around the deck beams and beam shelf as in these areas small unseen leaks through the deck planking can lead to rot in the top of the beams which cannot be sighted from inside. If rotting in the top surface of a deck beam is suspected then a hole should be bored in the timber up into the suspect area and the shavings from the timber removed and examined.

Dry and wet rot

It should at this stage be emphasised that there are two kinds of rot. Dry rot is relatively rare but when part of a hull is infected it can spread at an alarming rate. Symptoms of the problem are cracking both along and across the grain of the timber and a musty smell. There is never any doubt about the problem, as a spike will sink straight into the infected area with little pressure.

Immediate action is required once dry rot is found by cutting out all infected wood and that immediately adjacent to it. All such debris must be burned immediately to prevent further spreading of the spores. Even a few grains of infected sawdust falling into the hull can cause the problem to start up again somewhere else. Rot spores can be carried on the wind from boat to boat in a yard, or more likely, on the soles of the shoes of one of the yard workers. For this reason trays of disinfectant are often placed at the foot of the access to craft where rot has been found. The treatment of rot must be swift and decisive; delay will only mean a larger area has to be treated and this can involve major surgery in a wooden hull.

Wet rot is often found where there are leaks above the waterline particularly in the deck and cabin tops where minor leaks have remained undetected. Any area where there is a lack of ventilation will be susceptible to rot and good ventilation through all corners and lockers in a hull is essential. Wet rot again appears as cracks along and across the grain giving a crazed effect often with thin brown strands. There will be very little resistance when tested with a spike or other sharp instrument. Again, all infected timber must be removed and the area coated liberally with a rot preventive.

Rot is seldom found in bilges as saltwater is a mild antiseptic, though this cannot be relied upon 100 per cent as freshwater streams usually run to the sea via harbours and estuaries. Timber with a moisture content of less than 20 per cent is unlikely to decay through rot, and the drier the timber can be kept by good ventilation the less the risk.

Keelbolt inspection

Another item to which close attention should be paid in craft of all substrates is the keelbolts. In the traditional wooden hull these pass up through the ballast keel through the hog and are normally secured with nuts with good sized washers under the heads to spread the load. In GRP hulls the concept is basically the same, though this type of construction also allows for moulded keels to be fitted so no keel bolts are required. Many craft these days are fitted with bilge keels to allow them to remain upright when drying out on a mooring and such keels normally have short studs threaded into the top which then pass through the keel stub to be secured internally in a suitably strengthened location.

Sample keel bolts should be removed for inspection from time to time in order to ascertain any corrosion in this vital component. In the case of bilge keels secured with studs, the keels will have to be dropped off with the craft suitably blocked up on timbers or a cradle. Cases of keels falling off are, however, extremely rare, but corrosion of the bolts can lead to other problems. Bolts can be constructed from a number of materials—e.g., mild steel, stainless steel or bronze—though none of these metals is ideal for the purpose. The area most susceptible to attack is where the bolt passes out of the hull and into the keel. If a watertight seal cannot be kept between the hull and the keel, water will penetrate and act as an electolyte, thus setting up a galvanic cell between the bolt and the keel or some other item of metal in the hull. Equally, impurities in the bolts themselves can set up small cells all over the surface of the bolt. This action is continuous and soon leads to extensive pitting and corrosion.

Even stainless steel is not exempt from this action, as such material is basically an alloy of iron, chromium and nickel. The chronium acts with oxygen very rapidly to form an oxide film which is very durable and impervious to further attack. However where the keel bolt is embedded in the wood or GRP hull there is not a ready supply of oxygen to keep this protective coating intact and this leads to crevice corrosion of the material. Once this has started and the protective coating has been broken, pitting in the normal way starts up in the crevices and leads to further damage to the material. It should be pointed out that there are numerous types of stainless steel on the market, but those most usually associated with marine work are of the austenitic type and commonly referred to by their British Standards or American Iron and Steel Institute numbers. Types 316 and 304 are the most common marine grades, 316 being better than 304.

Double skin planking

A type of construction favoured in the 1940 era, particularly by the military services, was that of the double skin planking. With this method two layers of planking were used, running between 45 degrees and 90 degrees to each other. A membrane of calico or muslin soaked in boiled linseed oil was placed between the two layers of planking to provide a waterproof membrane. In later times this membrane was soaked in resorcinal or epoxy glue to bond the planks together. Planks are cut to fit tightly and no caulking is used apart from at the garboard strakes and the stem.

The older double diagonal construction can be a problem unless it is properly maintained. Any leaks between the outer planking will probably travel along between the planks and enter the hull elsewhere. Thus tracing and stopping leaks is difficult with these hulls. Over the years the membrane has dried out and rotted and any water leaking between the planks will lead to rot in both layers. Repairs then become expensive as, though its quite easy to reach the outer planking, a large area has to be removed in order to replace quite a small area of inner planking due to the overlap.

The principle of double diagonal planking has, however, taken on a new meaning with today's high-tech adhesives and resins, in the form of cold-moulded planking. By this method, a hull is formed by laminating several cross-diagonal layers of veneer over a mould or jig. The run of the veneers can be varied and if a varnished finish is required a final layer of laminates run fore and aft gives a pleasing effect. Each layer of veneers is cut and stapled into position before fastening the complete layer to that below with an epoxy resin. By this method layers can be built up to make an extremely strong hull entirely of timber and resin, without any mechanical fastenings. On completion of the hull, frames can be laminated or steam bent to provide additional strength, though because of the strength of the hull planking, the number of frames can be reduced or in some cases left out altogether. It is possible by this method to build a hull purely of timber and resin without any fastenings whatsoever.

Strip planked hulls

Another method of wooden hull construction is the strip planked hull. With a carvel planked hull each plank has to be spiled or shaped, each plank being wider amidship where the hull is more full and narrower at the stem or transom. With the stripped plank method the need for spiling is eliminated. The planking consists of approximately square cross section battens with one edge concave in section and the other convex. By this method the edges of the planks can fit together, the convex edge fitting into that of the concave. Glue is placed between the planks which are then edge nailed together using galvanised nails of length approximately two and a quarter times the width of the planks. Thus the nail passes completely through two planks and into the third.

As each plank is bent around the mould or frames it also has to be bent down into the preceeding plank. If no corrective measures are taken, the extra number of planks needed amidships to cover the wider girth of the hull will result in the planks running out along the deck edge at the forward and after ends of the hull. This is not a problem, the plank ends simply being cut off at the deck edge. If the hull is to be varnished such planking can look unsightly and spoil the line of the boat, although if it is to be painted no harm will result. It is possible to avoid this situation by building the hull upside down and planking from the sheer towards the bilge. Once the below-waterline area has been reached, feathered or tapered planks can be fitted to allow the planking to finally run out parallel at the keel, the garboard strake being fitted as for carvel planking.

In addition to the planks being nailed together edgeways, they are also secured with screw fastenings to the frames. This method of construction provides a structure which, due to the edge fastened and glued planks, allows fewer frames to be fitted internally.

Repairs to such planking, however, are far from straightforward as the edge nailing prevents the simple removal of individual planks. All internal hull fittings will require removal so that the damaged area can be cut out for rebuilding using similar strip planking and epoxy resin. Planks should be renailed edgeways where there is room, although inevitably this will not be possible with the final key planks. Once the planking has been replaced short ribs should be mounted across the planks internally and butt blocks secured over the joints and scarphs to provide backup and maintain local structural strength.

The final method of timber construction is that of the plywood hull. Such a hull is formed by cutting plywood panels to shape and then cold bending them over and securing them to the rigid frames and battens which form the hull framing. No caulking is used at the joints, epoxy resin or glue being used to seal any joints or seams. Only marine grade plywood should be used for boat building or repairs and this can be identified by the British Standard 1088 mark. Whilst delamination of marine grade plywood can be repaired, many boats have their decks and sometimes topsides constructed of inferior grade ply which is not so durable in the marine environment.

When surveying plywood boats great care should be taken in the examination of all edges and joins in order to see that no moisture has found its way into the material, thus weakening the joint or structure. Undulations on the surface of the plywood indicated by cracked or lifting paint indicate stressing of the material or possible delamination. A plywood hull is much lighter in both scantling dimension and weight, with frames being spaced much further apart than in a similar-sized carvel hull. It is therefore very necessary that the structure be maintained in first class condition in order to maintain its structural integrity.

Modern day technology has made it possible to coat the wooden parts of the hull individually with epoxy resin prior to assembling. The timber must be dry—i.e., below 10 per cent moisture content prior to coating—and the resin coat applied will act as a moisture barrier throughout the life of the hull.

Fibre reinforced plastic

Fibre reinforced plastic (FRP) is the method of construction which includes the well-known glass reinforced plastic (GRP), known to everybody as glass fibre. In recent years this method of construction has become very popular as it lends itself to mass production and the accompanying cost savings of large production runs. It was originally considered to be the ultimate maintenance-free material, but experience over the years has shown this not to be the case, although it is probably true to say that it needs far less maintenance than the alternative substrates.

For the designer it has the advantage that the material is actually formed during the construction, thus giving the opportunity to vary the thickness and strength in particular areas of the hull and form intricate shapes which would be imposssible with other materials without incurring considerable costs in time and labour.

Whilst the laying up of a glass fibre hull is a skilled job, it does not require the comprehensive skill and training of a shipwright. Additionally, the time taken to construct a hull in GRP is considerably less than that of timber or steel, so savings are made in both time and labour costs. GRP is a composite material consisting of a resin, reinforced with layers of glass material or cloth. The final lay up throughout the hull is decided by the designer.

The glass reinforcement itself comes in many types and weights so as to give the designer a full range of options in order to produce a strong but lightweight hull. These range from the chopped strand mat (CSM) through to the heavier cloths and woven rovings. Each comes in varying weights and weave patterns to give differing characteristics in use. The addition of specialised reinforcement, such as carbon fibre or Kevlar, can again provide different properties or characteristics to a particular area of a hull, reducing weight but providing the strength required. Such additions are usually confined to the specialised racing end of the market due to the high cost involved.

Polyester resins

Whilst a number of resins are available for use in laminating boat hulls, by far the most popular are those of the polyester type. These are generally easy to work with and of low cost for mass production methods. In order to make the resin gel and finally cure, the addition of a catalyst is required. Great care should be taken in mixing resin and catalyst as the addition of too much or too little will result in an incomplete cure of the finished product and a loss in its physical properties. Heat is given off during the curing process, so care should be taken by using large surface area mixing containers to allow the heat to dissipate. This should not be a problem in the laid-up hull as the large surface area will allow the heat to dissipate safely without any hot spots forming.

Temperature and humidity control of the workshop is important as this will affect the curing rate of the resin. An accelerator can be added to speed the curing process, or the resin can be purchased premixed with the accelerator. It is, however, vital that manufacturers' recommendations are carefully followed, as failure to do so will result in an incomplete cure.

The hull and deck will be moulded separately in either a male or female mould. In a female moulding the hull is laid up inside the mould and then lifted out leaving the outside of the hull in a smooth finished condition. In a male mould the hull is laid up over a plug. The outside of the hull will then require filling and sanding to final contours. The female moulding method is normally used for production line methods, while the male mould is generally used for the one-off design type of construction.

Sandwich construction

The majority of GRP hulls consist of a glass laminate made up of polyester resin and glass reinforcing material covered with a shining gel coat to finish on the outside. However, another type of construction is that of a sandwich method where two thin laminates of GRP are used to sandwich a balsa wood or foam core between. This method of construction gives a very lightweight but strong structure and is generally found, though not without exception, in performance craft. More generally it is found in the deck moulding of family cruisers where its light weight but strong properties stiffen the deck areas which undergo the most stress.

The main problems with sandwich construction is that of delamination or puncture. If the two laminates are not securely glued to the balsa core, then a weakened structure will result. With time, the continual flexing of the hull will extend the area of delamination, possibly resulting in structural failure. If one of the outer laminates is punctured, then moisture can enter the balsa wood core, causing it to expand and decay. This again will result in delamination and a weakening of the structure. In such cases the entire area of damaged and wet core will need to be replaced prior to repairing the damaged laminate. Great care should be taken in surveying hulls and decks constructed by the sandwich method as any delamination can spread and lead to weakening of the structure and possibly failure when under load or stress.

Impact damage

Any GRP laminate can easily be damaged by impact either internally or externally. This will normally show on the gel coat by a hairline crack. Such cracks generally become more visible as dirt becomes trapped in the surface break. It is the surveyor's job to establish whether these cracks are limited to the gel coat or extend through into the laminate.

It has not infrequently been known for fine hairline cracks on the gel coat to be the only evidence of what later turns out to be a major structural failure in the laminate once the cracked areas of the gel coat were removed. Surveyors, therefore, have to be most thorough in inspecting any damage caused to the hull, as what appears to be superficial damage can in fact hide much larger problems. Such problems may be the result of local stressing or flexing of the hull due to insufficient stiffening internally. It is therefore of little use recommending repair of the damaged area without removing the cause of the problem by the addition of stiffening within the hull.

If an impacted load is spread over a large enough area of the hull, the gel coat may remain apparently undamaged although delamination of the laminate internally could result. Such delamination will be caused by the glass material breaking away from the resin matrix, leading to a weakening in the structure. Many craft land heavily alongside the marina fendering in their life and all too frequently the internal damage to the laminate goes undetected. It is only when the surveyor is hammer testing the hull with a rubber mallet that he picks up the vibrating sound given off by the delaminated area. All delamination is serious and must be repaired without delay.

When GRP became the predominate boatbuilding material in the mid-sixties, it was acclaimed as being maintenance free and everlasting. However, this was found not to be the case when the initial gloss on the gel coat faded and chalked due to the action of sunlight and the harsh marine environment. Colour pigmented gel coats suffered particularly badly, resulting in the hull topsides being painted in order to maintain that pristine finish look. Once painted, however, it will require continual attention as the paint itself becomes scratched, worn and faded with time.

Osmosis

An even worse problem, however, is going on below the waterline in the form of degradation or rotting of the material. This is commonly termed as 'osmosis' in the UK or 'boat pox' in the USA. This term is used to cover a number of defects which can form in both the gel coat and the laminate due to immersion in water and usually manifest themselves as blisters in the gel coat. Many books and technical papers have been written on the subject of osmosis, which in itself is a complex problem, and it would be impossible to discuss the subject in detail here.

In its simplest form water permeates through the gel coat and into the many micro voids and air pockets left in the laminate. Within these voids remain very small amounts of acid or alcohol which has remained free from the curing process. The water gaining access to these chemicals causes a chemical reaction which leads to a build up of pressure within the void. Such pressure can be so strong that it will cause blistering to occur on the outside of the gel coat. Normally the boat owner will not appreciate that there is such a problem until the blistering reaches an advanced stage. The surveyor, however, should, by removing windows of anti-fouling and paint coating, be able to observe any gel coat disturbance at an early stage before any real damage to the laminate occurs. At its earliest stages it shows as little more than a nettle rash on the gel coat surface although it will be readily apparent to the practised eye.

The problem of osmosis is to be found in hulls which have been afloat for five to seven years or longer. Treatments are available and a cure is possible in approximately 90 per cent of cases. The causes of the problem are extremely complex both in the chemical breakdown and the quality control of the original construction and materials. Many different resins and reinforcing materials are being experimented with in order to solve the problem but at the time of writing a long-term solution has still to be proven.

Blistering within the gelcoat or between the gel coat and the laminate does not in itself cause a major problem. The real problem occurs when all the voids and blisters link up in order to extend deep into the laminate. Here the build up of pressure between the layers of glass material can lead to delamination and possible failure of the structure.

Wicking

An even more potentially dangerous problem is that of wicking. If the gel coat is applied too thinly within the mould then there is always the possibility of the glass strands in the first layer of reinforcing mat penetrating into or even through it. When the boat is afloat water can then travel up the glass strands by capillary action, through the gel coat and deep into the laminate. The moisture will cause expansion and this will cause the glass strands to break free of the bonding resin. It follows that this can eventually lead to a serious structural weakness. The only remedy for such a problem being to grind the damaged area out until sound material is reached before rebuilding the area with epoxy resin and glass laminating materials.

Much research is being carried out into resin systems which reduce the chances of 'osmosis' occurring, but in the end it comes down to cost. It is generally considered that epoxy resins perform better in a marine environment than the polyester type but as they cost three to four times as much, such increased cost will be reflected in the finished product. The leisure craft market is extremely competitive and any increase in costs is not welcome. Generally, however, GRP craft do require less maintenance than those of other substrates, though they are certainly not to be considered maintenance free. It is a prudent owner who lifts his craft out of the water for the winter layup. After all, it cannot get osmosis if it's not in the water.

Steel craft

This is the material with which all professional mariners or big ship surveyors will be most familiar. However, in the leisure craft field it is probably the most underestimated of all the materials available. Everyone knows that steel rusts and this generally causes people to turn to either timber or GRP for their choice of craft. In the UK relatively few builders produce steel boats, but in Holland they have been producing them in large numbers for decades and the Dutch are the acknowledged masters of this type of construction.

Any small steel craft will ultimately be of a stronger construction than would normally be required for its size, due to the fact that scantlings are increased in size to allow for any corrosion that may take place.

An oceangoing ship will only be strengthened sufficiently to take the stresses imposed during normal service. This will inevitably require web frames and heavier plating and scantlings locally, with doublers and other reinforcements fitted in numerous positions. In the small yacht or other leisure craft this is not necessary as the material used will generally be over-strength for the job it is doing anyway. Steel plate of less than 3mm in thickness causes problems in welding and after allowing a margin for corrosion it is easy to see how an overstrength structure is built up. Apart from its stiffness a steel hull has a high degree of residual strength making it ideal for the most arduous conditions.

A steel hull will be totally watertight and can take severe punishment and remain so. When building in steel it is preferable to have a steel deck fully welded to the hull rather than a timber or plywood one as again total watertightness will be achieved. Any small leaks are the prime source of internal corrosion or rotting in the timber deck itself. If a steel hull is damaged it is relatively easy to cut out the damaged area and weld in new plate or frames. Labour and materials for carrying out such work are readily available and the work can be completed relatively quickly. This would not be the case with timber or GRP construction.

Inherent strength

Steel, of course, has inherent strength and unless a major incident occurs it is only likely to be bent or indented, thus keeping the hull watertight. The construction of a steel yacht is basically on big ship lines, although much more detailed thought will be given particularly to the internals as not only will space be at a premium but all efforts will be made to prevent any possible pockets where areas of corrosion can form. Also the finish must be of the highest standard and here special welding techniques will be used to keep the distortion of material due to welding to an absolute minimum. It may be acceptable to see the frame outlines on the shell plating of a fishing boat or commercial craft, but on a pleasure craft it is definitely not.

Minor distortions and indents can be filled with epoxy fillers to give a fair line to the hull, although this is no substitute for a correctly thought-out weld plan which will result in a hull that does not require any filling. The surveyor should be wary when examining a steel hull that filler has not been used to hide a defective or damaged area, as if properly finished off and newly painted it can be difficult to detect. In the long term, however, if there is any corrosion on the steel surface when the filler is applied it will eventually loosen, look unsightly and possibly fall out.

The main problem with a steel hull is corrosion and its effect on the integrity of the hull. The application of quality paint systems to the finished hull will be the first line of defence against corrosion. Naturally, such coatings must be kept in good order for them to be fully effective.

Cathodic protection

Dissimilar metals should be avoided where possible, as this will lead to electrolytic action and corrosion. It is essential that the below-waterline areas be protected with a planned and maintained cathodic protection system. If this is correctly done, below-waterline corrosion can be virtually eliminated.

Internally, the framework should be designed so that any moisture from any source can be carried directly to the bilge well where it can be pumped out. Lightning and drain holes in frames and floors should be designed so that no small puddles can form. A hairline crack in a weld or even a badly finished weld which has left sharp peaks protruding through the paint coating is sufficient to allow water in and start corrosion of the steelwork.

Corrosion is not necessarily a gradual operation; given the right conditions corrosion can act locally very quickly. Hulls must therefore be carefully inspected at frequent intervals both internally and externally and any areas where action has taken place be attended to immediately. Naturally the source of or reason for the corrosion should also be corrected or the problem will reappear at the next inspection.

If the zinc sacrificial anodes are disappearing at an alarming rate then the reason must be ascertained and the problem corrected. Though there is no hard and fast rule on how long an anode will last, generally, they will need changing every 12 months or so. If a craft is moored alongside a steel pier on which electric power is available, it may be due to small earths in the pier cabling entering the water down the piling and setting up a galvancic cell between the boat and the pier. Close attention should also be paid to the electrical installation on board, as any small earths from cable junction boxes or instruments will eventually lead to corrosion at some point in the hull.

Any suspect areas can be hammer tested if their structural strength or thickness is suspect. The modern method of checking a hull for corrosion is by non-destructive testing (NDT) using an ultrasonic meter. This will give an immediate reading of the thickness of the material. The less sophisticated models do not take account of layers of paint and rust already on the surface, requiring it to be cleaned off to bare steel before a reading can be taken. The old method using an electric drill and a plate thickness gauge is just as reliable though few owners like having holes drilled in their boats. The real problem then is that such holes have to be sealed by welding and this is then a natural point where paint coatings have been disturbed for corrosion to start.

Aluminium construction

Aluminium is another suitable material for the construction of leisure craft and an alternative to steel for those who still require a strong but lightweight hull. Steel is one and a half times as strong, but three times as heavy as aluminium. Because of the resistance of modern aluminium alloys to seawater corrosion, it is possible to construct the hull with thinner plate than would be possible with steels and thus produce a much lighter hull. The reduced thickness does, however, mean that the hull is more prone to physical damage than a steel hull.

Aluminium can be welded by the MIG process or it can be bonded with epoxy resin, additional local strengthening being gained by the inclusion of aluminium pop rivets. An aluminium hull will not seriously corrode in saltwater provided it is not affected by any outside influences. The surface will pit and discolour, but this will stabilise once an oxide film is established over the untreated surface. The main drawback with aluminium is the cost, which can be more than ten times the cost of steel.

With an aluminium hull it is even more essential than with steel that the use of dissimilar metals is avoided in the construction, otherwise serious corrosion will result with surprising speed. Stringent precautions must therefore be taken to ensure that no electrolytic action can take place, as such action will lead to extensive corrosion in a comparatively short period of time. Anti-fouling paints containing copper should never be used on an aluminium hull as corrosion will quickly result. In fact with aluminium being less noble than copper in the galvanic scale corrosion will result due to electrolytic action if an aluminium hull is moored alongside another craft which has a copper-based antifouling or sheathing. Generally the welds will be attacked first, although the entire hull should be carefully examined for signs of corrosion.

Ferro-cement

This method of construction has been around for many years and used to be favoured by the self-build fraternity. Numerous part-finished and abandoned ferro craft are to be found, the builders having lost interest, run out of money, or just having realised that they had taken on more than they can cope with. As a method of construction it is well suited to those with little experience or expertise in boat building, but the construction of the steel and wire matrix is a long, time-consuming, and boring job, which causes many to lose heart before its finished.

The matrix which forms the hull shape and much of the internal stiffening such as floors and bulkheads is made up of steel rods, pipe and wire mesh. The entire basket then being impregrated and skimmed with cement. It is essential that the cement is pushed right into the mesh so that no voids are left to form a weakened area. The matrix is finally finished with a 3mm layer of cement to form a watertight barrier between the wire basket and the water. Decks can be built-in of ferro-concrete or a plywood deck can be laid on the ferro-cement hull.

Surveying a ferro-cement hull is difficult in that the surveyor can never be sure of the condition or construction both of the internal steel matrix or the quality and adhesion of the cement mix used. This becomes a particular problem when dealing with home-built craft, where the quality of both materials and workmanship can vary greatly. Insurance underwriters are particularly strict with their survey requirements on such a construction and a wise builder will have the hull inspected at regular stages throughout the construction so that a comprehensive and complete condition survey can be produced on completion of the craft. Such a survey will, of course, require updating every few years.

Cracking problems

The main problem to be found with a ferro hull is cracking in the thin surface layer of cement. Moisture can penetrate such cracks and cause corrosion of the internal steel matrix. Such cracks can be sealed after drying with an epoxy filler or resin. Larger cracks can be opened and dried before filling and sealing with a mixture of epoxy resin and ordinary Portland building cement. Where cracking has taken place on a ferro deck, care should be taken to keep such areas dry until a repair can be made, particularly during winter periods. Any moisture that gets into the crack can freeze and cause even greater damage and speed up possible corrosion of the reinforcing matrix. A ferro hull should have all metal fittings below waterline, such as rudder hangings, seacocks, etc., well protected from electrolytic action, so zinc anodes should be inspected regularly.

Checklists and insurance required

The subjects mentioned have only been briefly covered and an experienced surveyor will have his own checklist to follow to ensure that a complete and comprehensive survey is carried out. Most small craft surveyors operate as individuals and not as employees of a large company and their work is gained by their individual reputation or by recommendation from other satisfied clients.

However, with the increase of litigation in society today, it is inevitable that even in the yacht survey field claims will be made against a surveyor for damages should an owner consider than an error of judgement on the surveyor's part has resulted in a material or financial loss to such owner. It is therefore essential that surveyors be covered by professional indemnity insurance in order to meet the costs of any claim that may be successfully awarded against them.

Many organisations and individuals will not employ or recommend a surveyor who cannot produce evidence of such insurance. Failure to have such cover can mean that a genuine professional misjudgment can result in an extremely expensive negligence claim at the individual's door. Even if such claim is unusuccessful, it can result in expensive legal costs which will have to be incurred in order to defend such an action and protect the individual's reputation. To be sufficiently and effectively insured for such times is therefore very prudent and sensible planning as befits the reasoned investigative approach of a professional surveyor in the course of his work.□

PROFESSIONAL INDEMNITY INSURANCE

L. Ockendon
Assistant Director, Bowring Professional Indemnity Ltd

THE PROSPECT of having to 'buy' a Professional Indemnity policy and perhaps more importantly knowing whether the terms and conditions you intend to or have accepted are the most suitable for your particular 'service' is a daunting prospect to many. The advice given by or services provided for members of The Nautical Institute can vary considerably, and consequently so can their insurance requirements. The following notes are intended to provide basic details as to what a professional indemnity policy is intended to cover, why it is needed, and how you can seek such protection.

Why it's needed

A professional person is expected to exercise the skill and care of the ordinarily competent member of that profession. Precisely what is meant by this in some professions, of which this is probably one, can be very expensive to ascertain.

During recent years we have found ourselves in an increasingly claims-conscious climate. In view of this and the advent of the consumer protection age, the professions are now far more vulnerable to allegations of professional negligence. It should be borne in mind that the cost of defending such an allegation can be substantial, even though the allegations are ill-founded.

What is it?

There are a number of professional indemnity insurers and policy wordings that will therefore vary. Essentially, though, it will indemnify the insured for any sum (not exceeding the limit of indemnity), which they become legally liable to pay, whether by an award in a Court, or by a negotiated out-of-Court settlement, arising from any claim or claims made against them (for breach of professional duty) by reason of any negligent act, error or omission committed or *alleged* to have been committed in the conduct of their occupations. It is usual for insurers to require a description of the insured's occupation, where it is not, as in this case, necessarily clear from their qualifications. It is very important to get this right, and it is generally sensible to seek the help of an insurance broker who specialises in this kind of insurance.

In addition, the policy will indemnify the insured for the legal costs and expenses incurred by insurers in the investigation and defence of any claim, even although the insured may not ultimately be held legally liable. In some instances, such costs and expenses will be inclusive of the limit of indemnity; in others, in addition, depending on the particular insurer. Although it might be more desirable for the cost to be in addition to the limit of indemnity, this is not always negotiable.

One of the bonuses of professional indemnity insurance is that, if a claim is made, insurers will have a good deal of experience in dealing with such claims and in particular will know and use defence lawyers of considerable skill and expertise. It is not always easy for a professional man to find a lawyer with the right experience and reputation to look after him.

Who would claim?

The most likely claimant would be a person or company employing the services of the insured, alleging a breach of contract. In addition, actions in Court can be brought by third persons who may not have had a contract with the insured but may be 'consumers' of the product or may reasonably be expected to rely on the insured's expertise. The third party may not be known at the time the work is performed or the advice given, but if they can show that they could reasonably have been foreseen by the insured to exist, and to rely on his services, the insured could be held to owe them a duty of skill and care.

In order to bring an action of negligence against the firm, the plaintiff would have to establish that as a result of the firm's breach of duty the plaintiff has suffered damage whether physical or financial. Many claims against professional firms arise from inter-contractual disputes, in which the insured may have only a very small role to play, but lawyers are increasingly inclined to 'sue everyone in sight', and though very often it is possible to have an insured dismissed from the action at quite an early stage, the costs of doing so can still be quite sizeable, especially for a small firm.

Disclaimers of liability are often used within the contract agreements. Disclaimers in the UK are subject to a test of unreasonableness. They are disliked by the Courts in all jurisdictions, who usually try to overturn them if they can. Disclaimers may often be inappropriate and undermine the professional reputation or appearance of the firm. Again, substantial costs could be involved in defending any action, even with the existence of a disclaimer, and the final outcome could not be guaranteed in the Courts.

Potential liability

This will be dependent upon the services the insured, whether an individual or a firm, is providing or is *believed* to be providing. Allegations of negligence will often arise from the areas where you may least expect them to. Some potential problem areas of exposure or litigation could arise from:

(a) Pre-shipment responsibilities—e.g., the inspection of the hold to ensure the vessel is properly cleaned and fit to receive the cargo; or the inspection of the quality of the cargo in order to ensure that it is fit for shipment; or supervision of the loading of the cargo to ensure that it cannot become tainted or contaminated.

(b) Averaging adjusting—an error of judgment could be made in apportioning the rateable contribution. Whilst this can be corrected, it may not be possible should one of the parties go into liquidation.

(c) Marine surveying—as already mentioned, potential areas of the exposure will vary in accordance with the work performed. It should be stressed again that we are now in an increasingly claims-conscious climate and it can cost a great deal of time and money to defend an allegation of negligence, even though the firm may not ultimately be held liable.

How to obtain cover

The first step to obtaining such cover would be to speak to an insurance broker. They should have the knowledge or access to various possible insurers. Find one who specialises in professional indemnity if you can.

Potential insurers will probably require a proposal form completed in order to ascertain the details of the proposer requiring insurance. In addition, they may request curricula vitae of the partners or directors of the firm in order to establish the experience gained in the area requiring insurance. Examples of the contracts undertaken and any brochures might also be required. Such information would always be treated in the strictest confidence by potential insurers. The information requested is quite often very detailed but necessary in order that insurers may have an understanding of the risk and therefore their potential exposure.

Upon receipt of these details they should be in a position to consider whether they will be able to offer terms and at what cost. Quite often a professional indemnity policy will need to be specifically moulded to suit a firm's own individual requirements. The limit of indemnity chosen should be dependent upon the following factors: the potential exposure of the proposer; the limit of indemnity available in the insurance market; and the cost of the limit of indemnity required.

Again, a specialist insurance broker can help you to arrive at a sensible figure. Remember that the indemnity limit may be inclusive of legal costs. Quite often, these can be more than the damages. professional indemnity insurance is generally bought to provide catastrophe insurance protection. The insured will often accept a substantial self-insured excess and therefore reduce the cost of the professional indemnity policy, to protect themselves against only those losses which they could not bear themselves. Insurers will in any case stipulate a minimum excess that they require the firm to carry.

Upon obtaining quotations we would suggest that the following points be considered:

- Do the quotations offered indemnify you for all the professional services you provide or are certain restrictions imposed?
- Has a copy of the proposed policy wording been forwarded to you?
- Is the policy subject to a retroactive date—i.e., does it exclude any claim or loss arising out of work performed prior to a certain date?
- If this is the case can this retroactive date be deleted for an additional premium?

This is important, as professional indemnity policies are on a 'claims made basis'. This means that a policy must be in force at the time that a claim or circumstance that may possibly lead to a claim is notified in order for cover to respond.

Other factors which the insured will wish to consider will depend upon the firm's own particular requirements and the conditions imposed by insurers. An insurance broker would be able to discuss a firm's own particular needs in much greater detail.□

THE SURVEYOR AS A FRAUD INVESTIGATOR

Captain P J Rivers, FRGS, FNI, MRIN, MSNI, ACII, ACIArb

ONE AREA where the commercial nautical surveyor could become involved is that of investigating possible frauds which can take a variety of forms. I quote from a paper which I delivered at the end of 1981 when there had been a period involving an alarming number of suspected scuttlings and many millions of dollars. Insurers in Hong Kong, Singapore and Malaysia with interested underwriters at Lloyd's had set up their own commission of enquiry (as it were) called the Far East Regional Investigation Team, or Ferit for short, to review losses over the preceeding few years.

Ferit began to look into those losses firstly which occurred in the period 1978-9 and then back as far as 1959. They eventually short-listed 45 vessels from which a pattern emerged. Most of these vessels were over 15 years old, under 3,000 tons gross and flew the Panamanian flag. Surprisingly, only three of these ships were known not to have been classed. However, the report noted that the percentage of these vessels under the (Japanese) NK Class was 52.6. They elsewhere noted that only three of the vessels lost were registered in Japan.

A sub-standard vessel would obviously fit within this pattern, so a further look was required to see where the money was—that is, was hull or cargo over-insured or were values at risk excessive or any other anomaly (e.g., was there an undue amount of high valued goods in a low valued ship?). Some of the losses could be eliminated, but, in the end, 27 of the 48 losses considered were highly suspect or thought strongly probable of having occurred as the result of scuttling. Furthermore, common names of crew members, shipowners, agents and cargo interests were uncovered, so that it could be said 'that not only have a number of different small syndicates operated, but also there has been a passing on from one party to another of the latest "techniques".'

Variety of cases

This, then, was the overall picture in this region, which included sinkings not only in the South China Sea but in the Indian Ocean as well. But within this framework, there was a variety of different frauds. The ones involving the highest values were the 'rust bucket' type. An old ship with a low insured value of, say, $US250/300,000 would go down, supposedly carrying high-valued cargo. Electronic goods figured in most manifests (the total value of goods would vary from $US2 million to almost $US30 million on different ships).

Although a number of different names for the consignees and shippers might appear on the manifest, enquiry would show that they were connected. Bills of lading would be without letterheads or be those different from the purported carriers. Names of the ships or shipping companies often resembled more respectable ones, while the ships were often 'singletons (i.e., in a one ship company) and had been newly purchased. A variant was to scuttle a ship after stealing valuable cargo, such as consignments of tin ingots. This may have been done without the knowledge of the owners (in which case it would be barratry), but not very often. Other reasons for scuttling would be to hide the fact that inferior goods had been shipped. In one instance, with genuine cargo under some 250 bills of lading, a ship was sunk because there were four consignments of grossly over-insured goods on board.

Charter-party frauds form another type. In their simpler form, a charterer pays the first month's hire, puts the ship on the berth to load 'freight prepaid' cargo and absconds with the proceeds. The next month, the shipowner quite naturally wants his money and the cargo interests just as naturally don't want to pay any further freight. At one time, there were some 15 to 20 ships held up in various ports (mostly obscure) around the Indian Ocean with large and valuable amounts of timber destined for the Middle East. Charter-party frauds normally do not involve scuttling, but one shipowner, it is believed, tried to justify the legal abandonment of the voyage by setting fire to his engineroom. When this did not work, several attempts were made to sink or burn the ship before success was achieved.

Documentary frauds

Others have found the above frauds too cumbersome. With 'rust buckets', too, many parties have to be involved while stolen goods have to be disposed of. It is much easier to sell non-existent goods shipped on vessels which either do not exist or are in some other part of the world. These are generally termed 'documentary frauds' and are aimed at innocent consignees rather than at insurers. A genuine policy may be presented to the bank, but, if there were no goods, the risk never attached and it would be void. This type requires a sale against a genuine letter of credit, a false bill of lading with false invoices and packing lists. A false surveyor's certificate might also be included. A variant is to have a genuine surveyor's certificate of goods sighted at a warehouse but never shipped.

These are just some of the frauds that man can devise. There are many more. In considering any investigation into marine frauds, certain generalities must be borne in mind. Although suspicions of insurers may be aroused by certain peculiarities or by anonymous warnings, the initial enquiries must of necessity be carried out by the marine surveyor. Several surveyors can be involved, for example, for the hull underwriters, for the P & I Club who insure for carriers' liabilities and any number for various cargo insurers or interests. The investigative talents of these gentlemen can vary considerably and generally their findings are not shared even between the various cargo interests.

Surveyors are, of course, not detectives and are generally (but not necessarily) men with some sort of marine qualification, such as a master mariner's or chief engineer's certificate. They have no legal powers to demand any information or documents, although the insurers who employ them may decline to pay claims if full information is not forthcoming. Market considerations or pressure from insurance brokers may however result in claims being paid even though satisfactory proof has not been forthcoming. Some lawyers hinder enquiries by advising their clients for example, who may be shippers, that they need not supply any information whatsoever.

The best that can be hoped for is that the surveyor can show that goods were not in existence and therefore the insurers were not on risk. Although the identity of the culprits may be guessed at, satisfactory proof may not be forthcoming. If the police can be satisfied that the case comes within their jurisdiction, because their powers although greater can also be limited, a more detailed investigation can be carried out by them leading it is hoped to a successful prosecution. Even when the police do investigate crime a very high standard of proof is required. In one case, an owner was brought to trial mainly on the evidence of the master and chief engineer, but their testimony was discounted when it came out they had been promised a reward by the insurers.

Convictions obtained

In places where English law does not apply, it may be that convictions can be obtained. In Taiwan, some crew members and some cargo interests involved in two different losses were sent to prison for scuttling. In Singapore, a different approach was tried in a $S23 million fraud involving a loss. A long and thorough investigation carried out by the Commercial Crime Division enabled the Attorney-General's office to present such a strong case that, after a lengthy trial, an accused changed his plea to guilty on four charges with another 38 to be taken into consideration. He was sentenced to 11 years in prison, but scuttling was not among the charges which included conspiracy and the use of false shipping documents. It was to the latter charge that two other gentlemen pleaded guilty earlier and were sent to prison for up to six years. They were involved in three ships.

So how is this enviable result attained? Generally, in the normal course of events, an insurer presented with a notice of claim involving the non-delivery of a highly valued consignment of goods will appoint a surveyor to carry out enquiries. With the sinking or disappearance of the vessel the question will be: "Did the 'rust bucket' sink on her own or was she nudged?" With regard to fraud the main point is: 'Where's the money?' A 20 year old 3,000-tonner laden with $4 million worth of electronic gear would be of more interest than an old hulk laden with sand, both insured at market values. The same problem of unseaworthiness might have to be resolved, but the first postulates a more interesting possibility.

The attendant circumstances of the case may help resolve the quandry. As far as I know no one has attempted a case by collision. Fire is rare but not unknown. Strandings or sinkings in shallow water have been used, but leave the vessel concerned liable to examination. The preferred method is usually an unexplained sinking in deep water where the exact position is uncertain. It is one thing for a crew to ferry over to a rescuing vessel complete with neatly packed luggage and the pet canary in a cage. It is another if there had been heavy weather properly substantiated, a series of distress calls giving accurate positions, with the wives of the master and chief engineer on board and possibly, sadly, even the loss of life.

Discovering inconsistencies

In dealing with the circumstances of the loss the diligent investigator will of course rely upon discovering inconsistencies with the alleged cause of loss and events surrounding it. The ship's history. Has it been newly purchased or recently renamed? When was it last docked or repaired? What is its class record? Was it little below than scrap? Was it heavily over-insured? What do the trading certificates reveal? Lloyd's intelligence, classification records and market reports will all be of help, and even, as has happened in some cases, hotel registers. There is one case where the shipowner thoughtfully reserved rooms for his crew before the ship had sunk!

The extra question can reveal some astonishing results. In one case the master claimed that the log book had been lost. Speaking separately to the third mate I asked if he had made any record or note and he produced the 'lost' log book. This showed that portholes had been leaking prior to the loss. On another occasion, going ashore in a launch with the chief mate, I was asked what I thought about the casualty as reported. Wondering about his possible reaction I told him that the others put the blame on him. 'Is that so' was the response. 'Well, I'll tell you what really happened' and he did! It is as well to remember that in taking crew statements they will often invent a tale of the occurrence not because of skulduggery but to cover up some boneheadedness on their part.

In any question of fraud the major point is, as was mentioned earlier: 'Where's the money?'. An eminent London barrister, Charles Haddon-Cave, enumerated a series of questions when investigating a shipowner where scuttling is suspected.

- Do the master or crew have any personal incentive to sink the ship?
- Are there any unusual communications passing between the master and the owner?
- Does the loss look as if it must have been carefully planned with outside help?
- Is her owner financially solvent?
- Had he asked for advance freight?
- Is the vessel heavily over-insured?
- Is the vessel trading at a loss and a continual drain on her owner's resources?
- Had her owners been involved in other unexplained losses?
- Do they have any links with cargo interests?

A ready rule of the thumb is to check against the description of the vessel's possible market price as compared to the insured value. Similarly, with cargoes, a comparison of the market prices of various commodities and goods may show discrepancies.

Realistic cargoes

Is the voyage realistic? Could the vessel have carried the amount claimed? In 1974 the Somali authorities arrested the master of a ship which delivered only 678 tons of sugar out of 10,000 tons under a letter of credit for $US5.9 million paid out against 'proof of loading' (i.e. bills of lading). The ship had a cargo capacity of only 8,400 tons.

Captain H. P. Schulz, of the AIG Hong Kong, tells of shipments of colour TV sets to countries 'which did not have colour television at that time at all'.

An exporter from Singapore insured vast amounts of cloves carried on two ships (in which he had an interest) which sank within two days of each other. I totalled up the tons involved, checked against the trade statistics and found that the gentleman (who normally traded in textiles and automobile spares) had apparently cornered 80 per cent or so of the annual exports of cloves from Zanzibar to Singapore. This case illustrates the importance of checking out the parties concerned. I culled over 100 certificates from the registry of companies and managed to show a connection between the various 'different' assured exporters but also with the suppliers, transport forwarders and even the firm which issued certificates of weight and quality.

It is an amazing fact of life that under the rules governing letters of credit, the banks can accept at face value the bits of paper placed in front of them. They don't even have to check if the vessel named in the B/L was in port at the relevant time or indeed even exists. Eric Ellen, the well-known head of the International Maritime Bureau, pointed out that one buyer requested ten different documents to safeguard his letter of credit. All of which as it turned out came from the same source. I know of one case of a gentleman in Taiwan who even issued his own certificate to say that goods existed in Singapore.

Check back

No document should, therefore, be accepted on its face value. The questions are: Did the goods exist? Where did they come from? Who actually saw them? I approached one supplier who denied any knowledge of the transaction. On being shown 'his' invoice he pointed out a number of discrepancies. The form was counterfeit, the supplier was genuine but the sale wasn't. In checking back on warehouses and premises where goods were said to have been stored prior to shipment they may be found to be inadequate or unsuitable for the commodity. One warehouseman (whose records were incomplete) claimed that he merely provided transient facilities to any who required them. His address showed up in three fraudulent sinkings.

In another case through the lorry numbers on port dockets, and the registrar of vehicles, the route of some truckers was back-tracked to the premises where they had picked up some sealed drums whose contents of expensive ore had been substituted with black sand. In the warehouse was found a pile of sand and a number of seals and wire.

A visit to the assured's office and that of the ship's agents can be helpful. From the registry of companies it will be known if they were recently formed, what their paid up capital was (often $2) and what their objectives were. I found rubber merchants who overnight had become exporters of electronic components. One 'shipping agency' had started life running a night club. Not only can a feel of the company be ascertained—'fly by night', established active or dormant—but other links may be picked up—names of sister companies on the door, other vessels listed on a blackboard. (Other vessels may provide links with other frauds—so the port arrivals list should also be checked out.) Trade directories and market enquiries will provide yet further information, as may Lloyd's intelligence, Lloyds Register of Shipping and shipowners and even newspapers. Many sources will be utilised, many people contacted before the story will make sense.

A case history

In one case I was intructed by some of the cargo insurers to enquire into the loss. Little information was known other than the bare details given in the insurance policies—that is the name of the ship for a voyage from Singapore, the company who became the insured, and a brief description and value of the goods concerned, together with a bald statement that the ship had sunk.

A check with the port authorities showed that the ship had arrived in Singapore and had sailed presumably for an Indonesian port. The agent's name was given and when I called upon them I noticed that the same address was shared by one of the major shippers on the short list of insurance policies.

The managing director of the agent was a very charming gentleman who was apparently friendly and who talked a lot but said very little. He informed me that his company had recently purchased the vessel but had time-chartered it to another shipping company. We could get very little information from the owners who said that the ship was under the control of the charterer who very shortly disappeared behind the screen of his lawyers. The charterer proved to be very elusive indeed and his premises looked more like those of a fashion designer than a shipping company. Enquiries were to show that the charterer's company, after having been dormant, had been revived only quite recently.

When I pointed out to the charterer on one of the few occasions that he was available that it was rather strange that one of the shippers should pay him large freight rates to ship cargo in their own vessel he said that it was only after signing the charter-party that he became aware of the connection. The managing director of the owner also claimed that it was only after the ship had sunk that he became aware of the fact that his sister company in the same office had goods on the vessel.

The only information given to us at that stage was that the ship had been scheduled from Singapore for some intermediate ports and then to Hong Kong and Taiwan. It's not clear why as she was supposed to have been fully loaded. An SOS had been received that the vessel was leaking in the engineroom in the South China Sea and had been abandoned. Subsequently the crew had been carried to another port and repatriated from there.

The original homogenous crew had been paid off in Singapore, it was said at the charterer's insistence, and the crew for this particular voyage was a mixture. There was considerable confusion over the number on board. The vessel sailed from Singapore with, it was said, 13 on board but 21 were rescued. A name or two of these seemed to indicate that they had been on previous suspect vessels which had sunk. A smoke-screen had been raised by clearing the ship for Indonesia and a telex was produced saying that the vessel had engine trouble at a port there in the opposite direction from her scheduled voyage.

Port calls

I therefore asked correspondents in various ports to check out what they could discover. Despite later interviews with some of the crew in Taiwan, Singapore and Jakarta, the best that we could make out was that she had been said to have called at a number of ports, the names of which no one could recall. The names of some of the ship's agents given in other ports raised suspicions. One of these had some 30 vessels under their agency in two years of which 10 had become total losses. We also had consignees' names checked and found that in some cases they were secondhand dealers or operated from apartments or were even fictitious.

Various clearance documents and manifests were now becoming available and it seems that the master was not sure whether he had 4,000 tons of bulk ore or a full cargo of 4,900 tons of general cargo or even nil cargo on board. Fortunately the vessel which rescued the crew called at Singapore. Her master was positive that the ship had very little cargo on board and was still afloat when last seen. Fortunately a young officer had taken some photographs and he very kindly lent us the film. The photographs show (and the master of the rescuing vessel confirmed) that the weather was quite calm. The subject ship gave no appearance of distress. Although the depth of water was about 1,000 fathoms (the deepest water available) an anchor was lowered into the water!

We had by then managed to get hold of some manifests (there were several), bills of lading and a number of policies and found that there had been only a few shippers. Two of these companies were under the control of a certain gentleman who had been the only shipper on another vessel which had earlier sunk without the operation of a insured peril. He and a colleague were subsequently imprisoned for a fraudulent involvement in three sinkings. They need not concern us further.

The relationship of the other three companies were quite close. One was owned by the charterer and was also supposed to have supplied all the goods to the other two. Of these two one company as noted above was connected with the owners and the other shared a common director with the charterer.

Source of goods

The source of the various goods was never satisfactorily established. Certain of the items—wax, tin ingots and so on—were said to have originated in Indonesia and had either been loaded directly from small craft or had been stored for a time in an unidentified warehouse. Some were even said to have been kept in the store room attached to the offices of one of the shippers. As another surveyor pointed out in respect of his goods in the space available the stuff must have been bulging out of the windows.

We never really found anyone who had actually seen the goods although there were some survey reports issued by a local firm. They would not discuss these and indeed they later sent a telex withdrawing their certificates. They incidentally were also the surveyors of cargo on and agents for the other ship mentioned above.

An analysis of the various cargo documents that had been received revealed a number of discrepancies. For example the various dates on tally sheets and bills of lading did not coincide with the various mate's receipts which bore the initials of the chief mate for days before he had even joined the ship. The stevedoring firm who were supposed to have carried out the work were evasive in their answers and superbly vague about the operations which according to the tally sheets were supposed to have taken place over two distinct periods in two different anchorages in Singapore.

I managed to find a supply boat which had been alongside the ship at one of the relevant periods and the boat driver was adamant that no cargo had been worked while he was there. Even better from my view point was that earlier as proof that the ship had been in good condition on sailing, I was told that repairs had been carried out and was given some repair bills to prove it. In this instance the bills were genuine and the workmen recalled burning and welding in the holds which were empty at the time. The cargo I was interested in was supposed to have been loaded three days earlier! In addition I traced hull surveyors from reputable organisations who had been aboard over the 'cargo working' period who saw empty holds and no such activity.

Reports were made to the Commercial Crime Division of the Singapore Police and investigations led to a successful prosecuion. Imprisonment was imposed on the villains—not for scuttling but for fraud.

Attention to detail

In all cases and regardless of whoever the surveyor is acting for the approach should be the same. A careful attention to details avoiding short cuts, cross-checking all facts should be followed throughout. Anything which cannot be reconciled should not be ignored if it does not conveniently fit any particular picture. You may even come to the conclusion despite a number of pointers that there was no fraud! But remember inconsistencies point to irregularities. The surveyor has to 'start at square one' and throughout ask 'who, what, where, when and why', continually seeking corroboration of every detail. It is most helpful if a pattern can be found from previous losses and claims.

An almost embarrassing and confusing pile of paper will accumulate—claims documents, ship's documents, sales documents, company searches, telexes, letters and on and on. The method that I have found best is to write a summary based on the documents that have been acquired at various times, arranged according to subject matter rather than in a chronological account of investigations.

Some possible headings in a cargo claim are:

- The insurance.
- The assured.
- The subject matter (goods insured).
- The carriers/agents.
- The contract of carriage.
- The ship.
- The loading port.
- Subject voyage.
- Circumstance of loss.

From this summary, cross-checking can be carried out to see if names, descriptions, dates, etc., are the same. Anomalies will appear and missing papers or facts can be followed up. The points can then be marshalled and supporting documents arranged in order and indexed. At the end there should be nothing left loose or unaccounted for.□

PART III
SAFETY AND INSPECTION

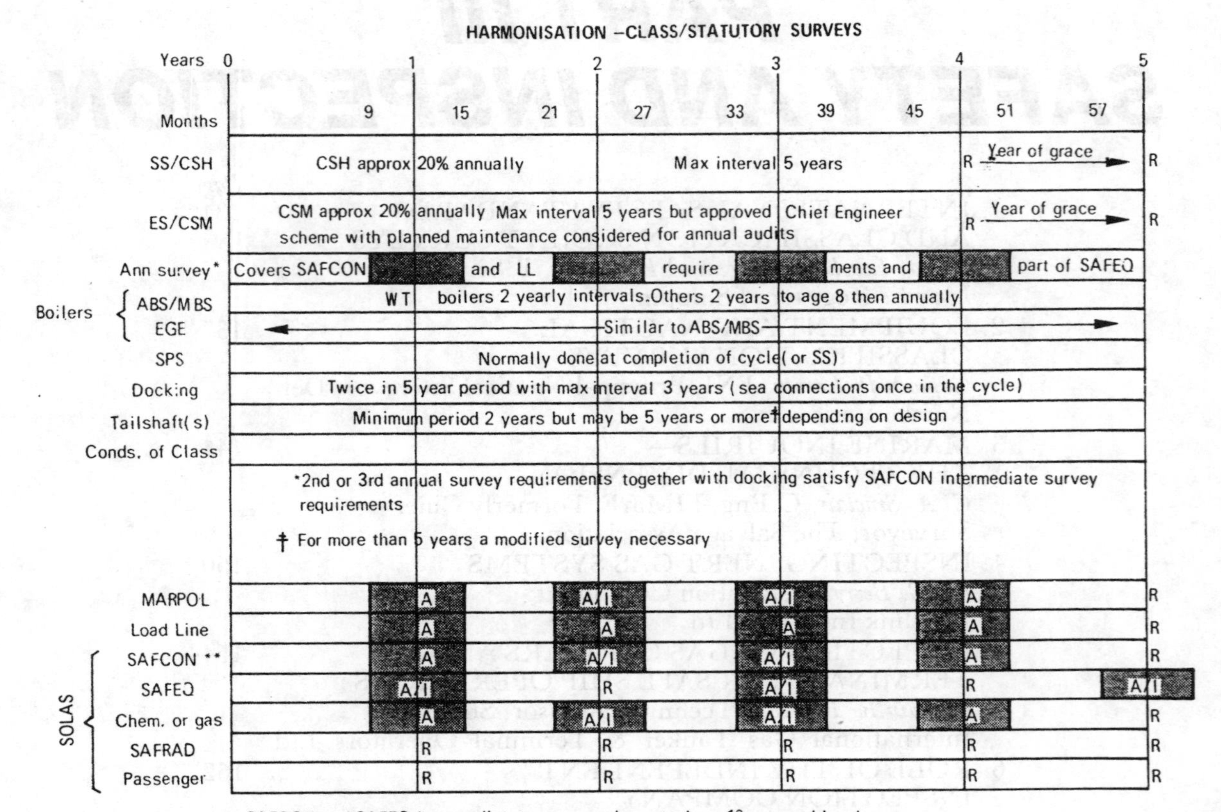
HARMONISATION –CLASS/STATUTORY SURVEYS
Years 0 1 2 3 4 5
Months 9 15 21 27 33 39 45 51 57
SS/CSH
CSH approx 20% annually
Max interval 5 years
Year of grace
ES/CSM
CSM approx 20% annually Max interval 5 years but approved Chief Engineer scheme with planned maintenance considered for annual audits
Ann survey*
Covers SAFCON and LL require ments and part of SAFEQ
Boilers ABS/MBS EGE
WT boilers 2 yearly intervals.Others 2 years to age 8 then annually
Similar to ABS/MBS
SPS
Normally done at completion of cycle(or SS)
Docking
Twice in 5 year period with max interval 3 years (sea connections once in the cycle)
Tailshaft(s)
Minimum period 2 years but may be 5 years or more† depending on design
Conds. of Class
*2nd or 3rd annual survey requirements together with docking satisfy SAFCON intermediate survey requirements
† For more than 5 years a modified survey necessary
MARPOL
Load Line
SOLAS
SAFCON **
SAFEQ
Chem. or gas
SAFRAD
Passenger
SAFCON and SAFEQ intermediate surveys only on tankers 10 years old and over
** Intermediate survey includes docking (done twice in the cycle with maximum interval of 3 years)

Fig 1

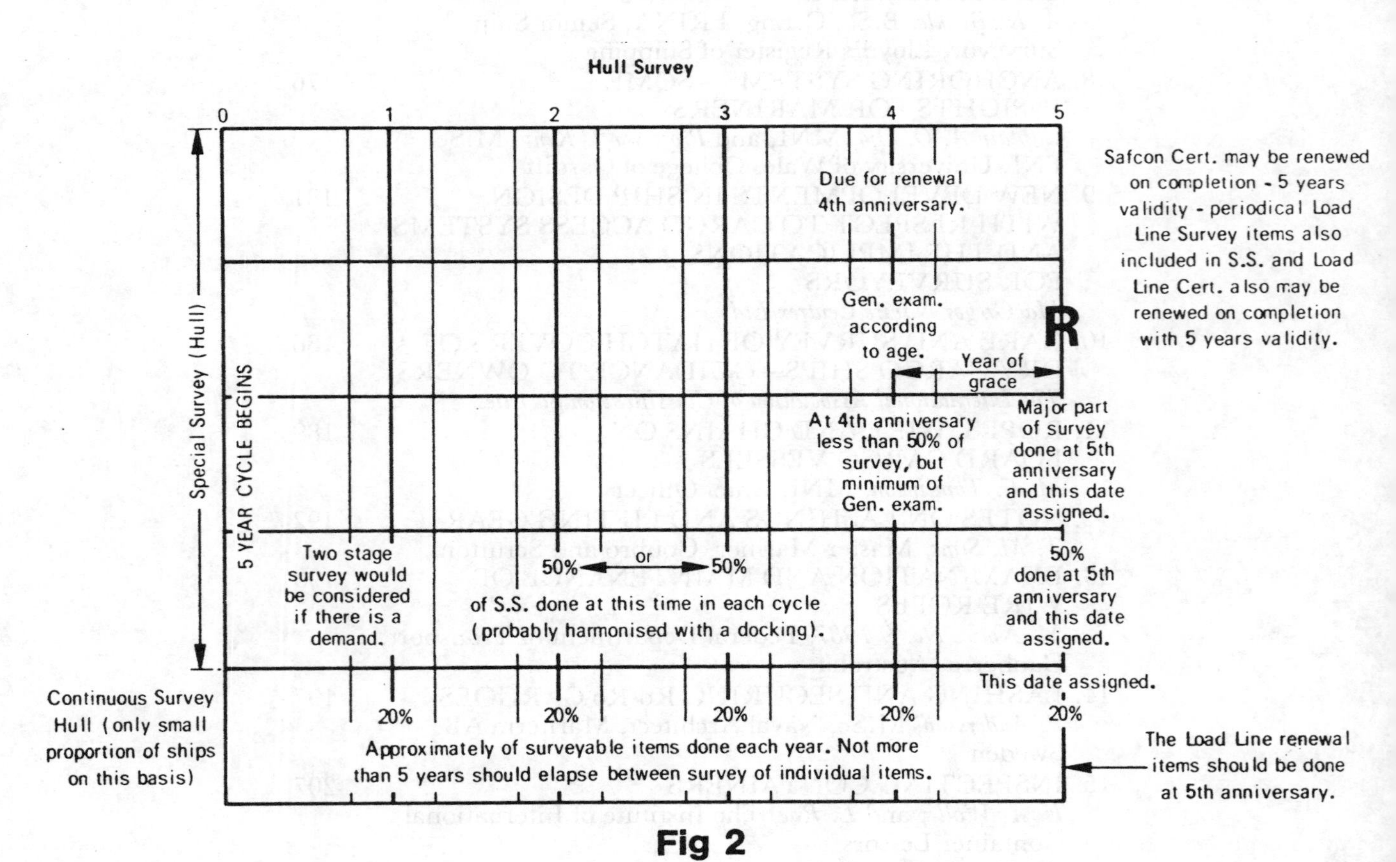
Hull Survey
0 1 2 3 4 5
Special Survey (Hull)
5 YEAR CYCLE BEGINS
Due for renewal 4th anniversary.
Gen. exam. according to age.
Year of grace
R
At 4th anniversary less than 50% of survey, but minimum of Gen. exam.
Major part of survey done at 5th anniversary and this date assigned.
Two stage survey would be considered if there is a demand.
50% or 50%
of S.S. done at this time in each cycle (probably harmonised with a docking).
50% done at 5th anniversary and this date assigned.
This date assigned.
Safcon Cert. may be renewed on completion - 5 years validity - periodical Load Line Survey items also included in S.S. and Load Line Cert. also may be renewed on completion with 5 years validity.
Continuous Survey Hull (only small proportion of ships on this basis)
20% 20% 20% 20% 20%
Approximately of surveyable items done each year. Not more than 5 years should elapse between survey of individual items.
The Load Line renewal items should be done at 5th anniversary.

Fig 2

INTERNATIONAL SAFETY STANDARDS AND CLASSIFICATION SOCIETY SURVEYS

R. M, Leach, C. Eng, and H. Nutter, B.Sc, C. Eng
Lloyd's Register of Shipping

CLASSIFICATION SOCIETIES came into existence in the 18th and 19th centuries to fulfil a need that was shared by all business interests connected with the operation of ships. Owners, underwriters, shippers and bankers accepted risks in varying degrees, and an independent technical opinion on the fitness of a ship to trade became a fundamental necessity in seeking methods of reducing some of these risks. Of all the parties concerned, the underwriters probably had the greatest total involvement in any particular venture and it is not surprising that it was they who took the first steps to 'classify' ships according to whether, from the construction and maintenance points of view, they constituted good or bad risks.

The first classification society to become established was Lloyd's Register of Shipping, which was constituted in 1760, followed by Bureau Veritas in 1828, Registro Italiano Navale in 1861 and the American Bureau of Shipping in 1862, Det Norske Veritas and Germanischer Lloyd closely followed in 1864 and 1867, respectively.

At the turn of the century, in 1899, the Japanese classification society Nippon Kaiji Kyokai was formed. More recently several other societies have been added to this list, including that of the Polish Register, the Register of Shipping of the People's Republic of China, the Bulgarian Register of Ships, the DDR Schiffs-Revision und Klassifikation, the Korean Register, the Hellenic Register of Ships, the Indian Register of Ships, the Jogoslavenski Registra Brodova, the Klasifikasi Indonesia, the Register Romano Navale, and the USSR Register of Shipping.

The first known register of ships was published in 1764 and this was a tribute to the enterprise of Edward Lloyd who owned a coffee house in the City of London and gave his name to Lloyd's Register of Shipping, the Corporation of Lloyd's and *Lloyd's List*.

When Lloyd's Register was reconstituted in 1834, a course was charted which is followed to this day. Briefly, this was that the governing body should be representative of all sections of the shipping community and that a uniform standard for ship construction and subsequent maintenance should be adopted in the form of written rules. As soon as these rules came into use, it was a natural development to establish permanent qualified staff both in outports and head office so that ships could be built under survey and maintained to the prescribed standard on a worldwide basis.

In the case of Loyd's Register, this new constitution ensured that overall management was exercised by duly elected representatives from the shipping industry with the prime object of providing an impartial professional service for the benefit of all sections of that industry. It also ensured that uniformity of practice was achieved by vetting reports technically in head office before outport surveyors' recommendations are confirmed.

Today, the main function of a classification society is to lay down standards for the construction and subsequent maintenance of ships and to ensure that these standards are fully implemented. These 'standards' are published in the form of rules. Lloyd's Register, in common with most of the leading classification societies issues a wide variety of rules, some of these are listed below:

1. Rules and Regulations for the Classification of Ships.
2. Rules for Floating Docks.
3. Rules for Inland Waterways Ships.
4. Rules for Mobile Offshore Units.
5. Rules for Refrigerated Stores (on land).
6. Rules for Refrigerated Cargo Installation (on board).
7. Rules for Ships for Liquefied Gases.
8. Rules for Bulk Chemical Tankers.
9. Rules for Submersibles and Diving Systems.
10. Rules for Yachts and Small Craft.

Whilst classification fulfils a very important function in maintaining standards of construction and securing the safety of ships and their cargoes, responsible governments for many years have been concerned about safety of life at sea. As the most important part of obtaining safety of life is to secure the safety of the ship, there is a large area of common interest which may cause confusion to the layman in differentiating between classification and statutory survey requirements.

Lloyd's Register, in the first half of the 19th century introduced a freeboard requirement of 3 inches per foot depth of hold, known as Lloyd's Rule, and this was used by responsible owners on a voluntary basis until 1880. In the United Kingdom, about that time, Parliamentary agitation was created by Samuel Plimsoll culminating in a Merchant Shipping Act 1876 requiring a 12 inch deck line and an 18 inch disc to be marked on the ship's side. It is still common practice to refer to the freeboard mark as the Plimsoll Line. Whilst freeboard tables for limiting draft had been proposed by Lloyd's Register, these were still only applied on a voluntary basis until they were used as a basis for the Shipping Act of 1890 which required freeboard to be calculated and marked accordingly. An international conference was held in July 1930 in London and the resultant recommendations were ratified by the majority of the nations attending, culminating in the 1930 Load Line Convention.

Similarly, further thought was given to passenger ship safety by governments mainly because of lessons learned from disasters, probably the best known being the *Titanic* in 1912. The first really effective international conference was held 1929 concerning construction, subdivision (water tightness and fire), pumping, fire appliances and radio, and it should be

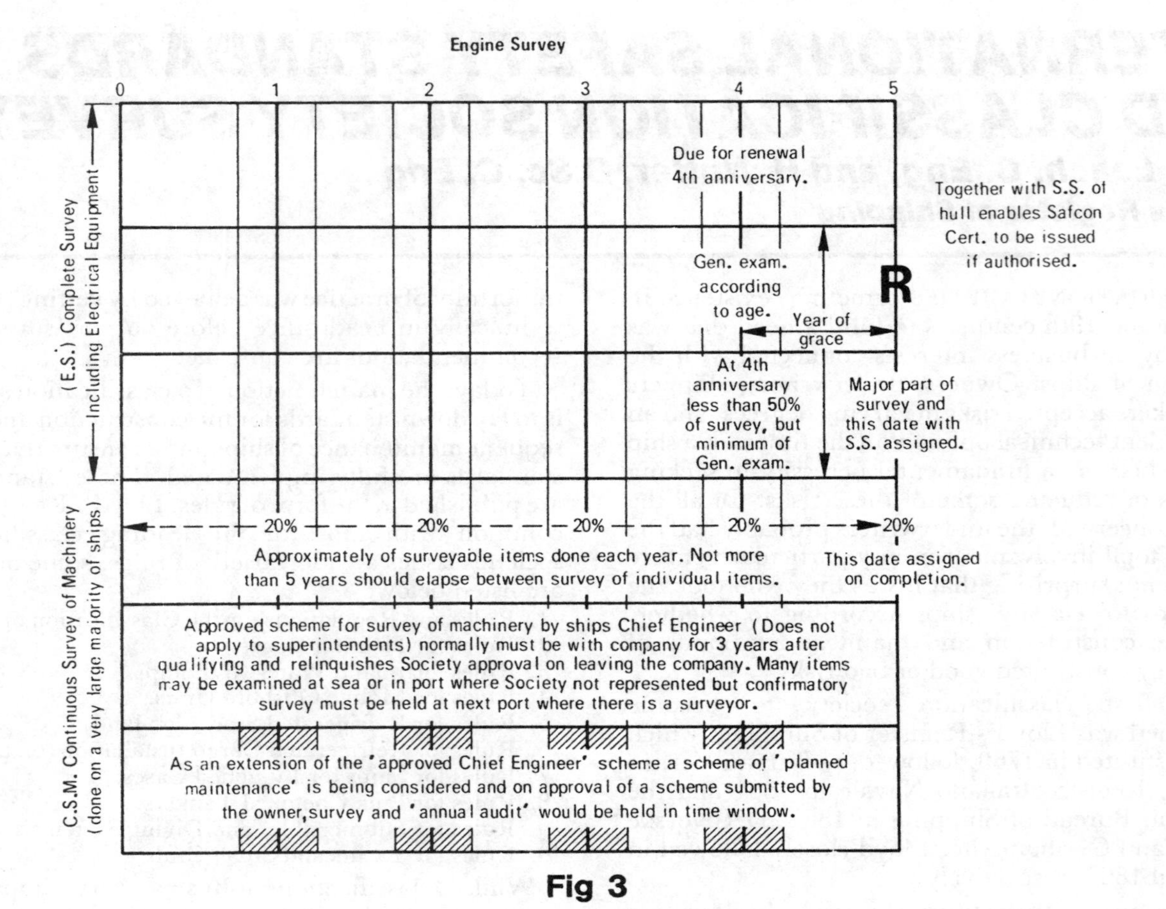

Engine Survey
0
1
2
3
4
5
(E.S.) Complete Survey
– Including Electrical Equipment
Due for renewal
4th anniversary.
Together with S.S. of
hull enables Safcon
Cert. to be issued
if authorised.
Gen. exam.
according
to age.
R
Year of
grace
At 4th
anniversary
less than 50%
of survey, but
minimum of
Gen. exam.
Major part of
survey and
this date with
S.S. assigned.
C.S.M. Continuous Survey of Machinery
(done on a very large majority of ships)
20%
20%
20%
20%
20%
Approximately of surveyable items done each year. Not more
than 5 years should elapse between survey of individual items.
This date assigned
on completion.
Approved scheme for survey of machinery by ships Chief Engineer (Does not
apply to superintendents) normally must be with company for 3 years after
qualifying and relinquishes Society approval on leaving the company. Many items
may be examined at sea or in port where Society not represented but confirmatory
survey must be held at next port where there is a surveyor.
As an extension of the 'approved Chief Engineer' scheme a scheme of 'planned
maintenance' is being considered and on approval of a scheme submitted by
the owner, survey and 'annual audit' would be held in time window.

Fig 3

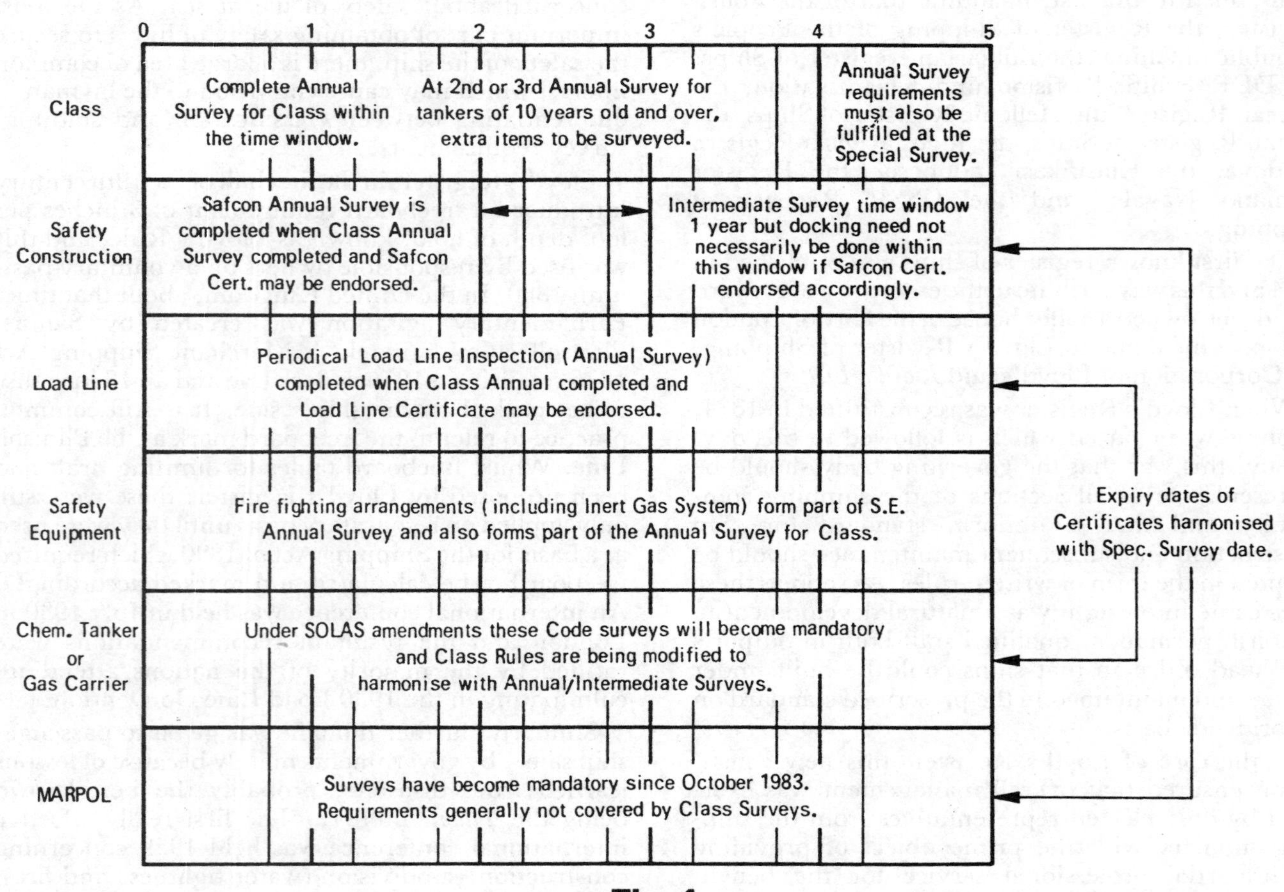

(MANDATORY) Annual and Intermediate Surveys
0
1
2
3
4
5
Class
Complete Annual
Survey for Class within
the time window.
At 2nd or 3rd Annual Survey for
tankers of 10 years old and over,
extra items to be surveyed.
Annual Survey
requirements
must also be
fulfilled at the
Special Survey.
Safety
Construction
Safcon Annual Survey is
completed when Class Annual
Survey completed and Safcon
Cert. may be endorsed.
Intermediate Survey time window
1 year but docking need not
necessarily be done within
this window if Safcon Cert.
endorsed accordingly.
Load Line
Periodical Load Line Inspection (Annual Survey)
completed when Class Annual completed and
Load Line Certificate may be endorsed.
Safety
Equipment
Fire fighting arrangements (including Inert Gas System) form part of S.E.
Annual Survey and also foms part of the Annual Survey for Class.
Expiry dates of
Certificates harmonised
with Spec. Survey date.
Chem. Tanker
or
Gas Carrier
Under SOLAS amendments these Code surveys will become mandatory
and Class Rules are being modified to
harmonise with Annual/Intermediate Surveys.
MARPOL
Surveys have become mandatory since October 1983.
Requirements generally not covered by Class Surveys.

Fig 4

noted that some sections of the resultant convention were made applicable to cargo ships. More recently there has been a great deal of emphasis on protection of the environment and this can also be related to the various governments' principal concern about safety of life, though in this case it is not solely human life that is involved.

In order to fulfil the main function of a classification society, which is to implement fully the standards of construction laid down in the rules and maintain the ships to a specified standard subsequently, it is necessary to have a procedure for building ships to class. This procedure is outlined below:

1. Request for survey, by the shipbuilder.
2. Submission of plans for approval/appraisal.
3. Survey of components—steel testing.
4. Construction at the shipyard.
5. Installation of machinery and equipment.
6. Tests and trials.
7. Documentation and submission of reports.
8. Issue of class and statutory certificates.

In general, the plans that are to be submitted under item 2 are the main construction plans of the hull, equipment and loading manuals and for machinery, plans for such equipment as boilers, shafting and gearing, pumping and piping and electrical and control equipment. A full list of the plans and supporting documents required are clearly set forth in the classification society's rules. In order to maintain the class of an existing ship periodical surveys are required to be held by the due dates, and these are shown in figures 1 to 6. In general, postponements beyond these dates are not encouraged, but short extensions are sometimes considered on application having regard to the circumstances of the particular case and whether or not there is a statutory involvment.

Having described the process for obtaining class certificates and maintaining class afterwards it may not be out of place here to state the circumstances in which class may be discontinued. The following paragraphs are extracts from Rules (Part 1) of Lloyd's Register of Shipping, but broadly represent the policy of all the leading classification societies.

1. The Committee has power to:
   Withhold, or if already granted, to suspend any class (or to withhold any certificate or report in any other case) in the event of non-payment of any fee.
2. Any damage, defect or breakdown, which could invalidate the conditions for which a class has been assigned, it to be reported to the Society without delay.
3. All repairs to hull, equipment and machinery which may be required in order that a ship may retain her class, are to be carried out to the satisfaction of the society's surveyors.
4. When at any survey the surveyors consider repairs to be necessary, either as a result of damage, or wear and tear, they are to communicate their recommendations at once to the owner, or his representative. When such recommendations are not complied with, immediate notification is to be given to the Committee by the surveyors.
5. When the Regulations as regards surveys on the hull or equipment or machinery have not been complied with and the ship is thereby not entitled to retain class, the class will be suspended or withdrawn at the discretion of the committee and a corresponding notation will be assigned.
6. When it is found, from reported defects in the hull or equipment or machinery, that a ship is not entitled to retain class in the Register Book, and the owner fails to repair such defects in accordance with the society's requirements, the class will be suspended or a corresponding notation indicating that class has been suspended or withdrawn at the discretion of the Committee because of reported defects will be assigned.

As stated previously, governments have been concerned for many years about safety of life at sea, and that gave rise to the first international conventions which came into force in 1932. However, much more emphasis has been placed on conventions and statutory certificates since the advent of the Inter-Governmental Maritime Consultative Organisation (now IMO) in March 1958. Ships are now unable to trade internationally without statutory certificates, and such certificates require initial, renewal and other surveys on board.

Leading classification societies are appointed by most administrations as recognised organisations to do surveys and issue or endorse many of these certificates on their behalf. As many of the surveys for classification overlap with surveys for statutory purposes, a more detailed consideration of the harmonisation of these is indicated.

Harmonisation of surveys

An International conference on tanker safety and pollution prevention was held by the then-named Inter-Government Maritime Consultative Organisation (IMCO) in 1978 and this resulted in two protocols, one relating to the International Convention for the Safety of Life at Sea 1974 (Solas 1974) and the other relating to the International Convention for the Prevention of Pollution from Ships 1973 (Marpol 1973).

A safety construction certificate for cargo ships came into being with Solas 1960 but no maximum validity was specified. When duly authorised by the administration concerned—i.e., the flag State—Lloyd's Register based the validity on the class special survey (SS) and as with the load line certificate, issued the certificate valid for five years from the assigned date of SS. The protocol specified that the Safcon certificate should have a maximum validity of five years.

As some concern was expressed at the conference regarding the condition of some ships and their equipment, it was felt that more control should be exercised by the port States and that 'unscheduled inspections' should be held by or on behalf of the flag State. Whilst such unscheduled inspections would fulfil the idea of additional control, there are many practical difficulties, particularly if the flag State delegates the surveys to recognised organisations. Mandatory annual surveys were therefore introduced and administrations could elect to require such annual surveys, in which case unscheduled surveys were not obligatory. None of the administrations has elected to do unscheduled surveys and therefore we need not dwell upon those requirements.

At the same time, intermediate surveys became obligatory for tankers of ten years of age and above, and these consisted of an examination of the items

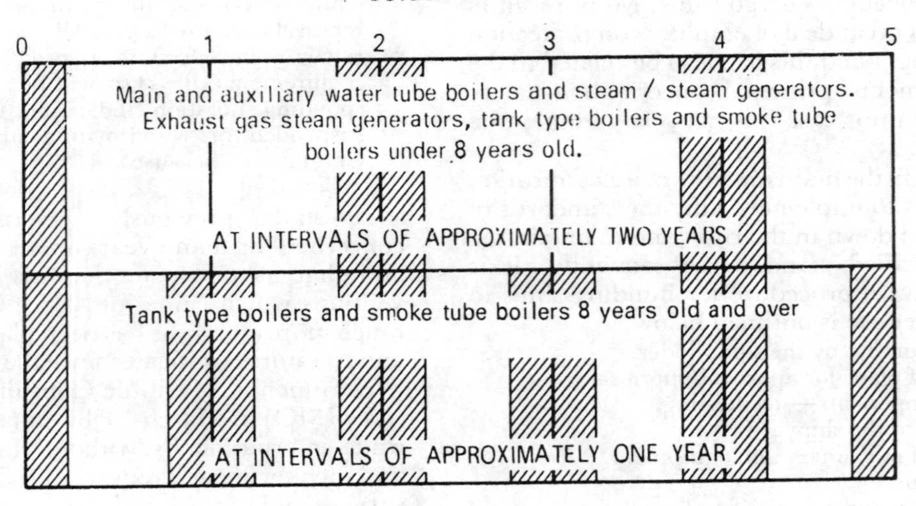

Fig 5

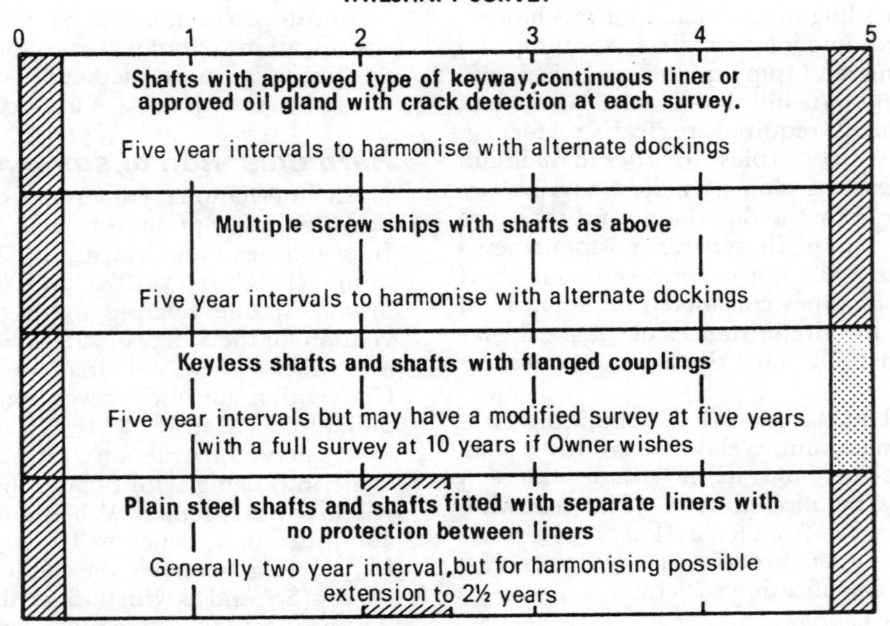

Fig 6

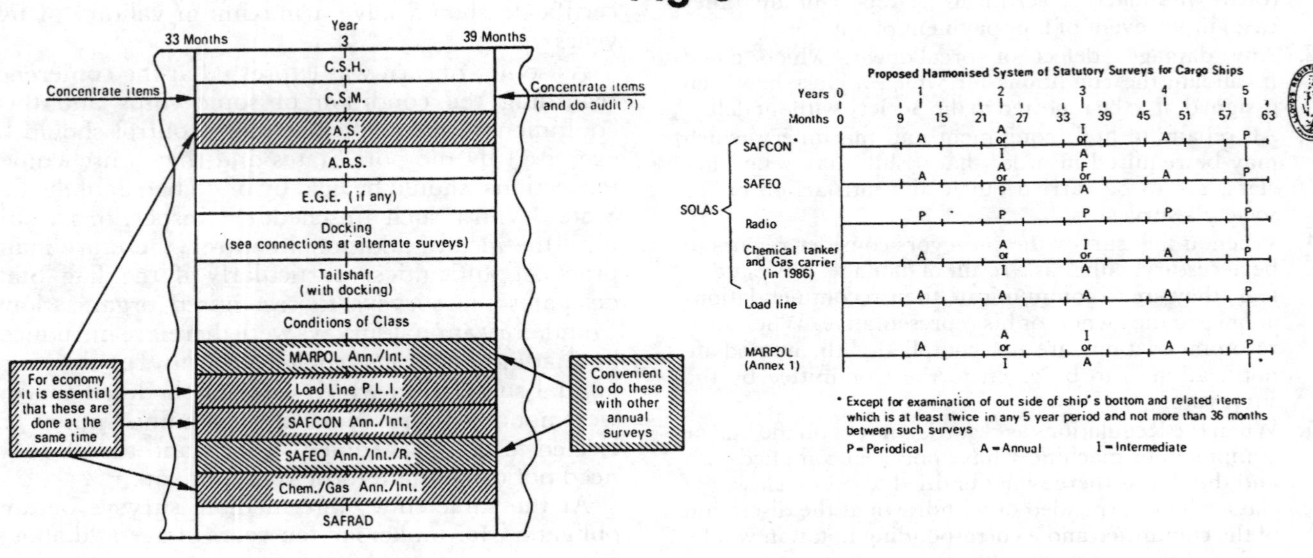

Fig 7

Fig 8

required for mandatory annual survey as a minimum, together with an examination of the ships bottom, stern frame, rudder, propeller and shaft seals and additional items such as two selected cargo tanks, anchors and mooring equipment and electrical equipment in dangerous zones. The mandatory annual survey is due at the anniversary date and is to be done within three months before or after that date. If only one intermediate survey is done, it is due at the half-way date of the certificates's validity and is to be done within six months before or after that date.

Before the Solas Protocol came into force 'Guidelines on Surveys under the 1978 Protocol' were produced at the now-named International Maritime Organisation (IMO) which detailed, *inter alia*, guidelines for mandatory annual and intermediate surveys. The society's rules for annual survey were amended to take into account these guidelines and to make the survey at the anniversary date with a 'time window' of six months. The rules for survey of tankers of ten years old and over together with docking survey of all ships ensure that the Safcon intermediate survey requirements are dealt with.

Lloyd's Register has always based the validity of the Load Line Certificate, when authorised for its issue, on the assigned date of the special survey and the requirements of a periodical load line inspection were included in the class annual survey. By aligning the validity of the Safcon and Load Line Certificates based on the assigned date of class special survey the anniversary dates coincide and enable the annual surveys to be carried out at the same time, all being within the same time window. This is essential if the owner wishes to avoid duplication of preparation for survey and attendance of different surveyors perhaps both administration and classification surveyors at different times.

It should be noted that mandatory annual and intermediate surveys concern safety equipment as well as safety construction, and safety equipment covers lifesaving appliances, fire-fighting equipment and navigation equipment. In addition there are other statutory surveys such as initial/renewal, annual and intermediate surveys for Marpol 73/78 and code requirements for chemical tankers and gas carriers which became convention requirements under Solas in 1986. As these came into force it became more and more important for the society to harmonise the validity dates of these certificates with the classification certificates so as to enable owners to combine surveys if they so wish. Due to the relatively generous time window of six months for annual surveys, the owner is not obliged to carry out several surveys at one time, but many are finding that it is much more economical to do so.

Figure 1 shows a chart of main surveys to be done, divided horizontally into a five-year cycle with months also marked to define the time windows. The surveys are arranged vertically with the classification surveys in the top group and statutory surveys grouped together according to their Conventions. Although the chart shows intermediate/annual surveys with full six-month time windows, the window for intermediate surveys for Safcon and Marpol is 12 months, from anniversary date to anniversary date. Safety Radio is an annual renewal certificate and the survey items are not class requirements so harmonising is not important. Passenger ships also have their safety certificates renewed annually, but being smaller in number and mainly on fixed schedules they receive special treatment.

Figure 2 gives details of special survey of hull which is due after four years, but most owners take advantage of doing a general examination at the fourth anniversary to obtain one-year postponement. This would then bring the SS date into line with the statutory five-year certificates. An owner may take advantage of the general examination to start the SS but care must be taken not to complete the major part of the survey as that governs the date assigned. Continuous survey (CSH) may be the preferred method in which case some 20 per cent should be done each year and not more than five years should elapse between survey of individual items. It should be noted that to comply with load line survey requirements, the survey items concerned should be dealt with at the end of the cycle.

Figure 3 is similar to figure 2 and concerns engine survey including electrical and remote control equipment. In most cases this is dealt with on a continuous basis (CSM) but a refinement is that survey of certain items may be done at sea or in port where the society is not represented, by the chief engineer if approved under the society's scheme. A confirmatory survey has to be held at the next port where there is a surveyor. As an extension of this a 'planned maintenance' system can be approved. This requires also that all chief engineers used be approved, and an annual audit is carried out on board most conveniently when the ship is doing an annual survey and certainly within the six-month time window.

Figure 4 brings together the various annual surveys, the important ones to harmonise being Class, Safcon and Load Line. The Safcon intermediate survey on tankers is covered by additional items to be surveyed for class at the second or third annual survey. The inspection of outside of ships bottom required for intermediate survey is covered by a docking survey for class, and in fact may be done outside of the time window, provided the Safcon certificate is endorsed accordingly. The only part of safety equipment which overlaps with class is fire-fighting arrangements, which includes the inert gas system if fitted.

Figure 5 for boiler surveys and exhaust gas economisers shows that in general these are only needed in alternate years, and class is flexible enough to allow the intervals to be adjusted slightly to fall in with the time window. **Figure 6** for screwshaft surveys shows the result of the recent modifications to the rules whereby most ships only need to have surveys at five-year intervals, which harmonises with the new requirement for two dockings in five years with a maximum interval of three years between alternate dockings. It should be noted that for keyless shafts and shafts with flanged couplings a modified survey which does not necessitate the removal of the propeller or complete withdrawal of the shaft may be done at five years with a complete survey at ten years.

Figure 7 is a section of figure 1 at the third year of the cycle and presents an ideal situation where survey dates coincide. If the hull and machinery are on continuous survey the due items should be arranged to fall within the window, and of course the audit under planned maintenance would become due. The sea connections which are done at a docking survey need to be done once in the CSM cycle. Conditions of class are generally arranged with the surveyor so that they would become due at a convenient time. Safety Radio has been added for completeness, but it would in general not fall due as shown since there is no time window.

The society sends a quarterly notice to all owners of classed ships which shows due dates of class surveys, also statutory surveys where the certificates are issued by us. Updated lists are also available at local offices. In addition, the master list of surveyable items supplied when the ship is first classed is updated as continuous survey is held and copies are available at owner's request both in numerical order and survey date order. These listings have been found to be most useful in the planning and harmonising of surveys.

Harmonising statutory surveys

Proposals at IMO for the harmonising of statutory surveys were submitted for full discussion at the Marine Safety Committee in 1988. The principal part concerning the carrying out of surveys can be summarised as follows.

Unscheduled inspections are omitted and intermediate surveys are extended to include all cargo ships instead of only applying to tankers ten years of age and above. The inspection of the outside of ship's bottom is extended to all cargo ships instead of tankers ten years of age and over. It should however be noted that all classification societies have had requirements for bottom inspection of all ships for many years, so this is not an extra and also is now at more flexible intervals.

Instead of the Safety Equipment Certificate expiring after two years maximum, the validity will be five years maximum and the periodical survey, equivalent to the present renewal survey, would be done during the 2nd or 3rd anniversary time window. Similarly, the Safety Radio Certificate, at present with maximum validity of one year, will be five years maximum and the periodical survey done each year within the time window.

If all the Solas certificates were issued for five years it would be possible to issue a combined certificate, though it would be necessary to have separate annual and intermediate survey endorsement spaces for Safety Construction, Safety Equipment and Safety Radio.

A limited extension of three months will be allowed on all five-year certificates, including Load Line and Marpol, but the certificate will be backdated to the expiry date of the previous five year certificate. Likewise, if the survey is done within three months before the expiry date, the new certificate may be valid from the expiry date of the previous certificate.

It should be noted that if certificates are issued for less than the maximum five-year period, they can be extended by completing the equivalent surveys due for a five-year certificate.

Though the above will not come into force for some time (not before 1 February 1992), mainly because modifications will be needed to the Solas and Load Line Conventions, either by Protocol or new Convention, it is considered there should be benefits for all concerned with surveys.□

EQUIPMENT AND NAUTICAL SURVEYS

Captain Per Larsen, FNI
Det Norske Veritas

WHY EQUIPMENT and nautical surveys?—In what way and to what extent can this type of classification service improve nautical safety?

This may be one of the first questions that should be answered, and the simple answer is: 'The aim of this classification service is to reduce the risk of nautical casualties by establishing operational requirements and assisting owners and shipbuilders in meeting them. By surveying ships in service, we intend to assist the users in maintaining the safety standard which is dependent upon the quality of the total system performance. Because, if relevant requirements for design and equipment are established on the basis of the functions to be performed on the bridge, and efficient surveys are carried out to ensure that the requirements are being complied with, there should basically be no other risk of nautical casualties than the types of "operator failures" which are related to operational procedures and operator qualifications, and, of course, failures which may relate to shortcomings in human quality.'

This statement also reveals that operational procedures and operator competence should be included in rules dealing with operational safety. But still, we would be left with the 'Operational failures' relating to 'human quality' or what we may name 'human errors.' For the purpose of this chapter, however, the influence of human quality on operational failures should be regarded as a problem of its own. The intention in rules for nautical safety, and the surveys implied, is to approach the areas of operational safety which can be influenced by rules and requirements. And, by doing so, there are reasons to believe the area of 'human quality' will be influenced in a positive way as well.

Ship safety

It is useful to focus on nautical safety by making ship safety the starting point, because ship safety is our overall aim and it is important to see any area of rules and requirements in relation to total ship safety. An overview of casualty risks and safety measures would assist us in establishing rules and requirements in a rational way. It would also enable us to judge the relative importance of casualty areas in question. Such information would also assist us in deciding where we should aim our efforts and to consider relevant adjustments or classification functions.

What are our efforts and the measures taken in relation to the apparent risk and consequences of ship casualties? Casualty statistics reveal the distribution of various types of casualties and the relative risks. The pictures given by the statistics may give reason to ask if we, classification and maritime authorities, address the total safety of ships in a systematic and adequate way.

Instead of elaborating on this subject, we may state that a classification society with the overall objectives to 'safeguard property, life and environment' needs to have a picture of the impact of classification rules on ship safety and take action accordingly. Statistics tell us that collisions, groundings and heavy weather damage account for about 50 per cent of all serious casualties, total losses inclusive. These types of casualties represent about half of the total, both with regard to tonnage and claims paid and also with regard to lives lost in ship accidents. Operational failures, collisions and groundings also account for the major oil spill incidents caused by ship casualties.

The statistics show that nautical safety is an important sector of the total safety of the ship. But today this sector is outside the traditional scope of classification. The traditional main rules for classification of ships stipulate technical requirements to hull structures, machinery installations and equipment with respect to strength and performance. The requirements are aimed at safety against hazard to the ship, personnel and environment by reducing the risk of structural damage, machinery damage, fire and explosions and other incidents which may be caused by technical failures. These rules do not influence the risk of collisions, groundings and heavy weather damage if we make exceptions for technical requirements aiming at reducing the risk of 'black out' and steering gear failures, and requirements regarding structural strength of the hull.

We also know that the effect of the traditional classification requirements are highly dependent upon the ship and its machinery and equipment being competently handled and operational procedures being established and complied with. Facing this situation, and with the objectives of 'safeguarding properties, lives and environment' in mind, it is our concern also to focus on the part of ship safety which deals with operational safety. In order to reduce nautical casualties, we have to focus on the factors influencing safe bridge operation.

Nautical safety and operational efficiency

Nautical safety can be improved by measures taken to reduce the risk of casualties originating from decisions made and tasks performed on the bridge. In order to improve operational safety, it is imperative also to consider operational efficiency. Operational safety requirements have to be based on analysis of tasks to be performed on the bridge with due regard to operational efficiency. Because, means to increase operational economy, like reduced manning, system integration and automation, influence the tasks to be performed and have implications for bridge design, instrumentation and operator qualifications.

Classification rules and requirements ought to be in the forefront of technical and operational develop-

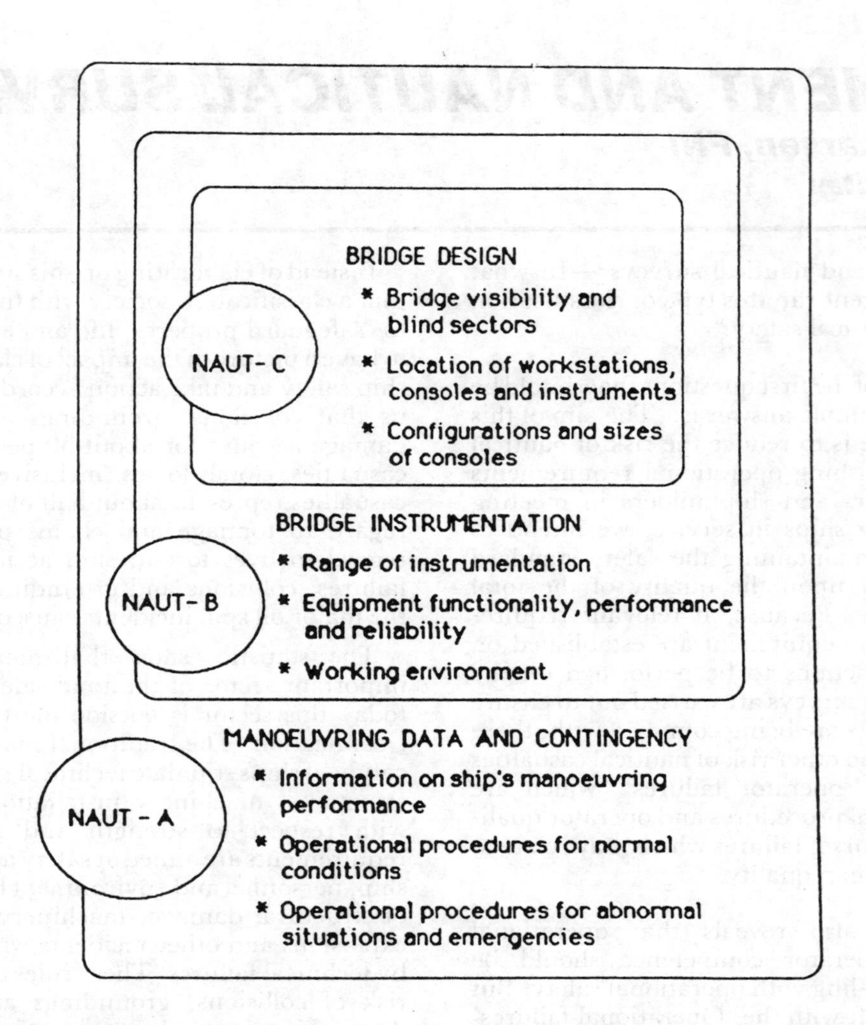
BRIDGE DESIGN
* Bridge visibility and blind sectors
NAUT-C
* Location of workstations, consoles and instruments
* Configurations and sizes of consoles
BRIDGE INSTRUMENTATION
* Range of instrumentation
NAUT-B
* Equipment functionality, performance and reliability
* Working environment
MANOEUVRING DATA AND CONTINGENCY
* Information on ship's manoeuvring performance
NAUT-A
* Operational procedures for normal conditions
* Operational procedures for abnormal situations and emergencies

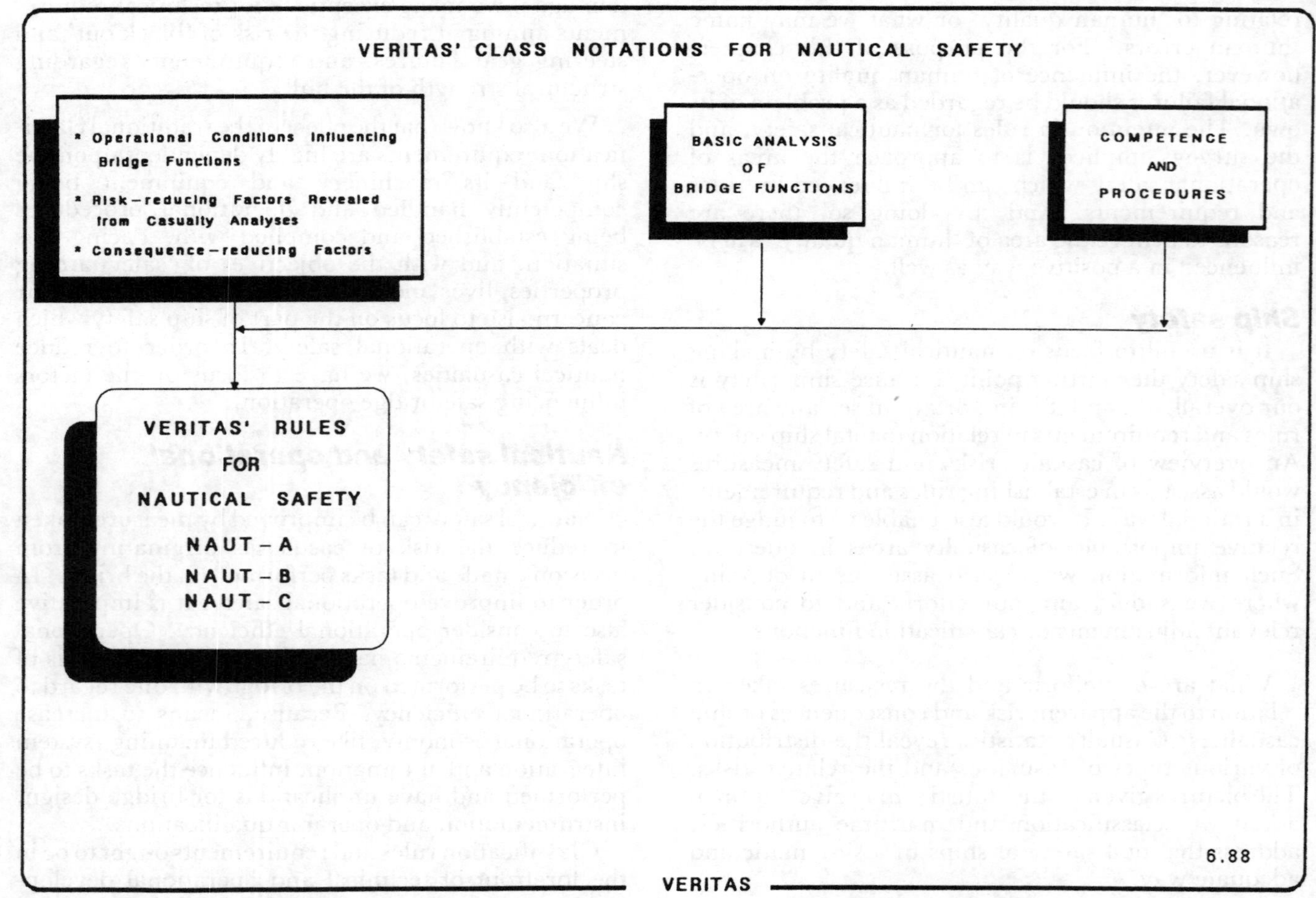
VERITAS' CLASS NOTATIONS FOR NAUTICAL SAFETY
* Operational Conditions Influencing Bridge Functions
* Risk-reducing Factors Revealed from Casualty Analyses
* Consequence-reducing Factors
BASIC ANALYSIS OF BRIDGE FUNCTIONS
COMPETENCE AND PROCEDURES
VERITAS' RULES FOR NAUTICAL SAFETY
NAUT-A
NAUT-B
NAUT-C
6.88
VERITAS

ment in order to serve their function. Therefore, it is vital to keep abreast of the latest technical possiblities and operational needs and to be aware of future developments. To achieve this, we should give our attention to the basic aims and means both with regard to safety and operational economy, as follows:

Increase safety

- Reduce operational failures through mutual adjustment between technology and the operator.
- Reveal operator information requirements through functional analysis,
- Ensure information quality through functional requirements.
- Ensure information availability through system performance and reliability requirements.
- Establish total system performance requirements considering the operator as an integral component in the total system configuration.
- Estimate operator workload in the system configuration through definition of functions, operational procedures and time available for carrying out the tasks.

Increase operational economy

- Reduce manpower requirements by arranging for safe one-man bridge operation.
- Reduce maintenance and repairs by improving equipment reliability.
- Reduce fuel consumption by increasing navigational efficiency.

In order to reduce operational costs, the general aim for most shipowners is—in addition to having established an unattended engineroom port to port—to achieve one-man cargo handling, no maintenance at sea and a one-man operated bridge, involving external communication and extended monitoring functions of engine, cargo and safety systems. It is important to know the operational aims in order to foresee the development and be prepared to evaluate new systems and take necessary actions to maintain an adequate operational safety standard.

'Future' developments

It is apparent that we are in the beginning of a new development process and must expect the proliferation of new types of instruments and system configurations. We know that a new generation of computer-based navigational aids and new satellite system for positioning including integrated grounding avoidance and collision avoidance systems together with satellite communication and data transmission facilities offer opportunities to increase operational efficiency beyond any standards of today.

The use of computers in automation systems serving primary bridge functions introduces the possibility of new types of failures related to software quality, to name one more reason to establish adequate operational safety standards. Extended integration of bridge instruments, further automation and utilisation of 'expert systems' to assist the operator in performing his tasks can reduce the workload and change bridge functions, operational procedures and bridge routines. Increased use of advanced technology, reduced manning and extended operational responsibility for the officer on duty emphasise the importance of mutual adjustment between the technology and the people who use it in order to safeguard ship operation.

Another factor influencing functionality, workload and safety is rational instrumentation. No matter the level of automation, it is of prime importance that the instrumentation is based on information needs derived from functional analysis rather than on what is technically possible. The result of this approach may also have a positive influence on capital investment—only buy what there is a need for . . . These are among the factors we have to take into account when we are dealing with nautical safety.

Veritas rules for nautical safety

Through the rules, we want to make our contribution in reducing the risk of collisions, groundings and heavy weather damages by aiming at improved bridge functionality and instrument reliability. Furthermore, we want to contribute in reducing the consequences to life and environment as well as financial losses if an accident should occur. We meet this aim by assisting the owners in providing the ship with effective procedures for different operational conditions and emergency situations. The rules give consideration to navigational efficiency and operational economy.

Put succinctly, safe and efficient navigation is dependent upon the ship proceeding on the correct course at optimum speed in relation to destination and environment. The wrong course or speed in relation to the waters may cause grounding. In relation to the traffic it may cause collision, and in relation to the weather it may cause structural damage as well as loss of time and increased fuel consumption. Lack of knowledge about the ship's responsiveness to an alteration in propeller revolutions and rudder angle may cause faulty manoeuvres in harbour and lead to contact damage, collision, grounding, explosion and fire.

Consequently, the overall aim of the rules is to focus on operational conditions that may cause the ship to proceed on the wrong course or at the wrong speed. Furthermore, they focus on conditions that may have an impact on the consequences of accidents. Accordingly, the Rules are based on: operational conditions influencing bridge operations; risk-reducing factors revealed from analysis of navigational casualties; and consequence-reducing factors.

Rules structure

The rules are based on functional requirements and give guidelines as to how the requirements can be met. They include relevant international rules and recommendations within the same subjects. The functional requirements are general and only to be adjusted if operational procedures and functions to be carried out are altered. The solutions and methods given as guidelines can be regarded as 'approved,' but will be amended whenever technological developments or operational experience form the basis for better solutions. The guidelines given do not exclude the application of alternative solutions, provided the functional requirements are met.

We have divided the Rules into three groups in

order to offer classification of nautical safety on different levels. Naut-C is the least extensive class notation and covers bridge design, comprising:

- Bridge visibility,
- Location of workstations,
- Console configuration,
- Location of instruments.

Naut-B will, in addition to items in Naut-C, also cover instrumentation as follows:

- Range of instrumentation.
- Instrument performance, functionality and reliability.
- Equipment installation.
- Bridge working environment.

Naut-A is the most extensive class notation and will, in addition to areas included in Naut-C and B, also cover:

- Information on ship's manoeuvring performance.
- Operational procedures for normal conditions.
- Operational procedures for abnormal situations and emergencies.

Classification competence

To develop rules with regard to safe and efficient bridge operation and to perform an efficient classification service in this area requires expertise in various fields, such as: ship operation, nautical experience, functional analysis, human engineering, bridge design, data technology, software quality, electronics, electromagnetic interference, radio communication, ship manoeuvring, contingency and international rules and requirements within relevant areas.

This classification service on nautical safety standards was established in January 1987 and we have since gained some experience by working with different owners and builders. With regard to training of surveyors, we are aiming that experts in automation, data technology and human engineering acquire pertinent knowledge in navigational subjects and gain practical experience by accompanying a nautical expert on ship trials. Furthermore, we are ensuring that everyone involved with this service attends courses in data technology. In order to obtain good team work and to have sufficient flexibility, everyone in the main team is to master two subjects in addition to his own expert competence.

Classification service

In order to achieve the goals of the rules and provide efficient assistance to owners, builders, manufacturers and the end user, it is imperative that classification starts at the planning/design phase. This will enable our experts to make their comments for further improvements of design and system configurations. We will comment on owners'/builders' drawings in relation to: rule requirements, intended instrumentation, flag requirements and ship types and trades.

In the building phase we will ensure that approved drawings and specifications are met and 'supervise' the building and installation. Before delivery of the ship, our experts will be observers at relevant tests and trials, ensuring that the bridge systems are functioning in accordance with specifications. We will follow up ships in service through a feed-back reporting system on instrument reliability.

The classification concept has not changed very much since its inception, in spite of the tremendous technological development that has taken place in the meantime. Apparently, the classification model has responded very well to the needs of the marine industry. Now, however, there are stong calls for reduced manning of ships to lessen operational costs. This will, in turn, stress the importance of human reliability in various operational environments as the means to maintain an acceptable safety level. Consequently, a need may arise for compulsory classification services which are extended to cover the vital operational aspects of ship safety.□

MARINE INQUIRIES: THE TECHNICAL DIMENSION

C.A. Sinclair, C.Eng, FIMarE
Formerly Chief Surveyor, The Salvage Association

WHENEVER THERE IS A CASUALTY within the shipping industry there will be several and, at times, many interests, all seeking to obtain details with a view to satisfying their particular principals. The vessel's master and other officers at times become rather confused by the number of people asking the same or similar questions. Also, they often do not know who is entitled to information; so that at times they withhold, and at others give, information to people who are acting for parties who are, or may become, in conflict with the owners.

Those inquiring into the cause of any casualty will have a distinct advantage if they are amongst the first to look into the case, are allowed to interview those directly involved and are able to set people at ease, so that they give honest answers to the questions raised. The only people who are in a position to demand co-operation are flag and, perhaps, port State officials but, because of their official capacity, those answering their inquiries are usually on their guard. Since there can be confusion as to the differences in entitlement to information, it may be useful at this stage to consider why these differences exist.

Formal inquiries into UK casualties are held under the Merchant Shipping Act; and, in other States, under its equivalent. These inquiries take place when State officials are seeking the truth with a view to avoiding a recurrence; or when they are dealing with persons responsible for casualties, be they owners or crew. It is usual in these cases for the inquiry to be official in character and, when the case goes before a court, there are generally technical assessors present to assist the judge in weighing the evidence. In some countries these official inquiries are inquisitorial in character. The tendency is for such inquiries to be held when there has been loss of life or loss of vessel; where severe environmental problems are involved, or in response to a public outcry.

When a serious casualty has been reported, it is usual for the flag State to send surveyors to conduct on-the-spot inquiries, interview crews or gather information from other available sources. This information will be used to decide whether to embark upon a formal inquiry; and, if so, should also be available at the time of the inquiry.

The disadvantages of this system of inquiry stem largely from the time taken up in setting up the court, accumulating evidence and instructing legal representatives. An additional disadvantage is that many crew members may not belong to the jurisdiction of the flag State and, due to lack of proper reciprocal arrangements, they cannot be forced to attend; indeed, at times they cannot even be located. A further possible disadvantage of this form of inquiry is that if a crew member loses his certificate, it will only be the one issued by the flag State, so the man may still be able to sail with the same rank in his own national fleet.

In taking testimony from those directly involved in the casualty, it is often found that their memory of detail will be defective: perhaps due to the passage of time, an inbuilt protective mechanism, the influence of discussion with others, or a host of other causes. Opinions will be generally available but facts will be more difficult to come by. Generally speaking, factual evidence of the type needed is destroyed by the casualty, so judgment must be based upon opinions.

Depending upon its own laws, the port State will be in a similar position to the flag State except that, quite frequently, they have the power of arrest immediately following the casualty.

It is clear that judgment of both flag and coastal States, at least as to cause should be about the same, given that both parties have the means to extract information and the advice of top experts in whatever field is necessary. The marine inquiry seldom falls into the staight black-and-white category: as may be noted from the *Betelguese* case, where the State inquiries of France and Eire both came to different conclusions, in spite of the employment of what they considered to be the best expertise available.

Some major disasters

To give some examples of the opinions of others, the comments on some typical cases are quoted below.

Titanic—'Although this inquiry was held with top level legal aid it would seem that after all these years the validity of some of its judgments leave room for doubt.'

Alexander Kielland—'Inquiry convened under Norwegian Maritime Law. The Commission of Inquiry was appointed by the Ministry of Justice and consisted of five men chaired by a District Judge.' (*Lloyd's List* 5 May 1981, page 14.)

Det Norske Veritas said that the Commission of Inquiry 'over-dimensioned and idealised its function and place in the structure of inspection and control which is based on supervision of control and repair procedures by the builder and operator rather than detailed direct involvement.'

(*Lloyd's List* 4 April 1981, page 1.)

Kurdistan—'Wreck Commission stated proceedings were taking too long with too much detail being given in primary and expert evidence.'

(*Lloyd's List* 7 July 1981, page 1.)

(In all the Inquiry lasted 10 weeks and cost over £1 million.)

Dr Ewan Corlett, giving evidence at the Inquiry, said the 'maritime world should take a leaf from the aeronautical industry's book in the way casualties are

The original version of this chapter was published in the Translations of the Institute of Marine Engineers, Volume 95.

investigated. More casualty information should be made available to owners and the procedures for investigating a crack in the hold of a ship or an aeroplane should be the same.

As soon as an accident occurs, the authorities in the region should carry out their own investigation as quickly as possible and send the preliminary information by telex to the appropriate administrations around the world. Then information would be given to classification societies for further investigation and information passed to interested parties and owners.' *(Lloyd's List* 15 July 1981.)

Betelguese—In a speech in October 1980 Dr Black said of the inquiry 'it seemed too much to have the character of a civil action rather than an inquiry into causes. The fundamental nature of the inquiry seemed to be lost sight of.'

Cuttlefish—Whilst noting the delays in most marine inquiries it must be admitted that the judgment on the skipper's certificate was delivered only about 10 months after the stranding. It is understood that the Department of Trade have a formula for deciding whether a casualty is of a serious nature or otherwise and thus merits the cost, time and trouble of an inquiry. It would be of interest if someone from the Department would outline their basis of decision-making. In the event that every accident were investigated, besides adding considerably to the cost, it would mean that the operating personnel's certificates and livelihood would be at stake and, where the loss was purely commercial, this may be too severe a penalty.

The collection of reliable data can now be implemented by making use of on-board computer printouts of a wide spectrum of operational data and the use of 'black box' techniques is also now a commercial possibility. Firstly, in relation to computer printouts, these can be log book-type information or may merely record readings of data which indicate operations outside of previously laid down parameters. Further to producing on-board records it is possible for the printouts to be conveyed by satellite to a printer in the owners' office thus providing a most valuable management tool.

The information available on a fully integrated recovery system may include much more detail than that required for the black box whose main function would be for casualty or defect investigation.

An indication of the data recorded for black box purposes on the vessel *City of Plymouth* was as follows:

Continuous Data

Channel	**Measurement**
Channel	Heave acceleration
1	Heave acceleration
2	Sway acceleration
3	Surge acceleration
4	Pitch Angle
5	Roll Angle
6	Bow acceleration
7	Stern acceleration
8	Midships stress 1
9	Midships stress 2
10	Midships stress 3
11	Midships stress 4
12	Bow stress
13	Compass
14	Wind speed
15	Wind direction
16	Rudder angle
17	Engine rpm
18	Air temperature
19	Barometric pressure
20	Sea temperature
21	Engineroom telegraph
22	Ship speed
23	Water depth

Event Data

Fire alarms (8)
Bilge level Alarms (6)
ME alarms (6)
SG alarms (4)
Bow thruster (1)
Nav. lights (5)
Bridge voice (1)
VHF radio (1)

The black box would carry suitable release gear and recovery beacons besides being suitably located and protected against fire on board. Whether considering computer printouts or black box information it must be realised that the data contained therein whilst the property of the owner may well be used against him by opposing parties in cases of legal conflict, because although confidential data could be obtained by subpoena.

If we re-examine the findings of inquiries into some of the disasters or the really terrible catastrophies in the light of the multiple causative factors, it will be clearly seen that the human error factor, whilst triggering off the event, is by no means the whole cause. Looking briefly at the tribunal's recommendations after the Widdy Island catastrophe, we note that there are 45 of them—some of which would have stopped the occurrence at failure level, others certainly before it was more than an incident.

Further, if we consider the IMO regulations on steering gears, we will note that the start of the *Amoco Cadiz* incident may well have been initially caused by a defect, but, had adequate design conditions made it possible for the damaged gear to become redundant and its functions taken over by other equipment, the stranding could have been avoided.

It is difficult to foresee the possibility of any vessel being designed where the built-in redundancy would take care of every possibility of human, electronic or mechanical failure. However, the pressure of environmentalists and others may well cause the industry to consider designs approaching this ideal. If we consider the ramifications of this we may well be required to duplicate the propellers, rudders, main engines control systems and other equipment. It must be realised that in the duplication of equipment, we are increasing the possibilities of failures, so that it is often better to simplify instead of adding to complexity.

Adversarial inquiries

Other inquiries are adversarial in character, in which the evidence selected for presentation to the inquirer is considered at the time to be in the best interests of the party presenting the evidence. The formidable list of inquiries may include: owners' management; owners' P & I club or surveyors; classification society surveyors; underwriters' surveyors; cargo interests; charterers; labour organisations;

and third parties.

Owners will expect to have full access to their vessels, crews and documentation so, given the goodwill that is necessary, their inquiries into causes of casualties should be straightforward. Even the owners' investigation can be rendered quite difficult if any crew member feels that his job, reputation or integrity is in question. In other cases, where the crew is under arrest by the flag or port State, owners may be temporarily denied access.

The owners' P & I club will usually be involved in investigations. It is their practice to appoint maritime law experts who, in turn, engage the technical expertise they consider appropriate. The P & I club inquiry is most searching and the lawyers seldom accept any statement on its face value; it is their normal practice to investigate in depth. Technical experts frequently find that the questions and probing of their legal colleagues open up quite unexpected avenues of investigation.

Classification society investigations

Classification society investigations have in the past been confidential between themselves and the owners; thus, the results were not released in a particularised form without the owner's prior knowledge and consent. Today there is considerable pressure being brought to bear for the publication of findings, especially in the effort to clamp down on substandard ships.

Under the present arrangements, the build-up of knowledge, and the further development of this knowledge necessary in the marine field, are promoted by the fairly free access given by most owners to the classification societies for inspection of vessels and records. Any change in this arrangement could well adversely affect the necessary flow of information. It must be admitted that the classification societies can and do withhold their certification where they find that they are being obstructed; however, the lack of co-operation can be much more subtle than is generally realised.

Whilst many outside interests consider that classification societies do not go far enough in their inquiries into the cause of damage, it must be realised that their function is to investigate with a view to ensuring that ships are initially constructed and then maintained to an acceptable standard. Thus, where there has been a human failure, they are unlikely to go into this aspect in any depth. The build up of knowledge within the societies stems not only from their internal expertise but from close co-operation with ship, engine and equipment builders and ship operators.

Underwriters

Dealing now with marine enquiries on behalf of underwriters, the owners will have many other matters to consider before they decide their basis of claim and they will certainly lay stress upon the factors which they, at the time, consider to be in their best interests. In every case of a marine casualty the onus of proof of cause rests with the owners: after all, the ship is their proprerty and any liabilities, penalties or compensation will accrue to them.

The owners are, nevertheless, responsible under policy conditions to provide full disclosure of all relevant information necessary in assessing the claim. In deciding the basis of claim and what information they will need to divulge, the owners will also have in mind the possibility of claims being made against them by charterers, cargo interests and third parties.

Underwriters have been criticised for insuring substandard ships but it must be stressed that the purpose for which classification societies were set up and the reason for their continued existence is to set standards and see that they are maintained, so that the vessels and their cargoes present a reasonable insurance risk. Thus, if substandard ships sail with classifaction certificates in order, the underwriters are being let down to this extent.

Where underwriters, or P & I clubs for that matter, call for pre-insurance surveys, it is a reflection on class and the underwriters often stress the fact that they are not in the ship management business. However, it should be stressed that, in the development of the marine industry, people other than underwriters, such as charterers and cargo owners, have placed considerable reliance upon the class maintained.

With the increasing degree of intervention by world bodies and States, it would appear that they could well ultimately take over the functions of class. If this were to happen, the policing of the regulations would still present the possibilities of lack of uniformity due to local political and business pressures. Even if an adequate organisation of an international character were set up, the interpretation of regulations would still be left in the hands of individuals, all with differing standards of knowledge and ethics. For these reasons it appears vital that the attempts of classification societies to unify their practices and standards are successful.

Human factors

It is not unusual for the cause of a casualty to have more facets than are obvious to the casual observer. Whilst human factors usually act as the trigger, there are other matters also worthy of mention. In a multidiscipline investigation it may well be found that the trigger is in a crew area, but that owners, managers, builders and even designers are not without fault.

The human factors connected with marine casualties go far deeper than are realised at first glance. There are the well-known and often-repeated factors, such as poor seamanship or lack of experience, but negligence and ignorance can also be involved. These factors apply to the on board situation. Short-based decisions also have a distinct bearing on causes. Cause or contributory factors can lie in the concept and design parameters, in ownership practices, communications, staffing, training, maintenance and supplies; and in charterer and other user aspects. No matter where we look, there is no room for complacency in any part of this great industry.

If we consider the chain of events in some major catastrophes, it looks rather like this:

- Catastrophe . . . Inadequate isolation;
- Accident . . . Inadequate containment;
- Incident . . . Inadequate safeguards;
- Failure . . . Inadequate redundancy;
- Defect . . . Inadequate load-time limitation;
- Error . . . Inadequate inbuilt safety factors.

When this chain of events is considered it will be realised that, to arrive at the proximate cause of any accidental occurrence, considerable thought and careful evaluation are required. Fortunately, the sifting of these matters in fine detail is not usually performed by the surveyor in the field but is entrusted to the average adjuster.

Prime and proximate causes

Before proceeding further it may be useful to consider the commonly used terms 'prime cause' and 'proximate cause'. Those of us who work in co-operation with insurance interests would define 'prime cause' as the first in order of time or occurrence; where as the 'proximate cause' is the 'dominant or immediate' cause, the nearest or next in a chain of causation. Where there are interacting causes, the efficient or dominating cause is deemed to be the 'proximate cause'.

It is clear that in all but the most simple cases there will be several causative factors and it has been held in law that insurance matters are not really concerned with the 'cause of causes'. Lord Bacon said:

> *'It were infinite for the law to judge causes of causes, and their inpulsion one on another, therefore it contenteth itself with the immediate cause'.*

In fact, an eminent judge decreed that even damage caused by negligence or skilful navigation could rightly be held as proximately caused by the perils of the sea. In this case—*Wesport Coal Co. v. McPhail* (1898)—the loss was regarded as caused by the perils of the sea and only remotely by negligence or unskilfulness of the master or crew.

In view of the complexity of insurance matters, and for other financial or liability reasons, it is becoming increasingly difficult for the owners' technical man in the field to be able to provide the underwriter's surveyor with the full documentation and information he needs to form a definite opinion as to cause. Apart from seeing the damage, it is frequently necessary to examine scrap and official logs, movement books, classification records, work books, cargo stowage plans and notes of protest—to name but some of the records involved. Where the owners fail to provide evidence which the surveyor considers necessary, he is quite within his rights to defer expressing any opinion. If the necessary documentation is not forthcoming, the surveyor should inform his principals so that they can, if they wish, bring pressure to bear on the owners.

Whilst the ultimate decisions as to where and with whom the fault or liabilities lie may well be made on the basis of records and other paperwork, these decisions will generally be affected by the descriptions of findings given by surveyors who have surveyed the damage. For this reason, it is important that the surveyor records a careful description of the extent of damage, condition of parts or surroundings, measurements and any other factual information. No damaged parts should be cut away, removed or repaired without first making a full record of the conditions found. To this end, it is often advisable to obtain photographic evidence. It must be remembered that, usually, when repairs are in hand the evidence is progressively being destroyed; and if those making the decisions have something to look at they are able to understand the paperwork somewhat better.

When presenting a report on the cause of the damage, the expert is fully entitled to express opinions on the basis of the technical evidence presented to him. However, he must make it quite clear as to whether he is presenting facts or opinions. Where opinions are expressed, they are quite likely to be challenged and, with this in mind, notes should be made of the basis upon which the opinions are formed, because any questions could well arise some considerable time later when the memory of the basic considerations has been dimmed.

Unsatisfactory investigations

Investigations into the cause of marine casualties, whether by governments or other interests, are often less than satisfactory for many reasons. It is probably relevant to mention just a few.

(a) Human nature being what it is, people have a built-in protective mechanism and whatever they reveal about the casualty, or its circumstances, will usually be affected by this—often quite unconsciously.

(b) The prime cause of the casualty is often masked by destruction of the necessary evidence. This could be due to the extent of damage, removal of vital evidence during subsequent operation or removal during repairs; or it could even be deliberate.

(c) There are occasions where, for commercial reasons, the owners reveal less than the full amount of information which is available to them; or hide behind classification certificates and the like.

(d) On other occasions, the expertise necessary to investigate a particular aspect of a casualty is either not used or even not available—after all, no-one can be expected to have all the answers.

(e) At times, cause aspects are considered in city offices far removed from the casualty and the people who were directly concerned. At times, opinions are arrived at without sufficient accounting being taken of those who saw the condition.

(f) Some damages defy cause definition, either because of their unique character or because all concerned are working at the extreme edge of the border between theoretical and practical knowledge.

(g) It is unusual to be presented with facts, except for the actual physical damage; opinions are usually provided and their influence will depend upon the skill of the presenter.

Underwriters—The depth of any investigation on behalf of underwriters will depend upon policy conditions, so the technical investigator needs to be properly briefed before the survey is conducted. For example, where a vessel is insured for total loss only, it is pointless agreeing to the cost of opening out machinery to investigate the cause of mechanical damage. Where vessels are insured on liner negligence clauses, there is little to be gained in trying to blame the casualty on the negligence of the owners' representative if he is alleging that the crew were at fault, as the damage will be covered under policy conditions if either type of negligence is the cause.

Hull underwriters—Generally speaking, in investigations on behalf of hull underwriters, if all the vessel's certificates are in order it is accepted that she was fit for the purpose intended at the time they were issued. This state of affairs is quite different from the attitude of cargo interests, be they on behalf of cargo owners or underwriters; and, because there is likely to be a conflict of interest, there should be complete confidentiality and separation between the two. Whilst it may be quite in order to reveal certain matters to hull surveyors, it may well be found that cargo interests, if informed, would latch upon the defect as proof that the vessel was unseaworthy before the casualty.

General average—Another area where there may be conflict of interest relates to general average, as cargo interests will have to pay their proportion of the repair costs. The law of general average derives from the principle that 'all must contribute to that which has been sacrificed for all'; or, in other words, 'when one who partakes in a maritime venture incurs loss for the common benefit, it (the loss) should be shared ratebly by all who participate in the venture'. Modern law and practice relating to the adjustment of general average is determined generally by the York-Antwerp Rules of 1974.

Labour organisations and third parties—When dealing with inquiries into casualties by labour organisations or other third parties, it is prudent to take the advice of the P & I club's legal advisor before revealing any details. In fact, as a general principle, in the case of without prejudice surveys, the extent of the physical damage and the time and place of the casualty are about as much as it is safe to reveal until after legal opinion has been obtained.

Collisions—In collision cases it is in the owner's interests to see that he is represented in surveys of the physical damage to both vessels. Thus the surveyor who sees his vessel, on behalf of underwriters, ideally should also attend the opposing vessel without prejudice.

Conflicting interests—When inquiries are being conducted, no surveyor should represent conflicting interests. It is quite difficult, but still necessary, to comply with this requirement in remote parts of the world where technical expertise is in short supply.

It is well known that in some ports one surveyor may find himself wearing many caps and called upon to represent conflicting interests. This is less than satisfactory but, if unavoidable, the consent of the parties involved should be obtained. A surveyor's duty is to his principals and he should never commit himself, either orally or in writing, to other surveyors or their principals.

Clarity of technical reports—Turning now to the technical content of reports prepared by surveyors and others who are experts in their own particular field, the value of any document will be in its clarity, relevance and simplicity. If any occurrence is difficult for the specialist to understand, the tendency will be to present a report which, because of its complexity, is of little use to the parties concerned in making the necessary decisions.

In these sophisticated days of automation, electronics, computers, hydraulic systems and satellite communications, there are bound to be many areas where those investigating casualties need specialist advice. My advice is if you don't know, find someone who does, but keep control of the case yourself. Use the expert to provide the specialist knowlege; but use your own maritime knowledge when arriving at decisions.□

TYPES OF CASUALTY INVESTIGATION

LATENT DEFECT has been defined as a 'defect which could not have been discovered by a person of competent skill and using ordinary care'. The general rule is that latent defect does not mean latent to the eye. It means latent to the senses—i.e., undetectable by physical testing or by any other practical means of examination, hammer, heel or forceful persuasion.

The criterion is that if a competent surveyor is employed he must also be diligent. It does not extend to carrying out a special survey at the commencement of each voyage. It is not obligatory for a ship's officer, at the start of a voyage, to go and tap every rivet to find if it has a defect or not (*Cranfield Bros.* v. *Tatem Steam Nav. Co.* (1964). LIR 264 270).

However, there has been considerable difficulty in interpretation and it is not the surveyor's job to interpret policy conditions. In view of the varying opinions of owners, their adjusters, underwriters and others, where a claim is made and it is alleged the damage has been caused by a latent defect, the surveyor should, if there is any doubt, and if possible, agree the manner in which the damage has been sustained without agreeing that the term latent defect applies, thus leaving the underwriters to interpret the meaning of the words within the context of the policy.

Where vessels are insured on the tramp-type clause, the cover is only for damage sustained to the hull or machinery through a latent defect, thus excluding the cost of replacing the latently defective part. Where, however, the line type clauses apply, the latently defective part (if it has failed), as well as the damage caused by the failure, is considered to form part of the claim.

In order to indicate further the types of defect which are or are not considered to fulfil the description, several quotations are included for consideration.

Materials defect

Judgment of Mr Justice Kennedy: 'The phrase "defect in machinery" in a business document means a defect of material in respect either of its original composition or in respect of its original or after acquired condition . . .'.

Judgment *Scindia Steamships (London) Ltd* v. *London Assurance:*' . . . the circumstances revolved round an old flaw in a tailshaft, the shaft breaking in drydock while the propeller was being wedged off. The end of the shaft, with the propeller attached, fell into the dock with resultant damage to the propeller. The crack in the shaft was treated as a latent defect, so that the damage to the propeller was allowed, but not the breakage of the shaft. As Mr Justice Branson said . . . "damage to hull or machinery caused thought a latent defect in the machinery is something different from damage involved in a latent defect in the machinery itself".'

Design defect

Judgment of Mr Justice Kennedy: '. . . but for the purposes of today it is sufficient for me without attempting to define its boundaries to say that the phrase at all events does not in my view cover the erroneous judgment of the designer as to the effect of the strain which his machinery will have to resist the machinery itself being faultless, the workmanship fautless, and the construction precisely that which the designer intended it to be'.

Error in design

Judgment *National Sugar Refining* v. *Las Villas:* 'Here there was damage to cargo by the entry of seawater because an improper design has been used in the manufacture of the port bilge suction valve. This was not apparent upon visual inspection nor discoverable by a reasonable inspection, and in the view of the court was a latent defect'.

Faulty workmanship

Judgment *C. J. Wills and Sons* v. *World Marine Insurance Co. Ltd:* '. . . the hoisting chain of a dredger broke owing to a defective weld in a link. The damage caused by the ladder and buckets falling was allowed by the learned judge who remarked "This case appears to me to afford a good example of the legitimate claims which the Inchmaree Clause was intended to cover". So, at least, in the mind of one eminent authority, there existed no doubt that imperfect welding could be classed as a latent defect.'

Crew negligence

Judgment *Dimitrios N. Rallias* (1922): 'With the operation of the ten per cent machinery negligence clause, it is, of course, important to distinguish between a 'latent defect' and negligence of the master, officers or crew. Lord Justin Atkin was quite clear that negligence is not a test of latency, it follows that if the defect could have been discovered by the ship's personnel using ordinary care, failure to detect the fault will probably result in a claim for negligence.'

Wear and tear

Section 55 (2) (c), The Marine Insurance Act 1906: 'Clauses are subject to the stipulation in regard to due diligence having been exercised by the owners and their managers, and in neither case does the cover afforded extend to include liability for loss or damage due to speed or "wear and tear". Marine Insurance Act, 1906 provides that "unless the policy otherwise provides, the insurer is not liable for ordinary wear and tear".'

'A latent defect is seldom related to wear and tear. It is more often of the nature of flaw in the material and so hidden that it cannot be discovered by a person of competent skill, using ordinary care.' (*Dimitrios N. Rallias* L1R 363.)

Corrosion

Corrosion is not a latent defect because it is a gradual deterioration and can be expected, but in '*The Bill*' (1942) AMC 1607, corrosion aspects appear to have troubled the Court.

It was said that 'a latent defect is one that could not be discovered by any known or customary test'. If in that ship corrosion has not been discoverable by any known or customary test, it could have ranked as a latent defect. Nevertheless, a competent surveyor should have been able to detect a wasted plate, since he would know the area where corrosion was likely to occur. He could have used a torch and hammer.

Where a pipe fails because of internal corrosion, the life span of the item is of importance. If it is of a design or of a material likely to suffer from accelerated corrosion, there is lack of diligence in not ensuring its frequent examination and replacement. If the pipe is relatively new and was defective in manufacture, causing it to corrode long before the end of its expected period of service, and if there is evidence that the defect was not discoverable by the use of due diligence, the owner could no doubt plead the exception of latent defect.

It is therefore indispensable to employ the owner's superintendent to check regularly all parts of the ship, so that it may be said later that the qualified people entrusted with maintenance and repairs did their job thoroughly, and that they were seen to do so.

A common cause of trouble is damage to cargo caused by a fractured pipe, the pipe being protected by a wooden or steel sheath. The fact that the pipe is not visible is not accepted as an excuse, since it has been held that there are many other ways of detecting the fault (*Sewaran* v. *Ellerman Lines Ltd* (1937), L1R 97). Dilligence must be exercised.

Fatigue

Fatigue, in an engineering sense, refers to those situations in which fracture is defined in terms of the number of cycles of loading sustained, as differentiated from failures which occur on the first application of load. Fatigue accounts for a vast majority of service failures encountered. Appreciation of the factors which influence it is often on a qualitative, rather than a quantitative, basis.

Fatigue is basically a surface problem. Under an applied bending load, the surface will be in a state of maximum tension so that the crack usually initiates from this position. Hence, the surface is the major weak link in the specimen. The average fatigue strength of a ferrous metal is about half of its ultimate tensile strength (at 10^7 cycles) provided that there are no surface defects present. The fatigue strength can be drastically reduced if surface imperfections are present. These can take many forms—e.g., geometric shape changes, notches, tool marks, pits, inclusions, etc.—but all have one factor in common: they provide a region of high tensile stress concentration, additive to the applied bending tensile load. The overall effect is to accelerate the initiation of failure time (reduced cycles to crack initiation).

The effectiveness of various corrosive media in decreasing fatigue strength is well known. Notched specimens of steel in ordinary tap water, for example, may exhibit less than one-tenth of the normal fatigue strength of the material. In fact, it is now recognized that 'air' fatigue is, in itself, a manifestation of corrosion fatigue since a substantial improvement is found when testing is conducted in vacuo. In the main, when the fatigue strength is lost, fatigue failure will eventually occur at any load in a corrosive environment.

Investigation into the cause of fractures, be they in hull or machinery, is a subject which needs much care. Certainly, the help of metallurgists and vibration experts can often provide valuable assistance. The fact that a shaft failed, or a plate was noted to be fractured, during heavy weather is not in itself any indication that the weather conditions were responsible.

Hull fractures

Recently, large areas of plating have, literally, fallen off the sides of vessels and further investigation has, almost without exception, shown that the initial failure occurred in the internal structure. It has usually been found that the corrosion of internal members has set up local stress concentrations, which are followed by corrosion fatigue. The weakening of one part of the structure will cause extra stresses to be placed on others; and so the progression continues.

Unfortunately, the corrosive effects of the working environment are such that it is seldom possible to get any positive indication of the growth rate, so it is usual to assume that the cause factors include all heavy weather since the internals were seen to be in sound condition. This assumption, whilst often suiting owners, is often wide off the mark. If the design factors were correct at building, it is frequently true to say that the vessel's pattern of corrosion, or even conditions of loading, render a problem quite different from that envisaged by the design parameters.

Fatigue fractures can be initiated by several basic factors, such as propeller-excited vibration, excessive local hull stresses and resonant vibrations, and it is quite clear that the investigating surveyor would need help if these aspects are to be considered in any depth.

Weather damage

Investigations into weather damage will involve an attempt to ascertain the actual severity of the weather conditions encountered on the relevant passage or passages. Logbook information may be the only evidence submitted but it must be conceded that what one man will describe as extreme another will merely say bad, so the logbook information does require some checking. In many areas of the world's oceans, government meteorological and commercial weather facilities can give reliable day-to-day information.

Having established that the weather pattern was as described, the next matter to consider is the condition of the vessel in so far as loading pattern and stability are concerned. It is clear that the types of damage likely to occur are different for loaded and ballast passages. The next factor to consider after the actual weather pattern is its effect upon the vessel. Wind force in itself seldom causes damage but pounding, racing or rolling may. Logbooks are frequently deficient in their watch-by-watch description of these effects, with the result that it is necessary to assume various voyage percentages; whereas if proper entries were consistently made, a better judgment would be possible.

When damage has been sustained the next question will be why, especially if the vessel has met similar weather conditions on other passages without sustaining similar damage. It would be wrong to assume that the cause or difference is due to bad seamanship without first looking at other possibilites. Perhaps the failed part has deteriorated to the extent that it is much more sensitive to the stresses caused by weather.

If a vessel has sustained severe pounding damage, the fact that at the last drydocking there was no notation of class is no proof that she was free of damage. The damage may have been progressive over a period of years but the severity had not increased to the extent where it required noting over the period to the previous inspection.

Structual damage to vessels, often ascribed to heavy weather but caused by shift of cargo or even carriage of unsuitable cargo, can hardly be blamed on bad weather even though sustained in such weather. Factors to consider are, for example, stowage, securing and also the metacentric height produced on the vessel. Many vessels, which were quite unsuited to the trade, have been lost when carrying ore cargoes having a very high GM, with consequent quick-rolling motion producing excessive wracking stresses.

Design factors also require close consideration. If we take for example some of the rudder failures recently ascribed to weather, it must be realized that design has had its part in the chain of causation.

Explosions

Investigating the causes of explosions will involve looking at each of the basic ingredients necessary—i.e. the fuel, the heat source and the oxygen supply. Where warlike activities are encountered, the three elements are introduced externally but, with the subsequent fire and further explosions, it may be impossible to judge with any degree of certainty whether the start was from an outside source or otherwise. Tanker explosions went through a phase where most investigators felt certain the cause would be found in some part of crew activities causing a heat source. Later everyone seemed to consider static within the tank atmosphere as being the possible cause.

The exclusion of fuel or a heat source from tanker operations has not been possible, the former by virtue of the nature of the cargo; and the exclusion of the latter has lacked certainty. For these reasons, today there is a general requirement to fit and use inert gas equipment and thus exclude the oxygen. Where explosions occur on ships fitted with this equipment it is usual to blame the crew for failing to operate the equipment properly. However, investigators should look at the whole pattern of operations. It would appear that all systems should have built-in redundancy, so that in the event of failure not only is the condition of failure indicated but also another plant or arrangement takes over.

In vessels other than tankers, a frequent cause of explosions is to be found in the existence of gas pockets. This has been especially true of coal cargoes and investigations generally show lack of proper ventilation of some part of the vessel. Surveyors and others need to take special care when investigating casualties on chemical carriers or vessels with dangerous cargoes and specialist advice is generally necessary. Because of the risks to the vessel and others, quite apart from the dangers to the surveyor's own person, close consultation with the crew and the IMO Code becomes imperative.

Fire

When trying to ascertain the cause of fire on board ship, it must be recognised that there is so much combustible material about that the area of greatest intensity may well not have been the source, but merely the place where the most combustible material is available . Spontaneous combustion is in itself not a cause of fire, the cause usually lying in the moisture content of some substance being carried in bulk. In any event, further investigation will be called for in each such case.

Accommodation fires, whilst usually caused by carelessness with cigarettes or other heat sources such as heaters or galley stoves, do sometimes turn out to have their origins in electrical failures. However, especially in port, welding or cutting splatter is a frequent cause. Because of liability factors it is usually necessary to establish the actual reason, first, for the start of the fire and then for the failure to extinguish it at an early stage.

Many materials used in accommodation and other areas of the vessel give off toxic fumes in the event of fire so care must be exercised to make sure fire-fighters are properly protected with suitable breathing apparatus before they enter such spaces.

Engineroom fires can be caused in the same ways as accommodation fires, but the more usual reason is failure to contain combustible material within its relative system. Generally speaking, heat sources and oxygen will be in abundant supply so investigations will usually involve looking at oil containment.

Where fire damage is extensive, close attention should be paid to the condition of the fire-fighting and containment systems as these affect seaworthiness and, therefore, liability. Where there has been damage caused by water or other means used for extinguishing the fire, the investigator needs to make a very careful separation between the fire damage and the damage caused in efforts to extinguish the fire

Machinery

The two most prominent causes of fractured shafts are bending stresses and torsional vibrations. The former are often attributed to bearing failures; however, it is sometimes found that, due to hull flexibility, the engine seating fails to provide the support necessary. Although shafting may be in proper alignment when static, the dynamic condition will require checking when shafting or gearing troubles are investigated. Torsional vibration can be caused by a design feature, such as ineffective vibration dampers or an unbalanced engine.

In studying machinery fractures it is useful to ask the metallurgist to prepare a fratography report, as this may give some indication of the number of cycles of stress causing failure. The value of this study will be in evaluating when failure can be said to have commenced, so that adjusters can decide upon which policy year the claim falls. The adjusters will also need to know when the component can be said to have failed.

Fortunately, with the wide application of torsion wrenches and pretensioning gear there has been a reduction in the number of smashed-up diesel engines due to bolt failures. However, in investigating such failures it is still worth ascertaining the age of the bolts and the maker's given expectant life, besides looking for notches and signs of abuse.

Fuel quality

Broken or burnt valves, fractured heads and liners all require careful checking to ascertain whether the operating temperatures and fuel qualities are satisfactory. Excessive wear of the running gear, fuel pumps and valves have all been occurring with increasing frequency and it is not good enough to accept that the blame lies with the operating personnel if in fact the shipboard equipment is unable to handle fuel supplied. It is well known that the old criteria of calorific value and viscosity are insufficient indication of fuel quality.

Unfortunately, until recently owners did not have a basic fuel quality specification to work to and refiners are now finding new and better ways of extracting by-products from the crude with the results that the ship-owner has to purchase less acceptable residues. Sometimes added to this problem is the fact that fuel from different sources is incompatible and the resultant sludge cannot be handled by the vessel's purifiers. At other times, due to the fuel's high specific gravity, the purifiers cannot remove the solids, including wax, at their operting temperature.

When considering fuel we could quote a new vessel which was in serious trouble after 20 days with worn fuel pumps and excessive liner wear caused by catalytic fires. Fortunately, the International Chamber of Shipping has drawn up a specification sheet giving maximum and minimum levels for most fuel parameters. Unfortunately to date there is no internationally accepted test for fuel ignition quality although researchers have come forward with a fair measure of agreement upon the subject. Ingition quality is a measure of the suitability for the engine to burn the fuel and some engines will be capable of burning lower quality fuels than others. Some engine makers give the limits of suitable fuel quality.

In investigating engine damage all these factors must be considered. It may be easy just to blame engineers for negligence but, unless the real cause is found, the repair will hardly be a lasting remedy.

Lubrication

Bearing failures are generally associated with lubrication problems, but it must be borne in mind that the fault may not be in the lubricating oil condition or quality. Historically, we used to get top end troubles due to flexing (bending) of the pins causing local high loading. The same condition can apply due to flexing of crankshafts, and not essentially on high load factors. It has been established that, under some circumstances, this can occur with medium-speed engines running on low load.

Where the lubricating oil is suspect, independent samples should be taken and analysed before forming firm opinions. Sometimes it is found that filter cartridges are at fault; purifiers should be checked and operating temperatures ascertained. Even so, no cause investigation is complete until lubricating oil analysis has established whether the condition of the oil is such as to render it unsuitable for continued service. Having once established that the oil is either diluted with fuel or dirty, the investigation should then be pursued to establish the cause. In other words, the dirty oil is hardly the prime cause.

Techniques have been developed whereby with careful monitoring of the materials in suspension in the lubricating oil it is possible to establish wear rates of bearings, piston rings and liners and thus decide upon maintenance programmes.

Damage to tailshaft seals and sleeves needs further cause examination than the mere acceptance of the possibility of damage through grounding and efforts to refloat. The cause is all too frequently concerned with the materials used in the seals or sleeves and, at times, is even associated with the actual head pressure of the sealing oil.

Corrosion

Within recent times there was a case where severe corrosion took place over a very short period within the hold steelwork of a sulphur carrier. It is interesting to note that microbiological attack was blamed, but even more interesting is that experts disagreed, some holding that it was purely an electrolytic condition.

Historically, many cases of crankshaft and journal corrosion have been blamed on engineers' negligence, without much attempt to prove such; the attitude being 'it must have been'. Now, of course, we know that many cases may indeed have been due to microbiologial attack. Fortunately, if bugs have been present they can be seen under the microscope so, where there is corrosion, samples should at least be examined. This type of attack was held to be responsible for attacks to heat exchanger tubes in some harbours of the world in World War II, so it is not new.

General average

When investigating matters involving general average, it is customary for surveyors to be briefed to investigate damages caused in 'efforts to' extinguish fire, refloat after grounding or whatever, thus in his brief no specific mention of general average is made. The decision as to whether or not a particular item falls under general average is the subject of discussion and negotiations with average adjusters and owners. In order that surveys be held with an intelligent approach to the subject and suitable investigations carried out, it is important to understand the basis of these claims.

Rule 'A' of the York-Antwerp Rules, 1974, defines a general average act as: 'when and only when any extraordinary sacrifice or expenditure is intentionally and reasonably made or incurred for the common safety for the purpose of preserving from peril the property involved in a common maritime adventure'. From this definition it is clear that the cost of refloating the vessel is one of general average. This cost is incurred for the benefit of both ship and cargo in that if the vessel were not refloated, a commercial loss would incur.

Rule 'C' of the York-Antwerp Rules, 1974, states: 'Only such losses, damages or expenses which are the direct consequence of the general average act shall be allowed as general average'. Therefore, it does not follow that the damage to the bottom is recoverable as general average. Further, Rule 'E' states: 'The onus of proof is upon the party claiming in general average to show that the loss or expense claimed is properly allowable as general average'.

The understanding of these matters will give those involved the chance to check out their findings before the evidence is destroyed, as after repair it is generally too late to get further evidence.□

INSPECTING INERT GAS SYSTEMS

M. G. Berry, Companion
Installation consultant, Interlink Inert Gas, Ltd

THE OBSERVATIONS HERE are extracted from vessels I have visited. The comments are not restricted to any one supplier or shipping company and have been selected to give a broad picture of inert gas (IG) system faults.

Case 1

The plant had not been operational when purchased by the new owner. The system was to be made operational with a list provided for updating the system.

1) Boilér uptake valve jammed with soft rubber actuator hoses perished.
2) Boiler uptake line filled with packed soot; it broke apart on being lifted out.
3) Scrubber filled with soot past bottom inspection hatch.
4) Scrubber internals disintegrated at touch.
5) Level switches without floats (corroded away).
6) Demisters broken and packed solid with soot.
7) All valves damaged with a high percentage being beyond repair.
8) Both fans perforated/patched.
9) Both fans drain lines blocked solid and corroded.
10) The two non-return valves (no deck seal) jammed.
11) Main IG pressure sample line blocked yet though the pressure transmitter had to be forced open and the case was badly corroded it functioned after adjustment.
12) The deck system required 150m of piping replaced (it could be kicked in with the toe of an unprotected shoe).
13) All deck valves and P/V units damaged/badly corroded. Yet when all the links/loops were removed from panels, the circuit and external wiring was proved as satisfactory with only the bellows of pressure switches and limit switches, etc., being found defective. This system had been left to sit and rot, which when considering that two-thirds of the cost of an IG system is in the installation of the piping is an incredible waste of the initial investment.

Case 2

On boarding, the chief superintendent had stated that no problems had occurred to date and that the crew were happy with the system. A quick check of the system suggested that the deck department would have gas flow/volume problems. A discussion with the mate revealed that they were in fact inerting the tanks after discharging the cargo. Adjustment to permit full volume to deck took less than five minutes. Further checks revealed that though the piping was in good condition the system had never been operated automatically and a high percentage of the interlocks, etc., were looped over (i.e., matches in relays, etc.) The system had to be (re?) commissioned from scratch, yet it was still in the guarantee period.

Case 3

I boarded the vessel for pre-commissioning checks, only to find that the plant was not ready for operation. Pointed out a large number of minor modifications required prior to starting the plant, including the fact that three of the instruments would not function as they were mounted vertically instead of horizontally.

When the sea trials were eventually carried out, the system was passed despite the fact that these instruments were still installed incorrectly and had to be nudged now and again to function. The yard was to 'modify at a later date.' The instruments? They were for temperature, scrubber water level and the oxygen/pressure recorder.

Case 4

The vessel had retained sufficient cargo for crude washing underway which was to be discharged prior to dry docking. Time was limited and, unknown to the writer, the discharge master had decided to try to shorten the gas-freeing operation by a quick hydrocarbon purge followed by opening the main hatches, anticipating that the ship's forward speed would quickly dissipate the IG plus any lingering hydrocarbon. Seeing the gas rolling across the deck, the writer lowered an oxygen probe 3-4 m into the tank, obtaining 21 per cent. The following MSA 40 full-scale deflection caused the writer to tip-toe back to cargo control and ask to leave.

Case 5

Being informed that the vessel was inerted, the writer checked the oxygen recorder. The time spent 'inerting' did not tally and an oxygen probe was lowered into the tanks. The reading obtained was 17 per cent.

Case 6

The vessel was an OBO and although it has not been used for oil charters, it had had a number of surveys carried out on its IG system as the owners wished to update it. None of the lists of modifications pointed out that the deck seal would only handle maximum back pressure of 1,250 mm wg on an even keel (the mechanical P/Vs being set at 1,400 mm wg) and that there were three other possible hydrocarbon sources being led back into the engineroom.

Case 7

The discharge had started prior to the writer's boarding. IG could be seen coming from all three mast risers. The answer to my query was that they leaked. Going further along the deck the writer noticed that the pressure transmitter take-off point was aft of the deck main isolating valve. The deck main isolating valve was partially closed.

As the pumps were picking up speed the writer asked that the discharging tanks be checked for oxygen content. Two tanks were found to be pulling air, yet the gauges gave a reading of +400 mm wg. There were two causes for this incident. One was that the chief engineer believed the fans capable of damaging the tank bulkheads and the second was his misunderstanding of the supplier's manual, which was pretty but did not inform the operator as to the reason and function of the plant's individual units.

Case 8

The shipyard had installed many systems and therefore opted to supply all the control equipment and panels with the IG manufacturer supplying the main units only. When inspecting the plant prior to commissioning, the writer found a mixture of pipes, valves and controls that did not appertain to the specific system installed, but that were easily recognisable as required for an alternative IG supplier's equipment. It took three days to modify the plant to an acceptable compromise.

Case 9

The writer informed the yard that a number of modifications were required to the system in order to prevent water from being carried-over from the deck seal. However, the yard stated that the vessel owners would use this factor as an acceptance trial delay and were not to be informed. Eventually the trial was stopped due to insufficient gas reaching the tanks. The IG main had a long 'dipped' section which was full of water. There was no drain valve and it took over an hour to drain the line from two flexible coupling location bolts.

Case 10

The deck personnel had stated that the deck seal was useless and that it constantly caused water to be carried over into the deck main. A check start showed that the problem was their method of operation and not the deck seal, although, again, the manual did not describe the operation clearly.

Case 11

The oxygen alarm was set at 12 per cent. 'Good inert gas, yes?' queried the second engineer. As the alarm came up the writer had to have a discussion with the master, chief engineer and 2nd before it was lowered, because the manual stated that 'Fire cannot be sustained at oxygen level lower than 12 per cent.'

Case 12

Both blowers were in the preliminary stages of corrosion. The rubber linings had obviously been damaged on assembly, with the studs being screwed through the steel substrate to push/fracture the lining (i.e., too long a stud and not using 'blind' holes). Both the inlet evases were holed to the extent of requiring replacement. These blowers were in an inaccessible position, yet the superintendent ordered them to be boxed up instead of making repairs. Estimated cost of replacing two very large 100 per cent blowers, cutting through bulkhead for access, etc., must be in the region of £10,000.

Case 13

The IG system had just been installed and the writer was training ship's personnel on the ballast voyage. The first mate had apparently (foreign language) informed the pumpman that they would be opening the purge pipes on five across later that day, when they had finished one across (experimenting with resistance, etc., plus checking a fault in this particular installation). While the vessel's master, first mate and the writer were checking one port, we turned to see gas rolling across the deck from No 5 tank. The pumpman had opened the main hatch (he was drunk).

Case 14

The writer asked the master for his company's 'Tank Entry Safety Regulations' to include them in the training. The captain laughed 'What do we need those for? We are tanker men.'

Case 15

Following the inspection of an IG system, a report was made to the owners detailing the items which needed repair or replacement to bring it into a correctly operating condition—amongst other matters all the system valves appeared to be jammed solid (and later had to be freed-off using pneumatic chisels), the scrubber internals and linings had been severely damaged by intense heat, and one tank was not even connected to the IG main.

A year later we were asked to visit the same vessel to effect a repair and were assured by the fleet manager that all the original faults reported by us had been cleared and the plant checked as fully functional. On boarding the vessel we found the system in the same state as when we had left, plus the interim year's deterioration! We cleared the faults, but apart from the chief engineer (who was heavily involved in other work) there was a distinct lack of interest.

Who's fault is it?

The above cases are by no means the worst being selected from areas of cause, plus the fact that they cannot be identified to any owner, supplier or shipyard. The writer has boarded vessels fitted with IG for over a year, yet the deck analysing equipment is still in its original packing and without batteries. The writer has repaired a system, made the total deck tight and inerted the vessel single-handed without the slightest interest or help from the crew, despite constant appeals. Chief engineers have stated that they spend more time on the IG plant than all the rest of the ship's gear put together. So who's fault is it?

Unfortunately the answer is nobody and everybody; unfortunate in that it means co-operation before legislation. To give an example, let's look at the initial commissioning of a plant. The 'trial' consists of checking all the required interlocks plus a volume or capacity check along the deck. Naturally it is a time of tension, with the commissioning engineer bowing to every whim of the purchaser provided it is in his company's interest.

The trial usually lasts a couple of hours though the writer has been given a one hour maximum to prove the system (without preliminary runs/checks) on two occasions. There is only one interest during this period and that is 'System passed' for the purchaser will have retained a percentage of the system. It is probably the only system that is not checked for its actual function, which is to inert the tanks. The excuse that people may have to enter the tanks later is understandable, but surely another duty is to gas free the tanks.

As case 1 demonstrates, there is ample experience to show that system breakdowns (irrespective of IG supplier) are created by two major factors and that is installation and operation. It is here that the co-operation of operating personnel must be sought after for they have a wealth of information on the problem areas. After all the wiring of relays in a logical sequence according to the governing rules is not likely to change once checked.

Optimum conditions

If ship's personnel were given the opportunity of saying 'no' to the positioning of equipment and pipe runs, what would they be looking for? First accept that you are dealing with a very corrosive gas that is in its saturated state after the scrubber and then:

1) Install valves that are internally stronger than any possible external pressure.
2) Valves to be installed in an accessible position with horizontal spindles and leading edge coming away from maximum build-up point.
3) Do not accept horizontal legs on any drain lines, expecially at fan drain outlet.
4) Avoid any dips and loops in pipelines including the pressure transmitter sample line as condenstion will occur and collect.
5) Ensure service and withdrawal areas, such as the fans split castings are clear.

Naturally the writer has his own additional pet theories; however, that would fill another chapter. Therefore the best way is to imagine that there is a small amount of water mixing with the unscrubbed particles swishing along the pipes. Where it comes to rest it will cause damage, for it is acid (i.e., look at dips and bends, etc.).

Cost? If it's done correctly, it should reduce the initial installation costs, but what is more important the system is likely to last twice as long provided it is operated in a sensible fashion.

Checks required

Earlier in this chapter, I suggested that a considerable number of the problems relating to the serviceability of inert gas systems could be significantly reduced with detailed attention to the installation by utilising experienced ships' operators in both the initial design and installation of the plant. Installation engineers/pipework draughtsmen cannot be expected to realise either the function of the equipment or the problems involved in operation, and the supplier companies are handicapped in that they gain very little feed-back once the vessel has sailed. Once installed, apart from minor changes, it is generally considered uneconomic to modify the plant and therefore the operation of the plant must be controlled to combat the corrosive/errosive effects of the inert gas.

There are many checks required for the safe operation of the inert gas system, such as ensuring the integrity of the deck seal and function of safety interlocks. However, although vital, they are not dealt with here. What we suggest is that the inert gas system can be operated in such a way as to reduce the maintenance to that of any shipboard equipment by accepting/understanding two facts. The first is to accept that inert gas can be extremely corrosive under certain conditions, and to protect against the effects as follows:

1. Exercise boiler uptake valves from open to closed daily when the system is not in use.
2. Check that air seal lines are free and that valves function.
3. Change blower seals as soon as they leak, otherwise the shaft becomes pitted, thereafter rapidly destroys new seals.
4. Ensure that fan drains are free.
5. Check the main and recirculating valves for correct action—the efficiency of the plant is dependent upon them.
6. Ensure that all valves open fully, otherwise the gas flow is restricted.
7. Check all deck drain valves are free and clear of debris.
8. Above all, ensure a clean system by:
   (a) Starting scrubber pump at least 15 minutes before the fan to allow all internal surfaces to be wetted before hot dirty gas enters the unit.
   (b) On shut-down run scrubber pump for one hour to flush out both the scrubber and effluent line.
   (c) Wash down the blowers.

Gas flow

The second fact to accept is that the system is designed for a gas flow that is seldom achieved in operation. The gas flow is designed according to the maximum cargo pumping rate plus 25 per cent to ensure that a pressure can be maintained on the cargo during the discharge. This positive pressure is essential and all the inert gas supply requirements are controlled by it. However the equipment once installed has set dimensions and is designed according to volume throughput or flow as follows:

1. The scrubber carries out three duties:
   (a) To cool the gas.
   (b) To absorb SO_2 SO_3.
   (c) To scrub or remove the particles within the gas stream.

   While the cooling and absorption duties do not deteriorate with a reduction of volume throughput (in fact, absorption increases) the scrubbing efficiency is primarily dependent upon the velocity through the unit. The idea is to give the particles in the gas stream sufficient velocity to impact or stick to a wetted surface and be carried away by the water while the gas continues onwards through the unit.
2. The demister is also sized for the designed volume throughput with excessive water carry-over being created with increased or decreased speed.
3. The fans used are at their peak efficiency at the designed volume and, while the fan normally chosen is very stable, any reduction in volume throughput brings a corresponding increase in temperature. This temperature increase can cause the fan lining to be subjected to temperature in excess of 100°C on low volume throughputs.

It is for the above three reasons that supplier companies require some form of controlled recirculation when the deck volume/flow requirement is low. However, as soon as the recirculation point is reached, there are two valves controlling the flow and being butterfly valves this means a large percentage flow for a small percentage opening. Therefore, these valves tend to hunt around the set point and cause the inert gas fluctuation. These fluctuations can cause problems with deck seals, particularly the semi-dry type that operate more efficiently on a constant volume within the designed flow range. Finally, prolonged recirculation increases the oxygen content of the inert gas delivered to the deck.

As previously stated, the inert gas supply is controlled according to the tank system pressure requirement and the actual controller or adjustable set point is normally situated in the cargo control room along with oxygen/pressure recording and alarm panel. Let us now assume that the system is to be used for a cargo discharge combined with crude washing.

Use in practice

The cargo control room inform the engine room who prepare the system while the tank system is checked. When the scrubber has been running for at least 15 minutes, the engine room inform cargo control that the system is ready to start (it takes seconds to start a fan and while the system should be run for a few minutes in recirculation to stabilise the check, it is a waste of energy to run for hours). We suggest that the system is started approximately 15 minutes before start of discharge, as follows:

1. Start the blower, thereby ensuring pressure in the engine room.
2. Open deck main isolating valves. This method will ensure that any gas flow (due to leaking valves, etc.) will be towards the tanks.

   At this point the oxygen content will probably be high due to insufficient boiler load. Also, the operation of the different types of deck seal must be taken into account. The two-tank dry type will require time to drain the bottom tank completely. This takes up to four minutes and even then there will be a small amount left in which will damage the internal lining if a sudden blast of gas is allowed to pass through it. Therefore:
3. Crack open the forward insert gas main valve to atmosphere.
4. Slowly crack open the safety bulkhead IG control valves, allowing the gas to pass slowly through the deck seal and dry it out.
5. Anticipate water carry over and open drain valves in the deck main.
6. Close the hydrocarbon check valve (aft of deck seal).

This method (4) can also be used for the semi-dry type deck seals. In fact, one system has retarded the opening of the main control valve automatically for start up. The gas is now passing along the deck IG main to atmosphere via the forward valve. It is vital to have back pressure to ensure that pressure is felt at the forward tanks.

When the cargo pumps start, the load on the boilers will reduce the oxygen content and the atmosphere valve can be closed down to pressurise the tanks with good inert gas. Ensure a high pressure during this period, for, with a loaded vessel, the inert gas plant sometimes trips out due to an excess of water lifting level switches, etc. With high initial pressure it gives time to restart the system quickly; also, it is easier to maintain tank pressure when following the cargo down.

As soon as the IG plant has settled down and there is reasonable ullage space, try to stabilise the IG volume (instead of just letting it follow the cargo discharge rate) by reopening the IG main forward valve when the cargo discharge rate is significantly reduced from maximum (i.e., when crude washing). This causes the following:

- Scrubbing optimum efficiency is gained, giving cleanest possible gas.
- Demister optimum efficiency means less water carry-over into the fans.
- The fans run more quietly, more smoothly and are at their coolest.
- The main and recirculation valves stop fighting each other and system pressure is easier to control.
- The deck seal and non-return valve are stable.
- The oxygen content is easier to control at the boilers.
- The gas velocity through the deck main reduces the settling out of particles and, therefore, the corrosion rate.

How can you tell if it is right? Of course if there is very little volume being used you can hear the fan reverberating through the vessel. However, an easier way is that the fan gas inlet temperature will be close to the scrubber water inlet temperature (2-5°C higher, dependent upon scrubber type). The fan outlet temperature will be just warm (inlet plus 10-15°C). A quick hand check will tell you without even looking at a gauge once you get used to it and you can check the scrubber for hot spots while you are there to see that all the sprays are operating.

The work requirement is negligible, yet the result is a cleaner system, less prone to corrosive attack—just by opening a valve and keeping an eye on things. Remember always to maintain a back pressure.

Having gained a cleaner gas by ensuring flow as well as pressure conditions on deck, try to maintain a tight deck to reduce topping-up operations. Here a major problem is leaking mechanical P/V valves caused by particles sticking to the valve seats during the hydrocarbon pressure release. The gas builds up en route during the day, so why not pre-empt the P/V by releasing the gas to atmosphere via the manually operated by-pass to atmosphere valve? □

INSPECTION OF GAS CARRIERS AND TERMINALS FOR SAFE SHIP OPERATIONS

Captain R. D. Izatt
Technical Advisor, The Society of International Gas Tanker and Terminal Operators Ltd

THE ACTION of the master of a gas carrier presenting his ship at a terminal to load or discharge a cargo implies that there is already some form of contract or arrangement established for the shipping or receiving of the liquefied gas cargo. The terminal is, of course, supporting that implication by the action of receiving the ship alongside at its gas handling berth. The commercial and liability aspects will be governed by actual contracts, charterparty conditions, bills of lading, etc., but inherent in all of these will be each party's expectation and requirement that the ship and terminal are operated in a safe manner.

Responsibility for safe operation rests with the master in the case of the ship and with the 'berth operator' (see definitions in reference [6] at the end of this chapter) in the case of the terminal. However, dependent on national or local laws or regulations, there is an overlap of this responsibility inasmuch that the master should always ensure that his ship is in a safe location and that the berth operator will always have a clear interest in, and perhaps external accountability for, the safety standards of the ships which he allows at his terminal. Regardless of the legal aspects, every prudent ship or terminal operator will wish to ensure that his own safety standards are complemented and not jeopardised by the actions or omissions of the other party. In a free market, or more especially FOB trading, ship and terminal operators may have but short advance notice of specific ship nominations against a given cargo and thus limited knowledge of the other's capabilities and limitations.

Recognition of a common safety standard is therefore required as an acceptable foundation upon which the commercial arrangements and subsequent practical operations may be based. In the past, various industry bodies have published guidance such as the *Tanker Safety Guide (Liquefied Gas)* ref. [1], *Safety Guide for Terminals Handling Ships Carrying Liquefied Gases in Bulk,* ref. [2], and *Liquefied Gas Handling Principles on Ships and in Terminals,* ref. [3], etc., all of which are recommended reading and are internationally recognised works. Happily such guidance is followed by the majority of ship and terminal operators, but not to the point that we would ever wish to dispense with inspections or safety checks! The International Maritime Organization (IMO) publish the IGC Code for ships carrying liquefied gases, ref. [4], which, by incorporation into Solas 1974, ref. [5], is now mandatory. Compliance with that code (or the appropriate earlier version for older ships) is taken by almost all parties as evidence of a satisfactory ship from a safety standard point of view. As yet there is no international code mandatory on terminals to provide a common base for safety requirements but, again, clear recommendations are published by IMO, ref. [6]. In the UK, of course, many of these provisions are enforced under the Dangerous Substances in Harbour Areas Regulations 1987, ref. [7], and Guidance Note CS40 from HSE, ref. [8].

Of the above references, [4] and [6] are particularly listed since a thorough knowledge of their content must be combined with the inspector's practical experience in order to carry out his function effectively. It may well be that an inspector's experience and intuition will be ringing warning bells in a general sense, but this must be translated into objective recognition of breaches of the above codes and clear recommendations for corrective action to be taken. This is even more important if the situation has deteriorated to the point where operations must be interrupted for safety reasons. Most tanker seamen will recognise that a much quoted first principle of safe operations is 'if in doubt stop!'

However, although the author very much subscribes to that principle, it will also be recognised that even in the best run operations a prolonged delay resulting in a serious shortfall against contractual obligations will eventually raise questions of liability, blame, claim and counter claim. In any case, therefore, an inspector who has felt it necessary to stop an operation would be less than professional if he could not justify his actions with evidence that can be readily substantiated and easily understood by commercial (rather than marine-technical) operator interests. The foregoing comments of course apply where the inspection of either side of the ship/shore interface is being carried out to establish that operations comply with safety standards which both parties accept, as opposed to an 'in-house' safety audit which may be carried out to seek improved safety of operations and risk reduction in accordance with a company's own operating policy.

Objectives of the inspection

Having touched on the pessimistic point that an inspection may result in stoppage of operations, it is important to establish the understanding of all parties that the function of an inspection is to confirm, support and expedite safe operations. It is infinitely better for the inspector to gain the confidence and co-operation of the party being inspected, on the basis of helping to solve problems, rather than cold formality resulting in a 'catch me if you can' contest. Such a contest could well defeat the whole object of the exercise.

This is, of course, a two-way relationship in that responsible operators will be pleased to co-operate, will actively seek advice and confirmation of safe practice and, ultimately, the inspector's willing endorsement of their standards and good reputation.

Inspection preparations

It is recommended that the inspector gathers as much advance information as possible on both the terminal at which a ship is to berth and the nominated ship. This is in order to minimise expenditure in 'data acquisition' of the valuable but limited 'on-site' time that will be available to carry out the physical inspection. The ideal situation is where a terminal and ship can be visited independently prior to the actual berthing of the ship so that staff are more readily available to discuss aspects of equipment and operations and can carry out any modifications recommended or improvements found to be necessary. This can then be supplemented by a short inspection of the cargo transfer operation if required.

Almost all liquefied gas ships are provided with a technical data sheet (commonly known as Gas Form C) which lists dimensions, equipment information and designed gas handling performance capabilities. Likewise, most terminals publish booklets of their operating regulations incorporating detailed berth information on designed mooring system, maximum and minimum ship sizes and cargo transfer equipment capabilities. It is not totally unknown for a ship to be commercially nominated and accepted, only to find that when the ship arrives at the terminal, it is physically impossible to connect up the transfer arms or, sometimes, even to bring the ship alongside!

The inspector should also plan his methods on his pre-obtained knowledge of the terminal and ship layout, immediate last cargoes carried by the ship, the cargo to be handled, the previous ship and cargo handled at the berth and the estimated time of arrival. In formulating the plan of inspection it should be remembered that regardless of whether a full inspection is to be carried out or not, a ship/shore safety check list will be carried out by the responsible operating staff prior to commencement of cargo transfer.

The IMO standard check list format is given in Appendix 4 ref. [6], and obviously any previous inspection of greater depth should give due attention to the requirements of that check list. OCIMF have compiled an aide-memoire for gas ship safety inspections with full guidance (currently under review) and further detailed check lists, ref. [9]. Terminal operation survey guidelines, ref. [10], are also available with relevant check lists. It is not intended to bore the reader by pointless repetition of the technical content of these check lists. However, it is intended that the advice given here should complement the use of these recognised lists and the universally adopted ship/shore safety check list is given as an example in the appendix.

The final preparation for the inspection is, of course, close liaison with the terminal scheduling departments, or ships' agents, as maximum flexibility is usually required to ensure that the inspector's schedule keeps pace with that of the terminal and ship. It might also be borne in mind that several agencies may have an interest in inspecting a ship, which occasionally unfortunately coincide.

As an example, flag State, port authority, current charter/shipper or receiver interests, similar interests for the next voyage, and the terminal representative may all require attention of the master in competition with the administrative requirements of port arrival such as Customs, immigration and port health formalities. This is part and parcel of his job, but most shipmasters would appreciate understanding of that situation. Obviously, if different inspectors know that they are coinciding, it would be beneficial if they can co-operate at least by sharing the inspection tour and exchange of straightforward information such as dates on certification sighted, etc.

General approach

In the above preparations the inspector will have obtained a clear brief on requirements such as whether the inspection is to be carried out as a precondition for cargo transfer or a routine inspection of operations. If the former is required, then the inspector will obviously arrive at the terminal a few hours before the ship is to berth so that the on-shore discussions are completed in good time to observe the ship's approach, berthing and mooring operations. Similarly, the most efficient use of ship's and berth time is to come to a conclusive opinion of acceptability for cargo transfer by the time both ship and shore are in fact ready to commence transfer.

The well-prepared inspector will therefore aim at being in a position to give a competent decision within one or two hours of boarding the ship. That is not to say that the inspection is necessarily complete, but that the inspector will be confident in either endorsing commencement of transfer or, alternatively, requesting a delay while he makes further investigations. It is not being claimed that a full function inspection of a ship or terminal is in any way possible in the stated period of a few hours. Safety audits can take teams of personnel some days to carry out, and even then may well carry reservations on areas not possible to cover in the time available. What is possible is that a competent inspector can assess the general standard of operations, personnel competence and, with the aid of the check lists, that the effectiveness of certain essential safety features is being maintained to the level that has been agreed by the parties involved or is required by regulation, statute or convention.

On the basis that safety and efficiency go hand in hand, and that efficient marine operations depend on professionalism, regardless of the commodity being transferred or transported, an inspector's assessment commences from approaching the terminal gates or ship's gangway.

Carrying out the terminal inspection

The layout of the check lists will, to some extent, guide the inspector to a 'route plan.' To supplement these check lists, this section is written as examples and suggestions of things that will influence the inspector's assessment of operations. Experience shows that particular items have been repeatedly found to be significant, or give the best insight of operational standards and it is on these points that inspectors commonly focus their questions.

On approaching the terminal, the method of

personnel and vehicle entry control, de-matching or ignition sources control, and identification checks will be observed. Explanations of hazards on site—e.g., from adjacent operations as well as that to be inspected—approved route instructions and information on alarm signals and responses expected of visitors will be included. On gaining entry to the terminal, the inspector will wish to introduce himself to the local supervisor and agree entry procedures, etc., for the areas that are to be inspected.

Having been advised of hazards, it will be noted if personnel are suitably equipped with the relevant hard hats, eye protection, ear defenders, suitable coveralls for the materials handled, etc. General housekeeping will be taken as a particular indicator of personnel attitudes. Roadways will be noted as clear of obstructions. Safety equipment sighted en route will be assessed on availability, signposting, evidence of maintenance and expiry dates, etc. Again, all of these items are looked at as indicators rather than for particular small faults which may have only recently appeared.

On approaching the waterfront area, the means of personnel control and preparations for receiving the ship will be reviewed. This includes methods of ensuring basic compatibility of size, draft, presentation of ship connections within the limiting envelope of shore equipment, etc. The 'work permit' system will be noted with a view to see that checks have been made on cut-off of non-compatible repair work on jetty mooring safety and cargo transfer equipment—e.g., that inshore fire pumps are in fact available and not under routine maintenance.

The re-setting of automatic mooring hooks should be established together with preparatory checks and procedures for loading arms or hoses—e.g., have they been purged of last cargo and depressurised? It is not suggested that the inspector checks all these individual items, but that he will assess the operational procedures that are established and the level of checking that is thus routinely carried out. For example, he would observe in passing that fixed firefighting monitors are effectively aimed for the ship in question and, if capable of remote or automatic start, that they are on 'spray' rather than 'jet' setting.

In terms of personnel control, the provisions for preventing ship visitors gathering in proximity to mooring operations is one example, together with control of access to exposed jetty faces and mooring dolphins. Are shore staff aware of quarantine and Customs aspects of ships arriving from abroad? It will be seen that the normal 'maritime' aspects should not be overlooked because of concentration on 'liquefied gas' concerns. As mentioned earlier, all of these indicators give valuable evidence on the basis that the gas transfer operation will be carried out with a similar level of efficiency.

On the matter of gas transfer, is the person who will board the ship as 'terminal representative' knowledgeable on the ship/shore check list application? Does he have a prepared draft cargo transfer plan for final discussion and *agreement* with the ship? Will the ship be adequately informed as to shore emergency procedures and *vice-versa* with joint agreement on these procedures? Are there adequate means of communication (including back-up) available between ship and shore control rooms? Having witnessed the mooring operation and that the ship is effectively moored in accordance with an agreed plan, boarding methods and procedures will be observed.

The terminal representative will agree with the ship's representative on cargo tank sampling and last cargo checks to be carried out. He will probably arrange for laboratory personnel to draw samples accompanied by a responsible member of ship's staff. Again, these activities should be adequately supervised and in accordance with industry-approved sampling procedures. The ship and shore connection points should be clearly labelled for the correct product to be handled—e.g., liquid/vapour, LNG, propane, butane, chemical gas, etc.—and connecting procedures will commence, again after joint agreement and by trained personnel. If cargo transfer is subject to a satisfactory ship inspection, it is during this time that the inspector will introduce himself to the master and, with his permission, commence the shipboard assessments.

Carrying out the ship inspection

Just as in the terminal inspection, the general housekeeping, safety equipment and crew personal protection can all be observed en route to the master's cabin. Additional items observed will be the way in which the moorings are handled, the gangway bulwark access arrangements, safety notices on materials being handled, portable safety barriers and signs guiding visitors to safe accessways, and the layout of portable firefighting equipment for easy access. The ship's emergency plan for advice of shore emergency services should be available close to the gangway and other items such as the turning out of the offshore lifeboat would be noted as indicators of active safety awareness.

On approaching the accommodation block, a quick look should confirm that all ports, windows and doorways facing the cargo area are in fact closed. In the discussions with the master, the ship's statutory certificates, such as Solas safety equipment, construction and load line should be sighted, and dates of expiry noted. The ship's IMO Certificate of Fitness should be examined and the date of the last intermediate survey noted. Having sighted the certificates, an examination of the cargo log book should indicate previous cargo and items such as how often routine tests of the emergency shut down (ESD) and deck water spray systems are carried out. Some evidence of last date of setting or adjustment of safety relief valves (SRVs), cargo tank instrumentation (level, temperature and pressure) and auto-gas detection system calibration might also be requested.

These initial discussions will be followed by a tour of the ship, certainly in the company of a member of the ship's staff, but preferably the chief officer or at least one of the officers responsible for cargo handling operations.

The ESD system will have a major priority with the objective of establishing not only that it is effective, but that it is fast enough to meet IMO requirements (i.e., not more than 30 seconds from initiation to total shutdown) and yet not so fast that it will create a pressure surge problem when considered against the planned cargo transfer rate. When physically timing the operation of manifold valves, the inspector may wish to establish repetition of valve closing times by several activations of the local individual valve controller before requesting ESD by use of a remote ESD initiation control. When checking the movement of the valve actuator it is also prudent to check that the actuator shaft is properly connected and actually moving the required valve.

In going round the cargo tank domes, operation of randomly selected valves might also be requested, the pilot valves or SRVs examined for intact seals or signs of recent disturbance in keeping with the documentary evidence requested above, the level-gauging equipment checked for indication of reference readings in the housed or maximum extension position, and back-up gauging devices such as slip tubes checked for proper securing. Local readings of temperature and pressure might be noted for later checking against remote readouts in the cargo control room, etc.

In estimating the effectiveness of the deck water spray equipment, the inspector may take into account other supporting evidence such as log book entries and other general standards in deciding whether an operational test is required while final preparations for cargo transfer are being made. If tank entry procedures permit, a void space may be entered for a short look in the drainage sump and underside of an independent cargo tank. This will give indications of any water ingress problems.

The general condition of deck service air lines should be examined, on the lookout for any significant leaks which may affect cargo valve control arrangements and similarly seawater condenser cooling and fire water supply pipes. The inert gas system will be checked to ensure that it is fully isolated from the cargo system, together with availability of instruments to check tank atmospheres and quality of inert gas output—e.g., are chemical reaction-type sensing tubes in date? In gas detection instrumentation, the fixed gas analyser is almost guaranteed to be sighted as operational and properly cycling, and a thorough inspector will trace a sample supply line to see that it is properly connected to the indicated detecting head. Likewise, any air lock arrangement to electric motor rooms will be checked for proper alarm indications that it is not being operated properly, with information requested on reactions (ESD?) if the protected space loses ventilation overpressure.

In electrical motor room spaces, drive shaft seals to compressors will be examined, preferably with the motor running. The general appearance of machinery will be noted for maintenance—e.g., are oil seal pots filled with clean oil? While noting the indicator boards, etc., in the electrical motor room for cargo machinery, opportunity may be taken to observe rear connections between alarms and instruments—e.g., explanations should be available for unusual wiring links, bridges or temporary repairs. In the compressor room the inspector's 'nose' will be particularly alert and, as example, cargo plant refrigeration condensers will be examined for level indications and venting arrangements.

In moving to the ends of the ship at least one random check at bow and stern will be made to establish that mooring manual brakes are fully applied and that systems are not left in automatic tensioning mode. The inspector's final area may be the engine room floor plates, to establish that overboard discharges are limited to segregated ballast and that any repairs that are being carried out do not impose on availability of propulsion or main supplies. Once again, the above wrinkles are suggested only as illustrations to supplement the full check lists and to indicate the type of approach that may be adopted.

On completion of the inspection, any deficiencies found should be fully discussed with the master of the ship or operations duty manager of the terminal. Equally, where no grounds for concern have been found, then this should also be clearly stated. Any experienced inspector will undoubtedly find very minor items, but these are not necessarily 'faults.' It may be that in his experience of different operating practices, he may have found one method to be better than another.

Inspectors should ensure that their findings are seen to be totally based on safety interests and devoid of commercial overtones. They will find that most personnel in the liquefied gas industry fully appreciate the potentials of the material they are handling. Consequently, provided inspections are carried out in an open manner and in the spirit of mutual safety enhancement, then active safety checking and subsequent advice will be welcomed. □

References

[1] *Tanker Safety Guide (Liquefied Gas),* published by ICS, distributed by Witherby & Co Ltd, London. ISBN 0 906270 01 4.

[2] *Safety Guide for Terminals Handling Ships Carrying Liquefied Gases in Bulk,* published by OCIMF and distributed by Witherby & Co Ltd. ISBN 0 900886 72 2.

[3] *Liquefied Gas Handling Principles on Ships and in Terminals,* published by SIGTTO, distributed by Whitherby & Co Ltd. ISBN 0 900886 93 5.

[4] *International Code for the Construction and Equipment of Ships Carrying Liquefied Gases in Bulk,* Published by IMO. ISBN 92 801 1163 9.

[5] *International Convention for Safety of Life at Sea,* published by IMO. ISBN 92 801 1200 7.

[6] *Safe Transport, Handling and Storage of Dangerous Substances in Port Areas* (revised 1983) IMO. ISBN 92 801 1160 4.

[7] *Dangerous Substances in Harbour Areas Regulations 1987.* HMSO, ISBN 0 11 076037 9.

[8] *The Loading and Unloading of Bulk Flammable Liquids and Gases at Harbours and Inland Waterways.* Guidance Note GS40. HSE, distributed by HMSO. ISBN 0 11 883931 4.

[9] *Safety Inspection Guidelines and Terminal Safety Check List for Gas Carriers,* published by OCIMF in 1979 (under review), distributed by Witherby & Co Ltd. ISBN 0 900886 43 9.

[10] *Marine and Terminal Operations Survey Guidelines* (1983), OCIMF, distributed by Witherby & Co Ltd. ISBN 0 900886 81 1.

The author wishes to thank Shell International Petroleum Co Ltd and The Society of International Gas Tanker and Terminal Operators Ltd for permission to contribute this chapter.

Appendix I Ship/shore safety check list

Ship's Name ______________________________

Berth ______________________ Port ______________________

Date of Arrival ______________________ Time of Arrival ______________________

INSTRUCTIONS FOR COMPLETION

The safety of operations requires that all questions should be answered affirmatively ☑ If an affirmative answer is not possible, the reason should be given and agreement reached upon appropriate precautions to be taken between the ship and the terminal. Where any question is not considered to be applicable a note to that effect should be inserted in the remarks column.

☐—the presence of this symbol in the columns 'ship' and 'terminal' indicates that checks shall be carried out, by the party concerned.

The presence of the letters **A** and **P** in the column 'Code' indicates the following:
A—the mentioned procedures and agreements shall be in writing and signed by both parties.
P—in the case of a negative answer the operation shall not be carried out without the permission of the Port Authority.

PART A *Bulk liquids—general*		*Ship*	*Terminal*	*Code*	*Remarks*
A1	Is the ship securely moored?	☐	☐		
A2	Are emergency towing wires correctly positioned?	☐	☐		
A3	Is there safe access between ship and shore?	☐	☐		
A4	Is the ship ready to move under its own power?	☐		P	
A5	Is there an effective deck watch in attendance on board and adequate supervision on the terminal and on the ship?	☐	☐		
A6	Is the agreed ship/shore communication system operative?	☐	☐	A	
A7	Have the procedures for cargo, bunker and ballast handling been agreed?	☐	☐	A	
A8	Has the emergency shut down procedure been agreed?	☐	☐	A	
A9	Are fire hoses and fire-fighting equipment on board and ashore positioned and ready for immediate use?	☐	☐		
A10	Are cargo and bunker hoses/arms in good condition and properly rigged and, where appropriate, certificates checked?	☐	☐		
A11	Are scuppers effectively plugged and drip trays in position, both on board and ashore?	☐	☐		
A12	Are unused cargo and bunker connections including the stern discharge line, if fitted, blanked?	☐	☐		
A13	Are sea and overboard discharge valves, when not in use, closed and lashed?	☐	☐		
A14	Are all cargo and bunker tank lids closed?	☐	☐		
A15	Is the agreed tank venting system being used?	☐	☐	A	
A16	Are hand torches of an approved type?	☐	☐		
A17	Are portable VHF/UHF transceivers of an approved type?	☐	☐		

PART A—*continued* *Bulk liquids—general*	*Ship*	*Terminal*	*Code*	*Remarks*
A18 Are the ship's main radio transmitter aerials earthed and radars switched off?	☐			
A19 Are electric cables to portable electrical equipment disconnected from power?	☐	☐		
A20 Are all external doors and ports in the amidship's accommodation closed?	☐	☐		
A21 Are all external doors and ports in the after accommodation leading onto or overlooking the tank deck closed?	☐	☐		
A22 Are air conditioning intakes which may permit the entry of cargo vapours closed?	☐	☐		
A23 Are window-type air conditioning units disconnected?	☐	☐		
A24 Are smoking requirements being observed?	☐	☐		
A25 Are the requirements for the use of galley and other cooking appliances being observed?	☐	☐		
A26 Are naked light requirements being observed?	☐	☐		
A27 Is there provision for an emergency escape possibility?	☐	☐		
A28 Are sufficient personnel on board and ashore to deal with an emergency?	☐	☐		
A29 Are adequate insulating means in place in the ship/shore connection?	☐	☐		
A30 Have measures been taken to ensure sufficient pumproom ventilation?	☐			

PART B *Additional checks—bulk liquid chemicals*	*Ship*	*Terminal*	*Code*	*Remarks*
B1 Is information available giving the necessary data for the safe handling of the cargo including, where applicable, a manufacturer's inhibition certificate?	☐	☐		
B2 Is sufficient and suitable protective equipment and protective clothing ready for immediate use?	☐	☐		
B3 Are counter measures against accidental personal contact with the cargo agreed?	☐	☐		
B4 Is the cargo handling rate compatible with the automatic shut down system if in use?	☐	☐	**A**	
B5 Are cargo systems' gauges and alarms correctly set and in good order?	☐	☐		
B6 Are portable vapour detection instruments readily available for the products to be handled?	☐	☐		
B7 Has information on fire-fighting media and procedures been exchanged?	☐	☐		

PART B—*continued* *Additional checks—Bulk liquid chemicals*		*Ship*	*Terminal*	*Code*	*Remarks*
B8	Are transfer hoses of suitable material resistant to the action of the cargoes?	☐	☐		
B9	Is cargo handling being performed with the permanent installed pipeline systems?	☐	☐	**P**	

PART C *Additional checks—bulk liquefied gases*		*Ship*	*Terminal*	*Code*	*Remarks*
C1	Is information available giving the necessary data for the safe handling of the cargo including, where applicable, a manufacturer's inhibition certificate?	☐	☐		
C2	Is the water spray system ready for use?	☐	☐		
C3	Is sufficient and suitable protective equipment (including self-contained breathing apparatus) and protective clothing ready for immediate use?	☐	☐		
C4	Are void spaces properly inerted where required?	☐			
C5	Are all remote control valves in working order?	☐	☐		
C6	Are cargo tank safety relief valves lined up to the ship's venting system and are bypasses closed?	☐			
C7	Are the required cargo pumps and compressors in good order, and have the maximum working pressures been agreed between ship and shore?	☐	☐	**A**	
C8	Is reliquefaction or boil off control equipment in good order?	☐			
C9	Is gas detection equipment set for the cargo, calibrated and in good order?	☐	☐		
C10	Are cargo system gauges and alarms correctly set and in good order?	☐	☐		
C11	Are emergency shut down systems working properly?	☐			
C12	Does shore know the closing rate of ship's automatic valves? does ship have similar details of shore system?	☐	☐	**A**	
C13	Has information been exchanged between ship and shore on minimum working temperatures of the cargo systems?	☐	☐	**A**	

	Ship	*Shore*
Are tank cleaning operations planned during the ship's stay alongside the shore installation?	Yes/No*	
If so, have the port authority and terminal been informed?	Yes/No*	Yes/No*

**Delete yes or No as appropriate*

Declaration

We have checked, where appropriate jointly, the items on this check list, and have satisfied ourselves that the entries we have made are correct to the best of our knowledge, and arrangements have been made to carry out repetitive checks as necessary.

For Ship	*For Terminal*
Name __________	Name __________
Rank __________	Position __________
Signature __________	Signature __________

Time __________

Date __________

ROLE OF THE INDEPENDENT INSPECTION COMPANY*

J. S. Hinde
Moore, Barrett & Redwood Ltd

CARGO INSPECTOR, SURVEYOR, INDEPENDENT INSPECTOR—the titles are largely synonymous: the organisations for whom they work are involved in a very large proportion of petroleum bulk liquid movements. Whilst the term 'inspection' has a well understood meaning in everyday English, it has a very specific meaning within the petroleum industry. It relates to the procedures through which the quantity and quality of a body of petroleum are established, normally as it is loaded on to, or discharged from, a vessel via a storage terminal.

Quantity' in this respect is self evidently related to the volume which is transferred; 'quality' is somewhat more complex to define and relates to one or more specific parameters—e.g. viscosity, density, cloud point, etc., deduced by laboratory test. The chosen quality parameters will vary depending on the petroleum liquid in question, and on the buyer/users intended purpose for the cargo.

To a greater or lesser extent, every inspection company should have staff with the expertise to establish the quantity of a cargo: they may also possess laboratories in which the quality parameters can be determined. Alternatively, they may have suitably qualified (chemist) staff, but not the facility, in which case the best they can offer is to be present to witness another laboratory doing the work.

The larger inspection companies, however, will have in their armoury of technology, a much wider spread of specialisations. They will include calibration for both land storage tanks and vessels, flow measurement calibration, analytical instrument expertise, engineering consultants, software skills, loss control analysts, claims handling specialists, and the management skills to organise all of these diverse specialists into an appropriately balanced team to support client companies in the business of moving petroleum around the world.

Need for independent inspection

The outsider or newcomer to the petroleum industry may imagine that crude oil produced at the wellhead is transported, refined, distributed and sold as motor gasoline by a single, vertically integrated oil company. This has, of course, become increasingly untrue; the majority of movements today represent a physical custody transfer between a buyer and a seller and the notion of an independent ascertainment of quantity and quality is seen to be essential. Even when buyer and seller are one and the same company, the carrier—vessel or pipeline—may be a third party and the need for independent inspection may arise.

Thus we can list those parties likely to need independent inspection:

1. Oil producers
2. Shipping, transportation and pipeline companies
3. Storage terminals
4. Crude, gas and product traders
5. Refiners and processors
6. Distributors
7. Insurance, banking and financial institutions

The principal motivation for using an independent inspector is self evidently to 'protect the client's interests'—the phrase is easily misunderstood and is discussed later.

There are many subsidiary reasons for using independent inspectors.

Geographical—Clearly every buyer of petroleum cannot be physically present in every port.

Speed of reporting—The inspection company will clearly be more strongly motivated to report on the movement to their principals than any other party involved. Speed of reporting, confirmation of loaded figures, etc., might well be the key to profitable success in some transactions.

Technical objectivity—Whilst the invoice for the inspection fee might be paid by one side only in the transaction, it is a generally accepted principle throughout the industry that all reputable inspection companies will act in a technically independent way in reporting their results, and all sides will generally accept the outcome.

Expertise—Very few oil companies, let alone oil traders, will possess in-house expertise to the same depth as the larger inspection companies within their narrow specialisation. Many of the traders and finance institutions will totally lack this expertise. Even the largest oil companies will not possess detailed local knowledge of practices at every installation they may use. The inspection companies, because it is their business, will certainly know.

Cheaper than in-house staff—In the halcyon days of yore, the major oil companies might have employed their own team of inspectors. Cost has long since changed the practice and in common with many other services, the industry finds it cheaper to contract out sevices for which it does not have intensive use.

Meaning of 'independence'

The concept of 'independence' is central to the role of the inspection company. How, it is asked, can you profess independence when your invoice is paid by only one side? With difficulty, is the glib answer! In reality, the major professional inspection companies will have less difficulty with the necessary balancing act than some of their less-than-reputable clients. Every inspection company will have stories of the client, perhaps through ignorance but perhaps also through deviousness, asking for less-than-objective results to be reported. Many companies will have gaps in their client lists which bear testimony to their

Reproduced by permission of Petroleum Review.

integrity on such occasions.

A more recent threat to independence arises from use, by oil companies, of marine specialists or marine expeditors. Their primary task is self-evidently to expedite the loading, or more usually, the discharge of the vessel with a cargo of crude oil, and enhance the out-turn to the benefit of the receiver. The companies involved in offering marine expediting services frequently claim that their presence at the discharge obviates the need for independent inspectors.

Some clients have already grasped the essential conflict of interest problem presented by such claims. Having the enhancement of out-turn as a major objective comes at best perilously close to negating an independent viewpoint.

The major inspection companies will be perfectly capable of ensuring the out-turn is as complete as possible, whilst their independent stance ensures that they remain acceptable to all sides in the transaction. Some of the larger inspection companies, having started marine expediting services themselves as a diversification quickly realised the difficulties presented within their own houses.

Whilst there is clearly a role for the marine expediting specialist, it is fundamentally different to that of the inspector.

Commercial considerations

The greatest danger faced by the nature of their work to the independent inspection company lies in the value of the commodities they handle and the errors consequent upon one of their staff making an error. Certainly in the last ten years, the increase in the value of the commodity, set against the level of fees, appears to be totally out of balance. With inspection fees typically in the low hundreds (of pounds) and cargo values in the hundreds of thousands or higher, and with every major inspection company undertaking some hundreds of such assignments each month (in the UK alone), it seems only common sense for the industry to recognise some upper limit of liability. Yet numerous attempts by the inspection industry to do so have generally met with failure. Professional indemnity insurance is prohibitively expensive and very limited in scope. The oil industry continues to treat the inspection companies as their own insurance policy against losses rather than a professional partner. Meanwhile the inspection companies remain nervous.

A further problem which faces the inspection industry is that of tumbling rates. It is exceptionally easy for a single individual to set himself up as an 'inspection company'. He needs pen, pencil, paper, calculator, telephone and (perhaps) access to a telex. He can offer services at excessively cut-price rates, which eventually work their way through the whole inspection industry. Recognising that competition is a healthy sign in an industry, the client should address himself to whether the rate for the job isn't just a bit too good.

It is an easy exercise: the fee is known and the hours involved for loading or discharging can easily be established. Subtracting any expenses from the fee and dividing the remainder by the hours gives an effective hourly rate. Whilst that should be lower than in-house hours costed by oil company accounts, a figure too low should arouse suspicion that

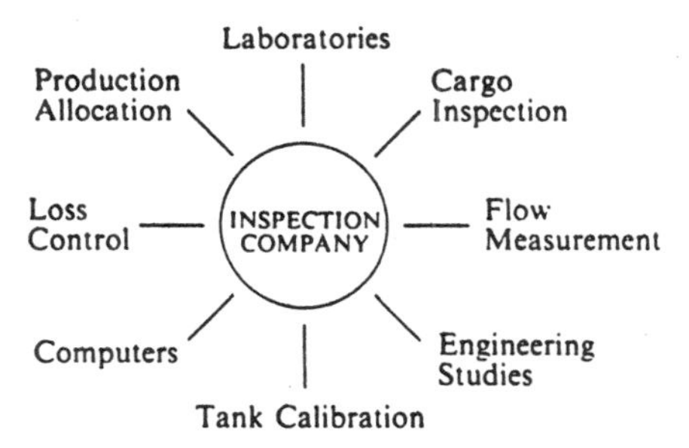

The structure of an inspection company.

either corners have been cut or, in the extreme, physical attendance never took place. The consequences upon the oil company of using a contractor who is too cheap can be serious, if it all goes wrong. Furthermore, such sharp practice has not and does not assist the image of the professional cargo inspector.

It is *de rigueur* to produce results to many significant figures, agreed by all parties who were witness to the cargo movement. The 'numbers' take on an exactness far beyond that which they deserve, because the banks can only deal with 'exact' numbers.

Firstly, there are a whole range of sources of error in physical measurement that cause the resulting calculation to be inexact. Then there are the variations in terminal practice of methods of arriving at composite densities.

There is a new generation of error arising from the 'numbers problem'. Computers on site at the terminal, may round or truncate differently than a hand-held calculator or computer or another computer offsite (possibly belonging to the inspection company). Much has been written about the statistical treatment of random errors which is largely accepted in theory by the technical branches of our industry. Is it not time the commercial branches found a way of coming to terms with it?

Crude oil operations

As an example of how the inspector performs his role, consideration will be given to a crude oil custody transfer movement.

Almost all companies involved in loading or discharging crude oil—whether inspection companies or buyers—have a set of procedures for the inspector to follow when loading crude oil. Attempts are being made to standardise internationally, but there are inevitably many variations for which to cater, including those created by differing terminal practice. Inevitably a balance has to be struck between what constitutes an exhaustive inspection and what time allows and what clients are prepared to pay for.

Furthermore, the most detailed procedures cannot supplement an experienced, intelligent inspector. Rather than list out a 'recipe', the following highlights some of the more common problems that arise.

(a) Clear, unambiguous communication between the inspection company and the client are essential and both parties should not draw back from pressing the other for fuller or clearer information in documentary or telex form. Their relationship will be the stronger for it.

(b) Where automatic sampler systems are permanently installed on pipelines or loading arms, inspectors must satisfy themselves of its cleanliness prior to start-up, that it is suitably adjusted to capture the correct volume of samples and especially that it is put into operation coincident with the start of loading and does not miss any initial water surge.

(c) Particular care must be taken by the inspector to ensure that only those tanks nominated to discharge crude oil are physically connected to the pipeline manifold and thus to the vessel and any adjacent tanks are adequately and safely isolated.

(d) Temperature should be establised wherever possible with electronic measuring devices rather than more traditional means using mercury-in-glass thermometers. Practical difficulties arise, however, from the former's combined disadvantages of being cumbersome yet easily damaged.

(e) Automatic gauging systems are frequently poorly maintained and calibrated and should never be taken on trust by an inspector, who should satisfy himself by comparing with manual measurements prior to each operation.

(f) Time pressures (e.g. early departure procedures) imposed by the terminal inevitably create difficulties for the inspector in the time-consuming activity of drawing representative samples. The inspector requires maximum co-operation from both ship's crew and terminal staff at such times and will need to draw on assistance from one or more of his colleagues if the task is to be completed satisfactorily.

(g) The inspector will require specialised instruments for sampling and temperature measurement to deal with vessels having inert gas blankets where the blanket cannot be released during measurements.

(h) The determination of on-board quantity (OBQ) can be the source of considerable discussion. A high degree of skill and judgement is required of the inspector to take the requisite measurement, select the most appropriate calculation or estimation procedure and report in a fair yet precise way on the nature and characteristics of the quantity on board on arrival.

Determination of quality

There are two general requirements to be satisfied by crude oil quality tests. The first is to provide data for the precise establishment of the volume or mass quantity: density and water determination are used in this. Other tests are designed to provide data on how the crude cargo will subsequently transport, handle or refine. The descriptions following are designed to provide an insight into the most common tests and their usage.

Density

Since the measurement of crude oil cargoes is almost invariably carried out volumetrically, the density must be determined before the cargo can be expressed on a gravimetric basis.

Density has been traditionally measured by the use of a hydrometer IP 160/82, but the digital density meter IP 365/84 is gaining acceptance and a portable meter for which the manufacturers claim an accuracy of plus or minus 1 kg/m^3 is available in both normal and intrinsically safe versions. Densities are often measured at ambient temperature and then reduced to a standard reference temperature of 15°C, 20°C or 60°F using internationally accepted tables.

Water—Some formation water is produced with almost all crude oils and although a variety of mechnisms are often employed to minimise the water exported with the crude, some will almost invariably remain in the cargo and its determination is obviously of interest to all the parties concerned.

There has been a slow trend in the UK to adopt the Karl Fischer method IP 356/84 as the standard. Unlike the distillation method IP 74/82 and the centrifuge method IP 75/82 in which the water is separated and then quantified, this method relies on any water reacting with iodine in the presence of catalysts. Since the water is not uniformly distributed through the bulk oil sample delivered to the laboratory, the accuracy with which it can be measured is largely dependent on the mixing and subsampling prior to analysis.

Salt—High salt contents in crude oil can have a radical effect on refinery costs, the saline water settling out of the oil can be quite corrosive, as can the hydrogen chloride formed when oils with a high acid number and a high salt content are heated. As the water is evaporated from high salt crudes the resultant crystals can cause plugging; moreover, sodium will be reflected, to their detriment in the resultant fuel oil products.

The preferred method for salt determination is IP 265/70 in which the electrical conductivity is measured of a sample of crude oil dissolved in a polar mixture of salts in the ratio in which they are often found in crude. As in the determination of water above, the precision of the salt determination is dependent on the quality of the subsampling.

Reid vapour pressure—The RVP IP 69/78 reflects the quantity of light ends, methane through to pentane, dissolved in the oil. The higher the RVP, the greater the probable loss of these light components during transhipment. As any loss of light components will radically affect the RVP determined, care must be taken during both the drawing and subsequent handling of the sample. It should be drawn through a cooling coil and remain cooled until poured into the oil chamber of the RVP apparatus. This is then connected to an air chamber fitted with a pressure gauge and the unit maintained at 100°F until a constant pressure value has been observed.

Pour point—The viscosity of crude increases as its temperature decreases, until the temperature is reached at which the oil has gelled and can no longer be easily handled. In IP 15/67 this gelling temperature is found and a 3°C safety margin is added and reported as the oil's pour point.

Sulphur—Although sulphur compounds are present to some extent in all crude oils, the total sulphur content can vary from 0.06 per cent to over 5 per cent. Most sulphur compounds are corrosive and can cause problems during refinery operations and there has been a trend to reduce the sulphur content of products for environmental reasons. For convenience, X-ray fluorescence IP 336/81 is the method most commonly used, but for high salt crudes there can be a problem with chlorine interferring and producing an abnormally high sulphur figure.

Distillation—The composition of a crude normally changes only slowly with time, but many loading terminals are now exporting blends or 'cocktails' of crudes from different fields and hence with different properties and in these cases an assay is sometimes required. These range from a preliminary distillation IP 24/84 to a full-scale assay based on ASTM D 2892. Similarly, the light-ends composition of the load can vary not only with differing proportions of crudes in the blend but also as a result of changes in operating practice such as stabilisation conditions or rates of condensate reinjection.

Sampling

Laboratory testing required relatively minute samples of oil drawn from the massive bulk of a liquid which can be anything but homogenous, as it resides in land or ship's tanks. Whilst static sampling is still the most widely used method, great progress has been made towards sampling dynamically at the ship's manifold, using transportable equipment. As well as its psychological advantage of being located at the actual point of custody transfer, technical advantages are also conferred, because at this point, the load or discharge stream is well mixed. Ownership and operation by an inspection company also removes from the equipment and technique the suspicion of bias which may accrue to a fixed pipeline sampler installation.

Use of statistics

Statistical analysis is used increasingly by inspection companies to assess the validity of measurements and control losses. In order to assess the accuracy of the measurements at both the loading and discharge ports, the quantity based upon the ship's tank measurements may be used as a monitoring tool. Although the ship's tanks are not calibrated as accurately as the shore tanks or meters and may in practice vary from −2 per cent to +2 per cent of the true value, they can still be employed as a powerful tool for monitoring purposes, provided the vessel is fully loaded.

The relationship between the quantity based upon the shore tank or meters and the ship's quantity is known as the ship's experience factor (SEF) and is conveniently defined as the ratio of the ship's quantity, expressed as gross volume loaded, to shore quantity again expressed as gross volume. Ship's experience factors may usefully be derived at both the loading and discharge port.

Studies of ship experience factors by many companies have shown that most loads result in SEF values greater than unity. This is primarily due to the variations in trim and list, which take place during the loading, which have the effect of indicating a higher liquid level than is real.

Random effects associated with individual transactions are particularly noticeable when shore measurement is based upon static tank figures. These result in random uncertainties associated with tank level and often lead to the common occurrence of one high SEF being followed by a low value on the next load. This may arise from the static measurement uncertainty, particularly where small changes in level are involved, or may arise from slack pipelines. In the latter case pipelines which have been allowed to drain between transactions result in measurement errors when they are refilled on the next load.

Other factors which affect the value of the SEF in loss control are the influence of oil grade, the on-board quantity (OBQ) when loading begins, the sediment and water (S & W) content of the oil and the degree to which each ship tank is filled. The calculation of SEF has been shown to have two major values. Firstly, it may be used to identify serious problems associated with any individual load or discharge, whilst the second use is in identifying any problem associated with individual port loading facilities. This latter use is much less widely appreciated and yet its value can be enormous. Overall the use of SEF values present a most valuable loss control tool in the analysis of voyage statistics.

After each individual loading, the ship experience factor, at the loading terminal should be calculated and filed. This procedure allows a data base of information to be established which enables the control limits for future SEF values to be established. The expected value will be the mean of the data base values, corrected for outliers, whilst the uncertainty band will be derived from the standard deviation of these figures. Provided at least 20 previous readings have been stored the uncertainty band may be established as twice the standard deviation, which is the 95 per cent confidence limit associated with a normally distributed variable. With a lower number of records the uncertainty band will be significantly wider, but nevertheless alternative statistical tools will provide a powerful control technique.

The SEF value associated with the latest transaction is then compared with the maximum and minimum limits. The most convenient means of carrying out this comparison is by means of a control chart. Such a chart allows trends or problems to be visually identified quickly. Based upon their wide experience across the world, the major inspection companies are ideally placed to provide extensive loss control services based upon statistical techniques, starting with simple control charts as described above, but extending to many other powerful techniques.

At a time when the operations and profits of the oil companies themselves are under severe internal scrutiny, it is inevitable that they will no scrutiny, it is inevitable that they will now seek to sub-contract many more operational services—laboratories, technical support, etc. This will create further opportunities for these inspection companies that have the resources, both in financial terms and in technical and managerial skills, to respond to the challenges. □

ANCHORING AND MOORING EQUIPMENT ON SHIPS*

A. K. Buckle, B.Sc, C.Eng, FRINA
Lloyd's Register of Shipping

WHILE THE WORLDWIDE number of failures of anchoring and mooring equipment due to bad design, installation or operation is now known, a survey made by Lloyd's Register revealed that no less than 38 ships exceeding 100 tons gross became total losses from this cause in the 4½ years ending June 1972. Of these, 13 dragged their anchors, one was run down by another ship whose moorings had failed and one capsized at its moorings, in a typhoon, while laid up. No ship of 16,000 tons gross or over became a total loss due to this cause but a number did break loose from their moorings, so large ships cannot be regarded as free from risk.

Many items of equipment are involved in the total system and a significant proportion are not covered by classification society rules or international standards. To keep matters in perspective, however, it must be noted that the losses from anchor and mooring failures given an average incidence of only 0.0143 per cent per ship year—i.e., only a fraction of the incidence from any one of the other causes under which losses are normally listed.

Classification equipment

International Association of Classification Societies' Unified Rules—In 1965 the classification societies unified the basis for anchor and cable requirements by accepting that the equipment number (EN) should be based on: EN = constant ($\Delta^{2/3}$ + Factor for windage).

The central feature was the concept that the ship would swing so as to tend to head into wind and tide. This meant that drag would be proportional to some actual or idealised cross sectional area of the vessel. Early calculations assumed a 25m/sec wind and 2.5m/sec current acting on a ship in a reasonably sheltered anchorage. Two snags immediately appeared. The first was that aerodynamic friction drag depends on the length as well as the cross section of the superstructure. Omitting to allow for this gave EN values a little too small for passenger ships and too large for tankers. This was put right by reducing the constant in the factor for windage and adding a third term to the formula; the term that now appears as A/10 in metric values.

The second difficulty was that windage was dominant in practice. This meant that if the formula truly reflected reality the equipment on a ship would have to be determined on the light displacement, not its load displacement. At the design stage, and during construction, only the load displacement is sufficiently 'fixed' to be used as a base for an enforceable classification rule. After some thought the constants were revised to give a swing of roughly 12 per cent from the h × B term to the $\Delta^{2/3}$ term. This had the result that a change in draft now has little effect on the value of EN but that, on cargo vessels in particular, any change that does occur will tend to cause EN to increase with displacement.

Table I Chain cable defects and losses on ships built 1965 to 1972 inclusive.

Grade of cable	I	U1	U2	U3
Number of failures	6	11	147	9*
Total of ship years	442	1471	11181	584

*Five of these nine failures were losses directly attributable to windlass malfunction in 1971 and 1972.

Table II Incidence of cable defects and losses

Year	Tankers % per ship year	Cargo ships % per ship year	
1954-1958	0.41	1.57	
1959-1963	1.45	1.53	
	Grades U1 and 1	Grades U2 and SQ	Grade U3
1965-1970	0.8	1.5	0.6
1971-1072	1.0	1.1	1.9*

*0.7% if windlass malfunction cases are omitted.

In the author's view this is realistic because wave induced snatch loads are now known to be a major item in anchoring and mooring forces and these do tend to increase with ship mass.

Having chosen the basis for the number the chain sizes needed to be selected. It was hoped to base these on a preferred number series but this proved impracticable because:

(a) No series gave suitable increments over the whole range of sizes,

(b) For metallurgical reasons the ratio of chain diameters, for the three grades of steel considered suitable, had to be 1.13 to 1 which was not coincident with the R20 increment factor of 1.12 nor with the proposed ratio of enlarged link to common link of 1.10.

(c) Flash welded chains tend to have link sizes about 2½ per cent less than the diameter of the bar from which they are made, so it is not possible to have both bar and chain to preferred sizes.

The final choice of sizes was therefore a compromise, the table values being rounded values derived from the formula $d = K_1\sqrt{EN}$ where K_1 = 1.75 for grade U1 chain, 1.552 for U2 and 1.375 for grade U3 chain. Giving priority to grade U1 values, the formula was used to obtain the mid-points of the range of EN values for each diameter.

These extracts are reproduced by permission of the Royal Institution of Naval Architects and Lloyd's Register.

either greater or less than optimum), and generally lies between 0.1 per cent and 1.0 per cent of the ship's displacement, depending on the location of the buoy and the weather conditions laid down for pumping to be discontinued.

Anchor testing

While the tensile proof test loads to be applied to anchors have been agreed by IACS the exact method of applying these loads is not part of the unified agreement and various systems are used to maintain the point of application of the load at one-third of the arm length from the extremity of the bill (fluke). As certain of these systems involve some form of wedging, the forces on the crown pin can be increased relative to tests where the palm is drilled to accept retaining pins or their equivalent.

However, as permanent set is not a problem, this variation seems to be more untidy than serious. Consideration is, nevertheless, being given to the possibility of standardisation of test methods.

In order that an anchor may be accepted as a 'high holding power' type full scale tests are required to be made on two sizes of anchor. The smaller of the two anchors is to be have a weight of not less than 1/10 of that of the larger and the larger is to have a weight of not less than 1/10 of that of the largest anchor for which approval is sought. The tests should be conducted on soft mud or silt, sand or gravel and hard clay or similar compacted material, and the results correlated to those of a previously approved anchor, preferably one of an ordinary standard stockless design. A scope of 9 or 10 is recommended but consideration will be given to a scope of 6.

IACS requirement 20 lays down that three tests shall be taken for each anchor and nature of sea bed. Tests should normally be carried out from a tug, but alternative shore-based tests will be considered, provided the anchor is fully submerged throughout. Measurements can be by means of a dynamometer or derived from the rpm/bollard pull curve of the tug. The tests should be conducted under the supervision of the classification society's surveyors.

Chain cable testing

Cracking has occasionally occurred in way of the weld behind the end of the stud on U3 chain. This cracking cannot be seen during normal chain inspections and one chain (not on an LR classed ship) with a high percentage of such cracks is known to have successfully withstood proof testing twice with a four year period of service in between. While it is not considered that ultrasonic testing to detect this fault is needed for ship anchor cables, manufacturers might consider it wise to make spot checks from time to time to see that the chain making process and machine settings are compatible with the steel being used.

When short link chain is permitted instead of studlink chain—e.g., on small craft—then sympathetic consideration will be given to proposals for chain to ISO recommendations R1835 and R1836 to be regarded as equivalent to the classification U2 requirement. In such cases all the tests required by ISO should be properly undertaken and supervised to the society's satisfaction.

In their negotiations with ISO, IACS agreed to recommend to member societies that, once ISO had published standard 1704 for anchor chains, then the stud link proportions and tolerances given therein should be adopted for classification purposes instead of the weight criteria now used by certain of the societies. This standard was published in April 1973.

Rope testing

The testing and inspection of fibre and wire ropes of the more common sizes has not posed many problems. The recommendation by IACS that the societies should amend their rules, where necessary, to align them with ISO Standard 2408 for wire ropes should help in the few areas of difficulty still existing in some countries. Difficulties have occurred with the very large synthetic ropes used for single point moorings where no machine seems to exist which combines the necessary stroke and power. This problem arises largely because of the need to splice the ends of the test sample. If some form of socketing could be devised, such as is used for wire ropes, then the test sample length would be greatly reduced and the extension at break would be within the capacity of existing chain testing machines. In the meantime, strength must be derived from extrapolated realisation factors and Lloyd's Register have agreed to the values in Table III for nylon rope using double braided construction.

Table III Realisation factors for double braided nylon rope.

Rope Circ. (inches)	Realisation factor
6	0.709
6½	0.697
7	0.685
7½	0.674
8	0.662
8½	0.651
9	0.639
9½	0.627
10	0.615
10½	0.610

Work has been carried out to determine the effects of dynamic loading on fibre ropes—(a) the effect of snatch loads; and (b) the effect of repeated loads at medium frequency. Both types of loading seem to result in reduced strength in some types of fibre. Based on the rather scanty rough water data available, Lloyd's Register were assuming that nylon ropes used for single point moorings may suffer up to 25 per cent reduction in their static test strengths due to dynamic effects. The deduction for polypropylene is 20 per cent, but both values are subject to revision.

Deck machinery—windlass design

The vast increase in ship sizes in the past ten years has left those windlass designers who relied on 'rule of thumb' methods on very uncertain ground. A whole series of failures of windlasses and associated structure and fittings has resulted in ISO, the classification societies and some governments taking a much greater interest in the arrangements for handling large anchors. IACS recommendations on a performance criteria not greatly different from their existing statement 19, the more important points of which are:

(a) A continuous duty pull of
$3.75\ d^2$ kg for U1 chain
$4.25\ d^2$ kg for U3 chain
$4.75\ d^2$ kg for U3 chain
where d is the chain diameter in mm.
This would allow a ship to weigh anchor in 100m depth of water or to pull itself up against a 3 knot current and Beaufort 6 wind by using the windlass alone, assuming a hawse pipe efficiency of 70 per cent.

(b) The mean speed for hoisting the anchor and cable from deep water should be not less than 9m/min measured over two shots of chain during trials. It is probable that such a trial would commence with three shots of cable fully submerged but this has yet to be finally agreed. The question of overload capacity has still to be settled but it is possible that large ships may be required to recover one anchor with 50 per cent of the chain on that side of the ship, from deep water.

(c) When a windlass is associated with a chain stopper the windlass, including the brake, is to withstand a proof load equal to 45 per cent of the breaking load of the chain. This proof load should be 100 per cent of the breaking load of the chain if an adequate chain stopper is not fitted.

(d) Only one anchor will have to be lifted at one time.

UK practice

Typically, a double-lifter windlass may be designed to raise two anchors simultaneously from a depth of 30 fathoms, a pull equivalent to raising one anchor from a depth in excess of 90 fathoms or 165m. Hoisting speeds around $7\frac{1}{2}$m/min are fairly common, the speed increasing as the anchor is raised and the effective load reduced depending upon the characteristics of the windlass prime mover. An overload capacity of 150 per cent is usual for assisting in the breaking out of an anchor.

Because of the wide range of chain cables a windlass manufacturer may produce 10 or 12 basic frame sizes from which he can select the appropriate windlass to cater for the specified requirements. In other words, the selection may be of a windlass one or two frame sizes above standard but modified to include the appropriate cablelifter. It goes without saying that the basic standard should cater for the vast majority of cases.

There seems to be a move towards designing single-lifter windlasses or windlass units powered by mooring winches. These anchor handling arrangements become more common largely because of bulbous bows and the necessity for spreading the hawsepipes to allow a clear fall of anchors. Here, the windlass designer has to decide whether to use the same total power for two single-lifter windlasses as for one double-lifter windlass or to double the power forward. The solution is usually a compromise, for example, each single windlass having 75 per cent power of a double windlass. The economics of this arrangement are justified in the case of a mooring winch/windlass unit arrangement since two functions are met with one power source.

For cable sizes above, say 84mm diameter, double-lifter windlasses are the exception rather than the rule and the mooring winch/windlass unit layout predominates. Ships requiring cables of this size will be quite large, in the order of 50,000—60,000 tonnes and here the desig[illegible]l not to reduce windlass p[illegible]t is not unusual for the ow[illegible]ies.

With regard tc [illegible] enerally assumed by British [illegible]ıction of the brake is to chec[illegible]ıd chain and to arrest the ca[illegible]able has been paid out. So[illegible] assume that cables will on[illegible]wer and downgrade their l[illegible]rdingly. For riding at anch[illegible]ıe that a bowstopper will be [illegible] engaged to remove the ridi[illegible]ass. The cable-lifter brake i[illegible]vithstand only a load equivale[illegible] a double-lifter windlass.

From the man[illegible]ew, any proposal which see[illegible]ss design on the basis of wh[illegible] or is not fitted cannot be v[illegible] aim is to eliminate unneces[illegible]rive at a common internati[illegible]

The question [illegible]hould be mentioned at this [illegible]te critical. Good practice wo[illegible]s induced under rated duty lo[illegible] per cent of the material yiel[illegible] yield on overload. Gearing[illegible] under full load horsepower [illegible]ed and for rolling element b[illegible]3.10 life is adequate. These l[illegible] but it must be borne in mind [illegible]nost windlasses are of co[illegible]ation and infrequent, proba[illegible]ion of time being spent in wa[illegible]

Windlass fail[illegible]

There are few [illegible]at have not failed at some ti[illegible]ps in recent years, but the m[illegible]l into eight groups:

1. Loss of power s[illegible]
2. Under powerin[illegible]
3. Slipping brake[illegible]
4. Inadequate lub[illegible]
5. Component in[illegible]ation.
6. Dynamic torq[illegible]
7. Shock loads fol[illegible]re.
8. Misalignment.

While any o[illegible] to cause a breakdown, the[illegible]in practice it is possible for t[illegible]ur together. For example, i[illegible]s it is quite common for th[illegible]p allowing a considerable le[illegible] through the hawsepipe. Th[illegible]es lost if the windlass powe[illegible]er it, so the master has to c[illegible] the ship may proceed. This i[illegible]to occur only on very large [illegible] found to be difficult to ob[illegible]xact circumstances of most [illegible]possibly come under this hea[illegible]

Table IV ...ndlass faults included in classification survey reports by Lloyd's Register's surveyors 1969 ...usive

Type of def...	...psy	Shafts	Bearings	Framing	Bolts	Gearing	Warping drum	Prime mover	Other
Broken		8	18	7	2	11	1	17	10
Cracked		4	15	16	—	20	1	9	18
Corroded		3	4	—	—	1	1	13	1
Scored		11	4	—	—	4	1	2	1
Overheated		2	3	—	—	1	—	45	2
Seized		—	5	—	—	2	—	1	3
Alignment		29	3	1	4	5	2	6	4
Flaking		1	—	1	—	1	—	—	—
Loose		3	7	1	1	3	1	3	4
Excess wear		1	3	—	—	9	—	2	3
Defective (type not stat...		3	5	—	—	9	1	67	8
Total		65	67	26	7	66	8	165	54

Attention ... ...lrawn to the high failure rate of windlasses fc... ...hain stopper failures. Damages reported in t... ...ımstances include:

- Gipsy fract...
- Gipsy drive... ...tured completely through.
- Gipsy drive... ...t.
- Key sheare... ...ass drive pinion.
- A frame col... ...forward edge.
- A frame fra...
- Both A fran... ...ed inwards.
- Both A fran... ...ed to starboard.
- Windlass se... ...rted.
- Brake lining...
- Brake pin fa... ...ue to vibration.
- Deck set do... ...windlass.
- Windlass ca... ...board.

Table V An... ... total number of faults to anchoring an... ...g machinery, ascertained as having occurr... ...ps visiting 130 selected ports during June 1... ...gust 1973 inclusive.

'A' frame defor...	...fracture	5
Seating collaps...	...ion	13 + (see Note 2)
Mainshaft fract...	...ortion	9 +
Motor failures		7
Brake failures		25 +
Clutch failures		9 (see Note 3)
Lubrication fail...		10 +
Other compone...		17
Chain stopper fa...		7

Notes: 1. Of ... ...orts contacted damages were repc... ...only 41.
2. Wh... ...ported 'a few' or 'several' cases with... ...ng numbers the values were reco... ...- for each port so reporting.
3. Incl... ...lures of rendering controls on auto... ...ring winches.

A very simil... ...damages has resulted from brake failures v... ...e prior failure of the chain stopper. Five ca... ...eported over a period of only a few months in... ...or port alone, but there has been some impr... The mechanism of failure is that first the cha... ...way until the cable locker is empty. At this p... ...length of cable between the gipsy and the b... ...in the cable locker comes under tension so... ...ultant force acts on the gipsy equal to almost twice the original chain tension augmented by some dynamic shock effect as the chain is suddenly brought to a stop. In the majority of cases, however, the bitter end is pulled out of the ship instead of, or in addition to, the windlass failure.

There seems little doubt that the cost of fitting better brakes would be far less than the cost of redesigning the windlass, its seating and the deck below to take the high loadings following brake slip. The loss of power is most commonly due, on steam and hydraulic windlasses, to corrosion especially corrosion resulting in pipe fractures. On electric windlasses causes are more diverse: thyrister control failures, burnt out motors following seizure due to poor lubrication and water entering through inadequately sealed covers have all been reported recently.

Lubrication failures are of two main types: (a) due to oil draining away from the bearings when the windlass is set at an angle other than horizontal, either due to the seating design or to ship motions or trim; and (b) due to lack of grease or oil grooves in the bearings. Failure due to this cause has occurred in various bearings, including the eccentric sheaves, and in this latter example caused the eccentric rods to break. Failure has also occurred when a steam windlass intended for use with wet steam (in association with graphite lubrication) was fitted to the ship's superheated steam system. Component inaccessibility comes in many forms—grease nipples that cannot be reached can cause lubrication problems, for instance.

A serious worry on some windlass designs is the detailing of the chain pipe connection to the windlass bedplate. On some designs the access to this joint has been so restricted that it has been impossible to weld the butt on board ship and water has flooded into the fo'c'sle in bad weather as a result. A more serious matter is when such bad access results in it being difficult or impossible to properly seal the chain pipes when the ship is at sea, and at least one ship has been lost due to flooding of the chain locker inducing progressive flooding of the other spaces.

It should be noted that on large ships, cement on its own, if used to plug chain pipes will often be broken up by the forces exered by the chain. Steel covers in halves, hooked over the chain pipe top (spurling pipe lip) are recommended. These steel covers should then, themselves, be sealed with cement and have a canvas cover lashed overall.

High torques can occur in windlass drive mechanisms due to inertial and similar dynamic forces arising when the machine starts or stops suddenly. The principal types of failure are the shearing of keys and the fretting of keyways but clutch problems run a close second. The main cause is the suddent deceleration when the anchor is housed in the hawsepipe or when the chain snags in the chain locker while being paid out under power. Occasionally, snatch loads also cause damage if the ship is being manoeuvred on the anchor in crowded harbours and the anchor suddenly holds after dragging. In Table Bay and some other ports rocky sea bed areas can snag the anchor when it is being weighed, but in such cases it is usually the anchor rather than the windlass that suffers.

Misalignment can take several forms. One of the more common is for slight transverse misalignment to permit the chain, when running out, to swing intermittently against the brake control gear. This can result in the circlips or cotterpins in the band link pins being broken so that the link pins can fall out. Other damages known to have occurred are:

(i) Over-running of the cable due to insufficient depth of chain locker giving lack of 'tail end' weight; or the chain pipe being too far aft so that the cable does not bed properly on the cablelifter. In general, the chain passing over a five pocket cablelifter should change direction through an angle not less than 120°.
(ii) Transverse misalignment of the windlass relative to other fittings which results in the chain slamming against, or rubbing on, guide bars or machine parts.
(iii) Misalignment due to wave impacts causing distortion of the windlass frame.

Wave forces acting on large ships can be almost unbelievably high on rare occasions. Damage which can only be caused by horizontal pressures in excess of 30 tonnes/m^2 occurs in the region of fo'c'sles at an incidence of about 1 in 5,000 per ship year, averaged over all ships over 100 tons gross. Most of this damage occurred, in fact on ships of more than 40,000 tonnes displacement. Five cases are known to have occurred since 1943 on ships over 15,000 tonnes displacement in which the pressures must have been in excess of 60 tonnes/m^2 and one case known and one heard of where the pressure was probably about 80 tonnes/m^2. In the second of these cases a windlass was reported to have been washed overboard off South Africa after shearing its holding down bolts, but the author has been unable to obtain adequate details of that case to enable him to check the claim properly. High pressures of this order can probably not be fully taken into account on an economic basis but lesser pressures can cause displacement of windlass seats and bearings. Displacements of 10mm have been recorded and in one case a main pinion was displaced 40mm. It is a question of judgment as to the loads that should be assumed for design purposes but no rational judgments can be made if actual loads are totally unknown to the designer.

Cable stoppers

Cable stoppers as used on ships usually consist of some means of obstructing one of the links of the chain as it passes over a grooved track or roller. Three types of mechanism are favoured:

(a) A bar which can be dropped across a horizontal link so as to block the movement of the following vertical link;
(b) A pawl, which may sometimes be hinged at mid length for quick release purposes; or
(c) A screw down, or hydraulic compressor mechanism.

Occasionally a Devil's claw or Blake slip is used as an alternative to track or roller type cable stoppers

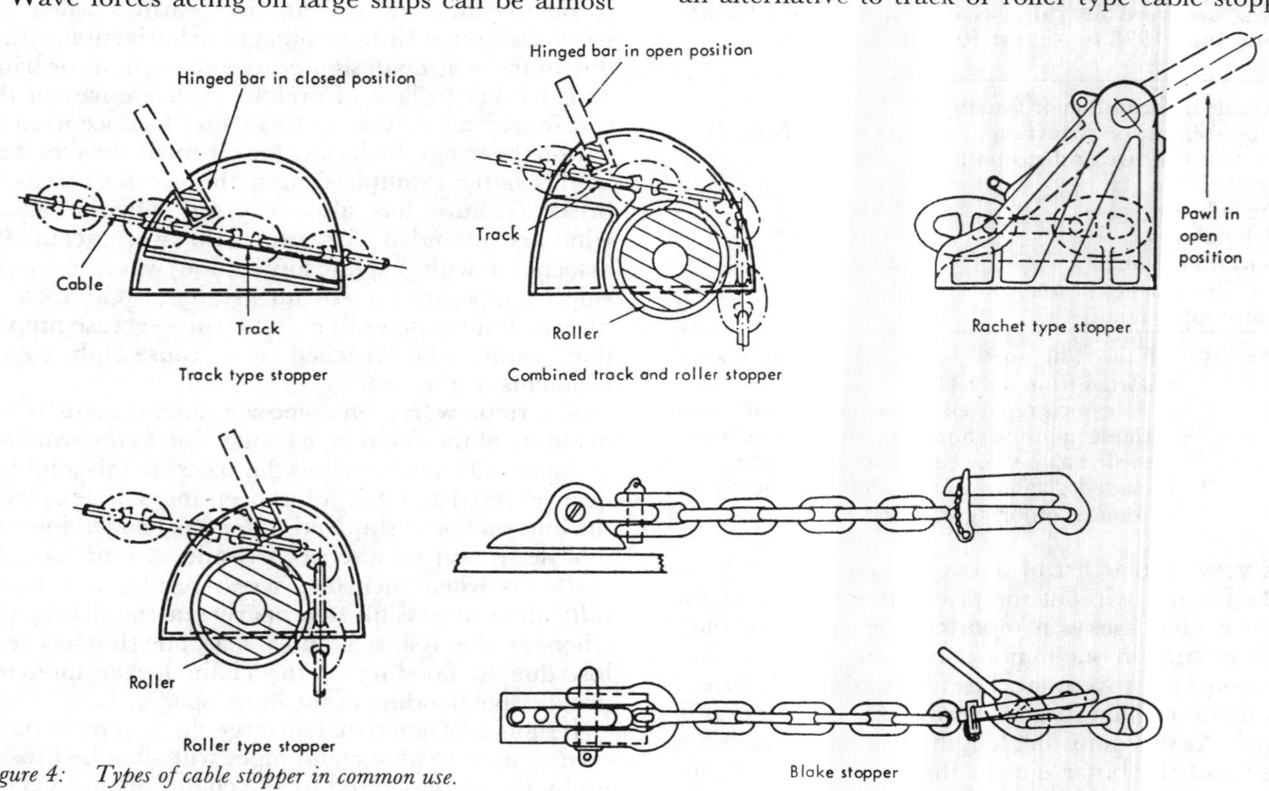

Figure 4: Types of cable stopper in common use.

although their use is more generally reserved for final adjustment to prevent movement of anchors in the hawsepipes. Cable stopper designs which use pocketed rollers with some form of locking device, instead of a bar or pawl are now occasionally seen, but mostly on drill rigs rather than on ships. Cable stopper design has tended to be the poor relation in the manufacturers' drawing offices and a considerable number of failures have occurred on large vessels in the last few years. Unfortunately no proper records have been kept and data could only be obtained when reference was made to cable stopper failures in reports relating to the loss of, or damage to, chain cables. It would appear that cable stoppers are generally designed to one of two levels of strength; 100 per cent of the rule chain breaking strength, or about 40 per cent of the rule chain breaking strength. Some stoppers are proof tested to the design loading and some to 70 per cent of the design load—i.e., in the case of the first type above to the chain proof test load.

Bar type cable stoppers are simple to operate and cheap to manufacture but the simplicity of design is accompanied by four deficiencies which must be recognised if accidents are to be avoided. These are:

(a) For normal rectangular section bars the chain link is put in single shear instead of the double shear for which it was designed. This means that if a standard anchor chain is placed in the stopper and a load applied, the chain will break at a load close to, or sometimes a little below its nominal proof load. At loads below this level it is not uncommon for the sharp corner of the bar to cause a notch to occur in the chain link with detrimental effect on the chain's fatigue life. Rounding the lower corner of the bar helps reduce chain damage but does not overcome the basic fault. Some manufacturers are now making bars specially shaped to fit the curvature of the link, sometimes with a lip on the lower edge that will provide support to the link on a basis of some compression rather than on pure shear.

(b) The working of the cable in a seaway can result in the bar gradually riding up out of its slot until it permits the cable to run free. As several examples of this type of failure have been mentioned in reports, it is obviously a matter needing attention, especially as in one case the bar was fitted with a locking pin 'which had bent through 90° to release the bar.'

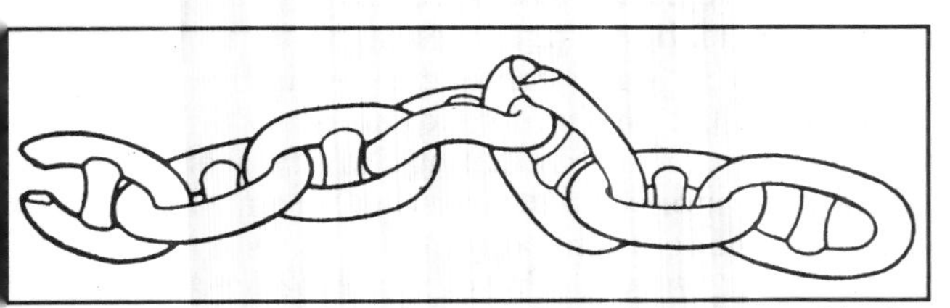

Figure 5: Typical shear damage after applying the proof load to a chain held in a bar type stopper.

(c) The bar is sometimes made of inadequate material so that the cable cuts a semicircular groove in the bar which is then no longer able to restrain the chain from running out.

(d) The stopper is subject to high glancing impact loads when the cable is being paid out under brake control only. Several cast cable stoppers have fractured under this type of loading. It is possible that heat gradient stress effects may also be significant in such cases.

A fault liable to be found in all types of stopper fitted with roller guides is the use of rollers which are too small in diameter. In such cases the links are subject to high bending stresses when the cable is under load. If the groove is deep enough to permit the 'vertical' links to ride clear of its bottom the bending movement will occur in the 'horizontal' links. If the groove is made shallower then the bending movement occurs in the vertical link. (Similar stresses can occur in the links in way of the shell/hawsepipe knuckle, especially if the ship overrides its anchor should the anchor be used for manoeuvring purposes. The designers of anchor pockets should bear this in mind.)

There is little doubt that in the field of cable stoppers the cheapest is seldom the best and shipowners might well consider that some thought should be given to cost effectiveness at this point when they are writing their specifications.

Mooring winch design

There is some controversy on safety considerations where mooring winches are concerned. One view is that a mooring rope is the cheapest component of the mooring system and should be designed to fail first. The other view is that means should be arranged such that the rope may be allowed to pay out before its breaking load is reached. It is agreed that the high elasticity of modern synthetic fibre ropes will cause them to whip back dangerously if they break (several seamen have been killed in this way), but it is not felt that this fact should result in a swing right to the other extreme at which the winch controls are set so feebly that after the ship has drifted away and been wrecked the ropes are in good condition and suitable for re-use. A number of ships have suffered that fate and several oil ports now refuse to accept ships using automatic winches and require positive attachment of the ropes to the ship.

Lloyd's Register introduced a rule requiring ships to be fitted with adequate bollards or with mooring winches with adequate brakes and also published advice on what is considered to be the absolute minimum that they would recommend as adequate. This is considerably in excess of what has been provided on many mooring winches. As over 20 ships had become total losses in $4\frac{1}{2}$ years due to moorings failing and considerably more suffered damage of various severity it was high time that something was done. In particular it is certain that brakes will have to be adequate to hold a load equal to 80 per cent of the nominal strength of the rope, compared with 20 per cent on some winches.

In addition to the requirements for brakes, which apply to all mooring winches, the render and recovery loads of automatic winches are of importance in ensuring safe mooring. Just how large the range of tensions between 'render' and 'recovery' should be depends on the use to which the winch will be put. For use with breast lines the ratio should be small with both values set as high as the controls will permit, but for use in conjunction with springs the render load should be set high and the recovery load quite low. The reason for this is that springs work against one another so that if 'hunting' is to be avoided the absolute difference between the render and recovery loads of the winch must exceed the sum of the recovery load of the opposing winches plus any normal dynamic loads acting on the system—e.g., a 200,000 dwt vessel in a deep-water berth liable to be affected

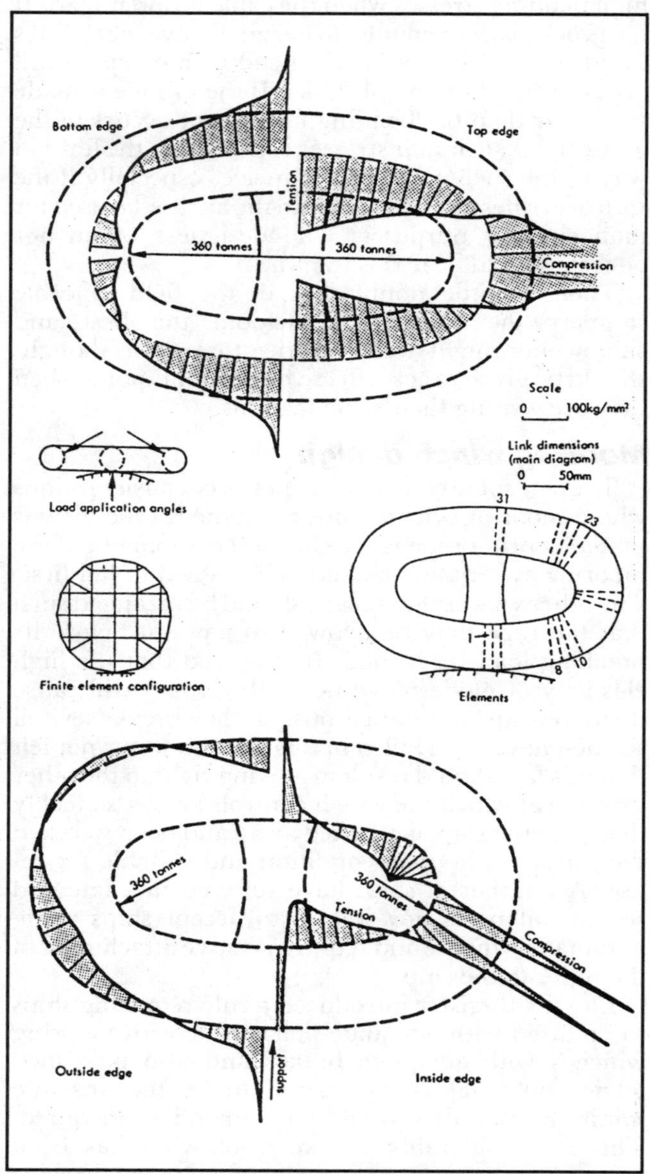

Figure 6: Stresses in 100mm U2 links passing over a 1,800mm diameter roller under proof load tension, calculated using Nastran solid element analysis program methods.

by ocean swell could be subjected to loads of the order of 150 tonnes perpendicular to the quayside and 200 tonnes parallel to the quay.

Assume a winch has a hauling load capacity of 30 tonnes if it were friction free; assume also that there is in fact 10 per cent friction. We then find that we would get a rendering load of $30 \times 1.1 = 33$ tonnes and a recovery load of $30 \times 0.9 = 27$ tonnes giving a difference of 6 tonnes. To withstand a 200 tonne loading we would need 200 divided by 6 = 34 (to the nearest whole number above) winches at each end of the ship—i.e., a total of 68. This is quite impracticable! By increasing the internal friction artificially, to, say, 70 per cent we get revised values of render and recovery loads of 51 tonnes and 9 tonnes giving a difference of 42 tonnes, so that now only $\frac{200}{42} = 5$ winches will be required at each end of the vessel to control the springs. Taken overall, it is considered that, in general, springs should not be controlled by automatic winches.

In the case of the breast lines the full holding load can be utilised in each case so only 150 divided by 33 = 5 winches would be needed in all for this purpose, say two at each end and one amidships, and if a gust did blow the ship off the fenders, there would be a total of $5 \times 27 = 135$ tonnes recovery tension available to pull the vessel back again during the subsequent lull. Statistical data on transient wind forces during stormy weather indicate that the recovery load should be at least 60 per cent of the render load if a reasonable chance of the ship hauling itself back alongside, after blowing clear, is to exist.

If the control of the winch was achieved by means of load switches instead of by the internal friction, then it would be possible in the example quoted above to set the render loads of the breast lines to 51 tonnes (the same as for the springs) and still retain the 27 tonne recovery tension, thus getting the best of both worlds. But such systems are relatively expensive and therefore seldom used.

Mooring winch failures

Apart from hunting and poor brakes, mooring winches have suffered from other failures including the following:

- Reeling the rope onto the drum the wrong way round. This usually results in a drop in brake efficiency and in one extreme case a 50 tonne breaking strength rope could be pulled off the drum, during tests, under a tension of only 2 tonnes.
- Lubrication problems similar to those on windlasses i.e., those due to oil failing to circulate when the ship (or winch seat) exceeds some angle of trim or list and those due to inadequate oil or grease grooves in bearings, etc. BSI are currently drafting a standard laying down heel and trim angles to be assumed for design purposes.
- Internal corrosion due to condensed water collecting in and being circulated from, hydraulic expansion tanks.
- Leaks in the seals of electric component covers.
- Fractures in the conduit piping of electric power leads due to winch and deck deflecting in opposite directions when loads are applied. This can be overcome by fitting an expansion joint or its equivalent.
- Failures of the thyrister controls.

Lloyd's Register have no reports of damage due to wave impacts, but as data are scarce, it cannot be said for certain that wave damage is negligible.

Some of the problems experienced in obtaining detailed data on winch damage can be seen from the following abbreviated extracts from replies to enquiries made by the author:

- '. . . no difficulties have been experienced. In gale conditions large tankers . . . have their moorings supplemented by tugs which thrust the ship against the quay.'
- '. . . mooring winches seldom give trouble . . . self tensioning winches are no longer used because of their inability to cope satisfactorily with surge conditions and a prohibition on their use . . . alongside refineries.'

In the original paper, the author thanked Mr H. Lee for information on design practices in UK industry, Mr G. Pumphry for his work and advice in connection with the diagrams, his outport colleagues in Lloyd's Register for doing most of the hard work in collecting the basic information on which this paper was based and the Committee of Lloyd's Register for permitting the work to be done and published.

This raises the question of when is a trouble not a trouble? It is obvious that, as with cable stoppers, it would be wise for shipowners to undertake a cost benefit analysis before writing their specifications. The alternative is to accept increasing onerous control by classification societies, IMO, etc., who give environmental protection a high position in their list of priorities.☐

Glossary and definitions

A = Area in m^2 in profile view of the hull and superstructure above the load waterline of the ship (a full definition appears in the unified rules of the classification societies).

B = Greatest moulded breadth of the ship in metres.

B10 life = The number of hours that 90 per cent of a group of identical bearings will complete or exceed before fatigue reaches the specified failure criteria, when the bearings are run at the specified load and speed. This life is also known as the basic life or the L10 life.

Chain grades = A means of indicating the properties of the steel from which anchor cable is made. The demarcation of grades has been internationally agreed and defined by the classification societies, and appears in their unified rules.

Common link = One of the ordinary links forming an anchor chain.

d = Chain diameter in mm.

Δ = Displacement of the ship in tonnes.

EN = Equipment Number. This is an empirical measure of the cross sectional area of the ship and superstructure and is determined from the formula:

$$EN = \Delta^{2/3} + 2Bh + \frac{A}{10}$$

h = Freeboard amidships plus the sum of the heights of major superstructure houses in metres. (A full definition appears in the unified rules of the classification societies).

Hawsepipe efficiency = (1.0—coefficient of friction between the hawsepipe and the anchor chain) × 100 per cent.

IACS = International Association of Classification Societies.

Preferred numbers = The conventionally rounded off term values of geometrical series including the integral powers of 10 and having as ratios the 5th, 10th, 20th, 40th and 80th roots of ten.

R20 series = The preferred number series with the ratio between consecutive numbers of $\sqrt[20]{10}$. This ratio is sometimes called the increment factor.

Realisation factor = The factor (always less than 1.0) by which the sum of the strengths of the individually tested strands or fibres of a rope must be multiplied to give the breaking strength of the made up rope as a whole.

Scope = Ratio of chain length to depth of water.

Tonne = 1,000 Kg = 2200 lb.

Wave group = The effect occurring when two wave trains of slightly differing frequencies are superimposed such that the actual measured heights of the waves increases and decreases in a regular manner. The waves occurring during one cycle of this variation are a wave group. The effect of wave groups breaking on a shore is known as surf beat.

ANCHORING SYSTEMS—SOME INSIGHTS FOR MARINERS

Captain A. O. Ojo, MNI, and Professor J. King, M.Sc, FNI
University of Wales College of Cardiff

THE PRACTICE of anchoring ships with a weighted line has been followed for centuries. Until recently, there has been little development in anchoring technology. As ships have become larger, anchors and cables have become heavier, windlasses stronger. But the standard stockless anchor commonly employed in merchant ships has been around almost unchanged for 100 years. In recent years, it has become increasingly clear that the capacity of conventional anchoring systems to serve large, modern vessels is now seriously in doubt.

Today, the anchor, cable and windlass of a ULCC or large bulk carrier must be regarded as an extremely fragile arrangement, for all that it may look massive when one is standing on the forecastle head. As ships have increased in size, anchors have become proportionately lighter, cables proportionately shorter, windlasses more vulnerable to shock loads. In consequence, the anchoring process must be conducted with extreme caution in such vessels lest the gear be carried away. The anchor, a monstrous weight often in excess of 25 tonnes, must be let go with utmost delicacy, with the ship travelling over the ground at no more than a few centimetres per second. There is no margin for error, and in consequence the notion that the anchors can be deployed in emergency situations is no longer tenable. Indeed, even the 34-tonne anchors fitted in very large vessels are totally in adequate to secure them in anything but the gentlest conditions. As Boylan has pointed out in the discussion to reference 1, the anchors of a 542,000-dwt tanaker are proportionately only one fifth as heavy as those of an 18,000-dwt vessel, and the cables proportionately only half as long.

Much experience has, of course, been gained in the offshore industry in recent years, where the securing of large fixed structures has presented challenging problems. Various new types of anchor have been designed and used with considerable success. There has, however, been little benefit to merchant shipping apart from these advances.

Unfortunately, merchant ships are supposed to be mobile most of the time. Being anchored is a temporary state. There, ease of deployment and retrieval is a crucial design consideration for their anchoring systems, to such an extent that weight and holding power may be sacrificed to achieve it.

Given their inadequacy, it is hardly surprising that anchoring systems should fail frequently. Anchors are lost or damaged, cables break, windlasses are unable to cope with the loads placed upon them. The extent of these failures has been highlighted by several recent studies (1, 2, 3, 4) which have shown, not only that the problem is a serious one, but that it is exacerbated by important technical and operational factors. There appear to be, for example, widely held misconceptions about the capabilities of anchoring systems and the operational procedures which should be adopted to meet them. There is, moreover, ample evidence of inadequate maintenance, poor design and, perhaps most significant to us here, inadequate training.

Table 1 summarises the major causes of anchoring system failures in VLCCs[1].

Table 1. Anchoring system failures

Background cause of failure	*per cent*
Engineering design	38.1
Operational practice	33.3
Combination of design and operational practice	14.3
Inadequate maintenance	9.5
Others	4.8

Although these categories of failures are very broad, it is clear that inadequacies of design and inadequacies of operation share almost equal responsibility for system failures.

There are several ways in which the present problem can be approached. We may attempt to design out weaknesses in the conventional system by better engineering: we may attempt to circumvent weaknesses by better operational procedures (remembering that seamanship has been defined as the art of overcoming poor design); or we may look for alternative ways to secure vessels. The first two of these are likely to be palliatives rather than solutions since it is becoming clear that for large vessels even the concept of anchoring in its traditional sense is unsound. The last alternative suggested above might be described as flying in the face of experience. Three thousand years of anchoring history must be telling us something.

However the problem is addressed, its solution will not be achieved without cost. Whether the failure of the system leads to the loss of an anchor or the loss of a ship, the costs involved are significant. There is now available sufficient information on such failures for their likelihood in any particular vessel to be assessed. This means that cost benefit analysis can be employed to help identify measures which might secure improvements in the operation of ships.

We are arguing here that there are grounds for concern that conventional anchoring systems are inadequate. But we are also conscious that these inadequacies only become unacceptable when there is sufficient pressure to do something about them. For individual ship operators this point is reached when it appears that there is some net benefit to be gained from taking action. For the shipping community as a whole it comes when the rules that they all follow are changed. this probably means that, notwithstanding all the technical weaknesses of modern systems, relatively cheap relief such as might be brought about by better training, is likely to be preferred to more radical measures.

What, then, are the areas to which attention might be given?

Improved engineering

Surveys (1, 2) published elsewhere have provided ample detailed evidence of areas of weakness in the engineering of conventional anchoring systems. Anchors, cables, hawsepipes and stoppers all give cause for concern. But the most vulnerable part of the system from the purely engineering standpoint is the windlass. Both during the anchoring process and when lying at anchor, the windlass components are subject to substantial dynamic loads which, in practically every case, must be resisted by a windlass brake of startling crudity. Table 2 identifies the principal compenents at risk.

Table 2 Windlass components at risk

Operation	*Component*			
	Motor	*Gearing*	*Brakes*	*Bearings*
Letting go		X	X	
Weighing	X	X		X
At anchor			X	X

Windlasses have always been regarded as major items of safety equipment. This is reflected in, for example, the prohibition of lap and lead in windlass steam chests which goes back to the nineteenth century. But at the present time very little is done to measure the performance of a windlass, even though there is ample evidence to suggest that in many ships they are often operating very close to their practical limits. This is unfortunate, since the information gained from better performance monitoring would provide evidence not only of the current state of the equipment for assessing maintenance needs, but also for improving overall design.

Windlass control is an area of significant concern. Traditionally the deployment of the anchor and cable is under the control of a manually-operated windlass brake. While this was no doubt acceptable in relatively small vessels (and carpenters developed considerable 'feel' for the task) it is barely so today in large ships. The anchoring environment is dangerous, the operation extremely sensitive to unskilled hands and it is probably unrealistic to continue to be satisfied with manual control in such circumstances, especially when braking arrangements are themselves physically inadequate. It is remarkable that disk brakes are still rarely employed in windlasses, and automatic control of their application is still highly uncommon. The performance of windlass brakes is also sensitive to the content of the lining materials and the considerable amount of heat generated when they are applied. Some attempts have been made recently to monitor heat dissipation in brake drums.

There is clearly ample scope for improving the windlass and several new designs have been proposed, including one hydraulically operated type which is claimed to be capable of absorbing the loads likely to be experienced in emergency stopping manoeuvres.

Improving operational procedures

Ships are generally not provided with information concerning the operational capabilities of their anchoring systems. This means that anchoring operations depend for their success very much upon the experience of the master. There is some data now available on such things as brake liner efficiency, cable scope for a given depth/draught ratio, limiting cable tensions, windlass overloads, and safe anchoring speeds, which could usefully assist masters, but at the present time these are not usually provided to ships.

In order to improve operational procedures there is a need for both better information on how the anchoring system actually performs in practice and for better training. Such monitoring of the system as is actually done at the moment is fairly rudimentary. At the very least, performance monitoring should include:

1. **Measurement of cable length deployed.**
   White paint and seizing wire is the rough and ready system of cable marking which history has bequeathed to us. It hardly needs further comment. Very few vessels are fitted with cable meters.
2. **Chain speed and acceleration.**
   In both manual and automatic operation this information is necessary if correct brake application rates are to be achieved. This is a vitally important aspect of anchoring, since correct brake application rates can reduce shock loads, and cable run away.
3. **Brake liner temperature gradients.**
   It is common practice for the windlass brake to be applied frequently at short intervals. This practice can lead to serious risk of failure because of the high temperatures that it induces in the brake liners. Heat fading is a well documented phenomenon. But without either the means to detect its onset or even an appreciation of its effects, operators are unlikely to apply the brake in the most effective way.
4. **Windlass bearing pressure.**
   Monitoring windlass bearing pressure would allow the operator to assess the level of stresses within the anchoring system.

Many seafarers recognise that anchoring arrangements are inadequate in large ships, although the precise nature of the inadequacies are not always clearly understood. This, in part, may be due to the fact that few seafarers have received anything more than the most cursory training in anchoring practice. Many seamanship text books, for example, do no more than promote traditional practices which are quite inappropriate for larger modern vessels. And their recommendations on, for example, speed over the ground when anchoring, would be suicidal if followed in today's larger vessels.

Anchoring is a practical operation for which practical training is necessary but rarely given. Clearly this is easier said than done. There is no question of training on the job (although *ad hoc* learning by doing, which is not quite the same thing, has been the normal way to acquire most seamanship skills). This applies to much maritime training of course such as watchkeeping and shiphandling, and in recent years there has been growing use of simulation techniques to provide opportunities for practice. Anchoring is ideally suited to simulation training.

Computer-based simulation training[6] allows the behaviour of the anchoring system behaviour to be demonstrated. It provides a means for realistic practice to be undertaken in a variety of circumstances without risk to either trainee or vessel and it also provides a means for examining the skills of trainees.

The value of such training depends very much on the validity of the mathematical models upon which it is based. Several such models are available and attempts are being made at UWCC to produce the necessary software to form the basis of an anchoring package.

An initial survey of seafarers has revealed that the essential consituents of any anchor handling training course are:

1. Practical anchoring operations
2. Anchoring forces
3. Operational planning
4. Cable deployment
5. Communications
6. Safety.

Alternatives to conventional systems

If the inadequacy of conventional anchoring systems is no longer acceptable, then alternatives have to be considered. Two alternatives suggest themselves. One is to employ the dynamic positioning techniques now well established for vessels in the offshore industry. Another is to so manage the operation of large vessels so that the need to anchor is reduced.

Dynamic positioning is expensive and has been installed to date in vessedls which are substatially smaller than those which we are principaly concerned with here. The technical problems associated with implementing such systems in large tankers are considerable. But that is not to say that considering them would not be rewarding.

Alternatively, solving the problem by avoiding it is a possibly fruitful approach for some trades, although it is not without cost. Since the present anchoring arrangements cannot be used for emergency operations, we need to be concerned only with planned anchoring. Good communications with some ingenuity might be exploited to plan voyages so that 'steaming-off' rather than anchoring is the usual way of waiting. The establishment of fixed, adequately-sized moorings at locations commonly used for anchorages would also help.

Both of these alternatives are more easily said than done. But after so many years of using anchors, no alternative which may be proposed is going to be simple—almost by definition. The inadequacy of anchors and cables for large ships will only become unacceptable, however, when there is sufficient discussion of the subject to generate the necessary concern.

Recommendations and observations

- In large ships, soon after brake release the rate of unrestricted descent of anchor and cable becomes excessive. Conventional band brakes operate at, or beyond, the limits of their capability in such circumstances.
- Repeated applications of the band brake after short lengths of cable have been paid out can keep the system under control, but overheating of the brake may still lead to a reduction in braking efficiency and subsequent loss of anchor and cable.
- Walking out the anchor can be recommended as a means of restricting velocities and loads in the system. Maximum benefit will be obtained if the anchor is walked out to the bottom or within a few fathoms of it. Atention must, however, be given to restricting the walkout speed so that damage to the windlass gearing and motor are avoided.
- Speed-limiting devices operating on the band brake help to relieve the windlass operator in the difficult task of controlling cable deployment, but undue reliance should not be placed on such systems.
- Even on windlasses fitted with auxiliary braking devices, or energy absorbers, it is advisable that the anchor should be walked out as recommended above.
- Once the required scope of cable has been deployed, the vessel should be allowed to bring-up on the bow stopper. In any manoeuvring on the anchor, the stopper should normally be engaged.
- The speed of the ship over the ground when the anchor is let go and whilst the cable is being paid out needs to be minimised. This requirement is more critical for larger ships, worsening weather conditions and decreasing water depth. The use of main engines may be required to achieve this. For a VLCC the permissible speed is of the order of 1/4 knot. Accurate determination of such low speeds over the ground is difficult.
- The energy absorption ability of a range of anchor cables indicates that increasing scope should be used with decreasing water depth. Higher scopes are required in ballast condition.
- On large ships, conventional band brakes frequently operate at, or near, their limits of capability. More frequent failures are probably avoided by palliative measure taken by sea staff, for example, walking-out the anchor prior to brake release.
- Designers of deck machinery are already considering, and in some cases have produced, alternative or auxiliary braking devices. At present, none of the devices considered is without some shortcomings, although disc brakes of one type or another appear to be the most promising for routine anchoring. Further development is needed.
- At present, most operators have very little information as to system performance. Instrumentation would be helpful which shows cable speed during free-running and walking-out, length of cable paid out and loads in the system. Limits of windlass speed during walking-out should be clearly indicated to the operator.
- On large ships the bow stopper should preferably be of the double-shearing type.
- Liaison between the shipbuilder and the manufacturer of deck machinery is vital at an early stage if leads of cable and deck layout are to be acceptable.
- Consideration should be given to suitable materials with which to coat the brake drum of the band brake to reduce efficiency losses caused by corrosion and contamination.
- Consideration should be given to water-lubricated synthetic bearings. These would help to reduce the on-board maintenance by sea staff.
- When brake liners are renewed it is important that at least the minimum recommended number of screws securing the liner to the band are fitted. The performance of the brake will be seriously affected if the liner is not properly fitted.

- The recommended mating surface for the brake lining should be maintained. Experience suggests that overlaid stainless steel will disintegrate after a few anchoring operations.
- The thickness of the brake liner is not a good indicator of brake liner effectiveness. new liners can be ineffective due to heat level, wetness and corrosion.
- New liners should be bedded-in before use.
- Training in the anchoring of large vessels is recommended. This should underline the differences in procedure required on large vessels compared to smaller vessels.
- Such training could be aided by computer simulation, but should also include supervised anchoring.□

References

1. King J. and Ojo A.: Some Practical Aspects of Anchoring Large Ships. Paper No. 8 RINA Spring Meetings 1983.
2. Ojo A., Byrne D. and Brook A. J.: An Investigation of Ships Anchoring Systems. General Council of British Shipping Technical Report TR/100 October, 1982.
3. Brinkmeyer H. and Russel K.: Dynamic Behaviour of Anchor Windlasses. Vol 1 & 2. Shell International Marine Ltd. London 1983.
4. Brook A. K. and Byrne D.: The Dynamic Behaviour of Single and Multiple Moored Vessels. Papers No 9 RINA Spring Meetings 1983.
5. Ojo A. O.; Survey of Anchoring Systems Journal. *Seaways* September 1983.
6. King J.: Applications of Small Computers to Shipboard Training. Paper No 7 RINA Spring Meetings 1983.

Bad or inadequate maintenance of hatch covers can result in: (1) Severe corrosion, as on this cross-joint end plate.

(2) Hogging, as on these (stowed) panels of a single-pull cover.

(3) Mismatching of cross-joints as shown here.

NEW DEVELOPMENTS IN CARGO ACCESS EQUIPMENT AND THE IMPLICATIONS FOR SURVEYORS

Contributed by the staff of the MacGregor-Navire Group, co-ordinated by the public relations department

THE DEVELOPMENT of the steel hatch cover over 50 years ago literally opened up cargo ship designs. It removed the limits on cargo hold access imposed by hatchways traditionally covered with wooden boards, laboriously laid and lashed down by tarpaulin sheets and ropes. The simple steel cover sowed the seeds of a new maritime technology discipline—cargo access and transfer systems—and an industry whose products and services have ever since played a key role in shaping the development of ships. New concepts and refinements of cargo access equipment (CAE) have progressively emerged to sharpen the safety and efficiency of merchant and naval vessels of almost every class.

The post-War birth and growth to maturity of a significant new fleet unit—the Ro-Ro ship—owes much to innovations by CAE designers. Shipboard technology has also moved shoreside to smooth the performance and widen the cargo-handling flexibility of established ports, and to allow the swift and cost-effective creation of new ports based on link spans.

Hatch cover designs

The range of hatch cover designs has evolved in scale and operating system options to suit different ship types and trades. The current portfolio of MacGregor-Navire embraces: rack-and-pinion driven rolling covers; Magroroll covers; hydraulic folding covers for weatherdeck and 'tweendeck hatches; telescopic covers; Rolltite covers; single-pull covers; direct-pull covers; piggy-back covers; weatherdeck pontoon or liftaway covers; and MacGregor Ermans flush sliding covers.

Cargo access equipment designers have played a key role in the swift rise to prominence of Ro-Ro ships and in the world merchant fleet. A portfolio of specialist Ro-Ro cargo access and transfer gear has been refined and expanded in the 1980s to serve vessels ranging from coastal cargo ships, rail and car/passenger ferries to large container and vehicle carriers. The MacGregor-Navire programme embraces: axial stern ramps; quarter and slewing stern ramps; side ramps; bow visors; wing doors; inner doors and forward angled ramps; varied types of cargo and stores lifts; and bulkhead doors. A range of shoreside link span designs for vehicle and rail ferry trafffic is also available for tailoring to the project.

Side loading and discharge of cargo, either exclusively or in combination with Ro-Ro and Lo-Lo handling, has increased in popularity in the 1980s. The mode is particularly valued for the efficient handling of palletised unit loads, crates and paper reels, and is thus attractive for operators in forest products and reefer trades, as well as the fishing sector. Side loading systems have been developed which either penetrate the hull or carry the cargo up and over the ship's side. They are normally located amidships and served by forklift trucks from the quay and within the ship.

Development of cargo access and transfer systems will continue to focus on improving the handling efficiency and safety of existing ship types, as well as promoting the evolution of new marine transport concepts. Advantage will continue to be taken of advances in electronics, hydrauliccs and materials technology, as well as computer-aided design techniques.

Cargo access equipment surveys

In this section the nature of the cargo access equipment survey is discussed and the main reasons for leakage are described. Maintenance procedures and recommendations are also given. The objective is to give a brief practical guide for surveyors.

Hatch cover surveys

Regular inspections of hatch covers are carried out by classification societies, both to satisfy their own rules and to fulfil their responsibilities as load line assigning authorities. The surveyors undertaking these inspections must be assured that the hatch covers are maintained in good condition and that they remain weathertight when closed. They may require chalk tests and hose tests to be performed, although in containerships these are usually dispensed with at all inspections other than those for four-yearly special surveys or when major structural repairs have been effected.

Hatch cover inspections are also carried out from time to time by cargo surveyors, usually, but not always, acting on behalf of charterers or cargo interests. These inspections are commonly conducted during loading or immediately before departure. Their purpose is to ensure that a particular cargo is properly loaded into the ship and adequately protected against sea water damage by the hatch covers. Hatch surveys are also made before discharging begins in order that the condition of the cargo, and the extent and cause of any wet damage can be assessed before it is disturbed. Wet damage is often concentrated on the surface of the cargo in patterns which directly mirror the arrangement of hatch joints and seals.

Pre-departure checks for hatch seals—Where the risk of sea water damage exists, it is necessary to ensure that all weatherdeck hatches can be tightly secured with their seals in good order. The condition of seals can easily be checked by spreading chalk along

By contrast, here are covers that have been well maintained (4) Chain drive perspective.

(5) Panel junction.

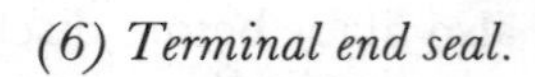

(6) Terminal end seal.

their compression bars and then closing the hatch. On re-opening the hatch, any ineffective seals will be rendered immediately visible by the absence of chalk marks transferred to the contact surfaces of their gaskets. The results of chalk tests must be treated with caution, however, because while they indicate whether or not compression exists they do not indicate its degree. Thus a hatch cover could successfully pass a chalk test even though the compression of the rubber gaskets is less than that required to effect a proper seal.

Whenever appropriate (at a cargo surveyor's request, for instance) chalk tests should be supplemented by hose tests. These are performed by directing a high pressure jet of water at the joints of a closed hatch. Any water penetrating the seals should be discharged through the drain holes. Hose tests should be carried out before loading has commenced to avoid damage to the cargo. But they can never reproduce the conditions that a vessel will experience at sea and satisfactory test results are no guarantee that hatch seals will remain tight during the ensuing voyage. For this reason some cargo surveyors doubt the value of hose testing.

Some reasons for hatch leakage

Failure of seals and joints is the most common of all the hatch defects occurring in general-cargo vessels and bulk carriers. There are numerous reasons for such failures, even assuming that covers and coamings have been installed to the correct tolerances, but often leakage is as much due to the shortcomings of human operators as to the failings of hatch equipment. Thus the majority of common causes of leakage fall into one or other of the following broad categories

Deformation of hull and hatch covers—Hatch covers receive rough treatment in service. Continual careless opening and closing leads to misalignment, localised damage and leaking seals. Careless cargo handling can have similar consequences.

Improper battening down—Battening down hatches with manual cleats is a fairly long job in medium-sized or large ships with small crews. It must often be done at night, in inclement weather with the ship proceeding to sea, and in such circumstances the crew may not be as attentive to their duties as they should be. Negligence is a major factor in practically every case of improper battening down: cleats left undone, eccentric wheels not turned up, multiple panels incorrectly aligned and cross-joint wedges not hammered up are among the most common forms manifested.

Permanent setting down of gaskets—There is a commonly held belief that the tighter a cover is secured to its coaming, the better the seal that is made and the smaller the chance of leakage. This view demonstrates a profound misunderstanding of the way in which hatch seals work.

In order to achieve an effective seal the gasket must be compressed—but never to the extent that the elastic limit of its material is exceeded. When a hatch cover is designed, the degree of gasket compression that is necessary to prevent the passage of water is determined, taking into account the nature of the gasket material and the acceptable contact pressure. Its cleats, whether of automatic or quick-acting manual type, have no effect on the amount of compression—the magnitude of the latter being determined by the height of the compression bar in relation to the rubber seal and the limiting effect of the hatch cover side plate support (or thrust) block in contact with the coaming. Once gaskets have been permanently set down in this way their characteristics are irretrievably changed and their sealing properties lost.

Some types of hatch cover employ highly elastic rubber seals which, besides keeping water out, also allows the cover and the ship's hull to flex independently.

Corrosion—Corrosion is a common cause of leaking hatches. Unless steps are taken to prevent it, compression bars become corroded, roughened and eventually waste away. Even minor roughening may seriously impair the effectiveness of hatch seals, yet it is not uncommon for ships to be sent for their periodic inspections and repairs with their hatch compression bars in very poor condition.

Hatch plating is also likely to corrode if it is not adequately protected. In some severe cases holes may appear, either directly as a result of corrosion, or as a result of fractures in steel weakened by corrosion. In combination carriers employed in oil trades, the inner surfaces of hatch cover plating often suffer severe corrosion unless properly protected. Internal corrosion is also often associated with the formation of sweat.

Blocked drainage channels—When sea water penetrates a hatch seal it is usually collected in drainage channels and discharged clear of the hatchway. If these channels are allowed to become blocked or restricted water may spill over on to cargo in the hold below. It is essential, therefore, that immediately before battening down, compression bar, gaskets and drainage channels are thoroughly cleaned. Drainage channels, particularly those serving cross-joints, are commonly found to be corroded after some years in service.

Seals and cross-joints—Seals that are continually being made and broken by bringing together and separating gaskets and compression bars are likely to leak eventually, no matter how well they are maintained. For this reason covers with the smallest number of cross-joints (e.g., side-rolling types) often perform best with regard to leakage. Where such covers are not suitable, permanent flexible membranes, linking adjacent panels, provide a rather better sealing arrangement than the conventional system.

Hatch cover maintenance

There can be no doubt that inadequate maintenancce is a major cause of many hatch cover defects. The marine environment is a harsh one. Damp salt-laden air, water on deck, and dusty abrasive cargoes all take their toll on a ship's structure and fittings which deteriorate rapidly if proper preventive measures are not taken. Yet it is too often evident by the condition of some ships when they are presented for survey and repair that such measures are neglected.

It is unquestionably more difficult for a crew to carry out extensive maintenance at sea today than it

(Left) The stern ramp on a Ro-Lo vessel—bearings, hoisting gear and cross-joints are all susceptible to the harsh marine environment.
(Below) Folding hatch cover on a Ro-Lo vessel—the cross-joints are particularly susceptible to corrosion.

was in the past when ships were smaller and slower. Crews have not increased in size as larger ships have entered service. Rather, the reverse is true. And not only has the amount of maintenance to be done grown considerably, but with the greater speeds of modern vessels, the crew has less time to do it in. The sheer size of modern hatch covers is another factor that makes their maintenance more difficult, while access to all parts is sometimes difficult to arrange.

Thus, with the limited resources normally available on board ships, it may be unreasonable to expect anything other than simple routine maintenance or essential repairs to be carried out at sea. Many ship operators have recognised this by making use of the maintenance services now offered by hatch cover manufacturers.

The ready availability of supplies of spare parts is an essential prerequisite for effective maintenance. In an international industry like shipping nothing less than a worldwide network of reliable supplies is acceptable and, in recognition of this, hatch cover manufacturers have established extensive aftersales service facilities. Besides spare parts, these services often include the availability at short notice of a trained maintenance engineer and, at major centres, skilled maintenance squads as well.

Steel work—Paint coverings naturally become damaged in service by impacts and abrasions, and corrosion can then be retarded only by repainting. Ideally this should be carried out in the manner prescribed by the paint manufacturer but in practice this is rarely possible—salt-laden air is not conducive to good paint adhesion! Moreover, some parts of hatch panels, such as cross-joints, may normally be inaccessible during a ship's operation.

The inner surfaces of hatch panels are also difficult to paint, especially in single deck ships, and the surfaces collect dirt and grime, so rendering a vessel unfit for the carriage of grain. For this reason some operators specify double-skinned panels with a rust inhibitor in the void space. This only partly solves the problem, however, because the outer surface of the lower skin must still be painted. Care should be taken to avoid painting over and blocking drain holes, and the surveyor should be alert for this defect.

Seals—Standard rubber gaskets can be expected to last from four to five years of normal service. Their life is severely curtailed by over-compression and contact with abrasive materials. Where local damage to gaskets occurs a common practice is to glue additional layers of rubber over the affected area. This practice is not effective and is to be deplored. Any section of damaged gasket must be renewed. In freezing conditions special grease or commercial glycerine should be spread over the surface of all gaskets in order to prevent them sticking to their compression bars. The roughening of mild steel compression bars can be alleviated to some extent by buffing. Where their contact edges are severely wasted, however, it may be necessary to build them up with weld or even renew them completely.

Cleats—Quick-acting cleats are probably the most common and most frequently handled of all hatch cover components. They exert the correct degree of compression through thick neoprene washers which, in time, tend to lose their elasticity. When this happens the cleat must either be adjusted or, more often, replaced. Galvanised cleats have been found to have a longer life than ordinary steel cleats. In order to cover replacements ships should carry a good stock of spare cleats.

Chains—On single-pull covers the connecting chains or wires between individual hatch panels must be inspected for signs of stretching. When the design tolerances are exceeded, and the chains on one side are not the same length as those on the other side, the panels tend to turn or 'crab' as they are pulled forward. In serious cases, panels may jam or become unshipped.

Hydraulic systems—Dust and fine particles, mainly from cargo sources, are among the principal enemies of hydraulic systems. They settle on lifting and hinge jacks during loading and discharging and lodge in their seals which eventually fail unless they are regularly inspected and cleaned.

The corrosion of steel hydraulic pipes is another common source of failures. Often this is due to inadequate paint protection (enhanced by the siting of hydraulic pipes too close to flat surfaces so that it is possible to paint those areas that are immediately visible). Because of the internal pressures that hydraulic pipes must withstand they have thick walls and it usually takes some years before they are sufficiently wasted to fail. However, when failure does occur it is invariably necessary to replace complete sections of pipe.

Ro-Ro access equipment maintenance

There have been comparatively few major claims for damages to cargo arising from the failure of Ro-Ro access equipment. This may be taken as an indication of the reliability of such equipment in service. Nonetheless, the consequences of a major failure could conceivably be very serious and it is essential that the maintenance of ramps, doors and elevators is not neglected. It is possible that the good record of Ro-Ro vessels is in some measure due to the high standards set by their operators, many of whom can be counted among the world's leading shipping companies.

It is not usually possible for the crews of Ro-Ro vessels to undertake anything but the simplest maintenance (such as routine greasing) of access equipment, both because of its complexity and because it is rarely accessible at sea, while port turnround times are short. Large items of equipment, such as quarter ramps, may have automatic lubricating and other facilities built in. Manufacturers also offer extensive after-sales service facilities for Ro-Ro equipment and are usually prepared to undertake regular maintenance supervised by their own engineers. Such work may be carried out under continuous classification society survey. The same general principles as described earlier for hatch covers apply. □

CARE AND SURVEY OF HATCH COVERS OF DRY CARGO SHIPS*

The International Association of Classification Societies

LOSS OF WEATHERTIGHT INTEGRITY continues to be a constant factor leading to cargo damage which could result in a threat to the safety of the crew, the ship and its cargoes, despite advances in modern shipbuilding technology, construction, navigation and means of preventing ingress of water into hold spaces. We need only to look at cargo insurance claims to recognise the continued prominence of this problem and the vigilance required from crews to ensure that cargoes spoiled as a result of water ingress through hatch openings is minimised. Although condensation can cause water to collect in holds, it is the problem of water entering into cargo holds through the hold openings as a result of badly maintained or damaged closing arrangements that this chapter addresses.

The classification societies are concerned with this problem in particular through the Load Line Convention. The experience of the classification societies confirms the importance of hatch cover weathertightness and the reports of survey—class surveys as well as load line surveys—contain frequent references to lack of maintenance of weathertightness of hatch covers. Where there is any doubt in applying this guideline, clarification should be obtained from the individual classification society. Attention is directed to the separate published rules of each. This guideline is not meant as a substitute for the society's rules or the independent judgement and experience of owners and surveyors.

Contributing factors

Lack of weathertightness may be attributed to several causes, which can be classed within two different types: Those which result from the normal use of the hatch cover system, such as deformation of the hatch coaming or hatch cover plating due to impact, or the normal wear-and-tear of the cleating arrangement, which may be corrected only through extensive repairs or overhauls; and those which result from the lack of proper maintenance—corrosion of plating due to lack of protection against corrosion, lack of adequate lubrication of moving parts, non-replacement of old gaskets, etc.

The classification requirements for the construction and inspection of hatch covers cannot, alone, ensure that the hatch covers will be adequate at all times; the improvement of the performance of hatch covers can be achieved only through an effort of all parties concerned. The hatch cover designers should, perhaps, give less thought to the sophistication of the systems and more to the sturdiness of the equipment, the ease of operation and the convenience of maintenance. Adequate protection of the steel work should be ensured with high quality coatings.

Shipowners and their crews should apply a programme of maintenance to ensure that the steel is not allowed to corrode, gaskets are periodically replaced, movable parts are kept properly lubricated and fittings periodically overhauled. The crew should also make sure, at each operation of the hatch covers, that they are sufficiently clean, especially at bearing surfaces, and that the drainage holes are clear. The crew should, however, bear in mind that in extreme cases hatch cover protection may have to be complemented by means of tarpaulins or adhesive tapes. This is a decision to be made by the master, taking into account the nature of the cargo, expected sea-conditions, uneven load-distribution on hatch covers and, above all, previous experience of the behaviour of the ship in similar circumstances.

Containers and cargoes should not be stowed on hatch covers unless they have been designed for such carriage. The crew should also make sure that not only the deck cargo, but also the derricks, etc., are properly stowed and secured so that they will not be dislodged in a seaway and cause damage to hatch covers. Recently, a ship has been lost due to the destruction of tarpaulins and hatch cover pontoons by the impact of blocks of a derrick which crashed on the hatch covers after being dislodged from its supporting crutch.

What is IACS doing?

Traditionally, hatch covers and hatch coamings are inspected every year, as prescribed in Article 14 of the Load Line Convention of 1966. This inspection is normally carried out by the surveyor of the classification society at the same time as the class annual survey, the scope of which is not much different as far as the closing appliances are concerned. These inspections are carried out in port, while the ship is in operation, and very often working cargo, their purpose being to ensure that no alterations have been made to the ship which may affect the load line, and that the fittings and appliances for the protection of openings are maintained in an effective condition.

This is only possible if a detailed examination of the hatch covers and fittings, which may necessitate the actual operation (closing and opening of hatch covers), is carried out by surveyors. Hose testing should be carried out whenever the surveyor is in doubt as to the weathertightness of a hatch cover. The surveyor can only assess the actual condition of the hatch covers at the time of survey and, in between two inspections, the proper maintenance of the hatch covers must remain the owner's responsibility.

Confronted with an increase in the number of reported cargo damages, the classification societies, members of IACS, decided to review the survey procedures issued to their surveyors for the survey of hatch covers and coamings. A unified requirement was consequently prepared and approved by the council of the Association during their June 1985

**Taken by permission from IACS Guidance to Owners No. 15.*

meeting; this unified requirement Z.4, *Survey of Hatch Covers and Coamings*, forms the appendix to this chapter.

Defects to covers and coamings

The purpose of this part is to review the various type of defects found in hatch covers, which for the most part are caused by a lack of proper maintenance or by improper operation.

Some defects affecting the weathertightness of hatch covers can be attributed to their design. Such defects, a very small percentage of all defects found, are not dealt with here. The crew should be attentive to them and report accordingly for eventual discussion with the designers.

Hatch covers may be categorised in three general types:

- Portable covers with tarpaulins and battening devices (usually wooden covers);
- Steel pontoon covers with tarpaulins and battening devices;
- Mechanically operated steel covers, of folding, sliding, rolling, etc., types, fitted with gaskets and clamping devices.

All types of hatch covers are basically of robust construction; under normal conditions of care and operation they are fully adequate for their intended purpose.

Wooden covers, steel pontoons and tarpaulins are affected:

- Mainly by poor handling and bad stowage.—
- Portable beams also are frequently deformed by rough handling and their locking devices have always been a source of considerable trouble due to lack of care in handling.
- By normal wear-and-tear.

Steel covers are affected by:

- Corrosion, which attacks the integrity of the cover itself and which also affects the moving parts (wheels and rails, hinges between panels, cleats and batten screws, etc.).
- Deformation caused by faulty handling, shocks from cargo being handled (especially logs and heavy loads), wear-and-tear on gaskets, and overloading with deck cargo. These deformations in turn, affect the weathertight joints and gaskets.

Hatch coamings are affected by:

- Corrosion aggravated by the presence of piping systems, utilising coamings as protection and support, thus preventing normal access to the plating for painting.
- Deformation, both of the plating and of the various supporting members and brackets, mostly due to cargo handling and aggravated by general corrosion.

All such defects can be considered as due to repetitive accidental causes, which for various reasons—cost or lack of facilities for instance—are not immediately repaired.

Checking at regular intervals (each loading and discharge operation) the condition of hatch covers, gaskets and hatch coamings is the responsibility of the vessel's owner and operator. The inspections and surveys carried out by the classification societies and/or the administration at yearly intervals are not intended to and cannot replace these regular checks and proper maintenance.

The hatch covers and coamings of ships engaged in the trade of carring timber or heavy parcels, also those of ships carrying deck cargoes such as containers on top of hatch covers, are likely to suffer very rapidly from excessive deformation. The owners should be aware that the weathertightness of such hatch covers is difficult to maintain in service, and that, in consequence, the greatest care should be exercised in the periodical checking of these hatch covers and coamings, in particular when the vessel changes its trading pattern.

Crew's examination of covers and coamings

A thorough examination should include:

General—Checking that the record of conditions of the freeboard assignment is available; the record itself is used when necessary to guide the checking. Checking that no significant changes have been made to the hatch covers, hatch coamings and their securing and sealing devices since the last periodical inspection of the classification society.

Hatch covers and coamings—When fitted with portable covers or steel pontoons, checking the satisfactory condition of wooden covers and portable beams, carriers or sockets for the portable beams, and their securing devices, steel pontoons, tarpaulins, cleats, battens and wedges, and hatch securing bars and their securing devices. When fitted with mechanically-operated steel covers, checking the satisfactory condition of hatch covers (corrosion, cracks deformations), tightness devices of longitudinal, transverse and intermediate cross junctions (gaskets, gasket lips, compression bars, drainage channels), clamping devices, retaining bars, cleating, chain or rope pulleys, guides, guide rails and track wheels, and stoppers; etc. Checking of the satisfactory operation of mechanically operated hatch covers—stowage and securing in open condition; proper fit, locking and efficiency of sealings in closed condition;—and checking the satisfactory condition of hatch coamings (corrosion, deformation of plating and supporting members and brackets, connection to deck).

Spare parts—Checking the satisfactory condition and number of spare wooden covers, spare wedges, and spare tarpaulins etc.□

Appendix Z.4

Z.4.1 Hatch covers, hatch coamings and their securing and sealing devices shall be submitted to surveys as detailed in Z.4.2 and Z.4.3.

Z.4.2 Annual surveys at intervals of about 12 months whether the ship is under continuous survey or not. They will normally be performed as part of the annual hull survey and for periodical load line inspection.
The scope of annual surveys of hatch covers and the specific requirements to be satisfied are listed under Z.4.4.

Z.4.3 Special surveys at maximum intervals of five years. They will normally be performed as part of the hull special survey and/or periodical load line survey.
The scope of special survey of hatch covers and the specific requirements to be satisfied are listed under Z.4.5.

Z.4.4 Annual survey
This survey shall consist of:

1. General
Checking that no significant changes have been made to the hatch covers, hatch coamings and their securing and sealing devices since the last survey.
2. Hatch covers and coamings
   - Checking the satisfactory condition of hatch coamings
   - When fitted with portable covers, wooden or steel pontoons, checking the satisfactory condition of: wooden covers and portable beams, carriers or sockets for the portable beams, and their securing devices; steel pontoons; tarpaulins; cleats, battens and wedges; hatch securing bars and their securing devices;
   - When fitted with mechanically operated steel covers, checking the satisfactory condition of: hatch covers; tightness devices of longitudinal, transverse and intermediate cross junctions (gaskets, gaskets lips, compression bars, drainage channels); clamping devices, retaining bars, cleating; chain or rope pulleys; guides; guide rails and track wheels; stoppers; etc.
3. Spare parts
Should the rules of the society require it, checking the satisfactory condition and number of: spare wooden covers; spare wedges; spare tarpaulins.

Z.4.5 Special survey (renewal of certificate)
This survey shall, as a minimum, consist of:

- A general inspection with the extent of the annual survey as stated in Z.4.4; and in addition:
- Random checking of the satisfactory operation of mechanically operated hatch covers: stowage and securing in open condition; proper fit, locking and efficiency of sealings in closed condition;
- Checking the effectiveness of sealing arrangements of all hatch covers by hose testing or equivalent;
- Checking the residual thickness of coamings steel pontoon or hatch cover plating and stiffening members as deemed necessary by the surveyor. □

ROPES, WIRES AND CHAINS ON BOARD CARGO VESSELS

M. F. Tomlinson, MNI
Chief Officer

A TYPICAL INVENTORY is as follows:

Deck items	Diameter (mm)	Construction	Length (m)	SWL (T)
Anchor cables, stud link				
Chain cable grade U3 (a)	50	chain	225	B/S196
Chain stoppers	8	chain	2	2.85
Union purchase chains, triangular plate with long links one end, cargo hook other	32	chain	0.76	6.0
Fork lift truck lifting frames: 4 leg chain sling with master link and couplers	10	chain	1.8	4.0
Plus 4 legs	10	chain	2@1.5	4.0
			2@1.8	4.0
Cargo gear extension chains Herc-alloy slings with master link and cee hook	12.5	chain	2.84	5.0
Towing wire	38	6×36	200	17.14
Mooring wires	28	6×36	165	10.06
Crane runners	32	17×26 dyform	110.3	13.0
Crane topping lifts (8T)	28	6×36	123	10.06
Crane topping lifts (12.5T)	32	6×36	123	13.14
Derrick runners	22	6×36	83	5.74
Derrick topping lifts	24	6×36	70	6.84
Derrick upper guy pennants	20	6×24	7	3.0
Derrick lower guy pennants	20	6×24	2	3.0
Derrick preventer guys	24	6×36	21.8	5.14
MacGregor hatch wires	20	8+1×36	55	5.0
Derrick bull wires	14	6×24	55	1.6
Hatch safety lines	10	6×19	various	0.9
Deck safety lines	12	6×19	various	1.28
Gangway falls	12	6×36	83	1.71
Lifeboat falls	16	6×36	137	3.04
Davit headspan	20	6×24	—	3.0
Mast stays	20	6×19	—	2.8
Forestay	20	6×19	—	2.8
Lifeboat gripes	14	6×36	5.23	2.32
20′ Container frame legs 2×2 leg wire rope slings with master link	28	6×36	5	13.18
Lifting legs: 4 leg wire rope sling with masterlink	26	6×36	7	25.0
Lifting legs: 4 leg wire rope sling with masterlink	19	6×36	3.5	10.0
Tirfor tackle wires	16.3	maxiflex	40	3.0
Whistle lanyard	5	6×19	—	0.27
Lashing wires	10	6×19	coil	0.9
Mooring ropes monofilament polypropylene	56	8 strand plaited	220	B/S36
Mooring wires tails monofilament polyprop	56	8 strand plaited	9.5	B/S36
Guy ropes: staplespun polypropylene	20	3 strand	—	B/S5.14
Pilot ladder manilla	20	3 strand	—	B/S3.25
Pilot ladder manropes manilla	20	3 strand	—	B/S3.25
Gantlines, staplespun polypropylene	20	3 strand	—	B/S5.14
Heaving line staplespun p.p.	12	3 strand		
Flag halyard, braided nylon	10	—		

Deck items	Diameter (mm)	Construction	Length (m)	SWL (T)
Lifeboat painters	24	3 strand	—	B/S7.6
Lifeboat lifelines	24	3 strand	—	B/S7.6
Lifeboat becket line	16	3 strand		B/S3.5
Lifeboat keelgrab lines	16	3 strand		B/S3.5
Lifeboat heaving lines	8	3 strand		
Lifebuoy lines	10	3 strand		
Sea anchor hawser	24	3 strand		B/S7.6
Sea anchor tripping line	16	3 strand		B/S3.5
Embarkation ladder	20	3 strand		
(All LSA cordage is DTp approved staplespun polypropylene)				
Engineroom items				
E/R gantry crane wire	16	6×36		3.04
Cross alleyway stores crane	8	6×36	28	0.9
Wire rope slings	20	6×36	2.44	4.76
	19	6×36	4	4.0
	13	6×36	1.83	2.0
	13	6×36	0.9	2.0
	11	6×24	2.0	1.0
	11	6×24	0.6	1.0
	10	6×36	1.83	1.18
	8	6×19	0.3	0.57
Braided wire rope slings		braid	5	3.5
		braid	2	3.0
Chain blocks			5.5	5.0
,, ,,			5.5	2.5
,, ,,			5.5	1.25
,, ,,			5.5	0.625
Yale Pullifit		chain	—	1.5
,, ,,		chain	—	0.75
Luffing tackles Sinal	24	3 strand	61	B/S4.07
,, ,, Sinal	16	3 strand	61	B/S1.80

Inspection and testing

There are various regulations governing the examination and testing of chains, wires and ropes. The Factories Act 1961 requires that chains and ropes be examined after testing and every six months and requires them to be tested before being put into use and after any repair. The Construction (Lifting Operations) Regulations 1961 have similar provisions.

The Docks Regulations 1934 required chains and ropes to be examined after testing and every 12 months. Also they must be inspected before use unless inspection has taken place in the preceeding three months, and the regulations require them to be tested before being taken into use and after any repair.

The Anchors and Chain Cables Act 1967 requires all chain cables to be tested unless they are under 12.5 mm diameter. The testing establishment takes three links from each shackle length and tests them to a tensile breaking stress. If this proves satisfactory, the length of cable is then subjected to a tensile proof stress. It is then inspected for flaws, weakness and deformation. The manufacturer has need therefore to provide three extra links with each length for testing.

Shackles and other cable accessories are subject to the same tensile proof loads as the cable with which they are to be used. One sample in 25 is also subjected to the breaking stress—1 in 50 in the case of lugless joining shackles. The chain cable is then awarded a certificate of test which shows the type and grade of chain, the diameter, total length and weight, the dimensions of the link and the loads used in the test.

Inspection of the certificate for my ship shows the cable to be Grade U3(a) stud link 50mm diameter, 550m long, weighing 30,643kg. A breaking load of 1,960 kilo-Newtons was applied to three links out of every four lengths and a proof load of 1,370 KN. It also shows the link dimensions, length over five links and the marks indicating the identity the marker, proving house, certificate number, date, proof load and grade.

Under the terms of the various regulations, the anchor cables should be examined every three months, though a thorough examination is not really practical except in drydock, an event which now takes place every two-and-a-half years.

Preparing for cable inspection

To prepare the anchor cable for inspection it should be ranged on the dock and descaled so far as possible. At a survey the joining links will be opened and examined, and the thickness of the cable determined (a 10 per cent weardown is allowable) and the cable links are inspected for distortion and loose studs. Every link will be tested with a hammer. If any links are replaced the cable is again tested to its proof load. At the intermediate inspection the joining links need not be opened but the cable should still be given a close examination, particularly at the 'working' end.

Other chains, such as chain-stoppers and chain legs used in lifting gear, can be examined more easily. If properly looked after they should not need scaling. They should be ranged out on the deck so that each link can be examined and the wear noted. This must be done at least every three months. No wrought iron chains are in use on my vessel so no annealing is required.

The chains used on the company's fork-truck lifting frames, for example, are 10mm diameter and are not therefore covered by the Anchors & Chain Cables Act 1967, so that links from each length do not have to be tested to breaking point. However, all chains have to be proved to twice their safe working load (in this case 'T' grade chain was proved to 6.4 tonnes and awarded a SWL of 3.2 tonnes) after examination.

After this proof test the chains do not need testing again unless they are repaired. However, as they form part of the lifting gear of the ship it is common practice to send them ashore for testing along with the blocks and shackles for quadrennial thorough examination by an approved examiner, usually a lifting gear specialist. In most cases the examiner will retest to the proof load and examine them before returning them to the ship recertificated.

Lifting gear wire

Each wire that is used for lifting gear is usually individually certificated, but in some circumstances a certificate covering a coil of wire from which several lengths are cut is sufficient. Wire rope is tested by breaking a sample and then if required it is made up by the supplier into lengths ready for use, spliced, with 'Talurits,' or having sockets attached, and supplied with a certificate for each length or one certificate for two or more similar lengths. It is not necessary to break a sample of every cargo runner, for example.

The certificate provides details of the diameter, number of strands, number of wires per strand, the lay, the type of wire in the strand (e.g., galvanised) and the length supplied. It also gives the tensile strength in tons/sq.inch or Newtons/sq.millimetre, the date of test, the test load under which the sample broke, the SWL and the safety factor. Wire rope will not normally be tested again. It must be examined every three months or every month if broken wires are known to exist.

To prepare for examination the wire should if possible be ranged on the deck and any hard old grease removed. The wire can then be examined for its full length looking for broken strands, kinks, excessive distortion, chafing or corrosion. If more than 10 per cent of the wires are broken in any length of eight diameters, then it must be condemned. It may also be condemned if the person inspecting considers it unfit through corrosion, wear, etc. For example, one strand only parting on a 17 strand rope would be within the 10 per cent limit, but clearly the rope could not be used through blocks for lifting purposes.

The internal structure of the wire should be opened up and examined at intervals for corrosion and wear. This can be done using two flat-ended splicing spikes to open up the lay. The wire should also be examined dimensionally. When new, wires are usually slightly oversize, but stretch to their normal working diameter with use. If a wire wears thin it will tend to cut a groove in the sheaves it runs through which will accelerate the wear of a new replacement. Wires should be renewed before this happens.

Fibre ropes

Most fibre ropes are supplied on board with test certificates although there is not a requirement for lifting gear. Ropes manufactured for LSA purposes, however, are required to be certificated. Small stuff comes on board without certificates. The certificate shows the diameter and length, the material (manila, staple spun, polypropylene, etc), the construction, specification and guaranteed minimum breaking strain.

To prepare ropes for inspection they should be coiled down on deck where there is sufficient room to examine the whole length of the line. Mooring lines should be flaked down for inspection. Some cordage cannot be conveniently displayed in this way and has to be examined in situ—e.g., some lifeboat lines. Ropes should be inspected for cuts, abrasion, melting-fusion, stretch, chemical attack (discolouration or loss of strength or flexibility) inside as well as outside. Look for powdering inside indicating heavy wear.

Three different types of mooring ropes are or have

been used on company vessels. One class of vessel uses Thor brand six-strand nylon line comparable in construction to wire ropes. This uses continuous filament cores with monofilaments making up the strands which are interspersed with filler yarns which produce a dense fibre pile to protect them against surface damage. This rope has an SG of 1.14 and good shape retention and is ideally suited to the single storage drum type of mooring winch on board; 44mm diameter rope is in use.

The other class of vessel started off life using Viking braidline—a braid on braid construction, again for use on single storage drum type of mooring winch; 56mm diameter rope is in use. The use of this rope was discontinued at the end of its life and the ship reverted to staplespun eight strand polypropylene ropes, 56mm in diameter.

Both classes of ship have the facility to operate their winches in the self-tensioning mode, though in practice this is seldom used. The trade in which we operate requires mooring lines which can be used for mooring in a closed dock system, which can be used for towage by harbour tugs when docking and undocking, and for use on berths in the Caribbean which are very low in relation to fo'c'sle height and which are open to the sea and thus influenced by wind, swell, and tide (not much rise and fall).

Roller fairleads

Multi-angle roller fairleads are provided, but the incidence angles on these is sometimes quite light—in the order of 70°-80° in the case of springs, and somewhat less, 60°-70°, in the case of the headlines. Sternline angles are usually less due to the poop being lower than the fo'c'sle, and the normal by-the-stern trim. The ship has to be heaved alongside against the wind, usually without the aid of tugs, requiring a fairly strong pull, and when alongside the movement of the ship due to wind and swell and cargo operations is sometimes considerable. Chafe at the roller fairleads is a major problem.

Another problem associated with the use of single storage drum winches is that the outer (tight) turns on the drum formed when heaving the vessel alongside tend to work through the inner (slacker) turns already on the barrel causing the line to jam on occasions when slacking out.

The Thor brand six-strand nylon used on one class of vessel has proved fairly good. Its relatively small size (44mm diameter) provides a rope which is easy to handle and its resistance to chafing is lighter than the polypropylene ropes found on this class of vessel. It does not reduce in diameter appreciably under tension, and so rarely jams on the barrel.

The braidline used on this class of ship fell out of favour principally because of the difficulty of splicing it! Its performance was good—it was more resistant to chafe than the polypropylene and its smooth surface made it jam on the barrel only rarely. It was a heavier rope to handle, 56mm diameter, but very strong and, but for the splicing problem, ideal. The need for splicing acknowledges that even with careful parcelling of ropes at the fairleads chafe still takes place, and eventually lines have to be cut and spliced rather than discarded for economic reasons.

Staplespun polypropylene ropes are now in use on this class of ship. Their main advantages over the braidline are ease of splicing and the fact that they float so that they can be seen to be clear of bow thruster and propeller and any underwater snags at the berth.

Their disadvantage is that they chafe more quickly and that they jam easily on the winch barrel. Surge is not a major consideration as they are normally worked off a down-end when heaving the vessel alongside. The down-ends are only used for last lines ashore to complete the tie-up after coming alongside.□

NOTES ON LASHINGS AND LIFTING GEAR

T. M. Sims, Master Mariner
Coubro and Scrutton Limited

Containers on deck

FOR THE UK, there are no statutory regulations for the securing of deck stowed containers other than the normal duty of a master to ensure the safety of the ship's cargo. Lloyd's Register of Shipping (and other classification societies) have requirements for container securing arrangements. These requirements lay out the strength of the securing systems and the determination of the forces acting on the containers. An annual survey of the securing systems should be carried out and the surveyor should see that the equipment remains in a satisfactory condition.

Quadrennial examination

A thorough quadrennial examination of the securing systems should be carried out:

(1) All permanent fittings attached to the ship's structure are to be carefully examined for excessive corrosion and their attachment to the ship's structure.
(2) All portable supports are to be erected and their attachments examined.
(3) All stacking devices, lashing rods and lashing terminals are to be examined for wear, corrosion and damage.
(4) All twistlocks, turnbuckles and other tensioning devices are to be examined for damage, corrosion and their mechanical operation.

In the case of turnbuckles and tensioning devices, particular attention should be paid to the condition of the threads and connecting pins. In the case of twistlocks, particular attention should be paid to the locking device to ensure that it will not work loose due to vibration, and that all the twistlocks in use have the same type of handle locking in the same direction. The handle should indicate that the twistlock is being used the right way up.

Lifting gear testing

Lifting gear on deck was covered by the Docks Regulations 1934 (S.R. & O 1934, No 279); on 1 January 1989 the Merchant Shipping (Hatches and Lifting Plant) Regulations 1988 came into force.

Lifting gear in the engineroom is not at present covered by specific regulations, but is under the general umbrella of the Health and Safety at Work Act 1974. In effect this means it is subject to annual inspections and quadrennial thorough examinations.

Other publications covering this equipment include *Safety and Health in Dock Work* ILO and *Code for Lifting Appliances in a Marine Environment* (Lloyd's Register of Shipping).

The general requirements for all lifting machinery is that it is tested and examined by a competent person before being taken into use, or after repair or modification. All derricks and permanent attachments including bridle chains, to the derrick, mast and deck used in hoisting or lowering shall be inspected once every *12 months* and thoroughly examined once at least every four years. All other lifting machinery shall be thoroughly examined once at least every *12 months.*

All loose gear must be thoroughly examined every *12 months* and inspected before being taken into use, unless it has been inspected in the last three months.

The new Docks Regulations will require testing of lifting appliances every five years. The ILO Safety and Health in Dock Work require testing every four years.

A thorough examination should be a detailed visual examination, carried out by a competent person, which may entail access to, or removal of hidden parts, or by such other means as non-destructive testing.

An inspection should be a visual survey carried out by a competent person to determine whether there is any visible defect liable to affect the safety of the gear.

For the purpose of the examinations and inspections a 'competent person' is one who has the requisite knowlege and experience, both theoretical and practical of the type of equipment under examination to certify with confidence whether it is free from patent defect and suitable in every way for the duty for which it is required. □

EXAMINATION AND MAINTENANCE OF WIRE ROPES

Australian Federal Department of Transport Marine Notice No. 9/1987

ON 14 AUGUST 1986 the heavy lift ship *Gabriella* capsized and sank in Port Kembla. Two lives were lost and the ship was a constructive total loss. The casualty resulted from the failure of a defective wire cargo runner at a critical stage of discharging a heavy lift. The runner wire was found to be extensively corroded internally due to failure of the applied lubricant to penetrate the rope. Outwardly, over most of its length, it appeared well lubricated and in good condition. The method of lubrication which had been adopted was ineffective.

This rope was of comparatively large size, 36 mm in diameter, and had been periodically examined a few weeks before the accident. It had not, however, apparently been examined internally. The lubricant had been applied manually. The cause of the failure is typical of a number of wire rope failures reported to this Department over the last decade. In all cases the common denominator is ineffective internal lubrication and ineffective, or no, internal examination.

Owners and masters of ships are required by law to maintain wire ropes in a safe condition. They must therefore ensure that maintenance and examination procedures are effective. To fulfill this requirement it is essential that, apart from external lubrication and examination wire ropes be lubricated internally, and wire ropes be examined internally to verify penetration of the lubricant and absence of defects.

It is appreciated that internal lubrication and examination of larger ropes present some difficulties. However, appropriate methods are available and are outlined in the following sections of Australian Standard AS 2759-1985—Steel Wire Rope—Application Guide. Advice is also available from rope and lubricant manufacturers.

Shipowners should ensure that those persons entrusted with the maintenance and examination of wire ropes on ships are provided with the instructions and equipment necessary to carry out those duties effectively. Australian Standard AS 2759-1985 contains good advice on the selection, storage, handling, maintenance, use, inspection and discard of steel wire ropes. All ships should be provided with a copy for use as a working manual. Where necessary, the tools and equipment referred to in the Standard should also be provided, particularly the clamps for opening up wires for internal examination.

Lubrication of wire ropes in service

Rope life can be reduced by a number of factors. A rope, during its service, is constantly undergoing changes, and eventually deteriorates as a result of abrasive wear, wire breaks, loss of lubrication, corrosion, and avoidable damage. Some of this loss of life is unavoidable in certain fields of service but often much of it is due to avoidable abuse, or to insufficient maintenance. One of the most common causes of neglect is inadequate lubrication.

The rope manufacturer should lubricate rope and its core with the correct grade and quality of lubricant to suit the size, construction and designed use of the rope. It is often incorrectly assumed that the rope manufacturer impregnates a rope with all the lubricant necessary to last out the life of the rope. Compression on the core while under tensile loading and while bending over sheaves and drums works lubricant out of ropes. A periodic application of suitable oil or grease must be made to ensure sufficient continuous lubrication. The frequency of application is dependent upon the lifting appliance, its use, the environment and the type of rope involved.

Inadequate lubrication can result in corrosion, heavy abrasive wear, fretting between wires and stiffening of the rope.

When used in a corrosive environment where service dressing cannot be applied, a shorter working life of the rope will result. Where corrosion, which can materially affect both safety and economic life of the rope, becomes a factor in rope operation, additional care and attention is essential. Inadequate lubrication also allows absorption of moisture by the fibre cores, which can result in a breakdown of the core and corrosion of the rope wires.

In operation a wire rope is a machine of many parts, and each time it bends or straightens under load, relative movement takes place between wires, strands and core in the rope, as well as between the rope surface and the drum or sheave groove over which it operates. If wear and bending fatigue failures of the rope wires and wear of drum or sheave wear are to be kept under control, frictional losses resulting from rope operation must be kept to the minimum by effective and regular lubrication.

Before a lubricant is applied to an operating rope, the rope should be cleaned by one or more of the following methods: (a) rotary or hand wire brushes or scrapers; (b) dry compressed air, if foreign materials or hardened lubricant fill the strand and wire valleys; (c) rubbing down with a light flushing oil (free from kerosine), where the surface lubricant is fluid enough to rub off.

Application

The method of application will vary with the type of equipment, service conditions, rope lengths and rope speeds. For short ropes under a machine operator's control, lubricants may be applied manually, by an oil can, by hand-brushing, or by swabbing.

For crane or hoist ropes, much of the rope length can be covered by applying the lubricant to the rope wound right up on the drum, leaving only the lengths

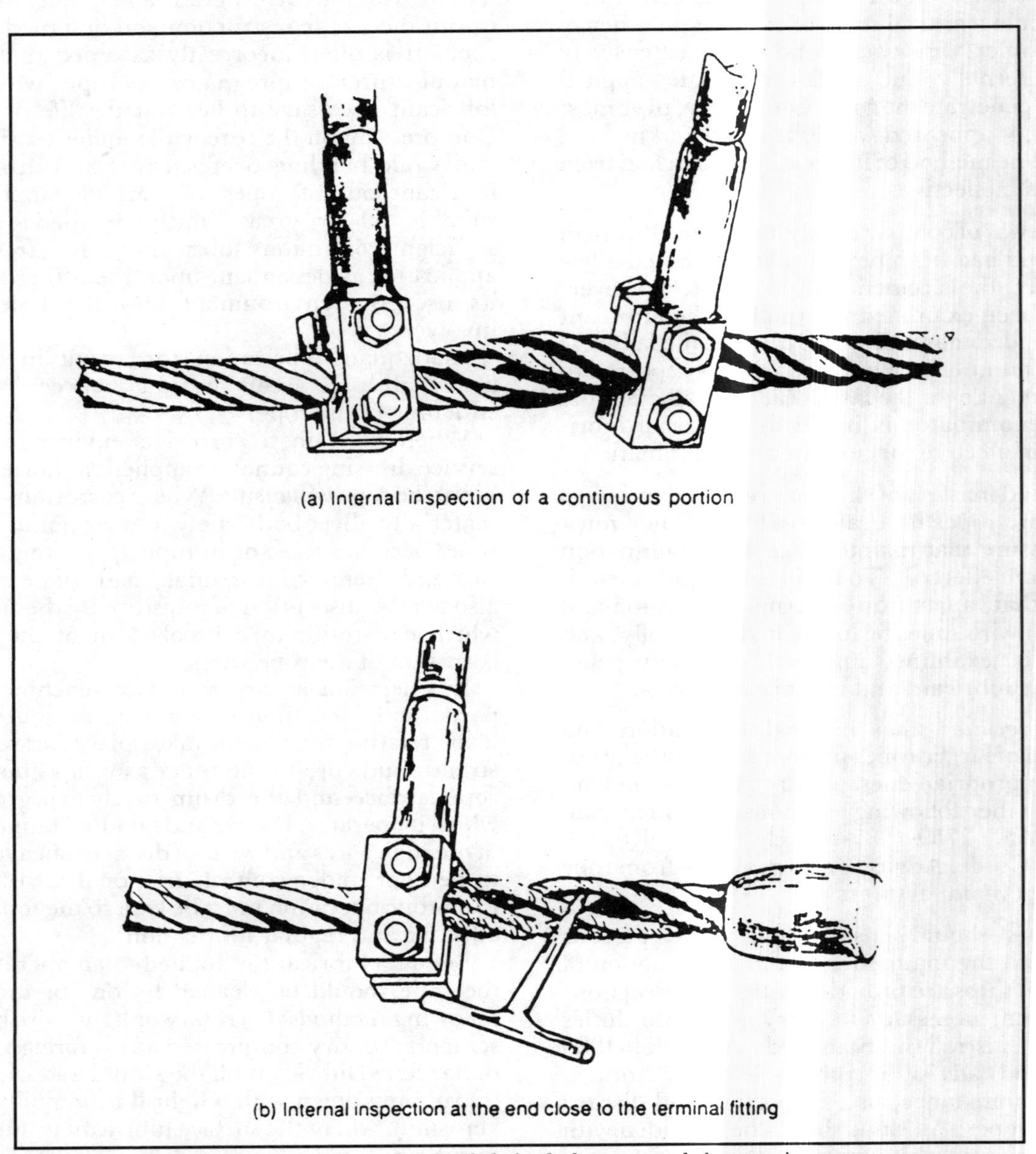

Internal inspection of unloaded single-layer stranded type wire rope

to block or hook to be lubricated by hand methods from the floor or ground. The drum section may be lubricated by hand-brushing or oil can, or by pressure-spraying which can be more economical both in time of application and cost of lubricant. As the lubricant is applied with the strands sprung slightly by being bent over the drum, penetration into the rope is improved.

Boom ropes on mobile, derrick or level luffing cranes, often out of sight and forgotten, should also be lubricated regularly to prevent corrosion.

Direct haulage and drum-type hoisting ropes may be spray-lubricated as in crane or hoist drums, usually as the rope comes on to the drum in a sprung condition, only the short length from the skip or cage to the drum remaining to be hand-lubricated.

Type of lubricant

Areas of wear that can be reduced by effective lubrication are essentially the following: (a) wear on outer strands caused by rubbing contact with sheaves and guides; (b) internal wear caused by rope wires rubbing together during flexing of the rope.

The best type of lubricant for a particular application will depend upon the type of rope and the conditions of use. There are many types of lubricants, each of which has a restricted range of operating conditions for which it is most suited. The advice of lubricant manufacturers should be sought when determining what type of lubricant should be used.

The type of lubricant used on a wire rope is governed by (i) size and construction of the rope; (ii) type of equipment on which the rope is used; and (iii) operating conditions of the rope. The type of rope construction will aid in the selection of a lubricant—e.g., coarse constructions such as the 6 x 7 rope groups with large valleys and bigger voids between wires can be penetrated by high-viscosity lubricants.

Ropes with more than 35 wires in each strand, which have very much smaller wire clearances, require low-viscosity lubricants which are able to penetrate into the core. These ropes and ropes with flattened strands require more frequent applications of these low-viscosity lubricants, which should contain water-repellent and rust-inhibiting additives.

The external working conditions have a bearing on the type of lubricant used. For example, where the rope is subject to water, grit and sand, where dredging, excavating and earthmoving, a low-viscosity lubricant with highly adhesive properties has proved most satisfactory.

Where the rope is subjected to the corrosive action of acid, brackish water, or fumes, or where high temperatures are met, a thick continuous coating of high-viscosity lubricant is recommended.

In special cases, a combination of initial lubrication with a low-viscosity lubricant to suit fine wire constructions is followed by an application of a high-viscosity lubricant as an external seal. The frequency of lubrication is mainly determined by the operating conditions—e.g., high speed heavy duty operations call for frequent lubrication, as also do wet, acidic, or other chemical conditions. Standing ropes call for less frequent applications of high-viscosity lubricants. Particular attention should be given to lengths of rope which bend when passing through pulleys.

Lubricants employed on steel wire ropes may require the following characteristics:

(a) Penetration—the lubricant should readily penetrate into both strands and core of the rope.
(b) Water displacement—the lubricant, particularly in the rope core, should contain an additive capable of repelling water, thus preventing the absorption of water which might lead to internal corrosion of the rope and deterioration of the core.
(c) Rust prevention—the lubricant, particularly for internal wire protection, should be rust-inhibiting.
(d) Pressure resistance—as wire ropes contain a large number of small bearing surfaces, for heavy duty operations the lubricant should possess good cohesion, even when submitted to high bearing pressures.
(e) Adhesion—while good penetration is required at all times, the lubricant should possess good adhesion properties, thus minimizing as far as practicable any throw of lubricant.
(f) Stability—the lubricant and its additives should be stable under working conditions.
(g) Compatibility—the service dressing should be compatible with the original lubricant used by the wire rope manufacturer.

Regular inspection

The safe handling of loads by ropes requires adequately designed and maintained equipment. The design of lifting appliances does not normally allow for an indefinite rope life. A rope should be regarded as an expendable component. There is danger in the belief that a rope does not need maintenance or that it will last forever. Thus, safe handling is also dependent upon regular inspections of the rope's condition. A deteriorating rope should be removed from service while still in a safe condition—i.e., before its strength decreases to a level where failure may occur. The rope shall be discarded immediately an inspection shows its strength to have diminished to an unacceptable level.

Inspections of wire ropes should be made by competent persons involved in the maintenance and inspection of lifting appliances. The inspection should include the following objectives:

(a) Record the rope condition so that the progressive deterioration of the rope can be monitored.
(b) Determine whether the rope retains an adequate safety margin.
(c) Particular attention should be paid to the rope at points of attachment to the appliance.
(d) Periodically check winding drums and pulleys to ensure that all these components are free to rotate. Stiff or jammed pulleys or rollers wear heavily and unevenly, causing severe abrasion of the rope. Ineffective equalizing sheaves can give rise to unequal loading in the rope reeving. With small rope movement, the equalizing sheave pins tend to wear out, restricting the movement of the sheave and causing the rope to skid under pressure.
(e) Investigate whether any defect in the appliance is assisting the deterioration. Any defect shoud be eliminated before a new rope is fitted.
(f) Check whether an improvement in the working conditions or increase of maintenance could reduce the rope deterioration.
(g) Investigate whether any kind of modification is necessary to reduce the rate at which the rope is deteriorating.

Much experience can be gained by opening up and completely inspecting discarded ropes, logging findings and causes found for deterioration, and displaying on boards samples cut from discarded ropes. Confidence in inspection can be built up by the submission of samples of discarded ropes of importance to a test house for testing to destruction.

Special inspection

Whenever any of the following conditions apply, the ropes should be given a particularly critical inspection before they are further used:

(a) Whenever an accident has occurred which may have caused damage to the rope or to its termination.
(b) Whenever a new deteriorating condition is recognised.
(c) Whenever a rope has been brought back into operation after dismantling followed by reassembly.
(d) Whenever a lifting appliance is reused after having been out of operation for a period.

It is necessary to inspect ropes regularly, and any equipment on which ropes are used. The frequency of inspection will be influenced by the nature of the equipment and its working conditions, and should be more frequent with advancing age. Depending upon the working conditions, wire rope may need to be visually inspected each working day with the object of detecting general deterioration and deformation.

In order to determine the frequency of inspection, consideration should be given to the following:

(a) Any requirements of the statutory authority.
(b) Environmental conditions.
(c) Severity of the operating conditions.
(d) Type of appliance.
(e) Classification of load application as specified by AS 1418, Part 1.
(f) Results of previous inspections.

Inspection log

Users should provide a log for the recording of information from each inspection of the rope. Regular recording of the condition of the rope should be made together with the service results obtained (i.e., in terms of work done and performance).

Accurate recording of information by the examiner can be used to predict the performance of a particular type of rope on a lifting appliance. Such information is useful in regulating maintenance procedures and also stock control of replacement rope. Any forecasting should not have the effect of relaxing inspections or prolonging the operating periods beyond that indicated by the criteria discussed above.

Although wire ropes should be inspected throughout their length, particular care should be taken at the following positions:

(a) The terminations at the end of both moving and stationary ropes.
(b) Any part of a rope which passes through a block or over a pulley.
(c) Appliances performing a repetitive operation.
(d) Any part of a rope which lies over pulleys while the appliance is in a loaded condition.
(e) Any part of a rope which lies in a compensating pulley.
(f) Any part of a rope which may be subject to abrasion by enternal features.

Internal inspection

Internal deterioration is the prime cause of many rope failures, mainly due to corrosion and the normal progress of fatigue. A normal external inspection may not reveal the extent of internal deterioration, even to the point where fracture is imminent.

Single layer stranded ropes can be opened up sufficiently to permit an assessment of their internal condition provided that they are at zero tension; however some limitation occurs with large rope sizes. Permanent damage may be caused to multilayer wire ropes if they are opened. Internal inspection should always be carried out by a competent person.

The method consists of firmly attaching two clamping jaws of suitable size at a suitable distance apart to the rope. During the inspection of portions of rope adjacent to terminations, it is sufficient to use a single clamping jaw, since the end anchorage system, or a bar suitably located through the end portion of the termination, can be used as the second clamp. By the application of a force to the clamping jaws in the opposite direction to the rope lay, the outer strands separate and move away from the core. Care should be taken during the opening process to ensure that the clamping jaws do not slip about the periphery of the rope. The strands should not be displaced excessively. When a limited opening is achieved, a small probe, such as a screwdriver, may be used to remove grease or debris which could hinder observation of the interior of the rope.

The essential points which should be observed are as follows: (a) state of the internal lubrication; (b) degree of corrosion; (c) indentation of wires caused by pressure or wear; (d) presence of wire breaks (these are not necessarily visible). After inspection, a service dressing should be introduced into the opened part and the clamping jaws rotated with moderate force to ensure correct replacement of the strands around the core. After removal of the jaws, the outer surface of the rope should be greased.

Since it is impossible to inspect the interior of the wire rope over the whole of its length, suitable sections must be selected. For wire ropes which wind onto a drum, or pass over pulleys or rollers, it is recommended that the lengths which engage the pulley grooves when the appliance is in a loaded condition be inspected. Those localized lengths in which shock forces are arrested (i.e., adjacent to drum and jib head pulleys) and those lengths which are particularly exposed to the weather for long periods should be inspected. Attention should be given to the length of rope close to its termination, and this is particularly important for fixed ropes, such as stays or pendants.□

LASHING AND SECURING RO/RO CARGOES

Peter Andersson, Master Mariner, M.Sc, Naval Architect
Project Manager, MariTerm AB, Sweden

IN MODERN CARGO HANDLING, with large, relatively heavy, cargo units, individually lashed on board, the shifting of cargo can be caused by one or several of the following factors:

- The cargo is not sufficiently secured inside the cargo unit.
- The cargo unit is not sufficiently secured to the ship.
- The cargo unit is not designed for sea transportation.
- The ship has unsuitable stability.
- Cargo which is difficult to secure is stowed where large acceleration may occur.
- The cargo is inappropriately packed.

If the cargo is not sufficiently secured inside the cargo unit, this can cause damage on the cargo itself but can also jeopardise the safety of the ship if the cargo leaves the cargo units. Loose cargo can damage the lashing equipment of adjoining cargo units or the ship. This problem has, in the first place, been highlighted by great damage to cargo transported in road trailers on ships.

A cargo unit, not sufficiently secured to the ship, can cause shifting of cargo in larger or smaller parts of the ship. Especially serious consequences can arise if the shifting of cargo occurs in a part where dangerous cargo is stowed. There is likewise a risk that cargo units, not designed for sea transportation, collapse and damage the cargo inside the cargo unit and the securing equipment of adjoining units. Also this problem has arisen during transportation of road trailers on ships.

The ship can either have too high or too low stability due to unsuitable design or unsuitable cargo distribution and considerable accelerations may occur on board. The accelerations which affect the cargo in the units are equal to those which affect the actual cargo units and are caused by the ship motions at sea. If the ship motions at sea can be mathematically calculated, the accelerations which the cargo is subject to can also be mathematically calculated and the correct requirement of securing equipment and cargo units can be determined.

In recent years there have been a number of accidents involving shifting of semi-trailers and general cargo in Ro-Ro ships, especially in short-sea traffic. These accidents have highlighted the problem of securing cargo on board ships. This has, among other things, led to research work within this field. Some examples of titles are:

1. *Safe Stowage and Securing of Cargo on Board Ships* (Ref. 1).
2. *Securing of Road Trailers on Board Ro-Ro Ships* (Ref. 2).
3. *Road Trailers Suitable for Sea Transportation* (Ref. 3).
4. *Securing Goods on Semi-trailers* (Ref. 4).
5. *Optimum Safety Factors for Securing of Cargo on Board Ships*(Ref. 5).

The shifting of cargo in Ro-Ro ships is the reason for many casualties. Cargo shifting, together with operational reasons, constitutes a larger part of the serious casualties than of all casualties: 16 per cent as compared to 12 per cent; 43 per cent of the total losses of Ro-Ro ships are caused by shifting of cargo or are due to operational reasons. These statistics, valid for Ro-Ro ships are taken from Det Norske Veritas' casualty statistics (Ref. 6).

Ship motions and accelerations

When the cargo is shifting on board a ship, this is normally due to ship motions at sea in bad weather and thus it is necessary to secure the cargo in one way or the other. In what manner the cargo is to be secured depends on the ship, the route and the type of cargo to be shipped. A ship at sea has six modes of motion; three rotational and three linear motions. The designations of these are pitch, roll, yaw, sway, surge and heave.

These motions can be combined into three orthogonal accelerations; the transverse a_t, the longitudinal a_l and the vertical acceleration a_v, which act across, along and normal to the deck of the ship.

The transverse acceleration is chiefly caused by roll, static and dynamic, but also by yaw and sway. The longitudinal acceleration is caused by pitch, static and dynamic, as well as surge and yaw, and the vertical acceleration by heave, pitch and roll. If the accelerations are known, the forces acting on the cargo can be calculated according to the formula $F = ma$ (kN), where F is the force acting on the cargo in kN, m is the weight of the cargo in question in tons and a is the acceleration in m/s^2. Besides being exposed to forces due to accelerations, cargo on deck is also exposed to forces due to wind and green sea.

Parameters influencing accelerations

When it comes to cargo securing the transverse acceleration is the most important. In principle, the transverse accelerations can be calculated according to the following formulas taken from the research report *Safe Stowage and Securing of Cargo on Board Ships* (Ref. 1):

$$a_t = \sqrt{0.01 + (0.0702\,\phi\,(Z - T)/T_r^2 + \sin\phi)^2}\; g \;(m/s^2).$$

$$\phi = 2865/(B + 75) \text{ (degrees)}.$$

$$T_r = 0.74\, B/\sqrt{GM} \text{ (seconds)}.$$

where

ϕ is the roll angle in degrees,
Z is the vertical distance in metres from the base line (BL) to the centre of gravity of the cargo,
T_r is the roll period in seconds,
B is the ship's breadth in metres,
T is the actual draft in metres,
GM is the actual metacentric height in metres and
g is the gravity constant 9.81 m/s^2.

These formulae are based on the different classification societies' rules for the securing of

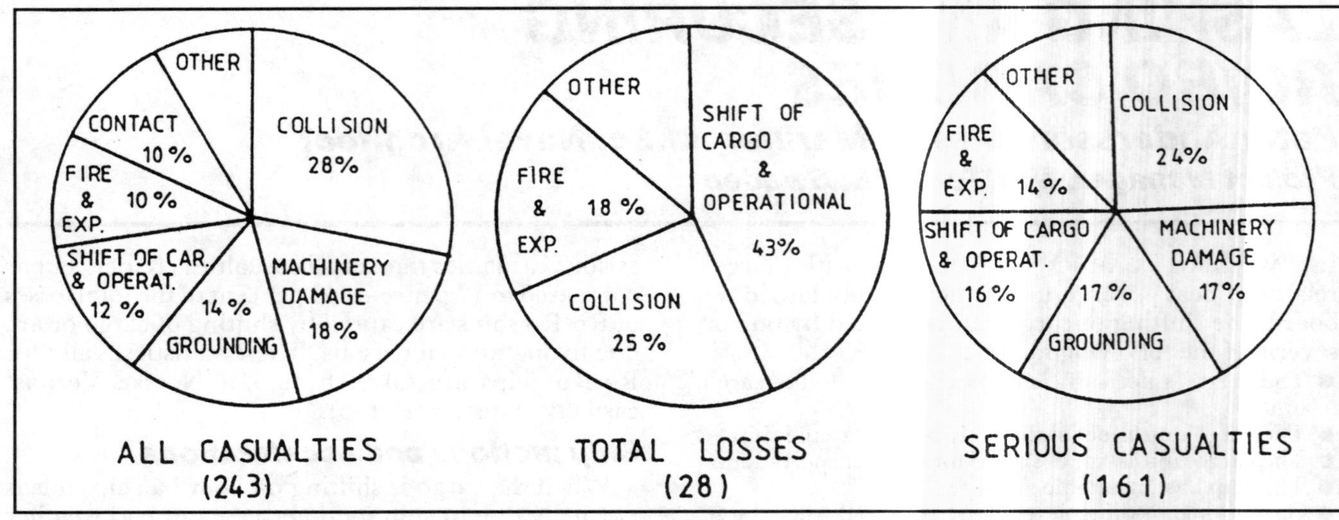

Figure 1: Ro-Ro casualty type distribution.

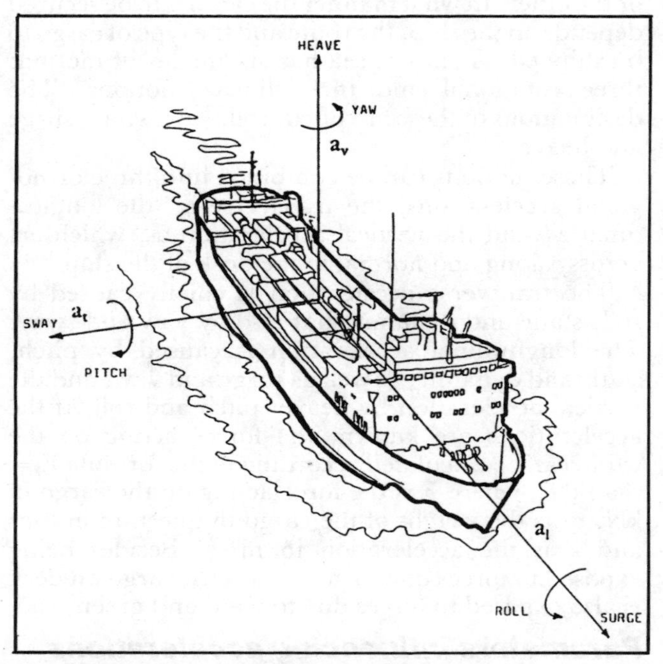

Figure 2: Ship motions at sea and three orthoganal accelerations caused by ship motions.

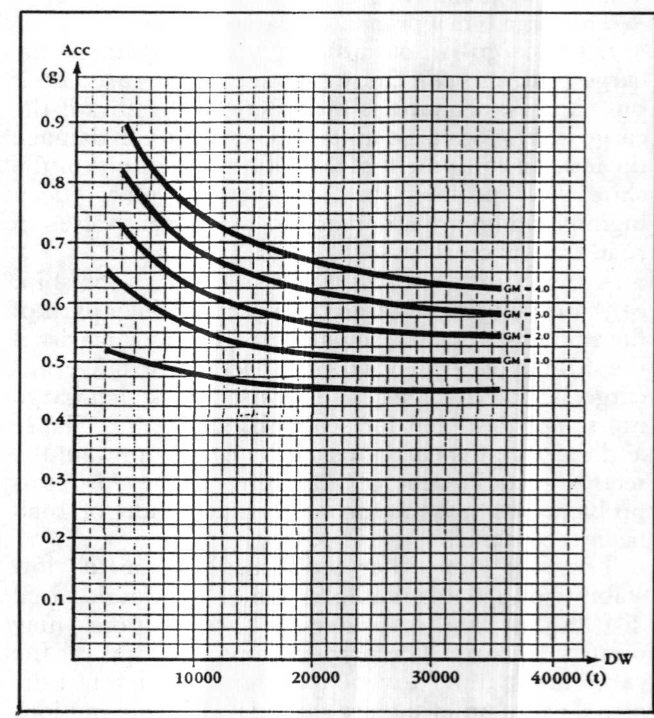

Figure 3: Transverse acceleration as a function of ship size and stability.

Figure 4: Securing of a truck with trailer in a file on a ferry.

containers on board ships, as well as results from full-scale measurements presented in the literature and two full-scale measurements, one on board the short-sea Ro-Ro *Tor Dania,* of 10,000 dwt, in North Sea traffic and one on board the jumbo Ro-Ro *Boogabilla,* of 32,000 dwt, in worldwide trading.

From the formulae it can be seen that the transverse acceleration decreases with increasing breadth, decreasing distance from the base line and decreasing metacentric height. These effects can clearly be seen in figure 3, where the transverse acceleration in typical Ro-Ro ships is shown as a function of the ship size. The four upper curves are valid for different GM values on upper deck (Z = D) and the lowest curve is valid at the ship's centre of motion (Z = T), where the acceleration is independent of the metacentric height. The accelerations are calculated with the main particulars for Ro-Ro ships of different sizes. The formulae for the vertical and longitudinal accelerations are set up in the same way.

Calculations show the largest accelerations as far from the ship's centre of motion as possible, which on a Ro-Ro ship means on upper deck, forward or aft in the outer lanes. The lowest accelerations on the other hand are found in the cargo hold near the ship's half length in the centre line.

How accelerations can be affected

The greatest risk of shifting of cargo occurs in beam sea conditions with a resonance between the period of the attacking sea and the ship's natural roll period. In a situation like this the roll angle can increase and reach values up to 30-35 degrees. The resonance situation can be avoided by changing course, which changes the period of the attacking sea.

Great roll angles can sometimes also occur in following sea conditions, especially to Ro-Ro ships with wide sterns. These situations are normally more difficult to get out of, as the course has to be changed considerably. The greatest longitudinal and vertical accelerations occur in head sea. These situations are not as dangerous for the cargo as for the ship's hull itself. The accelerations will decrease by reduced speed.

As can be seen from the above formulae, the roll period is reciprocally proportional to the square root of the metacentric height and the transverse acceleration increases rapidly with increasing GM. As the metacentric height can be affected by the cargo distribution, it is obvious that correct loading of the ship is of utmost importance to minimize the transverse accelerations. With a metacentric height of about 0.55 m the roll period in seconds will be equal to the ship's breadth in metres. A situation which should be aimed at, if allowed due to stability criteria. The risk of large roll angles will also be affected by a stabilising system. There are different types of stabilising systems on the market.

Bilge keels

A bilge keel is the simplest stabilising system with the smallest damping effect, but it is, at the same time, the safest system with no moving parts. The roll opposing force at the bilge keel is proportional to the bilge keel width and roll velocity. The result of this is, first of all, that the larger the bilge keel, the larger will be the roll damping and, secondly, that the bilge keel damping force will not be limited by the roll angle. As the wave height increases, so too will the damping force by the bilge keel. All other stabilisers, for one reason or another, will have some limit to the amount of stabilising force that they can supply.

Bilge keels do, however, increase resistance and, in cases where the bilge keel width has been made very large, there can also be a problem with the high stresses and maintenance of the bilge keels themselves. For the above reasons it is normal to install bilge keels which provide a degree of roll damping but also to minimise the effect on resistance. It is also normal to position bilge keels so as not to interfere with flow lines to the propeller in calm waters.

Uncontrolled passive tank

The passive system has the same stabilising effect independent of the ship's speed. The passive type is therefore ideal for ships lying on the same spot for some time or moving at reduced speed. Passive stabilising systems by means of anti-rolling tanks are not a new invention. The first tank was installed in a ship in 1874. Since then various models of such tanks have been adopted. The system consists of a rectangular tank, sometimes with a built-in damping device. The tank is filled with fluid and the fluid level is adjustable to suit various rolling conditions. The fluid can be sea-water, fresh water or fuel oil.

When the ship starts to roll, the fluid in the tank will also be set in motion. The damping device in the tank is designed to give the moving fluid a phase displacement of 90° compared with the ship—ending up in the opposite phase. The moving fluid will thus have a stabilising effect consisting of two components. One due to positive weight distribution of the fluid in the tank, the other due to the kinetic energy of the fluid moving from one side of the ship to the other. The tank extends across the whole breadth of the ship and is usually situated in the hull. Successful installations have been made in the superstructure as well.

Controlled passive tank

Also in this type of stabilising system the roll of the vessel causes an oscillatory athwartships movement of water in the tank system. Due to the U-shaped design of the tanks and by the automatic control which keeps the water cyclicly blocked on the upwards moving ship's side, the athwartships movement of the tank water is always tuned to counteract and reduce the roll. Thus, the sea in making the vessel roll delivers the necessary energy to reduce the roll.

There are two distinct main operational ranges: one purely passive operation at short periods and one actively controlled range at periods longer than the natural period of the tank system. As soon as the ship rolls with periods slightly longer than the natural period of the tank system due to reduced GM values or the effect of the waves, the tank water is immediately adapted to the changed roll motion by the automatic control.

Fin stabilisers

A pair of fins projecting from a ship's side can create a sizeable roll moment when tilted at full speed

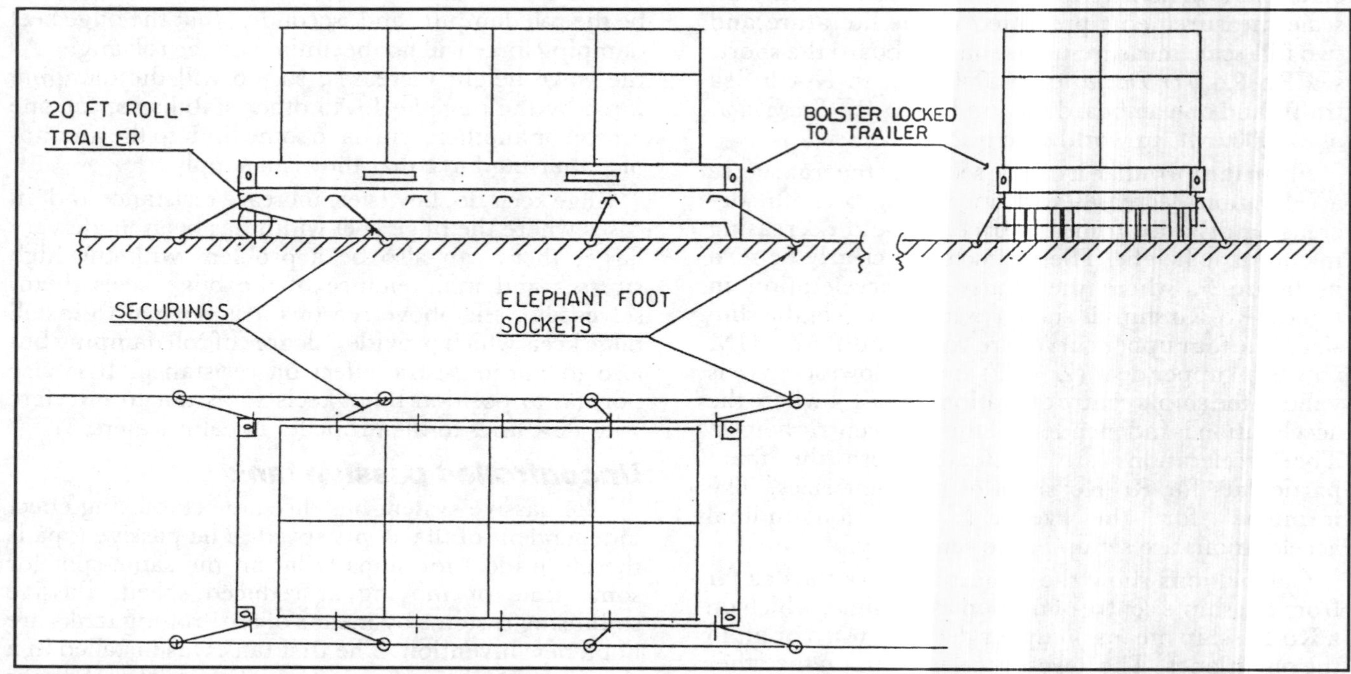

Figure 5: Securing arrangement of a 20-ft roll trailer.

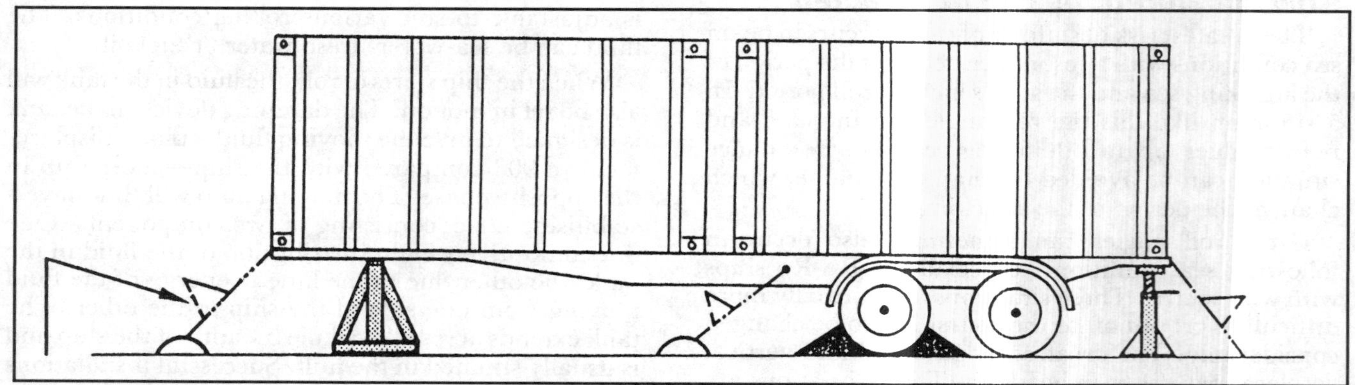
Figure 6: The lashing of a road trailer.

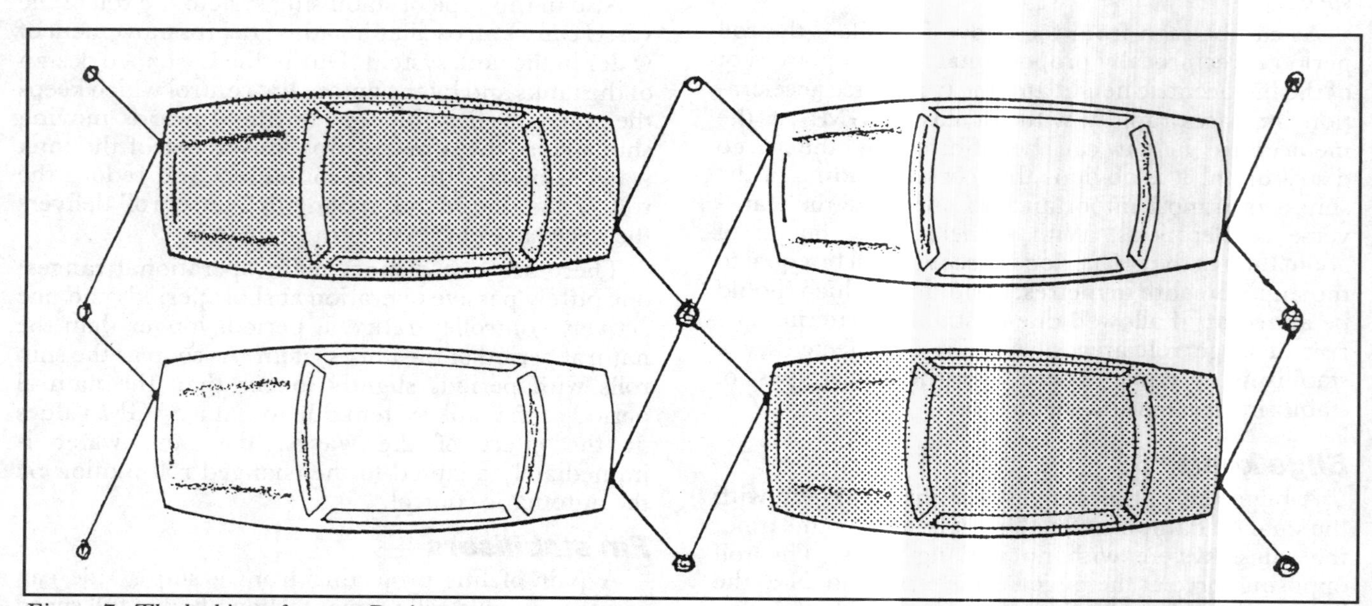
Figure 7: The lashing of cars to D-rings.

ahead. Essentially, activated fins provide the only practical means to counteract the actual wave moments attempting to roll the vessel. The use of control systems which activate (tilt) the fins in the shortest possible time period, and the use of sufficiently large tilting motors to also minimize the time lag after receipt of a control signal, combine to provide a system designed to sense roll accelerations and to try to minimise them. The reaction time is a fraction of one roll period.

Because fin stabilisers are designed as balanced foil sections, a relatively small torque is needed to operate the fins to angles capable of offsetting wave moments. In selecting a fin stabiliser, therefore, it is extremely important to estimate carefully the ship's operating condition (speed, GM, displacement, etc) and determine the fin physical size and fin tilt drive characteristics so that the system can respond quickly enough and with enough force to keep the roll within set small limits, even in the most severe seas.

Since the fin moment is proportional to the square of the ship's speed, it is important that the ship operates at or near the design speed in order to make the best use of the system. When the ship's speed is low or zero, the fins completely lose their stabilising effect.

Roll damping effect

The fin stabilisers can achieve a roll damping effect of over 90 per cent and reduce the roll angle to some 3°, as long as the ship runs at full speed while the damping effect is zero when the ship's speed is zero. The roll damping effect of the tank stabilising systems is 35-60 per cent, the higher value being obtained in more regular seas and the lower value in more irregular seas. These systems work independently of the ship's speed. The uncontrolled systems do not work properly in following sea conditions, while the controlled systems do not work in a blackout situation. The bilge keels give a roll damping effect of some 5-10 per cent. The effect is independent of speed and electric power.

According to a decision within the International Maritime Organization (IMO) stabilising systems, independent of type, may not be used to reduce the number of securings or their strength. Despite this many ships are equipped with stabilising systems as the ships, as long as the systems work properly, will move less at sea which in the long run will lead to less cargo damage on board. Small movements are particularly important on ships carrying passengers.

Securing cargo on board

When it comes to securing of general cargo and cargo units on board ships, up to now very few rules and recommendations have been published. One exception is, however, the classification societies' rules for container securing arrangements.

Some classification societies such as Det Norske Veritas, Lloyd's Register of Shipping, Germanischer Lloyd and Bureau Veritas have developed regulations for container securing arrangements. Within these rules, formulae for calculation of the vertical, transverse and longitudinal accelerations have been established.

Although there is a significant difference in the accelerations calculated according to the different societies formulae, it has been decided in the IMO sub-committee on containers and cargoes that any of these formulae may be used when designing cargo securing arrangements for any type of cargo on board ships. An attempt to develop internationally-harmonised formulae for calculation of the accelerations, based on the societies' formulae, has been presented in Ref. 1.

Activities within IMO

The international organizations ILO (International Labour Organization) and IMO (International Maritime Organization) have lately recognised the problems with combined transport of goods in cargo units in international traffic, where sea transport is included.

Especially within IMO, the main purpose of which is to promote safety at sea and to protect the marine environment, they intend to present a common international safety system. The work within the area of cargo safety has until now resulted in the following documents:

1 Recommendation on the Safe Stowage and Securing of Containers on Deck on Vessels which are not Specially Designed and Fitted for the Purpose of Carrying Containers, Resolution A 288 (VIII), November 1973.
2 Safe Stowage and Securing of Cargo Units and Other Entities in Ships other than Cellular Container Ships, Resolution A 489 (XII), November 1981.
3 Elements to be taken into Account when Considering the Stowage and Securing of Cargo Units and Vehicles in Ships, Resolution A 533 (XIII), November 1983.
4 IMO/ILO Guidelines on the Safe Stowage and Securing of Cargoes in Containers and Vehicles, MSC-Circ 383, January 1985.
5 Provisions to be Included in the Cargo Securing Manual to be Carried on Board Ships, MSC-Circ 385, January 1985.
6 Guidelines for Securing Arrangements for the Transport of Road Vehicles on Ro-Ro-ships, Resolution A 581 (XIV), November 1985.

The Sub-Committee on Containers and Cargoes is continuing its work in this field and is at present concentrating on two main areas. One is the development of a code of safe practice for the safe stowage and securing of cargo, cargo units and vehicles. The second and more important is the preparation of suitable regulations for inclusion in a new revised Chapter VI of Solas 1974 which will make elements of the code mandatory. The Maritime Safety Committee has recognised the pollution hazard, in addition to the ship safety aspect and the risk to the population living near coasts, of dangerous goods washed overboard as a result of inadequate stowage or securing.

The aim of the code will be to advise masters on the specific hazards and difficulties associated with the transport of certain cargoes; the stowage and securing of such cargoes; and associated ship handling measures. The basic principles of the code are that cargo should be stowed and secured in such a way that the safety of the ship is not put at risk by the cargo shifting.

The appendices to the code will give more detailed guidance and information to supplement the advice given in the main body of the code. The appendices to be developed by the sub-committee to the detail and extent as may be necessary.

Cargo securing manual

In the IMO resolution A 489, adopted in November 1981, governments are recommended to issue guidelines for the safe stowage and securing of cargo and to require ships to carry a *Cargo Securing Manual*. The information contained in the manual should include the following items as appropriate:

- Details of fixed securing arrangements and their locations (pad-eye, eyebolts, elephant-feet, etc.);
- location and stowage of portable securing gear
- Details of portable securing gear including an inventory of items provided and their strengths;
- Examples of correct application of portable securing gear on various cargo units, vehicles and other entities carried on the ship;
- Indication of the variation of transverse, longitudinal and vertical accelerations to be expected in various positions on board the ship.

Methods and equipment

As mentioned previously, cargo has to be secured due to ship motions at sea. For some types of cargo well formulated internationally-agreed requirements have been set up, while for others only general recommendations or no recommendations at all exist. In the following rules and recommendations for the securing of different types of cargo will be discussed.

Containers on Ro-Ro decks

On the upper deck in a Ro-Ro ship, containers are stowed three or four high and the securing systems are similar to those for the purpose-built container ships. Under deck in Ro-Ro ships, containers can normally be stowed one or two high. For flexibility reasons the containers are often stowed in blocks of ten.

Within the ten-block system ten 20 ft or five 40 ft units can be stowed longitudinally or transversely. Various combinations of transverse and longitudinal stowage of 20 ft units can also be achieved, as well as a combination of 20 ft and 40 ft containers.

The necessary number of fittings in a ten-block depends on the following factors:

- The ship's maximum motions.
- The stowage location in the ship.
- The stowage height.

A system with two fittings per transverse container position and three per longitudinal position has been developed. Should there be a risk of tipping, which means that four fittings are required per container position, it is customary to leave out the class nomination 'Container Carrier' for the decks in the ship rather than increase the number of fittings.

If a ten-block is equipped with lashing points for all container corners and row A is loaded first, the risk is evident that these units will be locked by four cones each. Then when row B is loaded, perhaps in another port, these units can be locked in three corners, which is often sufficient for the sea voyage. If the units in row A are later to be discharged in a port before those in row B, the problem is obvious with the locked cone in the inner corner. For these situations the lashing manufacturers supply a 4 or 6 m long twistlock actuator pole hook which can be inserted to twist the cone. This, however, does not work in practice due to jammed cones, ice, etc.

A situation of this kind is either handled by flanging the cargo in row B or by letting the cargo in row A accompany the cargo in row B to its port of discharge and be transported in a different mode back to A. There are many examples of cargo having to go on an extra trip because of being locked in.

Securing of 20-ft and 40-ft roll trailers

Roll trailers should be equipped with an adequate number of fittings for the securing of the units to the ships deck and the forward 'support leg' of the roll trailers should be supported to deck by a high friction sole of wood or rubber. Roll trailers should also be equipped with lockable cones for the fixing of ISO-flats or containers to the trailers as well as fittings for the securing of loose cargo to the trailer decks and the trailers to ships deck.

The gross weight of a 20-ft roll trailer is just above 26 tons when the trailer is loaded by one flat or container with a maximum weight of 24 tons. A 20-ft roll trailer with a maximum weight of 26 tons could be secured by four securings per side. Two of the lashings per side should be positioned in a fore and aft direction while the third and fourth lashings should be positioned as close as possible to 90° towards the roll trailer.

The gross weight of a 40-ft roll trailer loaded by two 20-ft flats or containers with a maximum weight of 24 tons is about 58 tons.

A 40-ft roll trailer with a maximum weight of 58 tons could be secured by six securings per side. Two of the lashings per side should be positioned in a fore and aft direction, while the four remaining lashings should be positioned as close as possible to 90° towards the roll trailer. More accurate information of how to secure roll trailers in a particular ship can be found in the ship's *Cargo Securing Manual,* if existing.

Wheel-based cargoes

Road vehicles of different kinds are shipped in an increasing number at sea, as well on short trades as on long international voyages. So, for example, are cargo-carrying vehicles with or without driver shipped over the North Sea and the Channel. A big number of brand new vehicles such as cars, trucks, buses, tractors and other wheel-based machines are also transported at sea.

When stowing and securing wheel-based cargoes at sea, the following points should be noted:

1 The cargo holds in which wheel-based cargo is to be stowed must be clean and the surfaces must be free from grease and oil.

2 Wheel-based cargo should preferably be provided with adequate securing points or other means on which lashings of sufficient strength may be applied.

3 Wheel-based cargo which is not provided with special securing means should, however, have those places well marked where lashings may be applied.

4 Wheel-based cargo which is not provided with rubber wheels or tracks with friction increasing lower surface should always be stowed on wooden dunnage or other friction increasing material such as softboards.

5 Securing means should be suitable for the wheel-based cargo and be of adequate strength to take up any of the forces which may arise at sea

6 Cargo stowed on wheel-based cargo should be adequately secured to its platform or where provided its sides.
7 Wheel-based cargo should be secured by lashings having at least similar characteristics as wire lashings.
8 Wheel-based cargo should preferably be stowed close to the ship's side or at such stowage places on board which are provided with sufficient securing points of sufficient strength.
9 In order to prevent lateral shifting wheel-based cargo, which is not provided with adequate securing points, should however always be stowed close to the ship's side and close to each other or be blocked of by other suitable cargo units such as loaded containers, etc.

In order to prevent longitudinal shifting all wheels and/or tracks should be blocked against movement in the forward and backward direction.

Road vehicles with drivers are mainly shipped on ferries on short routes. A semitrailer with a towing unit is secured in the same way as the truck with trailer.

Securing semitrailers

In the conventional way of securing a semi-trailer alone on board a ship, the front end of the trailer has to be supported by a trailer horse, and in the rear end trailer jacks could be used. The height of the trailer jack can be adjusted with a wheel-shaped handle. Against the wheels of the trailer, chocks can be placed. These could be made of wood or rubber.

Trailer lashings may consist of wires, chains or webs. Chains have traditionally been the dominant means of lashing but webs and winches have recently increased in popularity. Webs are lighter than chains and the handling time can be reduced. A chain lashing will get an elongation (L) of 50 mm when exposed to a force of 112 kN (11.2 tons), while the web lashing will get an elongation of about 240 mm at the same force. The difference in characteristics, of course, influences the behaviour of the trailer on a moving ship's deck at sea. This is one reason way IMO has recommended use of lashings made of material having elongation characteristics equivalent to steel chains.

IMO general recommendations

Within IMO the following general recommendations for the stowage and securing of road vehicles and semitrailers have been developed:

1 Lashings should consist of chains or other devices and be made of material having strength and elongation characteristics at least equivalent to steel chains. The strength of the lashings, without permanent deformation should not be less than 120 kN.
2 Lashings should be so designed and attached that, provided there is safe access, it is possible to tighten them if they become slack. Where practicable and necessary, the lashings should be examined at regular intervals during the voyage and tightened as necessary.
3 Lashings should be attached to the securing points with hooks or other devices so designed that they cannot disengage from the aperture of the securing point if the lashing is slack during the voyage.
4 Only one lashing should be attached to any one aperture of the securing point on the vehicle.
5 Lashings should only be attached to securing points provided for that purpose.
6 Lashings should be attached to the securing points on the vehicle in such a way that the angle between the lashing and the horizontal and vertical plane lies preferably between 30° and 60°.
7 Road vehicles should be stowed so that the chassis are kept as static as possible by not allowing free play in the suspension of the vehicle. This can be done, for example, by compressing the springs by tightly securing the vehicle to the deck, by jacking up the chassis prior to securing the vehicle or by releasing the air pressure on compressed air suspension systems.
8 The air pressure should be released on every vehicle fitted with compressed air suspension systems if the voyage is of more than 24 hours' duration. If practicable, the air pressure should be released also on voyages of a shorter duration. If the air pressure is not released, the vehicle should be jacked up to prevent any slackening of the lashings resulting from any air pressure leaking from the system during the voyage.
9 Where jacks are used on a vehicle, the chassis should be strengthened in way of the jacking-up points and the position of the jacking-up points should be clearly marked.
10 Special consideration should be given to the securing of road vehicles stowed in positions where they may be exposed to additional forces. Where vehicles are stowed athwartship, special consideration should be given to the forces which may arise from such stowage.
11 Wheels of road vehicles should be chocked to provide additional security in adverse conditions.
12 Vehicles with diesel engines should not be left in gear during the voyage.
13 Vehicles designed to transport loads likely to have an adverse effect on their stability, such as hanging meat, should have integrated in their design a means of neutralising the suspension system.
14 The parking brakes of each vehicle or of each element of a combination of vehicles should be applied and locked.
15 Semitrailers, by nature of their design, should not be supported on their landing legs during the sea transport unless the landing legs are specially designed for that purpose and so marked. An uncoupled semitrailer should be supported by a trestle or similar device placed in the immediate area of the draw plate so that the connection of the fifth-wheel to the kingpin is not restricted. The selected areas of the vehicle should be clearly marked.

When lashing cars, at least four lashings should be employed. The instructions of the manufacturer should be followed and cars may not be lashed together. The parking brake should be applied and the car should be in gear.

Securing break-bulk and unitised cargoes

As these cargoes can consist of any commodity in any shape or form, only general common sense recommendations can be given. The basic aim is to secure the cargo to prevent any movement in any direction to avoid damage to cargo or ship.

Securing is very much a consequence of stowage. A tight stow within a restricted compartment between rigid steel structures does not need any lashing at all. A way to simulate this situation in a large cargo deck is to make use of containers. Two basic methods are in use:

- Containers are stowed first and secured to deck by means of container stackers and then the unitized cargo is stowed in between.
- Unitized cargo is stowed first and then faced off with containers. The containers will then most likely not fit the deck fittings and must consequently be placed on antiskid material and lashed.

With both methods care must be taken to avoid damage to the containers. Only the container frames can accept any pressure from the unitised cargo. Other methods to secure cargo include lashing, skid prevention, tomming off and filling up. Lashing in general is done by means of chains and tension levers. To increase friction, cargo can be placed on antiskid material like rubber, plywood, timber or board.

Tomming off, if necessary, should be done in softwood for easier handling, whilst wedges, when used in tightening up a close weight cargo stow, should be of hardwood. Filling up void spaces in a stow can be done with just about anything available, but care must be taken to avoid damage to the commodity. Empty pallets, dunnage or, in damage-prone areas like reefer compartments or with sensitive cargo like paper reels, inflatable rubber cushions should be used.

Chains should only be attached to the ship in the permanent fittings. Chains should never bear over or round sharp structural corners. Tensioners should whenever practicable be placed so they are easily accessible for retightening at sea. Platform containers, flats and bolsters should preferably be stowed in container positions and thus secured by stacking fittings, but may as an alternative be stowed on antiskid material and lashed with chains.

When bolsters are stowed more than one high, chains should be used for lashing and if necessary walking boards for load distribution. The walking boards must not project out under the corner castings of the upper platform. Sufficient lashings must always be fitted to prevent shifting of cargo. Individual pieces may require cross-, vertical- and spring lashings. Blocks of cargo may require intermediate lashings as well as a secured face.

In a mixed stow, backlashings may be just as important as a lashed face. Lashings must only terminate in the specially designed lashing points in the ship. No matter how well the face of a blockstow is lashed, if gaps and holes between the individual items are not filled up or bridged, the cargo will start to shift at sea.

Where space or commodity does not allow tomming off a solution is to fill up the space with anything suitable. Hardwood wedges can efficiently be used in a stow of bundled copper ingots. An empty pallet can be used to bridge a gap and thus locking a stow. A couple of old tyres can fill a gap between horizontal steel coils.

Fixed securing equipment

When a ship is contracted to a shipyard and the cargo to be carried is specified, a stowing pattern, as well as amount, strength and location of fixed securing equipment, is decided upon. Fixed securing equipment are—e.g., lashing terminals and container fittings. Lashing points are usually painted in some contrasting colour to reduce the risk of tripping. The fixed securing equipment is designed for a certain strength, which can be found in the ship's *Cargo Securing Manual* or on drawings on board.

Loose securing equipment

Loose securing equipment includes lashing equipment, such as wire, rod, web and chain, equipment for the stacking of containers, tensioners, etc. Different types of hooks are used for rods, web, wire and chain. There are many types of tensioners on the market. Some of these can be used together with several kinds of lashing, others with one type of lashing only. The turnbuckle or bottle screw is the oldest type of tensioner and therefore known by operators all over the world. The turnbuckle can be threaded at one end or both, it can be open or enclosed and it can be equipped with a fixed or detachable handwheel for tensioning. Turnbuckles are simple and safe and therefore quite common, in particular when securing containers with rods. They are also flexible and can be used with chain, as well as wire.

Lever tensioners are used with chain only and they are the most common tensioners for trailer lashing. The weight is low, they are comparatively cheap and they do not need much maintenance. They require a lot of space and they are dangerous to tension in rough weather. The reason for this is that they first have to be slackened and then tightened. In the meantime, the cargo is not secured which is a risk. The lever tensioners can also cause injuries by springing when being released. They are furthermore quite laborious to tighten. Many types with different quality exist on the market.

The quick release tensioner is designed for use with wire lashings. It is quick to use and cheap to manufacture because all the components are cast. Some types have shown to be of poor design and, thus, have a tendency to spring and hurt the operator. Stevedores therefore prefer to strike the catch with a tool, a practice that wears out the tensioners in the long run. The quick release tensioner also requires more space than—e.g., the turnbuckle. As the lever tensioner, the quick release tensioner has to be slackened before retightened which can cause injuries in bad weather.

In recent years several new tensioners have been introduced. Among these hydraulic and pneumatic tensioners can be noted. These have become popular because they are light, quick and safe to use. They require less applied force and it is possible to apply the same tension in all lashings. They are, however, more expensive than traditional tensioners.

The quick tite tensioner can be tensioned pneumatically as well as with an hydraulic handpump. In comparison with a turnbuckle system, where the buckle weighs 20 kg and requires two stevedores during 5-6 minutes, the quick tensioner weighs only 11 kg and can be tensioned by one man in half a minute. The handpump can be set to give a preload of either 1.5 or 2.5 tons.

Turnbuckles, too, can be tensioned pneumatically—e.g., by using the speed lash tensioner. The speed lash system can tension up to 4-5 tons, it is safe, quick and requires little space. The source of power is often a problem. On many ships the power unit has to be pushed about on deck which is troublesome.

Tests have shown that lashings with wire and bulldog clips have very different strength depending on how the lashings are arranged. It is important to use ungreased wire together with bulldog clips with greased threads, as this combination gives the highest strength for a given number of clips.

Quality tests

In the Swedish research project 'Optimum Safety Factors for Securing of Cargo on Board Ships'[5], different types of loose securing equipment have been quality tested. Some 40 pieces of different types of equipment, webs, chains, rods, tensioners, car lashings and twistlocks were collected on board different ships and tested. The results of the breaking tests were then compared with the specified breaking strength of the equipment when new. The results were not the best.

These chains were taken from a large ocean-going Ro-Ro ship and the age of the equipment was estimated to be about four years, and all lashings were tested in the most stressed position, which is with the lever in 90° towards the chain. Two of the lashings had chain links of 11 mm with a specified breaking strength when new of 15 tons. Three of the lashings were made of 13-mm links with a corresponding breaking strength of 20 tons. Only one lashing showed a breaking strength of the same magnitude as the specified value. The other lashings broke at forces between 4 and 6 tons, which is 20-30 per cent only of the specified strength of new equipment.

One reason for the poor result of this test could be that some manufacturers of tension levers have restrictions for the use of the equipment. One restriction for example can be that the angle between the lever and the chain must be less than 45 degrees. If the equipment had been tested when mounted according to the restrictions, perhaps they would have managed 15 and 20 tons, respectively. Restrictions in the application of some levers are not known by the ship's officers and the stevedores and should thus not be allowed.

Chain lashings

Chain lashings with speed lash tensioners showed better performance. Two 11-mm lashings which had been used on board a North Sea Ro-Ro for two and four years, respectively, were tested. These lashings broke at a chain link at 15.1 and 14.9 tons, which is very close to the specified 15 tons.

The result of break tests of lashing chains alone without tensioner after 4 years use was as follows: One of the 11-mm chains with a specified strength of 15 tons broke at 6 tons due to one slightly damaged link. Another one broke at 9 tons due to a welding fault. The rest of the chains broke close to the specified values. Five web lashings for trailer securing were picked out at random on a Ro-Ro vessel also trading the North Sea. The lashings had a specified breaking strength for the assembled lashing with hook, web and winch of 12 tons when new.

The lashings had been in use between 1 and 11 months and were an average of the lashings in use on board. It was possible to determine the age of the lashings as the manufacturing date was still readable. The web broke at the hook in all the tests due to wear and tear, and results were obtained between 6.4 and 9.7 tons for the used equipment instead of the specified 12 tons.

As all five lashings in the test broke up at the hook, some of the lashings were dismounted and in a second step that part of the web, which during the first test had been wound up in the winch, was tested. This test showed a breaking strength of 10-11 tons instead of 17 tons, which is the strength of new web alone. Tests with new equipment confirmed the results that a web, which is loaded when it is wound-up, loses a significant part of its strength.

Different kinds of container securing equipment, such as twistlocks, rods, turnbuckles, etc., were also tested. These lashings had been in use on board an overseas Ro-Ro for about four years, and they all broke at loads close to the specified breaking strength for new equipment. This equipment had been classified by classification societies at delivery to the ship.

The quality tests of the shipboard lashing systems, which were carried out even though it was limited, clearly shows a demand of rules and regulations for classsification of all kinds of lashing equipment in the same way as for container securing equipment. The tests also showed a demand for clear marking of the equipment. The marking was proposed to contain information concerning manufacturer, batch number, safe working load and date of manufacture. To obtain the endurance of equipment in operation on board ships, regular tests and surveys of the equipment were also strongly recommended.

Calculation of lashing loads

To be able to calculate the forces in the lashing equipment all three orthogonal accelerations a_1, a_t and a_v, have to be calculated. The entire method developed in (Ref. 1) for the calculation of design accelerations and forces acting on the cargo on board is presented here.

The accelerations are calculated according to the following for two different cases; for maximum roll motion and for maximum pitch motion:

1) At maximum roll

$$a_{rl} = 0.6\sqrt{0.0064 + \left(\frac{0.0702\,\dot{\theta}\,(Z-T)}{T_p^{\,2}} + \sin\theta\right)^2}\; g \qquad (m/s^2)$$

$$a_{rt} = \sqrt{0.01 + \left(\frac{0.0702\,\phi\,(Z-T)}{T_r^{\,2}} + \sin\phi\right)^2}\; g \qquad (m/s^2)$$

$$a_{rt} \geq 0.5\, g \qquad (m/s^2)$$

$$a_{rv} = \sqrt{0.16 + 0.00493\left(\left(\frac{\phi Y}{T_r^{\,2}}\right)^2 + 0.36\left(\frac{\theta(X-LCF)}{T_p^{\,2}}\right)^2\right)}\; g \qquad (m/s^2)$$

2) At maximum pitch

$$a_{p1} = \sqrt{0.0064 + \left(\frac{0.0702\,\theta(Z-T)}{T_D^{\,2}} + \sin\theta\right)^2}\; g \qquad (m/s^2)$$

$$a_{pt} = 0.5\sqrt{0.01 + \left(\frac{0.0702\,\phi\,(Z-T)}{T_r^{\,2}} + \sin\phi\right)^2}\; g \qquad (m/s^2)$$

$$a_{pt} \leq 0.25\, g$$

$$a_{pv} = \sqrt{0.16 + 0.00493\left(0.25\left(\frac{\phi Y}{T_r^{\,2}}\right)^2 + \left(\frac{\theta(X-LCF)}{T_p^{\,2}}\right)^2\right)}\; g \qquad (m/s^2)$$

where

T_p = pitch period = $0.57\sqrt{L}$ (seconds)

T_r = roll period = $\frac{0.74\ B}{\sqrt{GM}}$ (seconds)

θ = pitch angle = $15\ e^{-0.0033L}$ (degrees)
$\theta \leq 10$

ϕ = roll angle = $\frac{2865\ C}{B + 75}$ (degrees)

$T_r < 20$ sec C = 1.1 without bilge keel
C = 1.0 with bilge keel

$20 \leq T_r \leq 30$ sec C = $2.3 - 0.06\ T_r$ without bilge keel
C = $2.0 - 0.05\ T_r$ with bilge keel

$T_r >$ sec C = 0.5 with and without bilge keel

X = longitudinal distance from L/2 to COG of cargo (m)
Y = transverse distance from CL to COG of cargo (m)
Z = vertical distance from BL to COG of cargo (m)
T = actual draft in m
LCF = distance to longitudinal centre of floatation from L/2 in m (positive forward, negative aft)
B = ship's breadth in m
L = ship's length in m
GM = actual metacentric height in metres
g = gravity constant 9.81 m/s^2

Index r = roll
Index p = pitch
Index l = longitudinal
Index t = transverse
Index v = vertical

In the accelerations calculated above, consideration is given to the static gravity force and the dynamic forces. For cargo exposed to wind, also a wind force has to be added. The wind force F is calculated from F = 0.125 × A (tons) where A is the projected exposed area of the cargo to be secured in (m^2). The cargo shall therefore be secured for the worst of the following two combinations of forces:

1) At maximum roll
Longitudinal force $P_l = a_{rl}\ M$ (tonnes)
Transverse force $P_t = a_{rt}\ M + F_t$ (tonnes)
Vertical force $P_v\ (1 - a_{rv})\ M$ (tonnes)

2) At maximum pitch
Longitudinal force $P_l = a_{pl}\ M + F_l$ (tonnes)
Transverse force $P_t = a_{pt} M$ (tonnes)
Vertical force $P_v = (1 - a_{pv})\ M$ (tonnes)

where:
—accelerations a_{rl}, a_{rt}, a_{rv}, a_{pl}, a_{pt} and a_{pv} are in (g)
—the mass of the cargo to be secured M in (tons)
—the wind forces F in (tons).□

References

(1) P. Andersson, B. Allenström, M. Niilekselä
Safe Stowage and Securing of Cargo on Board Ships
MariTerm AB 1982

(2) P. Andersson
Securing of Road Trailers on Board Ro-Ro Ships
MariTerm AB 1983

(3) P. Andersson, M. Niilekselä, A. Sjöbris
Road Trailers Suitable for Sea Transportation
MariTerm AB 1984

(4) *Securing Goods on Semi-trailers*
Swedish Transport Research Commission 1986

(5) P. Andersson, M. Koch, A. Sjöbris
Optimum Safety Factors for Securing of Cargo on Board Ships
MariTerm AB 1986

(6) B. O. Jansson
Safety of Ro-Ro Vessels—Casualty statistics
Ro-Ro 81 Proceedings

INSPECTING CONTAINERS

E. A. Woolley and L. Rae
Institute of International Container Lessors

Reasons for container inspection

AS CONTAINERS became the primary means for the transportation of manufactured goods, container surveying—or inspection—became an important field. Inspections were necessary to assure proper manufacture, safety to human beings and preservation of containers as assets of their owners.

The importance of container inspection is reflected in the increase in the number of containers and in containerised shipping. In 1973 there were approximately 1,000,000 20-ft equivalent units of containers (TEU). Fifteen years later, in 1988, there were probably 5.2 million TEU, a 500 per cent increase. In 1973, container movements through the world's 115 major container ports totalled 15,000,000 TEU. In 1986, such movements through 298 major container ports totalled 59,000,000 TEU(1), a 400 per cent increase (2). The number of inspectors had to keep up. In 1988 IICL's *Directory of Certified Container Inspectors* listed over 850 inspectors in 33 countries. There are perhaps another 2,000 uncertified inspectors around the world.

Inspections are necessary for the obvious reasons of safety and asset preservation cited above. In addition, certain specific developments have given a great deal of impetus to the field. These include the rise of container leasing in the 1960s and 1970s, the entry into force in 1979 of the International Convention for Safe Containers (CSC) and in 1975 of the Customs Convention on Containers, the publication of inspection and repair manuals and inspector certification. The first two of these developments created additional needs for inspection. The second two developments represented industry responses to these needs.

Increase in container inspection

A primary factor in the development of container inspection as a distinct discipline has probably been the development of the container leasing industry. Container leasing is one of those industries where the overriding criterion of safety to the public and in the workplace is reinforced and supported by commercial pressures. Under prevailing commercial practice, a leasing business cannot be operated without accurate inspections.

Accurate inspection

Firstly, like all other owners, leasing companies require accurate inspection at the time of purchase to warrant that their purchase specifications, manufacturers' guarantees and other contractual terms of the purchase agreements have been properly performed. While leasing company interests at the purchase stage correspond almost identically with those of ship line purchasers, leasing companies buy in much larger quantities, and these matters are of much greater significance to them since their only business is the supplying of containers.

Secondly, however, and more significantly, the leasing business is actually based in major part upon accurate inspection. When containers are purchased, the leasing company initiates the first lease by delivering the containers in mint new, first-class condition to the first customer. (At this time, as on each occasion when it accepts leased containers, the customer should confirm the condition of the containers by inspecting them.) Thereafter, during the lease, the customer subjects these mint new containers to normal wear and often to damage.

At the end of the lease, the leasing contract requires that the responsibilities of the parties for the condition of the containers be fixed as of the time that the containers are returned. The contract generally allocates to the leasing company responsibility for normal wear and to the ship line lessee responsibility for damage to the containers during the term of the lease. An inspector performs the essential elements of this allocation. He conducts an inspection to determine whether the formerly mint new container has acquired any blemishes during the term of the lease and, if so, which are normal wear and which are damage.

In order to conduct this inspection, the inspector uses an inspection guide. The condition of the container is noted, and whether defects are of the nature of damage or of normal wear. The cost of repair is calculated, and if it is substantial, sometimes the ship line will also inspect the container. The parties then agree on their respective shares of the cost of repair.

Whether or not the contract allocates financial responsibility in the usual manner, the leasing company must restore the containers to proper condition in order to make them marketable for the next ship line lessee. When the leasing company delivers the containers to that next lessee, the cycle is repeated. Somewhere between half a billion and a billion dollars of damage is estimated to be discovered and repaired in this manner annually.

Inspector's competence

The competence of the inspector plays a vital role for both leasing company and ship line in this process. The leasing company has two countervailing interests which can be balanced properly only by accurate inspections. For reasons of direct out-of-pocket cost and customer relations, the leasing company does not wish to repair too much. It does not want to repair normal wear needlessly, and ship line customers have no hesitation at complaining of being overcharged for repair of damage. Overcharging quite properly leads to looking for a new leasing company. The inspector must therefore avoid marking every minor blemish as damage.

On the other hand, the leasing company must repair the box properly to avoid safety risks, to return it to marketable condition for the next lessee and to preseve its life as an asset so that it can be re-leased many times. The inspector must therefore note every defect that really is damage.

The dependence of a leasing company upon accurate container surveying in order to maintain a marketable fleet and to keep itself in business is thus apparent. Too much repair will put it at a competitive disadvantage by increasing its costs of operation and causing its *per diem* rental rates to cover these costs to be too high; too little repair will result in rejection of its containers by customers, deterioration of the containers and early loss of income-producing assets. The inspector's accuracy thus directly affects the leasing company's profitability. These two countervailing interests tend to assure that the container is repaired properly—to maintain its life and to be marketable—but not too much—to avoid excessive costs.

International conventions

Container inspection was given a formal legal basis by the adoption of the International Convention for Safe Containers (CSC) and to a lesser extent by the Customs Convention on Containers. The CSC is intended to protect against danger to human safety, to the public and to the employee. While the container had increased safety enormously over the previous breakbulk methods of shipping, the threat of adoption of inconsistent safety regulation in different countries led to the ratification of the CSC by most countries with shipping fleets. The Customs Convention on Containers requires that a national inspection organisation certify that the container is secure for Customs purposes and that closing devices are as tamperproof as possible.

The CSC was signed and circulated for ratification beginning in 1972, and its first provisions entered into force in 1979. The CSC imposes essentially three obligations on an owner: first, to obtain manufacture of containers according to safe designs approved by approval authorities designated by government bodies; second, to examine containers for safety defects at least every thirty months; and third, to maintain containers in safe condition. Owner is defined as 'the owner as provided for under the national law of the contracting party or the lessee or bailee, if an agreement between the parties provides for the exercise of the owner's responsibility for maintenance and examination of the container by such lessee or bailee' (CSC Art. II (10)).

Enforcement schemes

Different countries enforce the CSC in different ways, but the function of the inspector is essential in all enforcement schemes. First, inspectors must give an assurance that containers are manufactured in accordance with the approved designs, a function also commercially desirable for the reasons referred to above. This is a procedure generally performed by the container divisions of the major ship classification societies, usually at the same time that the Customs certification is given.

Second, inspectors must perform examinations, whether of a 30-month periodic nature or more frequently under the Convention's ACEP procedures. The ACEP procedure permits a container owner to obtain approval of its frequent examination procedures and to place a permanent ACEP sticker upon its container certifying that it examines the containers under such procedures and at least every 30 months. The examination must 'determine whether a container has any defects which could place any person in danger' (CSC Annex I, Regs. 2(d), 3(c)). In some countries the inspector must certify that he has found no such defects. A proper CSC examination will also tend to reveal defects that would cause problems under the Customs Convention on Containers. Finally, inspectors are essential to the owner's programme for maintaining its containers in safe condition.

Industry manuals

In order to satisfy safety, operating, financial and legal requirements, the container industry has undertaken many formative steps in encouraging the development of a professional class of container inspectors. These efforts began with the publication of inspection and repair manuals and have culminated in the operation of the annual container inspector's examination. Since the inspector certification programme is based upon IICL's inspection and repair manuals, a word is appropriate to describe the nature of those manuals, first published in 1974 and now in third editions supplemented by technical bulletins.

IICL has issued a number of guides, manuals and other publications. The *Guide for Container Equipment Inspection* was first published in 1974 and is now in its third edition (1984). The *Guide* offers a set of criteria for inspectors to use in determining if damage to containers requires repair. Inspectors are to examine container components in a routine and complete procedure. Examinations should include rails, posts, panels, roofs, floors, understructure and other parts. The *Guide* is widely regarded in the container industry as the definitive work on the subject. In addition to tables of specific criteria, the *Guide* includes chapters on container design, types of damage, wear and non-conforming repair, and recommended inspection procedures.

IICL has also issued a number of technical bulletins revising inspection and repair criteria in its manuals. The last two technical bulletins, G1 of June 1986 and G2 of March 1988, came as a result of IICL's container testing programme and contained revisions to the third edition of the *Guide for Container Equipment Inspection*.

Technical publications

In addition, IICL has issued over a dozen other technical publications, including the *Repair Manual for Steel Freight Containers*, third edition (1984), a reference work for the repair of steel containers; the *Manual for GRP Freight Container Inspection and Repair* (1980), which pertains to containers of g + p reinforced plywood construction; *Specifications for Steel Container Refurbishing*, recommending instructions for surface preparation, coating and cleaning; the *Standard for 20 Foot Steel Freight Container Component Shapes and Profiles* (1985), providing designs for standardised repair parts; and *Specifications for Accelerated Laboratory Testing of Container Coating Systems* (1985), a test programme for evaluating the strength and durability of marine container coatings.

In 1982, IICL published a manual offering a code for describing damage to steel and grp dry cargo and open top containers and methods to repair such damage. This publication, *Guide for Container Damage and Repair Coding*, is considered the 'grandfather' of container codes. For inter-company communications, however, it is anticipated that the IICL code will be superseded by the proposed ISO code now in its final stage of preparation.

In 1988, IICL, the Steamship Operators Intermodal Committee (SOIC) and the International Container Repairers Committee (ICRC) published the *Guide for Container Depot Evaluation*, a workbook to be used in assessing the capability and performance of container repair depots. Inspection and repair of boxes being returned by ship lines to leasing companies are performed in depots, and the 'Depot Guide' should raise the standards of inspection in this industry. This publication marked the first time that the ship line, leasing and repair industries had collaborated on a publication.

As the sophistication of container inspection and repair continued to develop, voices were heard that the IICL inspection and repair manuals erred too much on the safe side and required too much repair. Certain ship lines formed the Container Lessees Commitee, which in 1987 published an inspection guide less stringent than the IICL recommendations. The dialogue between IICL and the CLC over this matter, which had begun much earlier, resulted in the adoption by IICL of a computerised container testing programme to ascertain the limits of damage which containers could accept and still operate in a safe and serviceable manner.

The testing programme was based upon finite element analysis, an advanced computer technique already used in the design of airplanes and automobiles. Today IICL's inspection guide and inspector certification programme are based upon the results of the computerised container testing programme, performed by HLA Engineers of Dallas, Texas, in collaboration with the University of Texas at Arlington.

Inspector certification

In 1982, IICL asked Educational Testing Services (ETS) of Princeton, New Jersey, to help develop a certification examination for container inspectors that would measure knowledge and skill in inspecting and repairing containers in accordance with IICL standards. Such a programme, IICL felt, would be helpful in encouraging professionalism and consistency in the container repair industry around the world.

Maintenance of a high standard of competency in the field of container inspection and repair, it was thought, would in turn contribute to greater safety and more efficient use and management of containers in general. ETS, as the leading American testing organisation with many years' experience in developing and administering national and international examinations, was highly qualified to develop a reliable certifying programme.

After consultation with shipping lines and other container industry representatives, IICL held an experimental test in 1983. When the results had been evaluated and modifications made, the IICL inspector's certification programme began on a regular basis with an examination held in October 1984. Five additional administrations of the test have followed. A total of over 850 container inspectors in 37 countries around the world have achieved passing scores and received certification. The examination is now given annually in at least 23 cities in 14 countries across the globe.

To make it possible for prospective candidates who live more than 250 miles away from the nearest test centre to take the examination, IICL offers such candidates the opportunity to establish special test centres in cities of their own choice. In the past two years special centres have been set up in Mombasa, Kenya; Bangkok, Thailand; Manila, the Philippines; Karachi, Pakistan; Jeddah, Saudi Arabia; and a number of other locations.

Examination contents

Originally based upon information contained in six technical manuals, the examination now tests knowledge of inspection and repair criteria in four manuals *Repair Manual for Steel Freight Containers*, 3rd edition, (*Guide for Container Equipment Inspection,* 3rd edition *Manual for GRP Freight Container Inspection and Repair*, and *Specifications for Steel Container Refurbishment*.)

The questions cover inspection, repair and refurbishment of steel containers, and variations in inspecting and repairing grp containers. The inspection criteria also apply generally to aluminium containers. The test consists of 100 multiple-choice questions, two and a half hours being allowed for completion. The test answer booklets are computer-scored by ETS in Princeton, NJ, USA. Those who receive passing scores are awarded certificates and photo-identification cards, and their names and addresses are listed in the *Directory of Certified Container Inspectors*. In addition, IICL grants an annual award, the Container Inspector's Award, to the inspector or inspectors who receive the highest score on the test.

The most noticeable change in the composition of candidates registering for the examination programme has been the growing number of candidates from the Pacific basin and Asia, including the US and South American west coasts. Whereas in October 1984, inspectors from this area of the world comprised 53 per cent of all registrants, the percentage has risen to 67 per cent of all candidates who took the test in 1986 and 1987.

Of the more than 850 inspectors who have passed the examination and whose names are listed in the IICL *Directory of Certified Container Inspectors*, 42 per cent come from the Far East and Asia (including Australia and New Zealand), 29 per cent from the USA and Canada, 21 per cent from Europe, 8 per cent from South and Central America, and less than

1 per cent from Africa and the Middle East. While some groups of non-English speakers do not score as well on the test as those who speak English as their native language, others actually do considerably better than their English-speaking counterparts.

Candidates for IICL certification come from a variety of different companies. The largest proportion, 49 per cent, work for container repair depots. The second largest group, the surveyors, or independent professional inspectors, comprise 26 per cent of the total pool of registrants. Leasing companies are represented by 16 per cent, followed by the shipping lines with 8 per cent of all examination registrants. Miscellaneous groups such as classification societies, railroads, and others make up the remaining 1 per cent of IICL test candidates.☐

References

1 *Containerisation International Yearbook 1976* at page 11; *Containerisation International Yearbook 1988* at page 7.

2 The number of inspections can be very roughly guessed by estimating that the 59,000,000 TEU of port movements in 1986 involved over 40,000,000 actual container units and that these were operationally inspected at at least one end of the trip or perhaps 20,000,000 times. More intensive inspections were conducted when containers were returned to leasing companies. If there were 1,800,000 actual units owned by leasing companies, and these were each returned 1.5 times in a year, 2,700,000 such inspections might take place in a year. Additional inspections were often performed on delivery to the next customer.

ENTRY TO AND EVACUATION FROM ENCLOSED SPACES

Captain F. G. M. Evans, BA, Cert Ed, Grad IFE, FNI

Senior Lecturer, College of Maritime Studies, Warsash

ANY ENCLOSED SPACE which has been kept closed may have an irrespirable atmosphere. An atmosphere may be irrespirable because it is:

1. Oxygen deficient.
2. Toxic, narcotic or intoxicant.
3. Injurious—i.e., corrosive, severely irritant, hot, or choking (dusts).

An atmosphere may be oxygen deficient due to:

- Oxidation of metals (rusting).
- Oxidation of oil residues in cargoes such as fishmeal.
- The presence of any cargo which absorbs oxygen.
- The drying of paint films.
- The use of de-oxidant chemicals—e.g. in boilers.
- Substances which give off vapours which may displace oxygen.
- The presence of an inert gas or gases which have been leaked or discharged from firefighting or refrigerating equipment.
- Biological activity—e.g., rotting.

An atmosphere may be toxic due to:

- Toxic vapours given off from cargoes.
- The rotting or fermentation of cargoes such as grain, edible oil, meat or fruit.
- The decomposition of residues from chemical cargoes previously carried.
- The presence of vapours from solvents such as those used in paints.
- Products of combustion.

Toxic hazards of dangerous cargoes should be made known to those likely to come into contact with them; such information should be readily available to the emergency response party, perhaps on a clipboard. But toxic gases may be evolved from the residues of previous and otherwise harmless cargoes. Rotting or fermentation may produce carbon monoxide or hydrogen sulphide and dangerous concentrations may not be released until the residues are disturbed.

Gases which are narcotic or an intoxicant, which may include hydrocarbons and solvents, may cause the victim to fall and injure himself. Injurious atmospheres may occur when, for example, water is added to a strong alkali or acid, when boiling may cause a corrosive mist and fumes to form.

The problem with many of the gases causing respiratory accidents is that they give no warning—no visible indication, no smell or in some cases the sense of smell may be anaesthetised at dangerous concentrations, and no taste.

The early symptons of oxygen deficiency and many types of poisonings are similar and may be confused with drunkenness. The victim himself may not be aware because his senses have been dulled. The person at the entrance must be trained to look out for these symptoms and raise the alarm at the earliest signs of unusual behaviour.

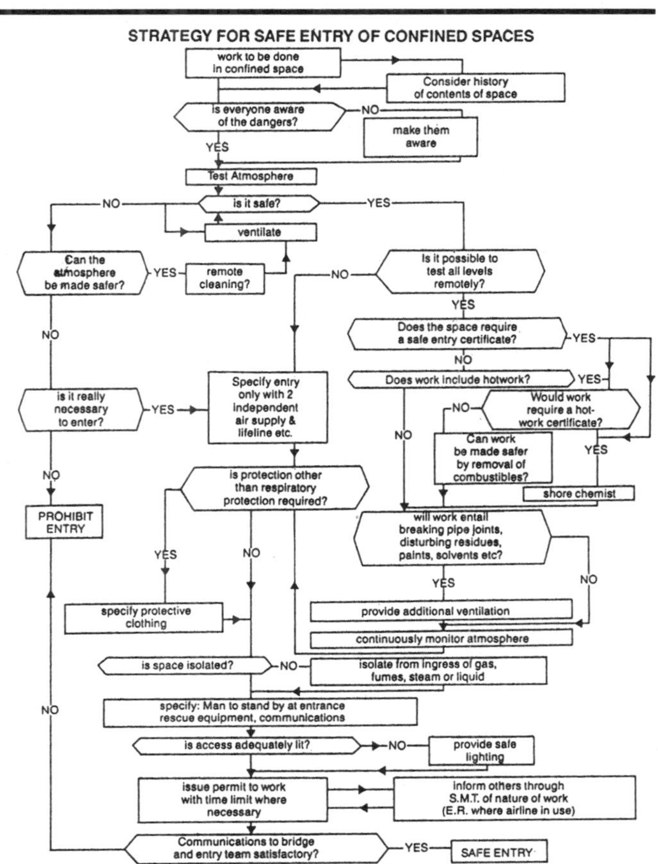

With oxygen deficiency there are no feelings of suffocation. Such feelings are caused by an excess of carbon dioxide, in the blood, operating sensors to tell the body to breathe out, in an oxygen deficient atmosphere this carbon dioxide is breathed out normally but not enough oxygen is breathed in. If the brain is not receiving enough oxygen it cannot operate properly, it speeds up the blood supply so that the victim may die of heart strain, or he may feel that, instead of climbing this ladder, when he lets go he will float . . .

Why TLVs?

The effect of poisoning may be chronic or acute. With acute poisoning, the effect is immediate or delayed only a few hours. Chronic poisoning may be either where a dose received now may cause chronic illness in later years, or regular small doses may be accumulative or have an accumulative effect which will cause chronic illness.

The threshold limit value (TLV) is the maximum concentration that can be worked in for an eight hour day. Sets of published figures are available from the Health and Safety Executive (see *Code of Safe Working Practices*). An STEL, short-term exposure limit, is the maximum duration that can be endured for 15 minutes with one hour's rest between.

An atmosphere may be dangerous by reason of being flammable, and in an enclosed space may form an explosive mixture with air.

This document is not meant to replace the various codes of safe practice (CSP) and should be read in conjunction with all the requirements of the codes which give information on the hazards of various cargoes, such as the IMDG code for packaged dangerous goods, the CSP for carriage of bulk dry goods and the international tanker safety guides. In particular reference should be made to chapter 10 of the *Code of Safe Working Practices for Merchant Seamen.* Couched in the officialese of nearly every paragraph of chapter 10 of the *CSWP* is a sad case history of death or injury involving needless waste of lives and skilled personnel and distress for bereaved relatives and friends.

Entry into dangerous spaces regulations

Every year a number of merchant seamen lose their lives through enclosed space incidents. This has prompted the British legislators to produce a Statutory Instrument (SI) and a revision of chapter 10 of the *Code of Safe Practice for Merchant Seamen.* It puts responsibility on people, including the master or person in charge of the vessel, lays down fines for summary conviction for a contravention.

A summary of the main points are:

- Entrances to unattended dangerous spaces to be kept closed or secured.
- There must be laid down safe entry procedures.
- The master must ensure safe entry procedures are followed.
- The principles and guidance contained in the *Code of Safe Working Practices* Ch. 10 must be followed.
- An offence will have been incurred by any person entering an enclosed space not following the procedures or without authority.
- Drills to be carried out for rescue from enclosed spaces, at periods not exceeding two months on tankers over 500 tonnes or other ships over 2,000 tonnes (entry in official log).
- Oxygen meters and other appropriate testing devices to be carried on vessels where entry into dangerous spaces may be required.
- Defences include taking all reasonable precautions and diligence to avoid the offence or that it was not reasonably practible to comply.

Prevention of enclosed space incidents

Rescue from an enclosed space is obviously very difficult and enclosed space incidents are best prevented. An enclosed or confined space should be the subject of a permit to work system as outlined in the *CSWP*. There was an incident in which an engineer entered an open-topped tank, and a cleaning cycle was put into operation which opened steam drains to this space. Obviously such a space should be the subject of a permit to work, which would include the posting of a notice on the controls of the cleaning cycle.

Men doing hotwork in an enclosed space where the only exit is upwards should avoid clothing of man-made fibre. Cotton or wool is preferable and will protect from flash fires but man-made fibres such as nylon may melt on the skin and cause third-degree burns.

Any space which has been closed up or is poorly ventilated must be ventilated and have its atmosphere tested prior to entry. Thought should be given to providing additional ventilation in the form of portable fans and plastic air ducting when work is being done in an enclosed space.

Instruments for measuring flammable atmospheres should be calibrated before use; remember they are only calibrated for a particular gas. Oxygen meters will indicate a respirable atmosphere when there are toxic gases present which are toxic at low concentrations. Other gas detectors will only detect a specific gas so the hazard has to be known before it can be sampled for.

The hazards associated with a particular cargo must be well known by all responsible persons. When persons are affected consult the *Ship Captain's Medical Guide (Chemicals Supplement)* or the *IMO Medical First Aid Guide*.

Emergency procedures

Unfortunately, there is not only the danger of making an entry into enclosed spaces, but case history has shown that when an incident occurs, inept and unpractised emergency response leads to multiple casualties. It is up to the master to ensure that, through drills and positive instruction, if there is an incident his crew will do the right thing and will be acting as an efficient emergency team because they will be doing what they have been 'conditioned' to do by training.

A few realistic drills involving the rescue of a live-weight dummy from an enclosed space, involving movement through lightening holes and up ladders, will soon convince people how difficult this is and make them follow proper entry procedures to ensure that such an incident could not happen.

Drills are needed to ensure that a proper entry procedure is followed even in an emergency. Many accidents which have occurred in enclosed spaces have been falls due to inadequate lighting, slippery surfaces, awkward access, etc., and when a man falls in a tank the instinctive thing may be to go and have a look at him; such an incident could be reported as an injury which may set the emergency response off on the wrong foot if the fall was caused by a foul atmosphere.

In the event of an incident in the enclosed space the person standing by should raise the alarm and must not make an entry until other persons arrive, and never make an entry without breathing apparatus. There should be one emergency signal which takes people to their emergency stations, whether that emergency is following collision, stranding, fire or an enclosed space incident.

Command communication and control

The emergency response station for the master is the bridge or other pre-arranged control centre. As a general principle the master should stay in the control centre; the bridge or control centre should only be left if:

1. A responsible person can be left at the control centre.
2. He can be sure that all the crew have been accounted for—there was an incident in which two persons were

rescued, and it was later discovered that a third person had made an unauthorised entry to the other end of the same space.

3. He has ascertained that the enclosed space incident is the only emergency occurring on board at that time.
4. Perhaps if he is receiving no communication.
5. There is no other responsible person available to take charge at the entrance.

The emergency response party is a small group of men whose muster point is at a place where the equipment necessary to deal with emergencies is stored. They are communicating with the bridge, mustering and getting their gear ready all at the same time.

There should be one responsible person in charge of the incident at the entrance with communications to the control centre (the bridge). If subsequent rescuers enter a space they come under the charge of this officer, even if they have come from a different party. There must be proper breathing apparatus (BA) control, with all entries logged. Backup relief rescuers will have to be made ready and enter to take over the rescue at about the time the first rescuers have to leave the space.

Priorities and training objectives

When there is an enclosed space incident, the first priority is to get air to the casualties. Second is first aid; removal of a casualty who has suffered a fall may be delayed to avoid compounding the injuries, but only if the atmosphere has been made safe or there is a limitless supply of air for victims and rescuers.

On-board training must ensure that, in the event of an enclosed space incident, nothing is done that is likely to produce more casualties. The multiple casualty situation gets more difficult and may go beyond the capability of the ship's resources to deal with it. Which is why the *CSWP* suggests that only the minimum number of people necessary to do the job are sent in to the space in the first place.

Without proper training, the instinctive thing to do is anything to save your shipmate. Lives have been lost:

- By removing a facemask to share the air with the casualty. Never remove a BA facemask in a suspect atmosphere.
- By would-be rescuers holding their breath and making a quick dash in to pull someone out or free a rescue line. Never enter a suspect atmosphere without a breathing apparatus.
- By rescuers staying to make that last effort and running out of air.
  Leave with enough air supply to gain fresh air, remembering that it takes more air to climb a ladder than go down it.
- By rescuers hurriedly donning BA and not going through the donning procedures and safety checks, or forgetting to switch to positive pressure.
  Every BA wearer, even in an emergency, must go through his safety checks and be checked by the control before entering a space. To do this quickly takes practice. A man who has stopped breathing needs to have his lungs ventilated within less than three to four minutes!

The above lays down the objectives of organisation and training of the crew for dealing with enclosed space incidents, plus the fact that rescuers need practise to be able to gain access to awkward spaces with a breathing apparatus on and to be able to move a dead-weight body through lightening holes or up ladder-ways. An effective training aid can be made from a plywood board with a hole in it of the same dimensions as a lightening hole. It can be used to develop the techniques for getting through with a BA set on.

The emergency response should see to it that air is got to the victim's lungs as quickly as possible. The rescuer must not enter the space without a breathing apparatus, but equally he should not enter the space without an air supply for the victim or victims. Also for a vertical rescue a lifeline needs to be taken in to haul out an unconscious person.

A STRATEGY FOR RESCUE FROM CONFINED SPACES

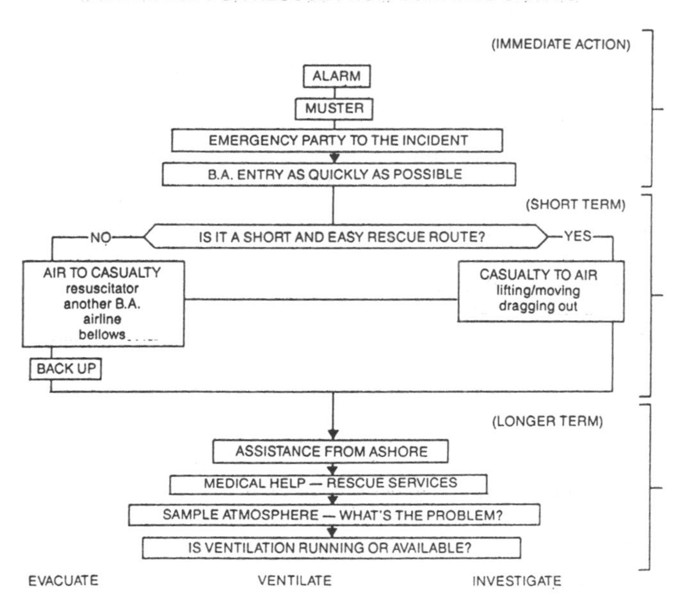

Procedures for rescue

Rescue equipment should be taken to the entrance when an enclosed space entry is to be made. The emergency response plan should be realistic and consider every eventuality in terms of what really may happen in an emergency. If a lot of equipment is taken up the foredeck, what happens if there is another emergency on board such as a fire and the equipment is not at the muster point? Certainly the one man at the entrance cannot carry it all back. There may be enough equipment on board for an attack to be made on a fire without needing the additional equipment except for backup. When giving instructions about taking equipment to the entry point, think where it should be taken from. Airline equipment is useful for rescue in that there is no limit on the duration of the air supply. As it is not likely to be needed for fire-fighting, and is heavy and difficult to move quickly, it would be a good idea to have that standing by.

But, it takes more time to bring into operation than a personal BA set so perhaps the initial rescue attempt should be made by BA wearers and the rescue taken over by men in airline equipment. Don't forget that the airline equipment should be plugged into a BA set and that the cylinder of the BA set should then be turned off so that the full contents of the BA cylinder remain if the airline fails. NB: If an airline is given to an unconscious person watch the pressure carefully.

The man standing-by at the entrance may don BA while he is waiting for someone else to arrive; he can then enter as soon as someone arrives and he has briefed them. Persons who arrive may be out of breath. The ship's management team may arrange to have others working nearby on radio call to assist. The *CSWP* simply says that a rescue should be based on a pre-arranged plan and that every ship will have its own individual problems, each of which may have a different rescue procedure.

It takes thought along the lines of the above to decide what are the best procedures in relation to: manpower availability, equipment availability, size of space involved, and ease or difficulty of access. When the procedure is decided upon, it should be practised as realistically as possible, making sure the space has been made safe. Afterwards the exercise should be debriefed and the procedures amended according to the lessons learned.

Some other points to be considered when laying down rescue procedures are:

- Rescue will be made easier if you have insisted that everyone entering a dangerous space was wearing a harness.
- Two men can handle an unconscious person better than one.
- Two men will be needed to move an unconscious person through a lightening hole.
- Going up a ladder, the men at the top do the lifting, the men below guide.
- Men in breathing apparatus may be needed to lift at a half-way stage.
- If there are two men in a space, air will have to be taken in for two.
- If there is no equipment available, concentrate on getting the men out as quickly as possible.
- Air shared is air halved; a device for sharing air from a self-contained breathing apparatus set must only be used for very short rescue routes.
- If bellows apparatus is being used, make sure the bellows is in fresh air, it might not be apparent in drills but there may be an irrespirable atmosphere outside a space, close to the entrance in a real incident.

Equipment

Members of the crew must be thoroughly familiar with all items of equipment. The best way of making sure that junior officers are familiar with equipment is: (a) arrange for them to give positive instruction to others, (b) see that they follow a planned maintenance schedule, and (c) give them opportunities to use the equipment in drills.

- It may be necessary to provide a portable sheerlegs and tackle for rescue from tanks on large vessels.
- Some companies are providing guide lines, as used by fire brigades, to get relief rescuers quickly to the scene of the rescue (tape could be used), but remember lifelines should be worn where appropriate.
- Ordinary breathing apparatus cannot be used as a resuscitation apparatus, but the victim may not have stopped breathing; if he has then the old 'Olga Neilsen' type resuscitation may be used to make him demand air.
- There are various types of resuscitation apparatus, some of which are not suitable for use in an irrespirable atmospheric as they use atmospheric air or oxygen enriched air. Some are switchable.
- If a resuscitator has a device which allows a recovering patient to breath in atmosphere air when the machine is in an exhalation cycle then it should be fitted with a non-return valve when used in an enclosed space. (Sometimes supplied as an extra, on a ship it should be left in place.)
- There is a danger in taking oxygen into a flammable atmosphere, where this is likely to happen, it is better to put air in a resuscitator, medical oxygen will be available after rescue if necessary.
- Where a facemask is strapped in place ideally it should have a quick release mechanism as a patient who recovers his breathing will probably vomit.
- Escape apparatus of the type worn by persons entering, which would give enough air to enable escape to fresh air in the event of the atmosphere becoming irrespirable, offer some security, but the persons may be affected and may lose the will or ability to survive before they have time to don them.

Body-handling techniques

All the body-handling techniques described in the *Ship Captain's Medical Guide* should be practised, plus any other technique you may feel necessary to surmount problems of awkward access on your particular ship.

The way to get a casualty through a lightening hole is face down (don't break his back); one rescuer goes the other side of the hole after the casualty is in position—arms through first, then the body, turn him slightly when necessary.

What is a confined space?

When gases evolved are heavier than air, an irrespirable atmosphere may exist in an open hatchway. In one incident involving tapioca root, which respires to give off CO_2, two men entered the hold to attach the crane wires to a grab which was sitting in a shallow depression in the surface of the cargo. They collapsed; two more went down to see what was wrong with them. Respiratory accidents have also occurred on the open deck where toxic fumes are coming from an opening to a dangerous space.

Proper entry procedures and pre-arranged emergency response is the only answer to the continuing loss of life from enclosed space incidents—plus the fact that every opportunity should be taken to make crews more aware, so that accidents do not happen due to ignorance.□

PART IV
MEASUREMENT AND CONTROL

AN INTRODUCTION TO THE METHODS OF CALCULATING A SHIP'S VOLUME AND DISPLACEMENT

Dr B. Baxter, RCNC
Professor of Naval Architecture, Strathclyde University

DISPLACEMENT AND TONNAGE are among the terms most widely used when considering ships and shipping and although there are different types of displacements and tonnages they are all related basically to the total volume or partial volume of the ship.

Displacement expressed as a volume (∇). This is the volume of the hole in the water occupied by the ship measured in cubic metres. There is no density correction.

Displacement as a weight (Δ)—This is the weight of water displaced by the ship and equals the volume displaced multiplied by a constant representing the density of water—i.e.,

In fresh water $\Delta = \nabla \times 1{,}000\ kg/m^3$
In sea water $\Delta = \nabla \times 1{,}025\ kg/m^3$

The displacement weight of a ship can vary according to circumstances and position in the world, although displacement weight and ship weight are equal when the ship is at rest in equilibrium in still water.

Displacement as a mass—This equals the quantity of water displaced and as the kilogram is the unit of mass and 1,000 kg = 1 tonne, the unit which is used when referring to the size of a ship is the tonne.

Displacement moulded—This is the mass of water which would be displaced by the moulded lines of the ship when floating at the designed load water-line.

Displacement extreme—This equals the moulded displacement, plus the displacement of the shell plating, bossings, cruiser stern and all other appendages.

Lightship displacement—This equals the extreme displacement of the ship when fully equipped and ready to proceed to sea, but with no crew, passengers, stores, fuel, water, or cargo on board. The boiler or boilers, if fitted, are filled with water to their working level.

Nautical surveyors of whatever discipline are constantly required to find the area, the moment of area and the second moment of area of figures bounded by straight lines and curves. In the case of ships, the curves will generally be fair and smooth. If the ordinates defining a particular curve represent areas, then the area under the curve will be a volume and if required the centre of gravity of the volume can readily be found.

If the equation of the boundary curve were known then the area under the curve could be found quickly and accurately using the appropriate mathematical formula for integration. In the majority of cases, however, the equations are not known and it is generally assumed that the curved boundaries consist of a series of parabolic curves. This enables approximate arithmetical, rather than mathematical, rules to be applied to calculate the areas and moments of areas about a particular axis. The values so obtained are normally sufficiently accurate for the great majority of practical purposes. Any errors which may remain after taking appropriate steps are of the order of less than $\frac{1}{4}$ per cent of the area which would be obtained using mathematical processes if this were possible.

Lines plan

Since this chapter is primarily concerned with the calculation of the volumes of ships, it is perhaps appropriate that consideration should be given as to the manner in which the form of a ship is normally defined.

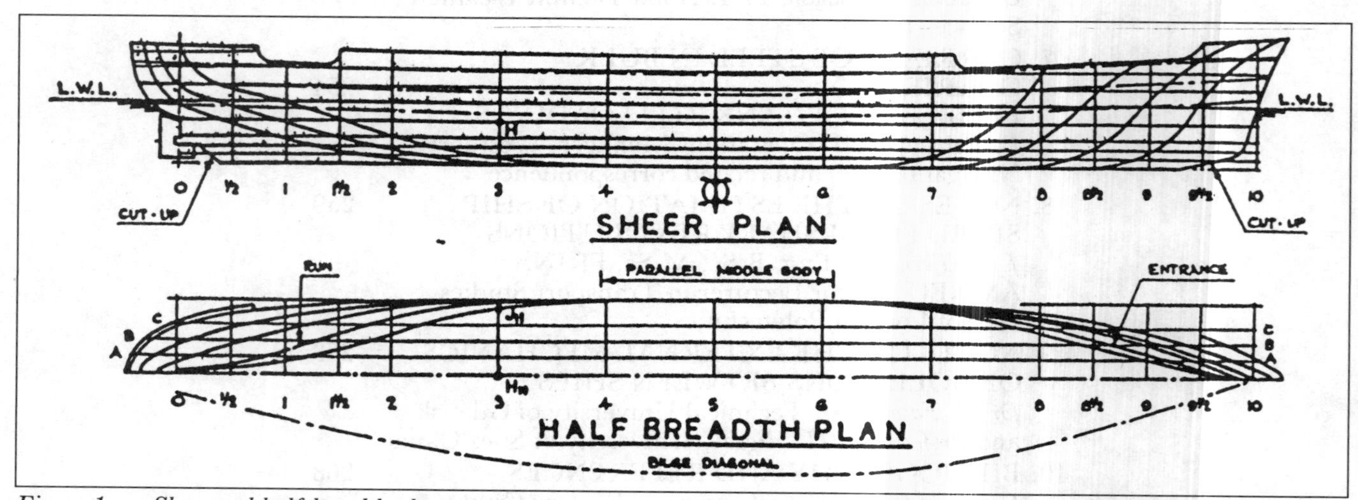

Figure 1: Sheer and half-breadth plans.

The form is usually shown on a scale drawing employing three planes of reference. The complete drawing is referred to as the lines plan, and the three views shown on it are:

(i) The sheer plan or profile, which gives the general outline of the ship, the position and sheer of the decks, and the position of the designed load waterline.

(ii) The half-breadth plan shows the shape of the decks and the waterlines which are formed by the intersection of the surface of the ship with horizontal planes.

(iii) The body plan gives the shape of lines formed by the intersection of the ship's surface with transverse vertical planes. The forward sections of the ship are drawn on the right-hand side, and the after sections on the left-hand side.

The lines plan represents the moulded surface of the ship—i.e., the inside of the inner strakes of plating for merchant ships—but the Admiralty lines plan represents the inside of the inner strakes plus the mean thickness of plating. For wooden ships the lines plan represents the outside surface of the planking. In the sheer plan the accepted convention is to draw the ship with the bow pointing to the right. Usually also, the sheer plan is drawn showing the ship as it would float in still water in its designed condition. The body plan is usually drawn to the left of the sheer plan and on the same moulded base line.

Displacement stations

For merchant ships it is usual to divide the length between perpendiculars (L_{pp}) into ten equal parts, and to number these from aft, with the after perpendicular (AP) as 0 and the forward perpendicular (FP) as 10. At the ends of the ship where the change of form is greatest, these divisions are often subdivided. At each of these divisions or stations a vertical line is erected in the sheer plan. Through each station a transverse plane is passed which intersects the moulded surface of the ship in a curve called a transverse section. These sections are shown in their true form in the body plan.

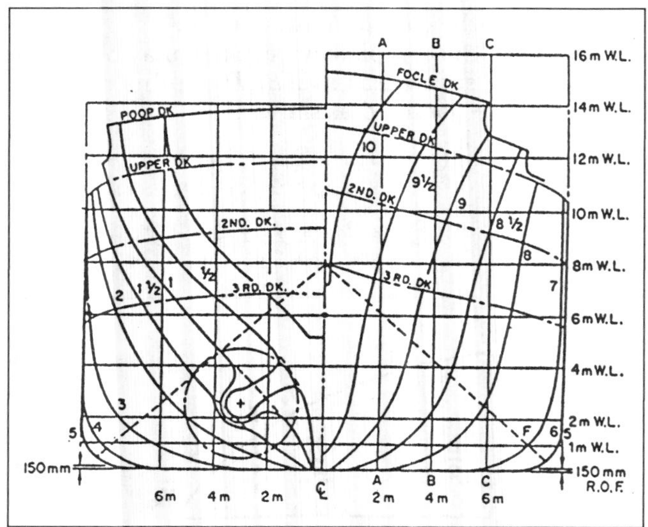

Figure 2: Body plan for a twin-screw passenger ship, 125m L_{pp} × 17.5m B × 10.5 m—to upper deck.

Waterlines—These are shown in their true form in the half-breadth plan, and are drawn parallel to the LWL at intervals above and below it.

Fairing—The finished lines plan as produced in the design office must be fair, i.e., all the curved lines must run evenly and smoothly, and there must be exact agreement between corresponding dimensions of the same point shown in the different views. The process of fairing consists basically of removing all unfairness from each line in turn, and then making sure that each line so faired fits in with lines already dealt with.

After the lines plan has been completed and faired in the ship design office, a table of offsets is prepared and sent to the mould loft. Offsets are the measurements made from the faired lines plan giving the distances of the points through which the curved lines must be drawn. For example, a table may be made from the centre line at each waterline for each section shown in the body plan. The lines of the ship used to be drawn full-size on the mould loft floor, but now are drawn to a scale of one-tenth full-size on a fairing table, and a final fairing is made. This is similar to the design procedure except that all the dimensions are now larger, and thus minor discrepancies may be eliminated. This tenth-scale drawing and fairing process is known as laying-off.

A corrected table of offsets is usually sent back to the design office and these may be offsets at displacement stations, frame stations or a combination of transverse bulkheads and frame stations.

Bonjean curves

The area of a transverse section of a ship to successive waterlines may be calculated and plotted in the form of a fair curve known as a Bonjean curve (named after the Frenchman who first used them.) The curves are often plotted on a profile of the ship as shown, and this enables the volume of displacement

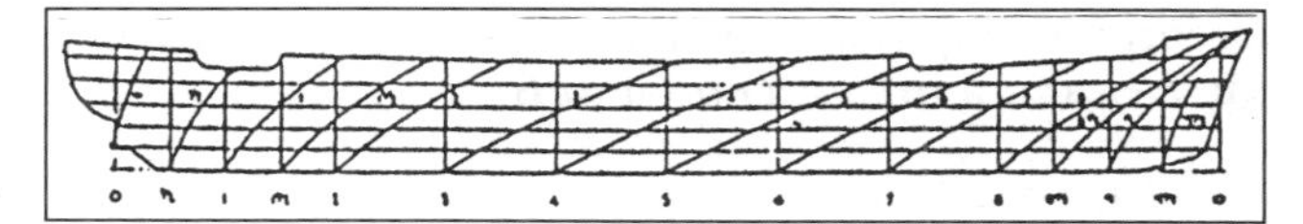

Figure 3: Bonjean curves.

and centre of buoyancy to be calculated to any waterline, trimmed or even keel. The curves are particularly useful for stability, strength, capacity and launching calculations.

The curves may also be plotted using a vertical line as a common base. This has the advantage that larger scales may be used for the areas of the sections, but if the ship is trimmed it is necessary to obtain the draught at each of the displacement stations before being able to read off the values of the areas.

Coefficients of form

Form is used as a general term to describe the shape of a ship's hull as defined by the lines plan. To compare one ship with another a number of coefficients are used. These may be obtained by calculation and are of value when making power, stability, strength, tonnage and other design calculations.

Block coefficient (C_B)—This is a measure of the fullness of the form of the ship and is the ratio of the volume of displacement to a given waterline, and the volume of the circumscribing solid of constant rectangular cross-section having the same length (L), beam (B) and draft as the ship.

i.e., $C_B = \dfrac{\nabla}{L \times B \times T}$

The L_{PP} is normally used in calculating the value of C_B, which varies with the type of ship.

Very fast ships (liners, destroyers)	0.50—0.65 (fine form).
Ordinary cargo ships	0.65—0.75 (moderate form).
Slow cargo ships	0.75—0.85 (full form).

Prismatic coefficient (C_P)—This is the ratio of the volume of displacement of the ship to the volume of the circumscribing solid having a constant section equal to the immersed midship section area A_M and a length equal to the L_{PP}

i.e., $C_P = \dfrac{\nabla}{A_M \times L}$

The C_P is a measure of the longitudinal distribution of displacement of the ship, and its value ranges from about 0.55 for fine ships to 0.85 for full ships.

Midship section area coefficient (C_M)—This is the ratio of the immersed area of the midship section to the area of the circumscribing rectangle having a breadth equal to the breadth of the ship and a depth equal to the draft.

i.e., $C_M = \dfrac{A_M}{B \times T}$

C_M values range from about 0.85 for fast ships to 0.99 for slow ships.

Waterplane area coefficient (C_{WP})—This is the ratio of the area of the waterplane to the area of the circumscribing rectangle having a length equal to the L_{PP} and a breadth equal to B.

i.e., $C_{WP} = \dfrac{A_W}{L \times B}$

The range of values is from about 0.70 for a fine ship to 0.90 for a full ship.

Tonnes per centimetre (TPC)—This is the mass which must be added to, or deducted from, a ship in order to change its mean draft by 1 cm. If the ship changes its mean draft by 1 cm and if A_Wm^2 is the area of the waterplane at which it is floating, then:

Change of volume	$= A_W \times 0.01\text{m}^3$
Change of displacement	$= A_W \times 0.01 \times 1.025$ tonnes in salt water
∴ TPC	$= A_W \times 0.01025$
	$= \dfrac{A_W}{97.5}$
or TPC	$= A_W \times 0.01$ tonnes fresh water
	$= \dfrac{A_W}{100}$

Centre of flotation (F)—This is the centre of gravity of the area, or centroid, of the waterplane of a ship. For small angles of trim consecutive waterlines pass through F.

Centre of buoyancy(B)—This is the centroid of the under-water form of a ship, and is the point through which the total force of buoyancy may be assumed to act. Its position is defined by:

(a) $\overline{KB}$ the vertical distance above the base.
(b) $\overline{FB}$ the longitudinal distance from the foward perpendicular.
(c) LCB the longitudinal distance from amidships.

Centre of gravity (G)—This is the point through which the total weight of the ship may be assumed to act. It also is defined by:

(a) $\overline{KG}$ the vertical distance above the base.
(b) $\overline{FG}$ the longitudinal distance from the forward perpendicular.
(c) LCG the longitudinal distance from amidships.

Calculation of areas

There are various rules in common use for finding the areas under curves. The best known and most widely used are named after Simpson, the mathematician who first published them in 1743.

Simpson's First Rule—This may be used to find the area under a curve defined by an odd number of equally spaced ordinates. In Figure AB is a portion of some curve such as a waterplane:

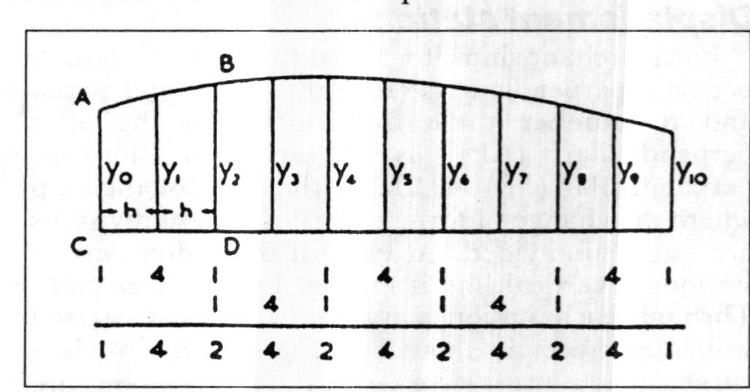

Figure 4.

Area $ABCD = \dfrac{h}{3}(y_0 + 4y_1 + y_2)$

where h = the common interval.

The area of the complete waterplane may be found by adding all such areas as $ABCD$. In this case the multipliers are obtained as shown:

$$\text{Total area} = \frac{h}{3}(y_0 + 4y_1 + 2y_2 + 4y_3 + 2y_4 + 4y_5 + 2y_6 + 4y_7 + 2y_8 + 4y_9 + y_{10})$$
$$= \frac{2h}{3}(\tfrac{1}{2}y_0 + 2y_1 + y_2 + 2y_3 + y_4 + 2y_5 + y_6 + 2y_7 + y_8 + 2y_9 + \tfrac{1}{2}y_{10})$$

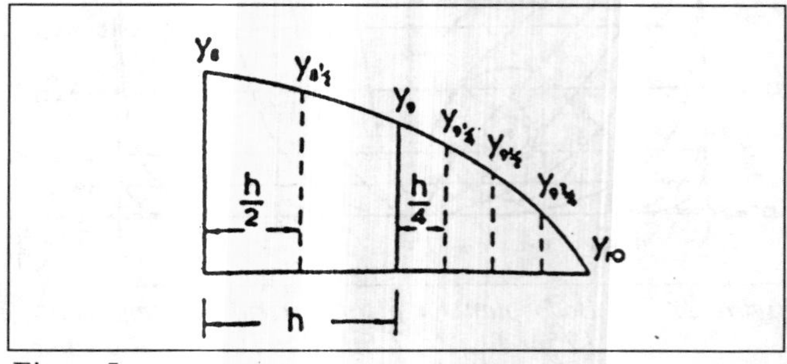

Figure 5.

Greater accuracy will be obtained when calculating areas if more ordinates are used where the curvature is greatest. At the two ends of a waterplane, for example, the common interval may be subdivided as shown in Figure 5.

$$\text{Area} = \frac{h}{2} \times \frac{1}{3}(y_8 + 4y_{8\frac{1}{2}} + y_9) + \frac{h}{4} \times \frac{1}{3}(y_9 + 4y_{9\frac{1}{4}} + 2y_{9\frac{1}{2}} + 4y_{9\frac{3}{4}} + y_{10})$$

In order to maintain the same value for h throughout the calculation, this may be written as:

$$\text{Area} = \frac{h}{3}(\tfrac{1}{2}y_8 + 2y_{8\frac{1}{2}} + \tfrac{1}{2}y_9 + \tfrac{1}{4}y_9 + y_{9\frac{1}{4}} + \tfrac{1}{2}y_{9\frac{1}{2}} + y_{9\frac{3}{4}} + \tfrac{1}{4}y_{10})$$

$$= \frac{h}{3}(\tfrac{1}{2}y_8 + 2y_{8\frac{1}{2}} + \tfrac{3}{4}y_9 + y_{9\frac{1}{4}} + \tfrac{1}{2}y_{9\frac{1}{2}} + y_{9\frac{3}{4}} + \tfrac{1}{4}y_{10})$$

Simpson's Second Rule—This may be used to find the area under a curve defined by a number of equally spaced ordinates equal to $(3n + 1)$, where $n = 1, 2, 3, 4$, etc.

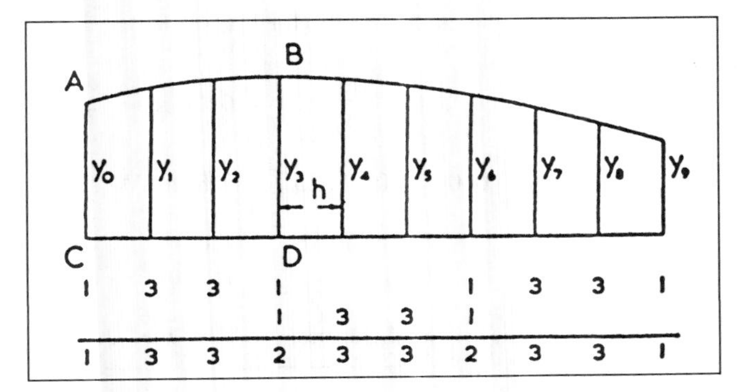

Figure 6.

In Figure 6 the curve AB is defined by 4 ordinates y_0, y_1, y_2 and y_3.

$$\text{Area } ABCD = \frac{3h}{8}(y_0 + 3y_1 + 3y_2 + y_3)$$

If seven, ten or more ordinates are used, then the continuous multipliers are obtained as before:

$$\text{Total area} = \frac{3h}{8}(y_0 + 3y_1 + 3y_2 + 2y_3 + 3y_4 + 3y_5 + 2y_6 + 3y_7 + 3y_8 + y_9)$$

The ends of a curve such as a waterplane may be subdivided as shown in Fig. 7

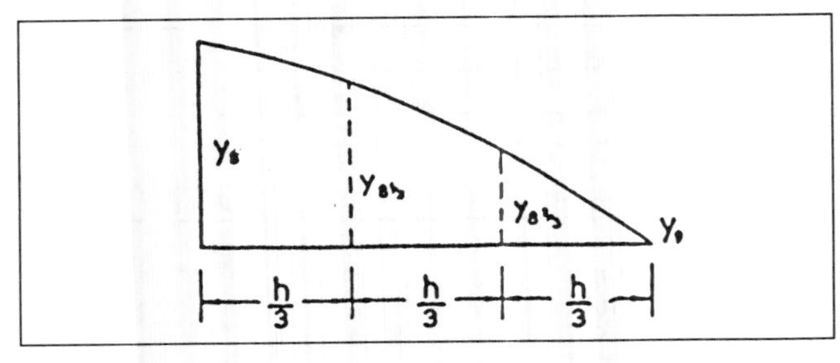
Figure 7.

$$\text{Area} = \frac{h}{3} \times \frac{3}{8}(y_8 + 3y_{8\frac{1}{3}} + 3y_{8\frac{2}{3}} + y_9)$$

$$= \frac{3h}{8}(\tfrac{1}{3}y_8 + y_{8\frac{1}{3}} + y_{8\frac{2}{3}} + \tfrac{1}{3}y_9)$$

5, 8—1 Rule—This may be used to find the area between two consecutive ordinates of a curve defined by three equally spaced ordinates.

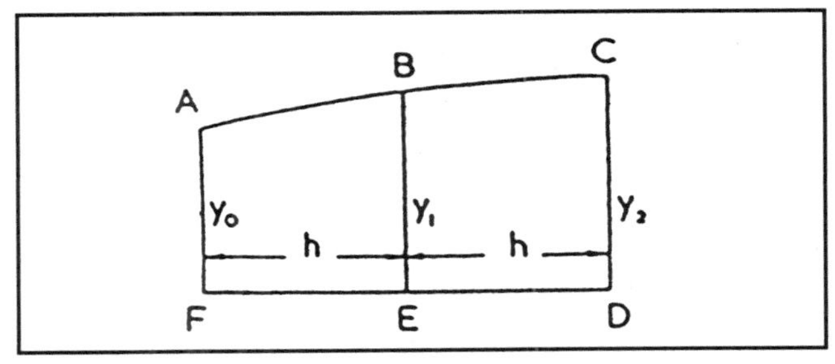

Figure 8.

$$\text{Area } ABEF = \frac{h}{12}(5y_0 + 8y_1 - y_2)$$

$$\text{Area } BCDE = \frac{h}{12}(5y_2 + 8y_1 - y_0)$$

$$\text{Area } ACDF = \text{Area } ABEF + \text{Area } BCDE$$

$$= \frac{h}{12}(5y_0 + 8y_1 - y_2 + 5y_2 + 8y_1 - y_0)$$

$$= \frac{h}{12}(4y_0 + 16y_1 + 4y_2)$$

$$= \frac{h}{3}(y_0 + 4y_1 + y_2)$$

(Simpson's First Rule)

Trapezoidal Rule—This may be used to find the area under a curve defined by any number of equally spaced ordinates. It assumes that the areas between successive ordinates are trapezoids, and the accuracy will therefore increase as the spacing of the ordinates decreases.

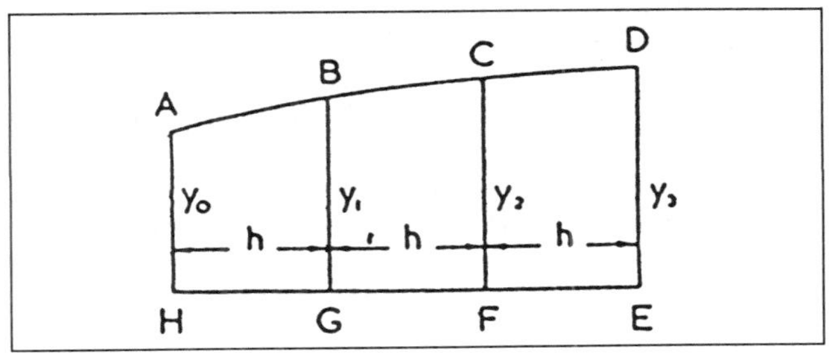

Figure 9.

$$\text{Area } ABGH = h\left\{\frac{y_0 + y_1}{2}\right\}$$

$$\text{Area } BCFG = h\left\{\frac{y_1 + y_2}{2}\right\}$$

$$\text{Area } CDEF = h\left\{\frac{y_2 + y_3}{2}\right\}$$

$$\text{Total area } ADEH = h(\tfrac{1}{2}y_0 + \tfrac{1}{2}y_1 + \tfrac{1}{2}y_1 + \tfrac{1}{2}y_2 + \tfrac{1}{2}y_3)$$

$$= h(\tfrac{1}{2}y_0 + y_1 + y_2 + \tfrac{1}{2}y_3)$$

Tchebycheff's Rules—All the rules so far considered have used ordinates at equally spaced intervals, and these have been multiplied by different multipliers. Tchebycheff devised a method which was published in France in 1874, whereby the area under a curve could be obtained simply by adding together suitably spaced ordinates. The area is given by:

$$A = \frac{\text{Sum of all ordinates} \times L}{\text{Number of ordinates}}$$

For example, in Figure 10 the curve is defined by two ordinates only and:

$$A = \frac{(y_1 + y_2)L}{2}$$

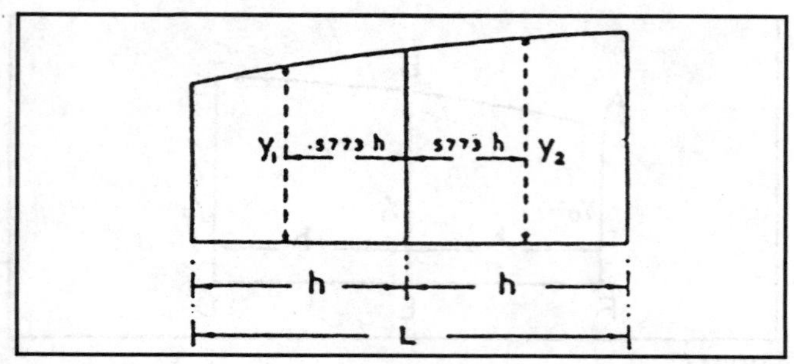

Figure 10.

Table 1 gives the position for a different number of ordinates equally spaced about the middle of the length. The fractions in the table represent fractions of the half length.

Number of ordinates n	Amid-ships	1	2	3	4	5
2		0.5773				
3	0	0.7071				
4		0.1876	0.7947			
5	0	0.3745	0.8325			
6		0.2666	0.4225	0.8662		
7	0	0.3239	0.5297	0.8839		
8		0.1026	0.4062	0.5938	0.8974	
9	0	0.1679	0.5288	0.6010	0.9116	
10		0.0838	0.3127	0.5000	0.6873	0.9162

Table 1

Simpson's Rules, particularly the First, are most commonly used in Britain, but the trapezoidal rule is used to a great extent in the USA and on the Continent. Tchebycheff's Rules are used for particular calculations in order to save long arithmetical work. The areas and moments obtained by using Simpson's Rules for normal ship work agree to within about $\frac{1}{4}$ per cent of the actual value, and this degree of error is usually acceptable. The error using the trapezoidal rule is slightly greater.

Associated with the rules for finding areas are rules for finding moments of areas, and thus the position of the centroid, using:

Distance of the centroid of an area from a given axis $= \dfrac{\text{Moment of area about the axis}}{\text{Area}}$

Moment of area using Simpson's First Rule—This is obtained in the longitudinal direction by multiplying each of the ordinates by its distance from the axis and then applying Simpson's Rules. If, as is usual, the position of the centroid is required, the calculation may be set out as follows. For simplicity in arithmetic, the distance between successive ordinates has been taken as unity, i.e., $h = 1$, and the actual value for h may then be substituted at the end of the calculation.

Example 1—The half breadths, in metres, of the load waterplane of a ship 240 metres in length, numbered from aft, are as follows:

Stations	0	$\frac{1}{2}$	1	2	3	4	5	6	7	$7\frac{1}{2}$	8
$\frac{1}{2}$-breadths	0	9.2	12.2	15.8	16.0	16.0	15.9	13.9	9.8	6.0	0

Find the position of the centre of flotation from amidships, and the value of the TPC in SW.

Station	Half-breadth	SM	Products for areas A	Levers	Products for moments M
0 (*AP*)	0	$\frac{1}{2}$	0	4	0
$\frac{1}{2}$	9.2	2	18.4	$3\frac{1}{2}$	64.4
1	12.2	$1\frac{1}{2}$	18.3	3	54.9
2	15.8	4	63.2	2	126.4
3	16.0	2	32.0	1	32.0
4 (Amids.)	16.0	4	64.0	0	277.7
5	15.9	2	31.8	1	31.8
6	13.9	4	55.6	2	111.2
7	9.8	$1\frac{1}{2}$	14.7	3	44.1
$7\frac{1}{2}$	6.0	2	12.0	$3\frac{1}{2}$	42.0
8 (*FP*)	0	$\frac{1}{2}$	0	4	0
			310.0		229.1

Common interval $= h = \dfrac{240}{8} = 30$ m

Area $= A \times \dfrac{h}{3} \times 2$ (for both sides)

$= 310 \times \dfrac{30}{3} \times 2 = 6200 \text{ m}^2$

TPC $6200 \times 0.01025 = 63.6$ in SW

Excess moment $= 277.7 - 229.1 = 48.6$ (aft in this case)

$$F \text{ from amidships} = \frac{\frac{h}{3} \times \text{difference in } M \times h \times 2}{\frac{h}{3} \times A \times 2}$$

$$= \frac{\text{Difference in } M \times h}{A} = \frac{48.6 \times 30}{310}$$

$= 4.70$ m aft

Moment of area using Simpson's Second Rule—This is obtained in a similar manner.

Example 2—Find the area and the position of the centre of flotation of the following waterplane, 120 m long.

Stations from aft	0	1	2	3	4	5	6
Half-breadths, m	0	8.2	10.1	9.3	7.8	3.9	0

Station	Half-breadth	SM	A	Lever	M
0	0	1	0	0	0
1	8.2	3	24.6	1	24.6
2	10.1	3	30.3	2	60.6
3	9.3	2	18.6	3	55.8
4	7.8	3	23.4	4	93.6
5	3.9	3	11.7	5	58.5
6	0	1	0	6	0
			108.6		293.1

Area $= 2 \times \dfrac{120}{6} \times \frac{3}{8} \times 108.6 = 1629 \text{ m}^2$

$F = \dfrac{293.1}{108.6} \times 20 = 54.0$ m from 0 station

Moment of inertia—The moment of an area about a point, or axis, is equal to the sum of all the elements of the area times the distance of each element from the point.

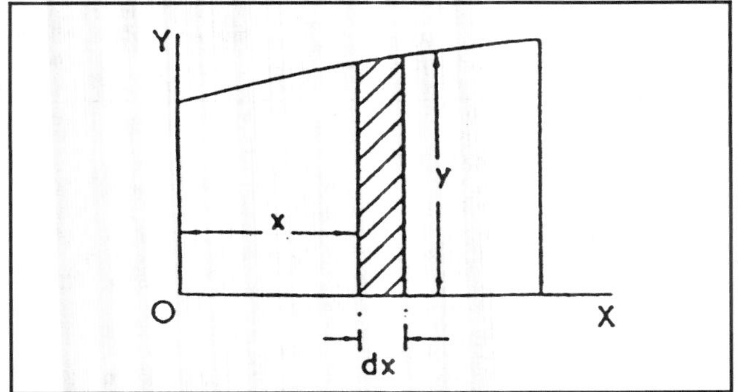

Figure 11

Moment $= xydx$ about OY

or $= \frac{y}{2} ydx$ about OX

where $xydx$ = integral $(xydx)$
= sum of $(xydx)$, and
ydx = element of area.

The second moment of area, more generally known as the moment of inertia, of an area about an axis is equal to the sum of all the elements of the area times the square of the distance of the element from the axis.

Moment of inertia $= x^2ydx$ about OY

or $= \frac{1}{3} \; y^3dx$ about OX

To find the transverse moment of inertia of the waterplane of a ship the calculation may be made as follows:

Example 3—The half-ordinates of the waterplane of a ship are spaced 20 m apart, and are 0, 6.9, 12.4, 17.0, 16.8, 15.4, 11.3, 6.2 and 0 m respectively. Find the transverse moment of inertia of the waterplane about the centre line.

Half ordinate y	(half-ordinate)^2 y^3	SM	Product for MI
0	0	1	0
6.9	329	4	1 316
12.4	1907	2	3 814
17.0	4913	4	19 652
16.8	4742	2	9 484
15.4	3652	4	14 608
11.3	1443	2	2 886
6.2	238	4	952
0	0	1	0
			52 712

$I_T = \frac{1}{3} \; y^3dx$

$= \frac{1}{3} \times 52\,712 \times 2 \times \frac{1}{3} \times 20$ . . . 2 for both sides
$\frac{1}{3}$ for SM
20 for common interval = dx

$= 234\,276$ m^4

This is a useful calculation since it enables the KG of a ship to be found using the formulae:—

$BM_T = \frac{I_T}{V}$ and $KG = KB + BM_T - GM_T$

GM_T may calculated using the results from an inclining experiment.

Volume of displacement

The areas of the successive waterplanes of a ship may be found by using the Simpson's rule appropriate to the number of equally spaced ordinates. If a curve is drawn in which the ordinates at equally spaced intervals represent the areas of the various waterplanes, then the area under this curve represents the volume of displacement of the ship. The centroid of this area will give the position of the centre of buoyancy from the design waterline or the base.

Similarly, if a curve is drawn whose ordinates represent the transverse sectional areas of a ship on a base of length, then the area under the curve will represent the volume of displacement. The centroid of the area will give the position of the centre of buoyancy from either amidships or the *AP*.

Example 4—A ship 120 m in length floats at drafts of 5 m forward and aft and has the following particulars:

Waterplanes	1 m	2 m	3 m	4 m	5 m
Areas, m^2	1579.4	1761.6	1881.8	1931.0	1943.2

The appendage below the 1 m waterline has a displacement of 1,063 tonnes with a $\overline{KB}$ of 0.55 m. The areas of the transverse sections of the ship to the 5 m waterplane at equally spaced intervals starting from aft are:

Section	0	1	2	3	4	5	6	7	8	9	10
Areas, m^2	0	33.1	68.6	89.8	96.2	96.4	96.3	95.1	78.4	40.2	0

Using the above information, find:
(a) The displacement when floating in SW;
(b) The value of $\overline{KB}$;
(c) The position of the LCB from amidships.

Water-plane (m)	Area m^2	SM	Products of areas	Levers about base	Moments of products of areas
1	1579.4	1	1579.4	1	1 579.4
2	1761.6	4	7 046.4	2	14 092.8
3	1881.8	2	3 763.6	3	11 290.8
4	1931.0	4	7 724.0	4	30 896.0
5	1943.2	1	1 943.2	5	9 716.0
			22 056.6		67 575.0

Volume of displacement $= \frac{1}{3} \times 1 \times 22\,056.6$ m

Displacement of main hull in SW = $\frac{1}{3} \times 22\,056.6 \times 1.02$
= 7536 tonnes

$\overline{KG}$ of main hull above base $= \frac{67\,575.0}{22\,056.6} = 3.06$ m

	Displacement tonnes	*KG* m	Moment tonnes m
Main hull	7536	3.06	23 060.1
Appendage	1063	0.55	584.7
Total	8599	2.75	23 644.8

Displacement in SW = 8599 tonnes
$\overline{KB}$ = 2.75 m

Section	Area m²	SM	Product of areas m³	Levers about amid-ships	Moments of products of areas
0(*AP*)	0	1	0	5	0
1	33.1	4	132.4	4	529.6
2	68.6	2	137.2	3	411.6
3	89.8	4	359.2	2	718.4
4	96.2	2	192.4	1	192.4
5 (Amids.)	96.4	4	385.6	0	1852.0
6	96.3	2	192.6	1	192.6
7	95.1	4	380.4	2	760.8
8	78.4	2	156.8	3	470.4
9	40.2	4	160.8	4	643.2
10 (*FP*)	0	1	0	5	0
			2097.4		2067.0

$$\text{Volume of displacement} = \frac{1}{3} \times \frac{120}{10} \times 2097.4 \text{ m}^3$$

Displacement in SW = 4 × 2097.4 × 1.025

= 8599 tonnes

$$\text{LCB} = \frac{2067.0-1852.0}{2097.4} \times \frac{120}{10}$$

= 1.23 m forward of amidships

It may be seen from the calculations that the displacement is the same using the two different methods. In this particular problem the areas of the waterplanes and the sections were given and the results were obtained by applying Simpson's rules to those areas. The method checked itself to some extent, since the displacements were found to agree.

It would be quicker, however, to find the displacement, $\overline{KB}$ and other particulars without the need to calculate areas of waterplanes and transverse sections separately before once again calculating the displacement.

This may be done using a displacement table.

Displacement table—Example 5 shows the use of a displacement table. The half-breadths of the moulded waterplanes are measured from the body plan and put into the table. The displacement is found by summing the areas in two directions—i.e., vertically and horizontally—and of course the two answers must agree. Only a few sections and waterplanes are used in this particular example, but the number may be extended in both directions as required.

The displacement multiplier

$= 2 \times \frac{1}{3} \times$ CI horizontally $\times \frac{1}{3} \times$ CI vertically $\times 1.025$

$= 2 \times \frac{1}{3} \times 6 \times \frac{1}{3} \times 1 \times 1.025 = 1.367$

Example 5

DISPLACEMENT TABLE

Section	SM	Levers	0 m WL, SM 1: Ord.	0 m WL, SM 1: Area product	1 m WL, SM 4: Ord.	1 m WL, SM 4: Area product	2 m WL, SM 1: Ord.	2 m WL, SM 1: Area product	Total products	Volume products	Moment of volume products
0	½	3	0	0	0	0	0	0			
(*AP*)			0		0		0		0	0	0
½	2	2½	2.56	5.12	4.86	9.72	7.46	14.92			
			2.56		19.44		7.46		29.46	58.82	147.30
1	1½	2	3.68	5.52	8.74	13.11	9.68	14.52			
			3.68		34.96		9.68		48.32	72.48	144.96
2	4	1	5.00	20.00	11.84	47.36	12.62	50.48			
			5.00		47.36		12.62		64.98	259.92	259.92
3	2	0	5.00	10.00	11.84	23.68	12.62	25.24			
			5.00		47.36		12.62		64.98	129.96	552.18
4	4	1	5.00	20.00	11.84	47.36	12.62	50.48			
			5.00		47.36		12.62		64.98	259.92	259.92
5	1s	2	4.12	6.18	9.16	13.74	10.00	15.00			
			4.12		36.64		10.00		50.76	76.14	152.28
5½	2	2½	3.16	6.32	5.42	10.84	8.12	16.24			
			3.16		21.68		8.12		32.96	65.92	164.80
6	½	3	0	0	0	0	0	0			
(*FP*)			0		0		0		0	0	
Sum of products				73.14		165.81		186.88		923.26	577.00
SM				1		4		1			
Volume products				73.14		663.24		186.88		923.26	
Levers				0		1′		2			
Moment products				0		663.24		373.76	1,037.00		
Displacement multipliers								1.367			
Displacement								1262			
$\overline{KB}$								1.123			

Displacement to 2 m WL
= 923.26 × 1.367 = 1262 tonnes in SW

$$\overline{KB} = \frac{\text{Moment of volume}}{\text{Volume}} \times \text{CI vertically}$$

$$= \frac{1037.00}{923.26} \times 1 = 1.123 \text{ m}$$

LCB from amidships

$$= \frac{\text{Moment of volume}}{\text{Volume}} \times \text{CI horizontally}$$

$$= \frac{577.00 - 552.18}{923.26} \times 6$$

= 0.161 m forward

It may be seen from the above table that 923.26 is the sum of the volume products, both horizontally and vertically. Further calculations should not be made unless these two totals agree exactly.

Computers

The preceding paragraphs have shown how arithmetical methods may be used to calculate the areas, volumes, displacements and basic form characteristics of a ship and how the coefficients associated with them may be used to compare similar classes of ships.

Modern design and drawing offices now make extensive use of computers, which because of their high-speed number-handling capabilities take much of the tedium out of the basic calculations.

The software programs use simple and multiple arithmetical calculations and are based on the principles already outlined; for a ship the input data are generally the offsets of the ship as measured from the waterlines or body plan. Separate programs are available from many commercial sources and can be used on small or large computers. Some programs have been derived to eliminate, in the design office, offset data errors by fairing mathematically the hull measurements made for use in hydrostatic calculations. These faired offsets can then be used to carry out the final calculations without recourse to the laying-off process.□

FACTORS AFFECTING DRAFT SURVEYS

J. Wolfram, B.Sc, Ph.D, C.Eng, MRINA
Strathclyde University

IT IS DIFFICULT to determine very accurately the mass of a ship under normal operating conditions. The traditional procedures employed by ship's officers are usually considered adequate; but there are many sources from which errors may arise. This note draws attention to some of these possible errors and in some cases suggests ways of eliminating or reducing them.

Determination of the equivalent level keel condition

To determine the ship's mass using the hydrostatic data normally available on a ship requires corrections to be made to determine the equivalent even keel draft. For accuracy this requires not only a trim correction but also a correction for list. When a ship of normal form heels it rises bodily in the water and its mean draft decreases. At the ends of the ship, where it is not wall sided in way of the waterline, the wedge emerging from the water is smaller than the wedge being immersed. To correct this difference, the ship naturally lifts out of the water until the emerging volume is equal to the newly immersed volume. For a small ship the difference in draft will not be large and the effect on displacement could probably be neglected provided the list is not more than a couple of degrees. (Fig 1.) However, for a large ship the error in the calculated ship mass when this effect is ignored may be significant. It would be a simple matter at the design stage when the hydrostatics are produced, using a computer program to include a sub-routine providing a correction table to account for the effect of list at any operational draft. With the data currently supplied to ships it would be difficult to make an accurate correction.

However, the following expression will give some idea of the error involved in the calculated ship mass if no correction is made: Error = $6\,(TPC_1 - TPC_2) \times (D_2 - D_1)$ tonnes, where D_1 and D_2 are the drafts at amidships on each side of the vessel and TPC_1 and TPC_2 are the tonnes per centimetre immersion corresponding to these drafts.

When corrections are made for trim the longitudinal centre of flotation (LFC) used is usually that which corresponds to the mean draft. But with large trims this may not be the axis about which the ship should be considered to rotate. The actual position of the axis will vary with the amount of trim and the hull form of the vessel. The trim correction tables produced for ships by some shipyards take this into account and it is unfortunate that such tables are not universally provided. Empirical formulae do exist to correct for the shift in LCF but these are unlikely to be accurate for all types of hull form. The calculation of the volume of displacement could be made directly from the Bonjean curves but this would be time-consuming and possibly inaccurate unless the curves were reproduced accurately and on a large scale.

Most calculations of displacement assume that the draft at the fore and aft perpendiculars of the design waterline are known. However, in many cases the draft read from the marks on the ends of a ship do not give these values. It is a simple matter to produce curves to enable the measured drafts to be corrected to equivalent FP and AP values. If no correction is made the calculated ship mass may be in error, in some cases, to the extent of several hundred tonnes.

Determination of density

Traditionally a bucket is thrown over the ship's side and the specific gravity of the water collected, as measured by a hydrometer, is used in the displacement calculation. However, in many rivers the density will change with water depth as different current layers are encountered. Therefore the density of a bucket of surface water will not necessarily be representative of all the water surrounding the ship and a displacement calculated using this one figure may be inaccurate. The mean of a series of samples taken over a range of water depths, extending to the ship's draft and at various points around the vessel, should give reasonable average density on which to base calculations, provided the variation in density is not too large. Simple devices exist for taking water samples from any depth alongside a vessel.

The hydrometers used to measure the density sometimes have scales which do not extend below '1', where '1' usually corresponds to the specific gravity of fresh water at 15°C. So when the hydrometer is placed in fresh water warmer than 15°C it is off the bottom of the scale. In tropical regions and close to warm water outfalls, the water temperature is often above 15°C and a 'guestimate' has to be made of the

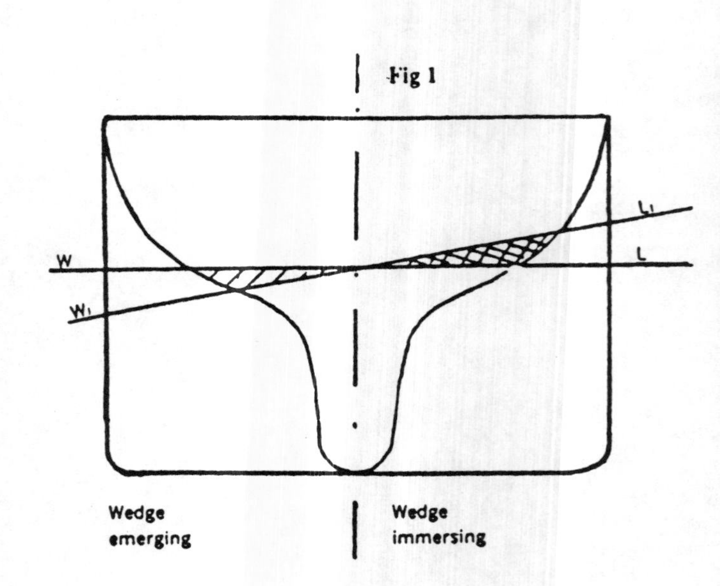

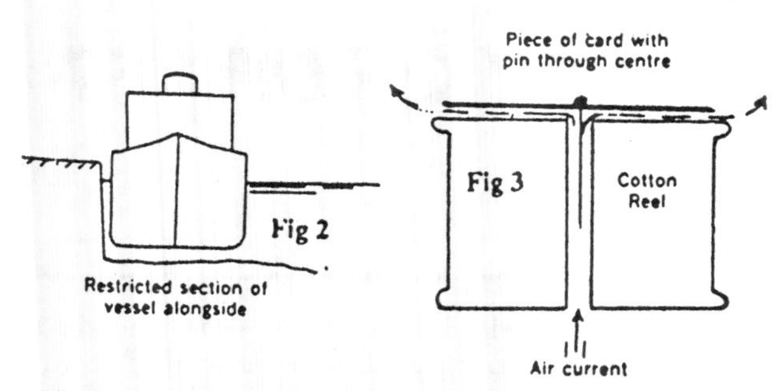

specific gravity by visually extrapolating the scale on the hydrometer downwards. It would help if the scale on all such instruments were extended down to, say, 0.995 which would cover the warmest fresh water likely to be encountered.

Another point worth noting is that the hydrometer measures specific gravity which, for the hydrometers commonly used, is defined as the sample water density divided by the density of fresh water at 15°C. Then the specific gravity of fresh water at 15°C, is '1,000'. However, the mass density of fresh water at 15°C is 0.9990 tonnes per cubic metre. For ship mass calculations it is the mass density that is required not the specific gravity. Assuming that the specific gravity and the mass density have the same numerical value involves a slight error (100 tonnes in the displacement calculation of 100,000 tonnes). Recalibrating an hydrometer to read density directly would be a simple job and a correction, which is often forgotten, would be avoided.

Sinkage and trim caused by currents and tidal streams

Most seafarers are well aware of the effect known as 'squat' which causes ships to increase their draft when travelling at speed in shallow water. What they may not be aware of is that a ship moored or anchored in shallow water experiences the same effect when there is a tidal stream or current running. The cause of both effects is similar.

Consider a ship moored in a river. (Fig 2.) When a current is running the ship constricts the flow. The water must then increase its speed in order that the same quantity passes through the restricted space as does through the unrestricted space, in any given period of time. The water flowing at higher speed under the bottom of the vessel causes a reduction in pressure on the bottom (this occurs by virtue of the Bernoulli effect) and the ship sinks deeper in the water.

The Bernoulli effect can be demonstrated by trying to blow a piece of card off the end of a cotton reel. (Fig 3.) It is impossible to blow the card off. The high air velocity on the inner face of the card causes a local drop in pressure relative to the outer face of the card, thus keeping it firmly pressed on the end of the reel. Bernoulli's equation, which governs this effect, is $P + p\frac{v^2}{2} + pgh = \text{constant}$, where P is the pressure, p the water density, v is the velocity and h the depth of water. Clearly as v increases, at a given water depth, P must decrease for the equation to remain constant.

The amount of sinkage caused by this effect will depend, therefore, on the water velocity. It will also depend on the depth of water beneath the keel and the

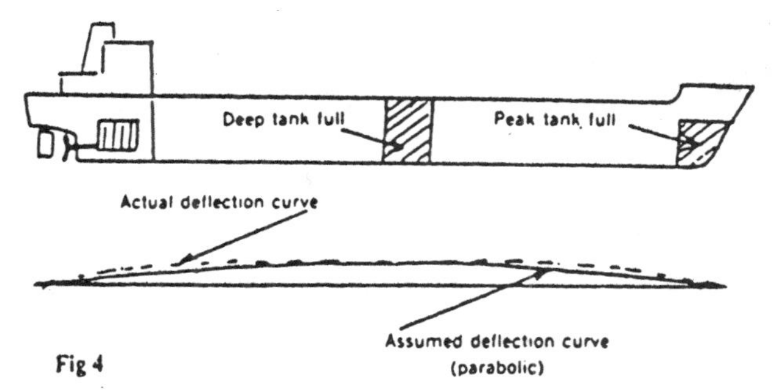

ship's length. The sinkage in some cases will be considerable. For example, a 1,600-tonne coaster moored in a river where the current is running at 4 knots will experience a sinkage of at least 5 cm where there is about 0.35 m of water under the keel. It is therefore desirable to wait until the depth of water under the keel is as large as possible before measuring drafts if there is any current.

Clearly in a tidal stream it would be better to measure the draft at slack water thus avoiding this sinkage effect if at all possible. With data currently available it would not be possible for the sinkage likely to be experienced to be estimated in all cases. An approximate theoretical estimate can be made but the procedure involved is relatively complicated (see Dand & Ferguson *The Squat of Full Form Ships in Shallow Water* TRINA Vol 115, 1973).

Further errors can occur when the drafts are read due to the wave pattern around the ship produced by the current. These waves are additional to those produced by the wind which also make an accurate reading of the draft marks difficult. Measuring draft with the help of a long transparent plastic tube, with its lower end well below the water surface with the opening pointing downwards, will largely eliminate the effect of these waves. The tube is held alongside the vessel and the water level in the tube read against the ship's draft marks.

Correction for hog and sag

When the mean of the forward and after draft marks is different from the mean draft amidships, a correction is sometimes made for hog or sag. For hog, two thirds of this difference is usually added to the mean of the draft at the ends, and for sag, two thirds of this difference is subtracted. The correction is based on the assumption that ships bend into smooth parabolic curves. This is quite reasonable in some cases, but not in all. (Fig 4.)

Consider a ship with engines aft, a large full deep tank amidships and with the forward peak tank full. The deflected shape will not follow a parabolic curve but a broken line. It should be possible for design offices to supply information concerning the form of the deflection curve for certain critical conditions allowing a more accurate correction for hog or sag to be made in cases where the deflection is significant.

Conclusion

There are many possible sources of error in the calculation of ship displacement, and this chapter touches on a few of them. The errors outlined here may not be significant in all cases, However, it is possible that one or more of these sources of error can result in considerable inaccuracy in the displacement calculation if appropriate steps are not taken.□

IMPROVING THE ACCURACY OF DRAFT SURVEYS

J. L. Strange, Ph.D, MNI

IN SOME BULK TRADES it is not possible to establish the weight of a dry or liquid cargo by direct measurement. In these circumstances the only possible method of finding the amount of cargo on board is by means of a draft survey, which entails taking the draft of a ship before and after handling the cargo and calculating the displacements. The difference between the initial and loaded displacement represents the cargo on board. This method is sometimes used in 'on' and 'off' hire surveys as a check on the bunkers remaining on board.

A commonly-held view is that draft surveys are not sufficiently accurate to be of any value. This probably came about because there is no universally recognised procedure for draft surveys and, although some surveys undoubtedly do an excellent job, in general the results are uneven. In an attempt to establish such a procedure, United Molasses has produced a manual on draft surveys and I am grateful for their permission to make use of it in this chapter.

Outline to follow

The outline procedure for draft surveys is:

(a) When loading

Read the drafts of the ship before loading and with a minimum of ballast on board—i.e., just enough to give a reasonable trim. Measure the density of the dock water and calculate the displacement. Subtract the weight of ballast (this will be discharged before the ship completes loading) to obtain the displacement of the ship before loading. When the cargo is loaded, read the draft again and, after taking the density of the dock water, calculate the displacement of the ship and cargo. After taking into account any bunkers and stores taken while loading, subtract the displacement previously calculated to obtain the weight of cargo loaded.

(b) Discharging

Read the draft on arrival before discharging any cargo. Obtain the density of the dock water and calculate the displacement. When all the cargo has been discharged and enough ballast taken to give a reasonable trim, read the draft again and after taking the density of the dock water calculate the new displacement. Take into account any stores ballast and bunkers taken or consumed when the ship is in port, and the difference between the two displacements will give the weight of cargo discharged.

By having draft surveys taken in both the loading and discharge ports and using the same procedure in each case the results can be compared and any errors will be shown up.

Ballast tanks

The aim should be to have as few tanks as possible containing ballast; such tanks should be full if possible to avoid the need for ullaging or sounding. In some ships where it is not possible to determine accurately whether a ballast tank is completely dry or if there is any sediment on the bottom, a more accurate procedure is to leave a measureable amount of water in the tank to be discharged after the ship sails.

Procedure in detail

It is important to realise that no matter how carefully the subsequent calculations are carried out, the accuracy of a survey depends primarily on the reading of the draft and the measurement of the density of the dock water.

Reading the drafts

Before reading the drafts ensure that the ship is not handling bunkers or cargo or carrying out any other operation likely to affect the draft. Try to have the ship upright, as the ship's stability information is calculated for this condition and any list will introduce errors. The stability information is specially inaccurate in the case of the empty ship with the bow out of the water, but as the measurement of ballast can also introduce errors the aim should be to only have enough ballast to provide a positive forward draft, but bear in mind that the trim corrections described in this paper are only accurate when the trim is less than 1 per cent of the ship's length.

When reading the drafts use a boat if possible and read the drafts on both sides of the ship, averaging them to allow for any remaining list. In the case of the amidships draft, it is usually more accurate to measure the freeboard to the top of the amidships deckline. When converting this freeboard to draft remember that the moulded depth is measured from the top of the statutory deckline to the top of the keel. The ship's drafts are, however, measured from the bottom of the keel.

The way to convert freeboard to draft is to add the deepest summer draft to the summer freeboard and

Diagram No. 1

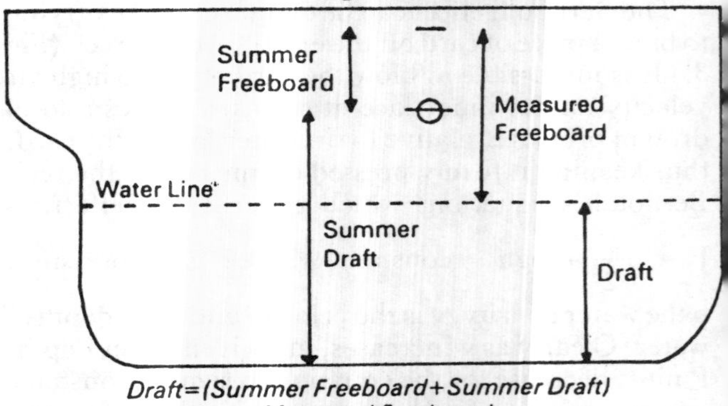

Draft = (Summer Freeboard + Summer Draft) − Measured Freeboard.

Diagram No. 2

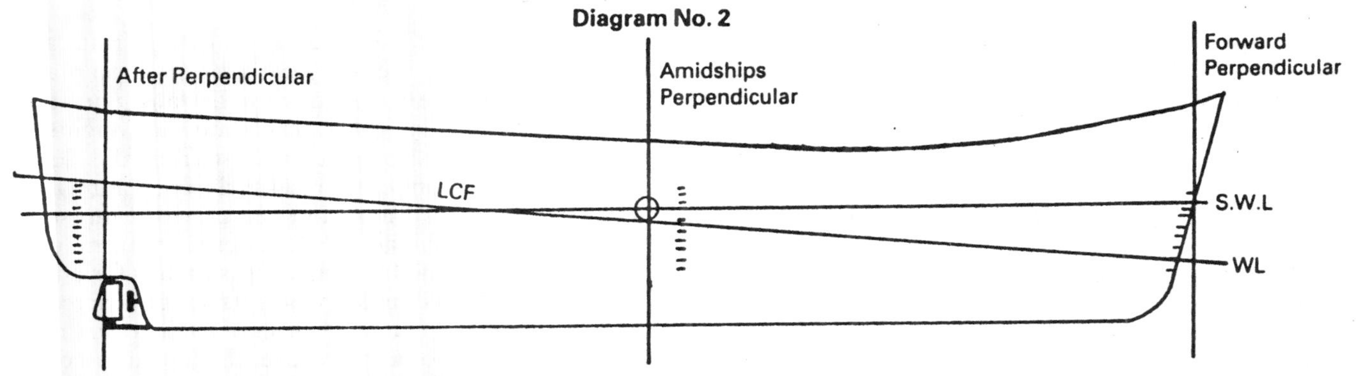

Position of the after, amidships and forward perpendiculars.

subtract the measured freeboard (diag. 1). United Molasses has designed a freeboard measuring instrument to smooth out the wave effect and make the reading more accurate.

Taking the density

As soon as possible after reading the drafts obtain the density of the dock water. It is important to take this reading without delays as the density varies with the state of the tide. When taking water samples use a container with a perforated lid; lower the container to a depth equal to the deepest draft and raise it up again at a constant speed so that it is not full when it reaches the surface. This ensures that a uniform water sample is obtained. Take three samples along the offshore side of the ship, as it is possible to have stagnant water trapped between the ship and the jetty.

When the sample is on board take the temperature before and after reading the density to ensure that it remains constant. There are a number of different instruments on the market for measuring densities; use one designed for water, not oil, as the surface tensions are different. Brass instruments are not accurate enough for this purpose.

Most instruments are calibrated for water in a vacuum and so 0.0011 and 0.002 should be subtracted from the glass and brass instruments respectively to allow for the different buoyancy of water in air. As the instruments are not being used at their calibration temperatures, further corrections supplied with the instruments must be used.

Correction to the perpendiculars

The forward perpendicular is a line, at right angles to the keel, cutting the summer waterline at the stem. The after perpendicular is a line, at right angles to the keel, passing through the after end of the rudder post. It is also the position of the frame marked 'O' in the plans. The ship's stability data is calculated for draft measured at the perpendiculars and, as the draft marks do not usually coincide with these lines, a correction must be applied.

To correct the drafts, if they are not already marked draw in the perpendiculars and the draft marks on the ship's capacity plan, and measure the horizontal distance between the draft marks and the perpendicular at the waterline. The correction is given by:

$$\text{Correction} = \begin{array}{c}\text{Distance of draft}\\ \text{mark from perpendicular}\end{array} \times \frac{\text{Trim read from draft marks}}{\text{Distance between draft marks}}$$

Inspection of the plan will indicate whether this correction is to be added or subtracted from the observed drafts.

In the case of the amidship draft, if it has been obtained by measuring the freeboard to the deckline, then no correction is necessary as the loadline disc can, in most cases, be considered to be at the mid-length of the perpendiculars, If, however, the draft has been measured from the draft marks then it must be corrected as the draft marks are not under the loadline disc. The correction is given by:

$$\text{Correction} = \begin{array}{c}\text{Distance of draft}\\ \text{marks from amidships line}\end{array} \times \frac{\text{Trim at perpendiculars}}{\text{Length between perpendiculars}}$$

Again, inspection will show whether this correction is to be added or subtracted. The amidships draft is corrected in this way because it is more accurate to use the 'length between perpendiculars' from the ship's particulars than it is to measure the distance between the draft marks on the ship's plan.

Correction for hull deformation

The amidships draft is only the same as the mean of the fore and after drafts when the ship is neither hogged nor sagged. There are a number of corrections for hull deformation. A suitable one for a ship shaped like a tanker or a bulk carrier is:

$$\frac{\text{Forward draft} + \text{After draft} + (6 \times \text{Amidship draft})}{8}$$

This correction is known as the 'mean of mean drafts' and will provide a more accurate mean draft when the ship is hogged or sagged. This demonstrates the importance of reading the amidships draft as accurately as possible.

Obtain displacement

Using the corrected mean draft extract the displacement from the ship's stability data.

Corrections for trim

When a ship is trimmed the calculated mean draft is not the same as the true mean draft measured at the LCF. To correct the displacement to that corresponding to the 'true mean draft' the following formula is used:

$$\text{Correction (tonnes)} = \frac{\text{Trim (cms)} \times \text{LCF (metres)} \times \text{TPC}}{\text{LBP}}$$

where LCF means distance of the LCF from

amidships and LBP is the length between perpendiculars. Both items can be obtained from the ship's stability information.

To apply the correction, known as the layer correction, when the LCF is in the same direction from amidships as the deepest draft it is added to the displacement, and when it is in the opposite direction it is subtracted.

This correction does not allow for the fact that, when a ship trims, the LCF moves from its tabulated position. Some ships are provided with corrections for this, but if they are not available the following correction must be added to the displacement:

$$\text{Correction in tonnes} = \frac{(\text{Trim in metres})^2 \times 50 \times dM}{\text{LBP in metres} \quad dZ}$$

where $\frac{dM}{dZ}$ is the difference between the MCT for a draft of 50 cm greater than the corrected mean draft and 50 cm less than the corrected mean draft.

Correction for heel

As previously explained, the ship should be upright for a draft survey, but if this is not possible then the following correction must be applied:

$$\text{Correction in tonnes} = 6 \times (T_1 - T_2) \times (D_1 - D_2)$$

where T_1 is the TPC for the deepest draft amidships (D_1) measured in metres and T_2 is the TPC for the shallower draft amidships (D_2) measured in metres. The correction is always added to the displacement because the effect of heel is to increase the waterplane area and so lift the ship out of the water.

Correction of displacement for density

The ship's stability information is calculated for a standard density, usually 1.025, but other standards do exist so it is important to check on the value used for any particular ship. The density of the dock water having been taken when the drafts were read the correction to the displacement is given by:

$$\text{True displacement} = \text{Scale displacement} \times \frac{\text{Density of dock water}}{\text{Density used for the displacement scale}}$$

The displacement now obtained is true displacement, within the limits of accuracey of the drafts and the ship's stability data. The weight of cargo on board is found from the displacement as follows:

(a) Before loading cargo

From the calculated displacement subtract the light displacement to obtain deadweight.

As soon as possible after reading the drafts and density, sound all the fuel, ballast and fresh water tanks. Correct the soundings for list and trim and using the calibration tables calculate the weights of fuel, fresh water and water ballast from the deadweight. The remainder represents the constant which in turn represents the difference between the scale deadweight and the actual deadweight that the ship can load. The value from the ship's data is for a new ship. However, as the ship ages its weight increases (partly due to the reluctance of ship's officers to throw anything away), so in many cases the figures used are too low. Therefore, a new constant should be calculated for each draft survey.

(b) After loading

Read the drafts and calculate the loaded displacement. Using the constant found previously and, having calculated the quantity of fuel, fresh water and ballast on board, calculate the quantity of cargo loaded.

At the discharge port the draft survey is repeated, first with the loaded ship, then with the empty ship, and the weight of cargo is calculated. It is now possible to obtain some idea of the accuracy of the surveys by comparing both the constants as well as the cargo figures.

Method of last resort

Draft surveys must always remain the last resort method of calculating the quantity of cargo on board a ship; but, if they are carried out carefully, it should be possible to produce more consistent results than would have previously been the case. It should be possible for a ship's officer to check his cargo figures where there is a dispute between ship and shore tonnages. The owner can assist in this by ensuring that the information required by the surveyor is readily available on board and in a form that can easily be used.

It is difficult to quantify the errors involved, as they must largely depend on the skill and care of the individual surveyor. With older ships the displacement scales were calculated using Simpson's first rule which is considered to be about 0.5 per cent low, so any displacement must be out by this amount. Even in an enclosed dock there is a possible error of ±2cm in the drafts and the error will depend on the TPC, while the measurement of density can introduce further errors.□

I am grateful to Mr J. E. Turner, lecturer in naval architecture at the City of London Polytechnic, for checking the formular given in this paper, and to United Molasses for allowing me to draw from work originally carried out by and on their behalf.

HOG/SAG AND TRIM CORRECTIONS FOR DRAFT SURVEYS

J. E. Turner, C.Eng, M.Sc, FRINA, AMNEC
Senior Lecturer, Transport Studies, City of London Polytechnic

IF A PRECISE CALCULATION of the displacement of a given ship in a given loading condition is required, then resort must be made to computer techniques as described in reference 1. However, such calculations require the surveyor to have access to considerable quantities of ship data and a cheap, easily accessible micro computer system. Thus, if such facilities are not readily available, draft surveys will continue to be made using the conventional approach based on the readings of observed drafts forward, aft and amidships.

Such an approach must incur an error in the result since the true nature of the deformation of the hull cannot be deduced from only three draft readings. In reference 1 it is suggested that at least five draft readings are necessary for this purpose, but until shipbuilders agree to mark the additional data lines on a ship in order that the additional drafts can be read with a reasonable degree of accuracy the surveyor must make the best use of the usual three draft readings.

Using the measured drafts, an allowance for the hog/sag of the ship may be calculated using the following simple expression:
Displacement
Correction = Hog/Sag × TPC × Coefficient.

If a computer is readily available, then a more precise calculation could be made using the three drafts on the assumption that the hull deformed as a three-point spline (Ref 1). However, since in general a quick and simple calculation is to be preferred then an expression as given above is required.

Over the years such an expression has been used with the value of the coefficient being taken as either 2/3 or 3/4. However, in reality the coefficient should vary and its value depends upon the form of the ship and its draft—e.g., ballast or fully-loaded condition.

Working in conjunction with United Molasses (Ref 2) a computer investigation was carried out regarding the variation of the above coefficient with draft ranging from lightship to full load condition with trim varying from even keel to 5 m by the stern for a typical 30,000-tonne-displacement tanker. The displacement correction according to the three-point spline method was equated to the expression given previously and the appropriate value of the coefficient deduced. The calculations were made using the traditional approach where the TPC is the value for the mean draft of the ship and then repeated using the true TPC—i.e., the value for the actual trimmed waterplane.

The computer output is shown in Table 1, which includes the waterplane coefficient values for both mean draft waterplane and actual trimmed waterplane. For the exercise, a sag of 0.2 m was assumed, but naturally the magnitude of the sag is of no consequence with the procedure adopted here in that it only affects the magnitude of the 'correction' figure in the table. The other columns would have exactly the same values irrespective of the hog/sag value.

Table 1

(Metric units apply)
CALCULATION OF DISPLACEMENT CORRECTION DUE TO HOG/SAG

Draft	Trim	Correction	WPM	CM	WPT	CT
2.60	0.00	465.34	0.776	0.738	0.776	0.738
2.60	1.00	465.13	0.776	0.738	0.774	0.740
2.60	2.00	464.92	0.776	0.738	0.772	0.741
2.60	3.00	464.70	0.776	0.737	0.770	0.743
2.60	4.00	464.09	0.776	0.736	0.766	0.746
2.60	5.00	462.45	0.776	0.734	0.757	0.752
3.90	0.00	472.47	0.795	0.732	0.795	0.732
3.90	1.00	472.93	0.795	0.733	0.796	0.732
3.90	2.00	473.40	0.795	0.733	0.797	0.731
3.90	3.00	473.86	0.795	0.734	0.798	0.731
3.90	4.00	474.30	0.795	0.735	0.799	0.731
3.90	5.00	474.40	0.795	0.735	0.799	0.731
5.20	0.00	477.38	0.808	0.727	0.808	0.727
5.20	1.00	478.23	0.808	0.728	0.812	0.725
5.20	2.00	479.08	0.808	0.729	0.815	0.723
5.20	3.00	479.96	0.808	0.731	0.820	0.721
5.20	4.00	480.95	0.808	0.732	0.825	0.717
5.20	5.00	481.78	0.808	0.734	0.831	0.714
6.50	0.00	481.73	0.824	0.720	0.824	0.720
6.50	1.00	482.87	0.824	0.721	0.831	0.715
6.50	2.00	484.01	0.824	0.723	0.838	0.711
6.50	3.00	485.14	0.824	0.725	0.845	0.707
6.50	4.00	486.09	0.824	0.726	0.851	0.703
6.50	5.00	486.85	0.824	0.727	0.857	0.699
7.80	0.00	486.27	0.847	0.707	0.847	0.707
7.80	1.00	487.16	0.847	0.708	0.853	0.703
7.80	2.00	488.04	0.847	0.709	0.860	0.699
7.80	3.00	488.90	0.847	0.711	0.866	0.695
7.80	4.00	489.60	0.847	0.712	0.870	0.693
7.80	5.00	490.25	0.847	0.712	0.874	0.690
9.10	0.00	489.77	0.868	0.694	0.868	0.694
9.10	1.00	490.43	0.868	0.695	0.872	0.692
9.10	2.00	491.10	0.868	0.696	0.876	0.690
9.10	3.00	491.77	0.868	0.697	0.881	0.687
9.10	4.00	492.35	0.868	0.698	0.885	0.685
9.10	5.00	492.80	0.868	0.699	0.889	0.683

In the above Table following meanings apply to columns:
Correction—Hog/sag correction based on 3 point spline
WPM —Waterplane area coefficient for mean draft waterline
CM —Coefficient 'C' in formula: C × hog/sag × TPC, using mean draft TPC
WPT —Waterplane area coefficient for trimmed waterline
CT —Coefficient 'C' in formula C × hog/sag × TPC using trimmed waterplane TPC value

It is obvious that taking a value of 3/4 for the coefficient, as mentioned in section 4 of reference 2, is strictly not correct, since, for the particular ship in question, using such a value would overestimate the magnitude of the correction. The actual coefficient clearly varies with both waterplane area coefficient and trim, decreasing as both trim and waterplane area coefficient increase. In addition, using values based on mean draft waterplane would also, in general, lead to overestimation of the correction.

Thus, to obtain a reasonable estimate of the hog/sag displacement correction using a simple and readily applicable formula, then both the TPC and the coefficient should be for the actual trimmed waterplane condition and not the mean draft condition.

It will be noted that if it is preferred to make the correction for hog/sag in the form of an 'an equivalent mean draft' as per section 4 of reference 2, then the expression given in that section should be written as:

$$\frac{(1-C)\,(\text{forward draft} + \text{aft draft}) + 2 \times C \times \text{amidship draft}}{2}$$

where C is the coefficient, strictly, for the trimmed waterplane. Obviously, to make such a calculation for a ship the variation of the coefficient (C) with trim and waterplane area coefficient is required. Such data could easily be produced by, say, the shipyard on a computer when the hydrostatic data were being prepared.

In reference 2, section 6, the corrections for trim were detailed which incorporated the conventional layer of LCF correction and a 'second correction' which takes into account the movement of the LCF as the ship trims. The combination of the two corrections is often referred to as the 'Nemoto trim correction.' However, it should be noted that such corrections cannot be applied for all trimmed conditions since the theory on which they are based only holds true for small changes in trim. If the corrections are applied to large trims a large error will arise.

Table 2 shows typical computer calculations of changes in displacement between even-keel condition and various trimmed conditions with constant mean draft. The calculations have been made using four methods. The first is the exact calculation based on the actual trimmed waterplane offsets, while the second method uses the total trim correction. The importance of the second trim correction is clearly seem from Method 3 which only takes into account the normal LCF correction. Method 4 is the same as Method 2 but uses the TPC and MCT values for the actual trimmed waterplane.

It is seen that when the trim exceeds 0.01 L the error involved with Methods 2,3 and 4 become large and the displacements based on such methods may be considered unacceptable for a draft survey. Thus, the best way to obtain precise results would be to supply each ship with a set of tables or curves giving displacement corrections for various mean drafts and trims. To obtain the required displacement for a given condition, the surveyor would only have to interpolate between the nearest mean draft and trim value.

TABLE 2

CORRECTION TO EVEN KEEL MEAN DRAFT DISPLACEMENT TO GIVE DISPLACEMENT TO TRIMMED WATERLINE USING VARIOUS METHODS

FOR MEAN DRAFT OF 4 METRE, CORRECTIONS ARE:

Trim metre	Method 1	Method 2	Method 3	Method 4
0.00	0.0	0.0	0.0	0.0
1.00	−60.1	−60.1	−63.6	−60.0
2.00	−113.2	−114.1	−126.5	−112.6
3.00	−159.1	−161.8	−189.8	−153.1
4.00	−194.9	−203.2	−253.0	−177.6
5.00	−217.7	−238.5	−316.3	−179.0

FOR MEAN DRAFT OF 5 METRE, CORRECTIONS ARE:

Trim metre	Method 1	Method 2	Method 3	Method 4
0.00	0.0	0.0	0.0.	0.0.
1.00	−51.6	−51.6	−54.6	−51.3
2.00	−95.4	−96.9	−109.3	−94.8
3.00	−131.1	−136.2	−163.9	−124.8
4.00	−155.7	−169.2	−218.6	−136.8
5.00	−165.1	−196.1	−273.2	−136.1

FOR MEAN DRAFT OF 6 METRE, CORRECTIONS ARE:

Trim metre	Method 1	Method 2	Method 3	Method 4
0.00	0.0	0.0	0.0	0.0
1.00	−36.1	−36.0	−39.8	−35.4
2.00	−62.5	−64.4	−79.6	−59.5
3.00	−75.9	−85.1	−119.4	−72.8
4.00	−75.8	−98.2	−159.3	−75.8
5.00	−61.6	−103.7	−199.1	−67.8

FOR MEAN DRAFT OF 7 METRE, CORRECTIONS ARE:

Trim metre	Method 1	Method 2	Method 3	Method 4
0.00	0.0	0.0	0.0	0.0
1.00	−11.2	−11.4	−17.7	−11.5
2.00	−9.7	−9.8	−35.7	−10.5
3.00	4.6	4.8	−53.5	2.0
4.00	31.0	32.3	−71.3	23.3
5.00	69.2	72.8	−89.2	49.9

Methods given in table are as follows:
1—Using body plan offsets
2—Mean draft plus whole trim correction
3—Mean draft plus LCF correction only
4—As 2 using trimmed waterplane data

The data required for such tables or curves are easily produced for the ship from a body plan using a computer.

Thus, it is seen from the foregoing that if more precise results for draft surveys are required using the conventional approach more information, all of which may be readily produced from a body plan, must be supplied for the surveyor's use. If such data are not forthcoming then it must be realised that the results will be in error, the error increasing rapidly once the trim exceeds 0.01 L in most cases. □

References

1 Wolfram, J. 'A Note on Corrections for Hog and Sag.' *The Naval Architect* July 1980.

2 Strange, J. L. 'Improving the Accuracy of Draft Surveys.' *Seaways* May 1980.

WEIGHT/VOLUME RELATIONSHIPS REQUIRED FOR DRAFT SURVEY CALCULATIONS

E. Stokoe, Companion
In collaboration with D.C. Marshall, J.L. Strange and J.E. Turner

ACCURACY in cargo measurement by draft survey depends on careful attention to detail, correction of draft mark readings to the perpendiculars, allowance for hog or sag and a variety of other measurements or corrections relating to the ship itself.

However, the accuracy of a draft survey also depends on a correct understanding of the weight/volume relationship of the supporting water and of any water ballast which may be on board. Errors in draft surveys frequently arise from the practice of using the terms specific gravity, density and weight per unit volume as if they were identical in meaning.

A series of tests was conducted in Antwerp so as to compare the readings obtained by the use of various different types of hydrometers in selected standard salt-water solutions. From these tests it became clear that the brass loadline instrument, calibrated in terms of specific gravity at 60°F/60°F, while probably quite adequate for the calculation of fresh water draft allowances, was not really suitable for draft survey work. After considering various possibilites, a new type of hydrometer was produced by G. H. Zeal, Ltd., London, covering a range of 0.990 to 1.040 kilogrammes per litre graduated in terms of kilos per litre in air—that is to say, in terms of commercial weights, and calibrated for use in sea water, a liquid of medium surface tension. This instrument is accompanied by a brief instruction leaflet indicating how the hydrometer should be used for draft survey purposes.

It is suggested that the introduction of this new type of instrument graduated in terms of commercial weights, in this case kilogrammes per litre in air, should do much to clarify the weight/volume relationship of the water in which the vessel is floating.

Since the new Zeal draft survey hydrometer was introduced, a certain number of questions have arisen with exprienced master mariners and draft surveyors either because they were using hydrometers graduated in specific gravity or density in vacuo or merely because they found it difficult to accept that the density of fresh water in warm climates may be less than 1.000.

The position is complicated by the fact that ship's displacement scales and ballast tank calibration tables are usually set out not in terms of volume, which would be clear and understandable, but in terms of fresh water tons or sea water tons without stating clearly the basis which has been used for their calculation. Furthermore, the tons may be long tons of 2,240lb or metric tons (tonnes) of 1,000 kg.

In this chapter D. C. Marshall, technical manager of G. H. Zeal, hydrometer manufacturers, J. L. Strange and J. E. Turner of the Faculty of Transport, City of London Polytechnic, and E. Stokoe, consultant to SGS-Van Bree, Antwerp, independent cargo inspectors, have joined forces in an effort to clarify the position and to dispel various misunderstandings which seem to have grown up over the years.

Assuming that the vessel's displacement has been correctly calculated from the draft measurements and displacement scales, practical experience shows that errors can arise from the following causes:

(a) Failure to draw proper representative samples of the surrounding water and the ballast water on board;
(b) Misuse of the water temperature to introduce 'corrections' which are unjustified;
(c) Use of density or specific gravity as if they were the same as weight in air per unit volume;
(d) Accepting ballast tanks as 'empty' when they still contain unmeasurable quantities of mud or sediment.

Definitions

Weight in air—If a quantity of liquid—for example, fresh water or sea water—is weighed in an open container on a laboratory balance or on a weighbridge against the equivalent of brass weights, then the atmosphere will exercise an upward thrust upon the water much greater than the upward thrust exercised on the smaller volume of brass weights (density about 8.0 g/ml). According to British Standard 718: 1979, for liquids over the density range 0.600 to 1.100, this 'air buoyancy correction' amounts to 0.0011 grams per ml of product. This correction represents the difference between the air buoyancy effect on the liquid and that on the brass weights.

It is commercial practice to make no allowance for air buoyancy so that commercial weights are normally weights in air. It is for this reason that the UK (Imperial) gallon used to be defined as being the capacity corresponding to 10lb weight of distilled water weighed in air under standard conditions. Weight in air per unit volume is known as apparent density in air and is usually expressed in kg/l. (In this paper the litre is used as a special name for the cubic decimetre. This (1964) litre should not be confused with the former (1901) litre which was equivalent to 1,000.028 dm^3.)

Weight in vacuo—If a quantity of liquid—for example, fresh water or sea water—is weighed in an open container on a laboratory balance or weighbridge and correction is made for the effect of air buoyancy, then for most practical purposes the resulting 'weight in vacuo' is equivalent to mass. Mass per unit volume is known as density, or 'true density', and is usually expressed in g/ml or kg/ml^3.

Specific gravity—Specific gravity may be defined as the ratio of the density of the substance at T1 to the density of distilled water at T2. It is most important that both T1, temperature of liquid and T2, temperature of water are clearly specified. The term specific gravity unqualified by both temperatures is meaningless and confusing. When specific gravity is used for draft survey purposes, sea water or river water is being compared with pure water and the two 'air buoyancy corrections' tend to cancel out. Over the specific gravity range 0.990 to 1.040, specific gravity results are therefore substantially the same, to four decimal places, whether in vacuo or in air. (In recent years the BSI and the ASTM have replaced the term 'specific gravity' by 'relative density'. The above definition still applies.)

Sampling the surrounding water

For various reasons there is frequently a difference between the density of the water taken near the surface and the densities at various levels below the surface. It is therefore necessary to take samples at various depths in order to obtain an average density. Furthermore, in special circumstances and particularly where large vessels are concerned, there may be differences between the densities of samples taken at the forward, midships and aft sections of the vessel.

Sampling at various depths—A simple means of sampling at various depths is to use a weighted sampling can provided with a cork. The cork is attached to the lowering cord at a point about 15cm above the the lower end. In use the can is lowered in the closed position by means of the cord and the cord is then removed from the can at any given depth by jerking the cord. The can should be flushed at least once before taking the first sample. After the can has been allowed to fill with water at the desired depth, it is raised to the deck level and the apparent density measured immediately. It is important to note that, once the sample can has been removed from the water, it should not be left on deck or in direct sunlight but should be placed in a sheltered position and the apparent density measured without delay.

Number of depths to be sampled—The number of depths to be sampled depends partly on the draft of the vessel concerned and partly on local circumstances, tidal waters, fresh water from rivers flowing into the sea and so on. The number of depths to be sampled is thus a matter of experience. Nevertheless, the following guidelines apply:

- For small coasters it is usually adequate to take samples from the open water side of the vessel close to the midships draft mark and at a distance below the water line corresponding to approximately one-half of the midships draft. At least two samples should be taken to ensure that consistent results are obtained.
- For larger vessels, at least three samples should be taken near to the midships position at a distance below the water line corresponding to approximately one-sixth, one-half and five-sixths of the midships draft.

Number of positions where samples should be taken—Whilst for small coasters it is usually sufficient to take samples at the midships position, further sampling positions are necessary where large vessels are concerned.

As a general rule samples should be taken at three positions, namely:

- Amidships;
- Midway between the midships and the aft perpendicular;
- Midway between the midships and forward perpendicular.

Care should be taken to ensure that samples are not taken near to positions where the vessel may be discharging cooling water or near outlets discharging water from the shore into the dock.

Weight/volume relationships

35 cu ft of sea water weigh 1 ton (2,240 lb)—For many years it was common practice in shipyards using the British system of measurement to use the above relationship for calculating ship's displacement tables. The underwater volume was calculated in cu ft, divided by 35 and the result was given as long tons of sea water.

A simple calculation will show that if 35 cu ft weigh 2,240 lb, then 64 lb is the weight of 1 cu ft of sea water. Using standard conversion factors it will be seen that 64 lb per cu ft equals 1.0252 kg/l, in other words 35 cu ft per long ton correspond to an apparent density of 1.0252 kg/l. The new Zeal draft survey hydrometer is graduated in apparent density in kg/l in air so that where older vessels are concerned with sea water displacement tables based on 35 cu ft per long ton, long tons could be otained directly by simple proportion using 1.0252 as a basis.

However, to avoid confusion between older ships and newer vessels with tables in metric units, it is proposed to proceed as follows:

Long tons of sea water × 35 × 0.028317 = cu metres
Cu metres × reading of Zeal hydrometer = tonnes (1,000kg).

Similarly, where fresh water tanks are concerned, the 'round figure' often used was that 36 cu ft of fresh water weigh 1 ton (2,240 lb). For tanks so calibrated the procedure would be as follows.

Long tons of fresh water × 36 × 0.028317 = cu metres
Cu metres × reading of Zeal hydrometer = tonnes (1,000kg).

1 cu metre of fresh water = 1 tonne (1,000 kg)—During the last 15 years or so many shipyards which previously used British units have changed over to using technical metric units.

Ships with tables in metric units usually have displacement scales based on the above simple relationship. In such cases metric tons of fresh water are equivalent to cubic metres and cubic metres multiplied by the reading of a Zeal hydrometer will give metric tons directly.

Ship's displacement and other tables are often stated to be based on 'SG 1.025', SG 1.000', 'Density 1.025' and so on. If these statements are taken literally, then much confusion and discussion can be generated. In most cases the figures 1.025 or 1.000 can be regarded as being straight forward weight/volume relationships—i.e., apparent densities in kg/l in air as indicated by the new Zeal hydrometer.

Care should be taken with certain vessels built in shipyards where 1.020, 1.027 or some other figure may be used for the sea water tables of vessels intended to trade in inland or confined waters. Fortunately, some vessels are now being provided with displacement tables in cu metres, thus avoiding the misunderstandings of the past.

Effect of water temperature

The new Zeal hydrometer is made of glass and has been calibrated so as to be most precise when it is used at a water temperature of 15°C. The coefficient of cubical expansion of hydrometer glass is usually taken as being 0.000025 per °C. If the hydrometer is used in water at some temperature other than 15°C then, for laboratory purposes, a small allowance should be made for the expansion or contraction of the glass above or below 15°C.

However, for draft survey purposes it should not be overlooked that the cubical expansion of steel is often taken as being 0.000033 per °C so that at high temperatures the ship will expand and rise slightly out of the water and at low temperatures the ship will contract and sink more deeply into the water.

If the hydrometer reading is corrected for thermal expansion of the glass then, to remain logical, the ship's draft readings or displacement should also be corrected for thermal expansion. This would be difficult to carry out in practice and the difference between the coefficients of glass and steel are so small that, for draft survey purposes, changes in the volume of the hydrometer and the ship due to thermal expansion or contraction, being of opposite signs, tend to compensate each other.

The point is illustrated in the following example: Let us take a ship floating in water at 15°C. The hydrometer reading is 1.0150 kg/l at 15°C. The ship's displacement is 60,000 m^3 at 15°C. Then $60{,}000 \times 1.0150 = 60{,}900$ m tonnes. Let us now move the ship by means of tugs to water, also of apparent density 1.0150 kg/l but at 30°C. The total weight of the ship remains 60,900 m tonnes. When the glass hydrometer is placed in the water at 30°C it will expand and float slightly higher in the water. The reading at 30°C will be: $1.0150 \times (1 + 0.000025 \times 15) = 1.0154$ kg/l at 30°C.

The ship will also expand due to the water temperature of 30°C and will float slightly higher in the water. If the average underwater temperature of the ship is 30°C, the displacement scale reading corresponding to the new draft will be: $60{,}000 \times (1 - 0.000033 \times 15) = 59{,}970$ m^3 then $59{,}970 \times 1{,}0154 = 60{,}894$ m tonnes. The difference is $60{,}900 - 60{,}894 = 6$ m tonnes, or less than 0.01 per cent.

From the above it will be seen that only the hydrometer reading and not the water temperature is required for draft survey calculations. If desired, the water temperature may be noted for reference purposes as for example when a sample of water is taken for subsequent checking in a laboratory.

It may here be noted that some draft surveyors seem to believe that because a hydrometer is standard at 15°C, then the corrected hydrometer reading should be further corrected to 15°C by means of sea water thermal expansion tables. This is a fundamental misunderstanding and if applied to the example given above for water at 30°C would lead to an error of about 170 metric tonnes. A golden rule is, therefore, 'measure the water temperature if you must but do not use it in the draft survey calculation'.

The coefficient of cubical expansion of the metals, mainly brass, used for making metal hydrometers varies slightly according to their composition. An average figure would be 0.000055 per °C. Neglecting the temperature correction would therefore give larger differences in displacement calculations. Thus also for this reason the use of a glass instrument is preferred.

Hydrometers and hydrometry

In draft survey work the hydrometer may be regarded as the apparatus used to weigh bulk cargoes of 50,000 tonnes, 100,000 tonnes or even more than 200,000 tonnes. In these circumstances it should be obvious that great care should be exercised in choosing the right type of instrument and checking its accuracy. The time has long passed when a brass instrument, possibly corroded and maybe deformed, is acceptable for draft survey purposes.

Those interested in the subject will find much useful information in British Standard 718: 1979—Density Hydrometers or in ISO 387—Hydrometers—Principles of construction and adjustment.

Hydrometers for draft survey purposes should be made of glass as they can then be certified by an official standardising body. Metal hydrometers cannot usually be certified because they may be corroded or deformed after they have left the manufacturers' premises.

The reading of a hydrometer is affected by the surface tension of the liquid in which it is used. High surface tension liquids will tend to pull the instrument more deeply into the liquid than will liquids of low surface tension. In practical terms if a hydrometer calibrated for use in petroleum is used in sea water, then the hydrometer reading will be too low by an amount depending on the size and shape of the instrument. Correction tables will be found in the relevant British Standard.

As previously mentioned, hydrometers used for draft survey purposes should have some form of certificate. The hydrometer scale should be numbered so that the number may be quoted in the certificate. Certificates are of two kinds:

(i) A certificate of conformity, which is issued by the manufacturer and which certifies that the instrument is accurate within a specified maximum limit.
(ii) A correction certificate issued either by the manufacturer or by some official body such as the British Standards Institution, giving corrections to be applied to the scale readings at, for example, four or five different points on the scale.

It is suggested that although hydrometers with certificates of conformity may be acceptable for practical work at the dock side, at least two instruments with official correction certificates should be available in the office for checking purposes and in case of dispute.

Although the use of a hydrometer such as the new Zeal instrument will avoid many errors and misunderstandings which occurred in the past, this does not mean that other types of hydrometer cannot be used, provided that they are acceptably accurate and that the necessary corrections are applied.

Such hydrometers should be made of glass, should be adjusted for the correct surface tension, should cover the range of about 0.990 to 1.040 and should preferably have a scale length of minimum 125mm. The instrument should be clearly marked to show the density or specific gravity basis used, the calibration temperature, the surface tension value, a serial number and the manufacturer's name or mark. It should also have a correction certificate or certificate of conformity.

Variety of scales

A bewildering variety of hydrometer scales is available depending on the purposes for which they are intended. Only a few will be mentioned below:

Density hydrometers graduated in g/ml at 15°C—These hydrometers are graduated in terms of density *in vacuo*. If they are used then 0.0011 g/ml should be deducted from the reading, so as to obtain the corresponding weight in air in kg/l.

Hydrometers graduated in Sp Gr 15°C/4°C—These hydrometers give specific gravity against water at 4°C and for most practical purposes the readings are numerically equal to the readings of a hydrometer giving density *in vacuo*. The procedure given above therefore applies.

Hydrometers graduated in Sp Gr 60°F/60°F—Hydrometers graduated as above were commonly used in the petroleum industry but should not be used for draft survey due to surface tension differences. If such hydrometers are found, graduated for medium surface tension, them 0.0020 should be deducted from the reading so as to obtain weight in air in kg/l.

Table 1: Specific gravity and density of air-free distilled water.

	Specific Gravity		Density	
Temp.	T°F/60°F in vacuo	T°C/4°C in vacuo	True Density in vacuo g/ml	Apparent Density in air kg/l
4°C	1.001 0	1.000 0	1.000 0	0.998 9
15°C	1.000 1	0.999 1	0.999 1	0.998 0
25°C	0.998 0	0.997 1	0.997 0	0.996 0
35°C	0.995 0	0.994 1	0.994 0	0.993 0
60°F	1.000 0	0.999 0	0.999 0	0.997 9

The above specific gravities and densities are those which should be obtained by using a correct hydrometer, after correcting for thermal expansion of the glass or metal of which the hydrometer is made. The table has been calculated to four decimal places, from BS 718: 1979, Table 11, the litre being taken as $1 dm^3$.

Sampling and sounding ballast tanks

When vessels arrive at a loading port they frequently contain substantial quantities of water ballast in order to maintain the vessel in reasonable trim. This water ballast may have been taken on board at various stages of the incoming voyage and the various ballast tanks may contain water of differing apparent densities. It is therefore most important that each ballast tank should not only be sounded but should also be sampled and the apparent density determined. This is not always easy because some vessels would seem to have been so constructed that it is difficult to take samples of adequate size.

Similarly, when the vessel has been loaded the ballast tanks will normally have been pumped empty and sounding the ballast tanks is an essential part of the loaded ship survey. Practical experience shows that ballast tanks regarded as empty may contain appreciable quantities of unmeasurable mud or sediment. When circumstances permit it is therefore preferable to arrange with the chief officer that a sufficient quantity of water should be left in the ballast tanks to cover the tank bottom, so that the tank can be sounded. These small quantities of ballast can be pumped out after the loaded survey has been completed.

The above procedure will not measure the volume or weight of the mud or sediment. It will, however, ensure that the volume of ballast water pumped out is measured by the difference between the horizontal water levels before and after pumping. This volume difference is not affected by any mud or sediment remaining below the water levels before and after pumping. As a last thought in connection with ballast tanks we would like to ask shipbuilders to consider whether clear identification of each ballast tank could not be welded on to the structure near to the sounding point. Experience has shown that removable identification plates attached to the structure by bolts or screws may be transposed, accidentally or intentionally, during drydocking or at other times.

The sampling and measurement of ballast tanks is often a vital part of the draft survey procedure. Co-operation between shipbuilders and experienced draft surveyors to ensure that dip pipes are of adequate diameter, are perforated throughout their length and convenient for sampling and measurement would thus be most constructive.□

TANK CALIBRATION

R.I. Wallace, BA, FInst Pet, MCMS, MNI
Chairman, The Institute of Petroleum Calibration Panel

ON FIRST CONSIDERATION the need for any land tank to be accurately calibrated may appear unnecessary. Modern practice is to measure the volume of bulk liquids transferred from one container to another by means of an 'in line' metering system. The need for tanks to be calibrated is connected with the word 'transferred.' Meters measure transferred volumes; they do not and cannot measure static volumes, and hence the need for accurate calibration of tanks. Calibration not only serves this need, but also provides a means of verifying meter performance if statistical analysis methods are applied to the metered quantity and the quantity delivered 'ex tank.'

Why calibrate ship's tanks? In some trades—e.g., liquefied natural gas, the ship's calibration is used as the basis for the bill of lading quantity. In other trades the need may not be so obvious, but it has become the rule rather than the exception that the vessel is on charter to the oil company and not owned by them. It is also likely that the cargo has been purchased from a third party rather than coming from an owned oil production source. The ship has the advantage of being the only part of the transportation system which is common to both load and discharge port and when accurately calibrated provides an ideal measurement station.

I quote from the Institute of Petroleum's *Measurement Manual:* 'Accuracy in measurement is essential in the sale, purchase and handling of oil. It not only obviates possible disputes between buyer and seller but it also provides the only reliable means of maintaining adequate control over storage and distribution losses . . . The calibration of storage and transport tanks is the basis of all measurement (other than metering) which are required in the handling of oil. No bulk quantity in a container can be determined with a greater precision than that inherent in the calibration of that container. Any error made in the calibration acts constantly in one direction and during the long period in which the tank tables remain in force such errors can involve very large quantities of oil.'

Recognised standards

Three internationally recognised bodies prepare and issue standards for the measurement, calculation and production of calibration tables: the International Standards Organisation (ISO), the Institute of Petroleum (IP) and the American Petroleum Institute (API). The three bodies attempt to maintain a common standard and approach but, unfortunately, differences do arise in the issued procedures.

Land tank designs conform to prismatic shapes—e.g., cylinders, spheres, rectangular boxes, etc. The methods of calculation are all based on this fact and simple geometric formulae are used in the procedures. The methods of measurement are all designed to provide dimensions which will enable the simple formulae to be used but also to give an acceptable level of measurement redundancy—i.e., too many measurements rather than too few.

Methods of measurement available are gravimetric, volumetric and dimensional. Gravimetric methods entail weighing the container both full and empty. The liquid used to fill the container must be of known density and temperature so that the volume which it occupies in the container can be calculated. This method is not in general use, but is used in the calibration of tanks which are mounted on strain gauges. Effectively the gauges are calibrated and volumes of liquid, subsequently stored in the tank, are calculated by pre-programmed, on-line, computers.

Volumetric calibration is carried out by introducing known volumes of a liquid into the tank to be calibrated. After each batch of liquid is introduced into the tank the liquid surface is allowed to settle and its height, above a datum point, measured. (These heights are usually refered to as 'dips'.)

Both gravimetric and volumetric methods are usually only used in the calibration of comparatively small containers. The difficulty with both is associated with potential variations in liquid temperature. However, both methods offer a lesser uncertainty (more accurate calibration), if properly carried out, than the third method, dimensional measurement.

Measuring dimensions

Dimensional measurement methods are those which are normally used in the calibration of large storage containers. The vertical cylindrical tank is the most common type of bulk liquid storage container. They vary from 3 metres to in excess of 250 metres in diameter and from 3 metres to in excess of 30 metres in height. Some have fixed roofs, others have floating roofs and some have both. All are calibrated using the same methods.

A vertical cylindrical tank is constructed of a series of curved plates welded together to form a ring or course of plating. The courses are erected one atop the other to form the tank. The tank is measured and calculated as a series of cylinders mounted one above another, measuring and calculating each course separately. The reasons are two fold: it is easier to carry out the measurements on a course by course basis; and secondly it is likely that the thickness of the plating used in the construction of each course will differ, the courses getting progressively thinner with increasing height.

The volume of a cylinder is given by the formula $\pi r^2 h$, where r is the radius of the cylinder, h is the height and π is a known constant. Measurement of height, h, is a simple matter of measuring each plate's vertical height and recording it. Measurement of radius, r, cannot be carried out by direct measurement. All oil storage tanks conform to a construction standard which allows some deviation from a true circular cross-section, and thus the centre of the 'circle' could not be found. The normal method of obtaining radius is to measure circumference and derive r mathematically. The circumferential measurements are made at positions laid down in the particular standard method being used. The IP lays down three measurements per course, API lay down either one or two per course dependent on the accuracy and purpose of the calibration, and ISO adopts both, the number and position being dependent on the reasons for which the calibration is being undertaken.

The original means of obtaining the circumferential measurements was to encircle the tank with a steel tape especially manufactured for the purpose. The tape is called a strapping tape and the method known as strapping. Other methods such as optical reference line, triangulation, and electro-distance ranging are increasingly being brought into use. These methods are not necessarily more accurate but they require less manpower and are considered to be intrinsically safer. Whichever method is adopted, the resultant dimensions are circumferences or diameters.

Optical method

The optical reference line method is based on measuring the distance between the tank shell plating and optical lines of sight set up vertically around the tank at a predetermined number of stations. The measurements are termed 'offset' measurements and are taken at the same heights, on the course of plating, as that at which a strapping measurement would be taken. One set of offsets, referred to as the reference offsets, are taken at a position where a circumferential measurement has been taken previously. The reference offsets are averaged as are each set of offsets at the higher levels on the tank. The difference between the average of each set of offsets and the average reference offset is then multiplied by 2π and applied to the circumferential measurement taken at the reference position to give the circumference at the higher level. By this means the necessary measurements of the tank barrel are obtained.

The triangulation method is based on either a single station or a two-station triangulation. Basically it relies on taking angular measurements of fixed points on the tank shell and then mathematically converting them into circumferences. The mathematics involved are fairly simple trigonometric relationships based, in the case of the single station method, on a reference strapping and in the case of the two station technique on the known measured baseline length.

The electro-distance ranging is a method based on the same principles as radar. A beam of short-wave energy is aimed at predetermined positions on the tank shell and distance derived from elapsed time between emission and reception of the reflected beam. As the method is carried out inside the tank from one fixed position the distances derived from the ranging device at each level will give a measure of the tank's radius at that level.

Another, infrequently used, method is that of photogrammetry, which is however more frequently used in ship tank calibration.

Other measurements

Whichever method is used, other measurements are necessary to complete the work. These include the measurement of any items which either add to or reduce the available capacity of the tank. Such items as heating coils, stirring paddles, manhole openings, etc., are examples. The tank bottom must be calibrated either by survey techniques or by liquid calibration methods and most importantly the position of the dip datum in relation to the other measurements determined. The dip datum is the point on which the calibration chart will be based. Any error in its position will lead to a systematic error in the calibration table.

Calculation of the table is then a careful process of applying the routine method laid down in the standard. Calculated manually the process is long and exacting and great attention has to be paid to detail to ensure that the resulting table is correct. It is obvious that a properly programmed computer is a great asset.

The other types of land tanks—spheres, horizontal cylinders, noded spheres, etc.—are all measured in a similar manner to that applied to vertical cylinders—that is, they are viewed as mathematically perfect solids and the measurements taken allow the calculation of a calibration table. The methods of obtaining the measurements are tailored to the type of tank but basically the same rules apply, sufficient to ensure a redundancy of measurement. Calculations follow the accepted formulae for the tank being calibrated.

Ship's tanks are slightly more difficult to measure than land tanks primarily due to the shaping of the hull which in most crude and products tankers forms the containment system.

Some liquefied gas carriers have cargo tank systems which are self-standing and mounted in the holds of the ship. Such tanks are calibrated in the same manner as similar land-based tanks, although they tend to be larger and do present problems of scale when measuring.

The determination of the volume of liquid in a ship's tank requires a knowledge of both the capacity and geometry of the container. In some cases, particularly liquefied gas carriers, it is also necessary to know the temperature of the tank containment system and that of the measuring equipment used. This knowledge permits calculation at a known base temperature, enabling subsequent cargo calculations to be carried out at any temperature, adjusting container size for thermal effects.

The selection of the calibration method will depend on the size and shape of the container, the facilities available and the requirement to include trim/list correction tables in the completed calibration tables. Tanks can be calibrated by either the liquid method or by the internal physical method. It should be noted that the former method does not provide any information on the geometry of the container and cannot therefore be used in the calculation of trim/list corrections.

Choice of method

The choice of method is dependent on:

(1) Whether the vessel's structure will allow the tank to be filled with liquid, usually water, in dry dock. It is essential to maintain even-keel conditions during calibration.
(2) Whether the vessel can be liquid calibrated while floating as an even-keel condition must be maintained during the whole operation.
(3) Whether an adequate water supply is available at the required flow rate and without entrained air (meters will measure air just as efficiently as other fluids).

If any of the above points cannot be adquately met, then accurate liquid calibration is not possible.

Corrections for the effects of trim/list can only be calculated from data obtained from internal measurements. It is obvious that the shape of the container will affect the level of the liquid, measured at the dip point, when the container's attitude is altered. In a simple rectangular box shape the liquid level will rotate around the point defined by the intersection of two horizontal diagonals, but consider what happens to the level in a foreward or after shaped wing tank. Even in the simple box the calculation of trim/list corrections at the top and bottom of the tank, where the liquid surface is not in contact with all four bulkheads, requires a knowledge of the geometry and dip datum position.

Ship's tanks vary from the crude oil/product carrier where the hull constitutes the containment system through double-hulled LPG/chemical carriers to self-standing, rigid and membrane systems fitted in LNG vessels. They are built in a variety of shapes and sizes to give the vessel her maximum carrying capacity consistent with good, safe design.

Tank types

There are five main types of tank:

Shell or hull type—This is the most common type in which the hull forms the outer containment and internally is divided by longitudinal and transverse bulkheads. Calibration can be achieved by either liquid or internal measurement methods.

Rigid self-standing prismatic type—Tanks of this design are either built and then lifted into the holds or built within the hull as the vessel is constructed. The systems were first built for LPG/LNG carriage and the standard materials were cold temperature steels or aluminium alloys. Liquid calibration of such tanks may not be possible as they are designed to contain liquids with far lower density than the normal calibrating liquid, water.

Cylindrical or prismoidal type—These tanks are also rigid self-standing types, but normally built to pressure vessel standards, if somewhat larger than the normal shore-based pressure vessel. Calibration of this type can be either liquid or internal physical.

Spherical type—Spheres are rigid and free standing usually mounted in a chined saucer-like construction or in an 'egg cup' (more properly a steel hoop mounted on pillars fixed to the hold's tank top). Calibration can be either liquid, internal physical or photogrammetric.

Membrane type—So far this type of construction has only been used in LNG carriers due to the extremely high capital outlay involved. LNG is kept liquid by insulation systems not by pressure, having been liquefied in a shore-based plant. The containment system is a barrier mounted on insulation which is affixed to the inner hull of the vessel. One system uses two 'Invar' metal membranes separated by plywood boxes containing purlite insulation, the outer box system being mounted on the inner hull. A second system uses waffled stainless-steel plates fixed to a composite insulation system mounted on the inner hull. Both systems are built to exacting standards as they have to operate in an extremely hostile environment at temperatures at or below −163 C°, the boiling point of LNG. Calibration has to be by internal physical means as liquid calibration would impose too great a stress on the structure.

There are a number of problems in carrying out an internal survey to achieve the high levels of accuracy required in large tanks (20,000 cu m) due to:

- Measurements between tank walls in excess of 30 metres where corrections for tape sag may be subject to errors.
- The large number of measurements required in order to obtain a capacity which includes the plate undulations.
- The distortion or 'strain' on the tank due to the pressure of liquid which cannot be evaluated in terms of additional volume.

Free span

The 'free span' method of measurement with a tape is accurate up to approximately 10 m. In excess of this errors in both measurement and sag correction become too important to be disregarded. New techniques had to be devised to overcome these problems.

Traditionally a ship's tank was measured by taking transverse cross sections at known distances along the tank's length. These sections were accurately measured and defined the area of the section as exactly as possible. The section areas were calculated by either integration or more normally by use of Simpson's rules to obtain areas defined by height. The results of this calculation were then integrated a second time to obtain volumes relative to height, The 'open' volume was then adjusted for the internal structure present in the tank and further adjusted so that the heights were relative to the datum from which the liquid level in the tank would be measured when loading or discharging.

A 'new' method known as the internal alignment method uses a very low divergence (and very low power) laser beam aligned horizontally and approximately parallel to the tank side. The distance from the tank side to the beam is measured with a metre rule graduated in millimetres; the deviations of the tank walls from a straight line, the laser beam, can then be determined. The distance between the tank walls is measured along each tank wall where the tape can be fully supported throughout its length and the correct tension applied. In effect the method is to build a 'laser beam box' and measure from that 'box' out to the tank walls at various levels in the tank. Knowing the size of the 'box' and the offsets measured from it to the tank walls, the distance from wall to wall can be then calculated.

Photogrammetric methods have been used, especially in the USA and Japan. Photogrammetry is a method using overlapping photographs taken with special cameras whose lens systems are as near perfect as can be made. Targets are placed on the tank walls at positions which define the tank shape; reference tape is also placed in camera shot. The photographs are then placed in a stereoscopic viewer and the dimensions of the tank determined by scaling from the targets using the reference tape as the base for measurement. The method is extremely accurate but the equipment is prohibitively expensive unless used for purposes other than ship tank calibration.

Other methods have been investigated, such as a radio frequency system of direct distance measurement, but problems rapidly became apparent. It is hoped that electro-distance ranging methods at present under development for shore tank calibration will eventually be adapted for ship tank calibration work.

VLCC problems

A further aspect of ship calibration is in the measurement of the tanks in VLCC/ULCCs. Such large ships have a tendency to hog or sag to a quite alarming extent, their bulkheads also tend to bow considerably when subject to unequal loading in adjacent tanks. In addition to these points it is not unusual for bulkheads to be misplaced in the hull by anything up to 10 to 15 cm. This makes any attempt to calibrate from drawings, the normal method, subject to large uncertainties. Uncertainties of 0.5 to 1.0 per cent are not unusual, but in terms of cargo volumes such percentages are unacceptable. It is considered highly impractical to carry out a full calibration of such vessels. Each tank would have to be scaffolded to allow measurements at the higher levels in the tank, the tanks would have to be cleaned and gas freed, be equipped with lighting plus all the other expenses which would be incurred by taking the ship out of service for a period of up to three months, not to mention the loss of earnings involved.

A simple inspection of the calibration tables will usually show that the wing tanks share the same calibration. This is not possible. It is more than likely the quantity below datum—i.e., that quantity below the point at which the dip tape weight strikes the tank bottom—will be zero. Again it is not possible unless the ship has a completely flat bottom. It may be found that the tables come complete with trim and list corrections, but these are a single line of corrections based on various trims and lists, no account being taken of varying liquid levels. Again this is not correct and is calculated from a simple tangent formula using tank length, LBP and the dip position.

A different approach is required and this involves what has been termed 'enhanced drawings calibration,' which is a way of saying that certain basic measurements are taken in each tank and then the drawing dimensions adjusted to the physical dimensions taken 'in tank.' Careful scrutiny of most ship's drawings and plans will either show them to be 'as designed' or 'as built.' Whichever is the case it can normally be shown that neither type of drawing is what it purports to be in every respect. Careful measurement of the main dimensions of each tank, measurements such as lengths, widths, dimensions of the main items of internal structure and the exact position, in terms of position and height above base line of the dip datum, will show differences when compared to those measured on the drawings.

The calculation is carried out using the drawing dimensions but adjusting them so that they agree with the actual dimensions measured 'in tank.' By this means a revised calibration table can be prepared which will reduce the uncertainty of calibration to acceptable levels in the region of 0.05 to 0.1 per cent.

Calibration is an exacting job. Some would have that it is a science, but there is an element of art in its practice. A knowledge of the different methods of measurement, calculation and preparation of final tables is required. A knowledge of computers and computing is a distinct advantage but more important is a sense of humour when climbing out of a nice warm bed at five o'clock on a cold, damp November's day knowing that there is a tank out there to be calibrated.☐

OIL, FAT AND GAS CARGO INSPECTION AND THE DETERMINATION OF LOSS FACTORS

N.C.I. de Spon, MNI
Caleb Brett Services Ltd

WHILST THE TITLE of this chapter covers the vast range of products from crude oil, through refined products, chemicals and gases to edible oils, it must be understood that one chapter could not cover every particular requirement for each commodity. It can therefore be only a guideline to general good practices within the inspection field. Specific publications on gases, chemicals and edible oils are shown in the appendices. Throughout this section the word oil therefore should be taken to mean the cargo under survey. In many cases, parts of the survey will be superfluous, but it should be remembered that following the outline will ensure that a full and thorough survey is undertaken.

If the purpose of this article is to give guidance to cargo inspectors for measurement procedures, it is necessary to define what is a cargo inspector? A cargo inspector may be regarded as a person who, by reason of his knowledge and practical experience in the field of bulk oil cargo measurement and analysis, is competent to provide impartial judgements, reports and recommendations on matters relating to the quantity and quality of these cargoes.

It is also important to specify the purpose of a cargo survey, which is to provide a form of certified statement of the quantity and quality of oil loaded or discharged and to highlight matters which may be relevant to the protection of the client's interests.

To achieve this, the cargo inspector has a number of responsibilities. Some are of a general nature. Others are highly specialised and clearly defined. The general are listed below and the specialised form the subject matter of the rest of the article. Safety matters and related responsibilities are not included, but the need for cargo inspectors to be conscious continually that safety requirements take precedence over all other considerations must be emphasised.

When is a survey needed? When a cargo of oil is transported by ship from one terminal to another a survey is undertaken to:

(a) Establish the quantity and quality of oil loaded (i.e., to confirm, compile, or indeed to dispute the information shown on the bill of lading);
(b) Establish the quantity of oil received by the receiving terminal;
(c) Establish the outturn difference—i.e., between the quantities established under (a) and (b) above;
(d) To provide a time log of the events;
(e) To identify other conditions at either the terminal or the vessel which may affect the above;
(f) To provide certified documents which may be used as a basis for the recovery of losses, the settlement of demurrage and despatch claims and assist in arbitration or litigation settlement.

As with any industry there are a number of terms used in the oil survey business peculiar to this field and a glossary of these is included in the appendices. These terms should be clearly understood by the inspector as confusion between ship, shore and inspector has in the past led to a lot of time and money being wasted.

General principles

There are a number of general principles that should be understood and followed by all cargo inspectors. In doing this the inspector should be aware of the principle of independence and must protect the confidentiality of his client.

An inspector should:

(1) Understand for which party he is acting and be properly nominated to do so;
(2) Be aware of the terms of the charter party/contract which may affect his client;
(3) Fully understand those elements of quality/quantity that relate to the cargo and may be of particular importance;
(4) Be properly equipped to undertake a survey and be aware of his client's instructions regarding the use of this equipment;
(5) Be aware of international and local standards for inspection, testing and equipment;
(6) Be aware of the safety requirements pertaining to inspection work in general and the particular cargo being inspected;
(7) Make himself available in time to carry out (or witness) each stage of the survey which he has been appointed to attend;
(8) Perform the survey he is instructed to perform noting where he is unable to comply with instructions;
(9) Keep the client informed before, during and after the survey of relevant details and issue a report of his findings promptly.

These general principles give a general outline of the inspector's duties. However the following give a more detailed description of a cargo survey and identify a general path that can be followed. Every survey is different and it is impossible to deal with all contingencies. Therefore these notes can only be used as the basis of a survey and the cargo inspector will rely on his own judgment to build a survey around them.

Cargo inspection

Key meeting—Before any cargo operation commences, the cargo inspector should meet all key personnel concerned with the operation to discuss operational plans. The approval of the master should be obtained for survey procedures performed aboard and that of the terminal operator for those ashore.

Naturally any applicable government, local port authority, and terminal regulations shall be complied with.

Loading—In general cargo will be delivered to a vessel from either standing tankage or via a meter bank. With standing tankage it is necessary to determine the quantity and quality of material in the shore lines from the tank to the vessel, the quantity and quality contained in the shore tank, and to obtain samples of this material as appropriate. Where the line volume represents a significant proportion of the quantity to be loaded the line contents should also be sampled for analysis.

When loading is via meters and the quantity figure is to be based on meter figures it is necessary to determine the type, size and maximum flow rate of the meters together with the position and accuracy of the temperature probe. In addition the average flow rate should be recorded for the intended cargo plus the temperature, viscosity and grade. The monitoring of meter performance during loading should be properly understood by a cargo surveyor.

Shore side

Shore tanks—Shore tanks should be examined for noticeable deformities which might affect the tank calibration data. Prior to gauging they should be isolated from other systems by closing and sealing valves on the filling, crossover and drain systems. Supply valves may be left open to the line depending on the system in use. Where tanks have floating roofs these should be free of debris, snow, water, ice, etc., and should neither be grounded nor in the critical zone.

The calibration tables for the tank should be checked to record the last calibration check date together with the issuing authority. In addition, record data regarding the measurement point and referenced height. It is advisable to record when the tank was last cleaned and inspected and if any repairs have been effected. If the tank has been in recent use or if mixers have been used, a period of 30 minutes should be allowed for settlement before any gauging is performed.

The tank reference height, which should be prominently marked at the reference point, should be compared with the calibration table. It should then be confirmed by measurement. The tank dip or ullage should be measured using approved equipment and the measurement should be checked until two consecutive measurements agree within 3mm.

Where tanks have sludge or debris present on the bottom, ullage measurement is preferred. This measurement should be related to the tank reference height. Where water is present in the tanks this should be gauged using water-finding paste and a steel tape or a portable sonic tape.

The tank temperature should be obtained using an electronic device or mercury-in-glass thermometer. Temperatures should be measured at a number of levels in the tank to obtain a more accurate assessment of the temperature profile and these should be averaged. Tanks should be sampled as required by the client and as specified within the industry.

Measurement procedures

Where parties concerned mutually agree, automatic tank level gauging and temperature measurement systems may be used for custody transfer. Wherever possible, the surveyor should take his own measurements and compare these with those recorded by the automatic gauge system. Where terminals do not allow surveyors to take these, the surveyor should satisfy himself from the terminal's gauge-proving records that the gauges are satisfactory, and should make an appropriate note in the general comments of his report.

Where the difference between the change in *tank level* measured by automatic gauge and by manual gauge is less than 7mm, the use of the automatic level gauge is acceptable. Note that level comparisons become unreliable in high wind conditions with floating roof tanks. The difference in *tank temperature* measured by automatic temperature gauge and a manual electronic thermometer shall be less than 0.5°C.

Before gauging, the surveyor must determine the nature and quality of material in the shore lines and the total capacity of the lines, from the vessel's manifold flange to the shore tank(s) in use. Record what steps were taken to determine that the shore pipeline was full of liquid. This check may take the form of a physical line displacement at the beginning of loading from a single shore tank to a single ship's tank. Alternatively, the line may be proved full or empty by inspection of high or low points on the line. Pressurising the line by using a shore tank opened on the system is often an effective method.

The terminal should arrange for lines and valves to be set so as to prevent cargo being contaminated or lost through other lines and tanks and this should be confirmed in writing. The inspector must satisfy himself as to the system integrity and report his findings. Attempt to ascertain the previous cargo in the line and if possible the density and temperature of this.

Meters

Before loading, the inspector should record meter data as previously described. Meter readings should be recorded after the correct line-up procedure is completed and where meters are zeroed the master counter reading shall be recorded.

Automatic sampling equipment may be employed during loading, and this should be checked prior to the start. Details such as the make, type, sample frequency, control settings, etc., should be recorded. Amongst the points to check are the siting of the sampler to ensure that proper mixing takes place and that the equipment is properly set up to accurately sample a representative quality of the material passing the sample point.

Ship inspection

Where possible, record the vessel's draft, trim and list. Check these against automatic readouts in the control room. It is advisable to make a quick visual inspection of the deck at this time to familiarise yourself with the vessel's general layout, the position of tank hatches and dip points, the cargo system on deck,

COW and IG arrangements, venting systems, inspection ports, etc.

Further to this it will be necessary to study available ship's drawings and plans to record details of the vessel. These will include the general arrangement plan. From this the position of gauging points, the length and breadth of tanks, the layout of pipework and the pipeline quantities may be determined. The surveyor should also check the ship's tank calibration tables to obtain the tank heights and whether the pipeline quantities are included in the tank capacities.

Whenever possible, tanks should be visually examined from deck level so that an accurate picture of the interior can be built up. Where the appropriate safety precautions have been taken, tanks may be inspected by entry. A physical inspection will normally be necessary to examine the condition of the tank surfaces, heating coils, pipework, submerged pumps, etc., and to assess more fully the cleanliness of the tank, the absence of odours and the integrity of the coating.

Where physical inspection is not necessary, the amount and nature of any OBQ should be determined prior to loading. This should be described as either liquid, non-liquid, free water or sediment, and where possible the temperature should be measured and a sample obtained. It is convenient at this time to check the tank reference height and this should be compared with the tabulated height in the calibration tables.

The inspection of a ship is a complex undertaking and cannot really be taught in a classroom as it is often a matter of experience. Whilst guidelines can be written for it, there is one aspect that cannot be covered. That is the feeling of knowing that something is not quite right. This feeling only comes with time and experience and may be ill founded in many cases. A full inspection of a vessel should include a check of the ballast system, the void spaces, the pumproom and the bunker system. This will depend on the time available, the nature of the survey and other factors.

When the surveyor is satisfied that he has obtained the necessary information to report on the vessel's condition prior to loading he should allow loading to commence. In some cases this may require his advice and supervision. It may also require the taking of check readings, samples, etc., etc.

During the loading the role of the surveyor can vary. It may be that he will play an important part in the loading being involved at all stages. It may be that sampling and analysis will need to be performed. In addition, he may have to monitor the operation of meters, samplers and the like. It is at the completion of loading that the greatest involvement of the inspector will be required.

Ship after loading

The survey of the vessel after loading will necessarily be both rigorous and extensive. The cargo tanks on the vessel will be checked for ullage and temperature and should be gauged for water. If this can be done on an even keel this makes calculations easier. It is important that all tank valves on the vessel are closed prior to this survey and that they remain closed during the survey. Not only should the loaded tanks be checked, however, it is important to check non-loaded compartments, void spaces, ballast tanks, etc., and a further bunker survey may be undertaken at this stage. Sampling of the loaded tanks will be important and proper procedures should be understood by a competent inspector.

When the physical inspection and gauging on board has been completed, the correct calculation of the vessel's loaded quantity should be made using the appropriate corrections. Application of the vessel's experience factor should, if properly prepared, provide a reasonable check on the quantity stated on the shore loading certificate. If it does not, then an investigation should be undertaken to find out the reason for the discrepancy.

The shore tank check after loading where applicable will be undertaken in the same manner as prior to loading. It must be remembered that the shore figures are generally the most important figures there are. They are the ones on which are raised the bill of lading. Whether this is by tank dipping or by metering they should be carefully checked as errors are very difficult to rectify and considerable sums of money depend on the reliance all parties can place on the information on a bill of lading.

This, then, should conclude the general conduct of a cargo inspection at a load port. There may well be a requirement for analysis of samples, but this is too complex a subject to cover in a single article. It is sufficient to state that analysis should always be performed by competent personnel working to approved methods and using standardised equipment. It can only be witnessed using this same criteria.

Determination of loss factors

The inspection of a cargo at a discharge port follows to a great extent a loading survey in reverse, and the care and attention to detail shown at the loading should apply to the discharge as well. However, there will come a time after the discharge when the quantity and quality of the cargo discharged will be compared with that at the load port. It is for this reason that the inspector's impartial report of both operations is of vital importance as rarely, if ever, will the quantity loaded be the same as that discharged.

With a general cargo this does not often apply, as shortages or loss are generally easily noted. After all, ten packing cases are ten packing cases, five cars are five cars and damage to these can be noted with some ease.

A bulk cargo, such as coal or grain, is more complex as the measurement of these relies on either a draft survey, some kind of weighing scale or both. The quality however should not change radically provided the carriage has been properly undertaken. This is not the case with a liquid cargo for a variety of factors and these are outlined below.

Measurement of volume

All bulk liquid cargoes are measured by volume. Many cargoes will also be bought and sold by volume whilst others will be bought and sold by weight. Whilst the *volume* of a solid bulk cargo is not a factor in most cases, it is of vital importance therefore in all and every liquid cargo.

Unfortunately the measurement of volume is extremely difficult to determine accurately and it is further complicated by temperature. Liquids expand and contract with temperature changes and unless volume is stated at a standard temperature it is largely meaningless. Measurement of temperature is an inexact science. A medical thermometer placed in the mouth will record different temperatures in different parts of the mouth. Imagine the difficulties of accurately determining a single temperature to apply to a tank which may be 100ft (30m) deep, 150ft (50m) long and 100ft (30m) wide, with the bottom layers immersed in sea water of varying temperature and the top heated by the sun. Additionally, tanks may be affected by the temperature of other surrounding tanks and their contents. The temperature of liquid in a tank will vary throughout the tank both vertically and horizontally and yet we, as inspectors, must average these out and provide a single temperature to correct a volume that may be in excess of 1 million barrels. The effect of a difference of 1°C in the amount of the standard volume is shown in appendix 1.

Measurement of volume can be done in two ways. One is to pass the volume through a meter, the other to measure the contents held in a tank. Whilst metering is in general more accurate, physical tank measurement is more common. It is therefore important to ensure that the physical measurement is accurate, and yet trials conducted by several major oil companies have shown that different people measuring the same tank, at the same time, with the same equipment, will rarely arrive at the same result. The variance between their results will be in the order of 2mm however careful they are and in most cases they will average the figure between them. If this 2mm difference is applied to a large shore tank it will make a significant difference to the figures.

As can be seen therefore the determination of volume alone is fraught with difficulties. It is further complicated by the next category.

Disparity of systems

As if the job of the inspector was not complicated already by the difficulties of accurate measurement, he is also faced with two different systems in use. Oil cargoes were traditionally measured in barrels and long tons and corrected to a standard temperature of 60°F using the API gravity. This system is still regularly used throughout the world. A more logical system, the metric system, is also widely used where measurement is expressed in cubic metres and metric tonnes at 15°C (sometimes 20°C) using the density of the liquid. As the two standard temperatures 15°C and 60°F do not correlate, direct conversion between the two systems is not possible except at the same temperature.

As if this was not enough, let us further complicate the process. Given that the volume of a liquid changes with temperature and that this change is not linear, it has been found necessary to provide correction tables showing the correction to apply to correct the volume to a standard temperature.

These tables were recently revised (in 1980) to more accurately reflect the new crude oils on the market and the different behaviour of products. The effect of the new tables compared to the old is to reduce by a small amount the volume at the standard temperature. As this in effect is good for the purchaser but bad for the seller, these tables have not gained universal acceptance however genuine their aim may be and we now have the problem of two sets of figures being produced from the same basic raw data both of which are correct. This difference is shown in appendix 1.

Given that there are two systems with two sets of tables for each, it is hardly surprising that errors and paper losses occur. They are not physical losses as yet, merely technical ones. This leads us on to the next area of possible loss.

Disparity of testing

Despite an effort at standardisation on the part of the ISO it is a fact that the majority of the testing undertaken on oil products is to standards set by the various national or business organisations. These are the IP, API, ASTM, ANSI, DIN, etc. Therefore, a single test may have a variety of different methods by which a result may be obtained. For not only are individual organistions involved. There are also different methods issued by the same organisation for determining a result. An example of this confusion is shown in appendix 2. This problem can be compounded by two other factors, repeatability and reproducibility.

Repeatability

A chemist repeating analysis in a laboratory using the same method, equipment and, of course, sample of material may get two slightly different results. He may get a different result every subsequent time or he may get the same. As this is recognised to be normal, each test will have a range of repeatability of results. Provided that the results found fall within the range of repeatability then that result is acceptable. And yet so is a result at the other end of the range. As you can see it is getting more complex.

By the same token there is reproducibility. A correct result found in one laboratory may be compared with that found in another. The same test method may have been used, and the same type of equipment, but a different result is found. Provided this result is within the range of reproducibility then this result is also correct. This apparent contradiction is shown in appendix 3 by a worked example. However, it should be realised that whilst a loss of some kind appears to have occurred it is still only a technical loss.

We have now looked at the areas of loss or error that will arise at random on any shipment. The fourth area where a loss may occur is at the point of supply, and it is here that we start to get an actual physical loss of oil, although there are still a variety of technical or paper losses that can occur.

What are the factors that will produce errors or losses? Well, *What about the tanks that contain the cargo?* As a company that undertakes tank calibration work we will draw a veil over the tank that has been improperly calibrated. That was done by someone else. But let's face it whilst shore tanks are calibrated by a variety of methods these come down to strapping or diameters in the long run. Both these methods can suffer from inherent errors and are only as accurate as the person who calibrated them.

A tank that is calibrated when empty will be slightly incorrect when full. A tank that is calibrated when full will be slightly incorrect when empty. A tank that is calibrated when two-thirds full, a quite common practice, will be slightly incorrect when full and empty. You just can't win. Then there is bottom movement. Most tank bottoms move as the weight of liquid on them increases. It's called *springing* and can occur quite suddenly. It is difficult to predict or measure when this happens, but it will affect the volume in the tank.

Temperature

What about temperature? The expansion of a tank due to temperature change is small in the ranges that we are dealing with, but with heated cargoes it can become significant. After all the tanks are calibrated at a fixed temperature. If you are dealing with high temperature products in large tanks this factor should be borne in mind. I said earlier that tank calibrations would be incorrect at different stages of filling. What I mean by this is quite simple. The volume in the tank will distort the tank by the effect of pressure on the sides. The more pressure the greater the distortion. The distortion may have been allowed for in the calibration tables or it may not.

In a fixed roof tank the effect of this distortion is different. Here the ullage or dip may be affected as well as the indicated volume and this factor is never allowed for. I will not dwell upon the possible inherent errors in the tank anymore.

What about the actual physical problems? There are floating roofs with debris, rain water, etc., on them which affect their weight and therefore the ullage; tanks with stand pipes that are not perforated through the full height allowing a thorough mix of the contents of the pipe (this can result in inaccurate readings); tanks that are deformed or distorted due to accidents or mishaps; tanks that are gauged when their floatng roofs are in the critical zone. All these should give you pause for thought.

The operational factors which govern oil movements within marine terminals and shore installations are also subject to audit. For the purpose of custody calculations, it is often assumed that large pipelines are perfectly filled with the same grade of dry oil before and after transfer. In practice, however, there is concern towards the incidence of voidage pockets and possible vapour locks in elevated sections of piping, as well as the presence of free water tending to accumulate in lower sections. With products it is often slightly easier as line pigging is more common and free water is less.

Following from this there are two ways in which a loss can occur when considering pipelines. The first is fairly simple and is actually fairly easy to check. That is that the line is not in the same condition before and after use—i.e., that it contains air or water in different places before use, but contains oil afterwards. Obviously if it is your oil going into the line and you get out a mixture of your oil, water and air you will not be too pleased. The way to check this is by performing a line push or displacement, or for some products a 'slopping' operation to a road tanker or similar.

A displacement is particularly valuable with crude oil where SBMs or long pipeline systems are in use, as a 30-in pipe will contain about 45 m^3 per 100 metres. It does not take much of a percentage in a long pipe to lose a lot of cargo.

The second error in pipelines is again that the pipeline contents are different before and after. But whilst they might be full of oil before, it is not known what the temperature or density of the material is. Obviously this density difference would have to be quite large to affect the volume, but the affect of only a small change in temperature is different. The affect of a 1°C change is about 0.01 per cent of the volume of the pipe. If, therefore, there is a change of about 10°C in the temperature on a pipeline of, say 5 km, the difference in volume would be:

(1) Volume in Pipeline

$$= \frac{0.762^2 \times 11 \times 5{,}000}{2}$$

$$= 2280 m^3$$

Before volume at 68°F 20°C = 2,280m^3 × VCF .9962
After volume at 86°F 30°C = 2,280m^3 × VCF .9877
2,271 = Vol at 15°C
2,252 = Vol at 15°C

As can be seen this is a significant loss—and a 5km pipeline is the norm for offshore loadings and discharges.

Terminal systems are by no means as efficient as they would appear. A modern control room with electronic diagrams indicating what is going on might hide a multitude of sins, whereas a ship (whilst in many cases appearing to be far less sophisticated) can often be in far better condition. Detailed investigations over recent years have demonstrated that measurement errors and operational mishaps at terminals frequently accounted for the greater proportion of official outturn losses.

The fifth area where a loss can and will occur is in the handling of the material. The amount of this loss will depend on a variety of factors. However, this is a genuine physical loss although it is difficult to attribute accurately the portions of the loss to any particular factor.

Losses due to tanker operations

During the course of shipments of oil by tankers, it normally happens that some operational losses occur whereby the quantity delivered at the discharge port is somewhat less than the total supplied on-board at the point of loading. It has been found, however, that in the majority of cases the losses can be subdivided into the following categories.

Real losses occur, to a greater or lesser extent, over each stage of the shipment, and may include evaporative losses of the most volatile fractions or 'light ends' during loading, carriage, and discharge operations; additional evaporative losses during COW operations for crude oil; build up of oil on tank surfaces and internals as clingage; increase in ROBs in relation to the initial OBQ; unaccounted hold up in vessel's cargo lines and pumps; and accidental spillage and leakage or diversion to non-cargo spaces.

Evaporative loss

There are four distinct stages during carriage where the loss occurs. Firstly there is the amount lost during loading. Even with today's closed loading systems, this loss is considerable. The cargo is entering the tanks under high pressure through a small aperture and immediately has a big empty tank to fill. Vapour generation on the surface is very rapid, particularly when the temperature of the air or inert gas in the tank is high. And all the time this vapour is being forced out of the tank as the vessel loads. It is hardly surprising that the old practice of topping off tanks with sticks through open tank ullage ports has been done away with. The losses, quite apart from the dangers, were considered too significant.

Secondly, there are the losses occasioned during the voyage. These will naturally depend on the weather conditions and temperatures experienced and the duration of the voyage, but are caused in the main by the sun. During the day the sun warms the deck of the vessel and the cargo. The gas and air mixture in the ullage space heats up. Obviously there is a resultant rise in pressure which is partially contained by the PV valves. But eventually this pressure will be released to the atmosphere allowing further generation of gas to occur in the ullage space. If the vessel is moving to any sea this effect is increased as surges in the tank force the pocket of gas in the ullage space either out of the PV valve or through other points. The tank starts to breath with the movement. Vapour is therefore continually being lost to atmosphere and is continually being replaced with more vapour from the liquid surface. This breathing does not only occur during movement. As the night comes the tank cools down and the ullage space cools with it. A vacuum is formed which activates the PV valve. Air enters the tank and in turn becomes saturated ready to be emitted the next day.

Thirdly, there are the losses occasioned during the discharge. By this I mean that vapour that is lost when, having completed the discharge, the vessel sails. The empty tanks will be full of a mixture of vapour and inert gas, much of which may well be lost when ROB measurement is taking place. The rest will then be lost as the vessel will sail with it.

Fourthly, there are the increased vapour losses due to COW operations. It is obvious that spraying a jet of crude oil around inside a cargo tank will generate vapours. These on occasions will be vented to atmosphere due to over pressurising of the tanks. Obviously this vapour will also be lost during the ROB survey and this loss is increasingly being looked at against the benefits offered by COW.

What sort of figures are we looking at with evaporative loss? It is difficult to arrive at a figure for the losses incurred during loading as you are trying to compare a shore figure with a ship figure. There are enough problems built into that comparison without adding a further one. However, on the passage loss side there is an amount of research that indicates a figure of between 0.1 per cent and 0.15 per cent over an average 20-day passage at a temperature in excess of 25°C should be expected.

When the question of losses during a discharge are considered, again we are faced with the problem of with what to compare the before and after figure. It is generally held that the highest vapour loss occurs during discharge and this is further aggravated by COW. Estimates ranged from 0.05 per cent to 0.15 per cent, although one study came up with a figure of 0.4 per cent. One oil company's rule of thumb was that vapour loss from COW amounted to as much as 1 per cent of the throughput of the guns. This figure should be borne in mind when COW is being monitored.

Clingage

Well, what about clingage? What is it and how do you estimate it? Clingage is the material clinging to all surfaces both horizontal and vertical within empty cargo tanks other than the bottom surfaces. That is the official definition of the ISO and they also include a guide as to how to estimate it. It is based on the area of the horizontal surfaces other than the bottom within a tank. Provided this is known, the estimation of clingage can be related to the volume of non-liquid material found on the bottom of the tank. It is normally a problem only with crude oil and other heavy or sedimentary products. Unfortunately this ISO document was written without taking into account the effect of COW. Whether the equation has altered is unknown, but in my opinion it should be changed, as from inspection of tanks in VLCCs which have been crude washed and of those that have not I would say that you can only relate the two areas in the latter. Tanks that have not been COW'd exhibit a build-up of sludge almost uniformly, although there is normally more on the bottom.

Tanks that have been COW'd tend to be totally free of sludge in the upper areas, on all verticals and on most horizontals down to the bottom. The bottoms tend to be free of sludge to a varying degree, but the distance from the nozzle of the COW gun and the effect of shadow sectors is apparent. Clingage must be seriously considered as a factor in oil losses where COW has not been undertaken, particularly if the vessel had undertaken COW prior to loading.

The increase of ROB after discharge against OBO prior to loading will, of course, be a loss that is easy to quantify and one that is easy to demonstrate to a receiver. What is not so easy or at least is not so well looked at is the proper estimation of these figures.

An inspector in one area may look only at discharges, whereas in another area he may look only at loadings. It is unlikely that a proper OBQ or ROB survey is undertaken by either. (That is a statement that might not be acceptable, but I feel it is the truth.) The proper evaluation of the ROB, for example, requires that a number of soundings be taken at different points in a tank to attempt to build up a pattern of the bottom. But a sounding rod has an area of only one square centimetre, so four dips of an average tank on a large vessel, say, 40 metres by 20 metres, represents a sample of only 0.00005 per cent. This is hardly representative is it? And yet most times the tank only gets a single dip. If this is taken at the middle of the tank the presence of oil at the after end might go unnoticed. If it is done at the after end, after a discharge there may be a considerable amount of liquid oil held in pockets at the for'd end which will only reach the after end during the course of the ballast passage.

Whilst a vessel will arrive at a loading port with all tanks well stripped and drained, the OBQ might (and only might) be a fair guess of the quantity in the tanks. The ROB found after discharge is unlikely to be even that.

As for the material that might be held in the pipelines and pumps, there are further problems. During the ballast voyage tanks may well have been washed and cargo pipelines and pumps will have been flushed with water. The vessel will therefore arrive at the loading port with little, if any, cargo in these. At the discharging port it is a different matter. Even with the use of efficient eductors, stripping pumps and the Marpol line I think it is fair to reckon that only perhaps 80-90 per cent of the line and pump capacity will be stripped out and that remaining amount cannot be measured. On an average sized vessel it might amount to a considerable quantity which together with the underestimation of ROB in the tanks should be borne in mind.

Accidental spillage, leakage or diversion is actually less common, but normally far better documented as a source of loss. Vessels that have had a loss due to spillage or leakage are normally the subject of exhaustive investigation by everybody concerned. Leakage from cargo tanks either into other spaces or to the sea can normally be spotted either by a proper survey of the vessel at the discharge port or by examination of the ullage reports. Void spaces and ballast tanks that have inadvertently been filled with cargo are as much a matter for concern to the ship's crew as to an inspector. They should be investigated where and when they arise and should not be discounted. They should not, however, blind you to the effects of other loss aspects.

Deliberate diversion—It has taken a long time to get to this point and I'm afraid that I am going to have to disappoint you. This subject is far too large for the space available and far too complex. Should readers be interested in the various forms of diversion carried out by a very small minority of companies, they should look carefully at the opportunities that arise for this form of fraud. They are numerous and I firmly believe that a competent and experienced inspector should prove a deterrent to this type of loss.□

Appendix 1

True average bulk temperature	31°C
Gauged average bulk temperature	30°C
Density @ 15°C	0.8500
Observed volume of product	= 1000m^3
Correction factor for:	
31°C	= 0.9867 Table 54B
30°C	= 0.9875 Table 54B
Gross standard volume @ 15°C	= 986.7m^3 or 987.5m^3
The difference is	0.08%
Observed volume of product is	1000m^3 @ 31°C
Therefore the true gross standard *volume @15°C using Table 54B is*	*986.7m^3*

Observed volume of product is	1000m^3 @ 31°C
Correction factor for:	
31°C	= 0.9872 Table 54
30°C	= 0.9880 Table 54
Therefore the true gross standard *Volume @15°C using Table 54 is*	*987.2m^3*

If both the temperature is incorrect and the old tables are used, the gross standard volume found will be 988.0m^3 against the true figure of 986.7m^3. Therefore by applying a slightly incorrect temperature from an old table, still widely used, an error of 0.13 per cent has arisen.

Appendix 2

As stated, there are a variety of international and national standards organisations issuing a bewildering number and variety of test methods.

For example:

Determining the flash point of a petroleum product by Pensky-Martens closed tester is a recognised method and is issued as:

ASTM	D93
IP	304
ISO	2719
FTM	791-1102
EN	11
BS	2839
DIN	51758
AFNOR	M07019

and yet all of the tests are identical and a result found using one can be compared with one found using another.

Similarly there are six methods approved by the ASTM for determining sulphur in Petroleum Products. These are:

ASTM D129	Bomb Method
ASTM D1552	Induction Furnace Method
ASTM D1266	Lamp Method (Liquid)
ASTM D1551	Quartz Tube Method
ASTM D2622	Xray Spectrography Method
ASTM D2784	Lamp Method (Gas)

These methods are all different and may produce different results from the same sample. It is therefore difficult to compare one with another or evaluate the accuracy found.

Appendix 3

Repeatability and Reproducibility—A sample is drawn of a cargo of gas oil. This product has a specification for the cold filter plugging point of − 12°C. This sample is tested in laboratory A. The result found by the analyst is −10°C. The repeatability of the test is 0.033 (30 − x) where x = − 10°C (the result found)—i.e., both results are acceptable if within ±1.32°C of the result found (i.e., average 10°C).

Therefore, a second result can be between − 8.68°C or − 11.32°C but as whole numbers only are reported, this could be averaged with − 9°C or − 11°C:

− 9°C + − 10°C/2 = − 9.5°C = − 10°C

or

− 11°C + − 10°C/2 = − 11°C.

Both the above results show that the material is *off specification.*

The same sample is tested at another laboratory. The reproducibility of the result is 0.092 (30 − x) where x = − 10°C (the result found at the first lab.)—i.e., both results are acceptable if within ±3.68°C of the result of the previous *results* found (i.e., average 10°C). Results found could be − 13.68°C (− 14°C) or 6.32°C (− 6°C).

If, therefore, the second laboratory finds two results (within the range of repeatability, 1.32°C of each other) that are also within the range of reproducibility, − 13.68°C or − 6.32°C, then this result is also valid. If those results are − 13°C and 12°C, then the average of the two figures is −12.5°C or −12°C. This means that the material is within specification.

Which laboratory do you choose to believe?

Bibliography

The following books and publications provide source data for a more detailed study of this subject and should be consulted freely:

1. *Institute of Petroleum: Petroleum Measurement Manual.* Part XVI Procedures for Oil Cargo Measurements by Cargo Surveyors. Section 1. Crude Oil. March 1987. Published on behalf of the Institute of Petroleum, London, by John Wiley & Sons Ltd, Distribution Centre, Shripney Road, Bognor Regis, W. Sussex PO22 9SA, UK. ISBN 0 471 91656 0

2. *Manual of Petroleum Measurement Standards.* Chapter 17—Marine Measurement. Section 1—Guidelines for Marine Cargo Inspection—2nd Edition, January 1986. Published by the American Petroleum Institute, 1220 L Street, Northwest Washington, DC 20005, USA.

3. *Significance of Tests for Petroleum Products.* Prepared by ASTM Committee, D—2 on Petroleum Products and Lubricants. ASTM Special Technical Publication 7C. Published by the American Society for Testing and Materials, 1916 Race Street, Philadelphia, Pa. 19103, USA. ISBN 0 8031 0767 6.

4. *Compilation of ASTM Standard Definitions.* Sponsored by ASTM Committee on Terminology. 6th Edition 1986. PCN 03-001086-42. Published by the American Society for Testing and Materials, 1916 Race Street, Philadelphia, Pa, 19103, USA. ISBN 0 8031 0928 8

5. *Criteria for Quality of Petroleum Products.* Edited by J. P. Allinson, Technical Secretary, Institute of Petroleum, London. Published by Applied Science Publishers Ltd, Ripple Road, Barking, Essex, UK on behalf of the Institute of Petroleum. ISBN 0 85334 469 8.

6. *EPCA Quality and Quantity Control—A guide to good practice during the distribution of petrochemicals.* Published by the European Petrochemical Association, Avenue Louise 250, Bte 74, B-1050 Brussels, Belgium.

7. *1987 Annual Book of ASTM Standards.* Volume 00.01 Subject Index; Alphanumeric List. PCN 01-000187-42. Published by the American Society for Testing and Materials, 1916 Race Street, Philadelphia, Pa. 19103, USA. ISBN 0 8031 1079 0.

8. *Liquefied Gas Handling Principles—On ships and in Terminals.* McGuire & White. Published by Witherby & Co Ltd, 32-36 Aylesbury Street, London EC1R OET, UK. ISBN 0 900886 93 5.

9. *National Institute of Oilseed Products Rules Book.* Published by Administrative and Conference Office, 2600 Garden Rd, 208 Monterey, CA 93940, USA.

10. *FOSFA International Manual.* Available from 24 St Mary Axe, London EC3A 8ER, UK.

11. *American Fats and Oils Association, Inc, Trading and Arbitration Rules—By Laws—Roster.*

12. *International Association of Seed Crushers, Oilseeds, Oils and Fats.* 3rd Edition. Available from IASC, Salisbury Square House, 8 Salisbury Square, London EC4P 4AN, UK.

13. *National Institute of Oilseed Products, 1986-87 Trading Rules.* Available from NIOP, 111 Sutter Street, San Francisco, California 94104, USA.

Terms and definitions

This glossary of terms incorporates terms which have been approved by ISO/TC 28 for use in connection with crude oil measurements.

API
— American Petroleum Institute.

API GRAVITY
— An American unit used in the petroleum liquids.

$$\text{API Gravity} = \frac{141.5}{\text{Relative Density } 60/60^{\circ}\text{F}} - 131.5$$

ASTM
— American Society for Testing and Materials

CALIBRATION TABLE
— A table, often referred to as a tank table or tank capacity table, giving the volume of material held in a storage tank for various liquid levels. Frequently, a subsidiary or proportional parts table is also included to simplify interpolation if required.

CERTIFICATE OF QUALITY (CQ)
— A certificate stating the quality of the product. It generally accompanies the B/L.

CHECK LISTS
— These are lists of questions concerning operation, safety, experience and communication presented by shore terminals to vessels. These lists are normally completed and signed by both ship and shore before each cargo transfer. They are designed to ensure that misunderstanding leading to dangerous situations does not occur.

CLINGAGE
— Oil residues which adhere to the surface of tank walls and structures on completion of discharge.

COFFERDAM
— Is the isolating space between two adjacent steel bulkheads or decks. This space is commonly void, but may be used as ballast in some vessels.

CRITICAL ZONE
— The volume close to the bottom of a floating roof tank in which there are complex interactions and buoyancy effects as the floating roof comes to rest on its legs. The zone is usually clearly marked on tank calibration tables and measurements for custody transfer should not be made within it.

DEEPWELL PUMP
— A multi-stage centrifugal pump used in some vessels for cargo discharge. The impellers are situated within the cargo tank itself and are driven by a shaft from an intrinsically safe motor situated on deck.

DENSITY
— The ratio of the mass of a substance to its volume. Since density is dependent on temperature and pressure these should be stated.

DIP
— The depth of liquid in a storage tank measured from a dip reference level. Also known as innage.

DIP PIPE
— A vertical pipe installed in a tank. In well designed tanks it should be at least one meter from the tank wall, well slotted to provide easy access to the bulk fluid, located immediately above the dip reference level and of sufficient diameter to accommodate a sample cage. Other names are: gauging, sample, stand, still and ullage pipe.

FLOATING ROOF
— A tank roof which floats freely on the surface of the liquid except at low levels when it is partially or wholly supported by legs. This must not be confused with a lightweight floating cover frequently installed in fixed roof product tanks to minimise evaporation loss.

FLOATING ROOF CORRECTION (or Roof Correction)
— The correction which allows for the oil displaced by the floating roof.

INERT GAS
— A gas or vapour which will not support combustion and will not react with the cargo.

INNAGE
— See Dip.

IP
— The Institute of Petroleum

LIGHT ENDS
— The low density constituents which may be easily lost by evaporation.

LOAD ON TOP (LOT)
— The procedure of comingling the recovered oil slops with the next cargo by loading the cargo on top of the slops.

MANIFOLD
— The final pipe of a cargo system before the shore connection. The pipe through which cargo is discharged into the loading arm ashore and from which loaded cargo is distributed to the various cargo tanks.

METER FACTOR
— The ratio of the actual volume of liquid passed through a meter to the volume indicated by the meter.

ON BOARD QUANTITY (OBQ)
— All the oil, water, sludge and sediment in the cargo tanks and associated lines and pumps on a ship before loading a cargo commences.

OUTTURN
— The quantity of cargo discharged from a vessel, measured by a shore terminal.

OUTTURN CERTIFICATE
— A statement issued by a receiving terminal certifying the outturn.

QUANTITY REMAINING ON BOARD (ROB)
— All the oil, water, sludge and sediment in the cargo tanks and associated lines and pumps on a ship after discharging a cargo has been completed, excluding vapour but including clingage.

SEDIMENT
— *Suspended sediment:* Non-hydrocarbon solids present within the oil but not in solution.

Bottom sediment: Non-hydrocarbon solids present in a tank as a separate layer at the bottom.

Total sediment: The sum of the suspended and bottom sediment.

SLOPS
— Material collected after such operations as stripping, tank washing or dirty ballast separation. It may include oil, water, sediment and emulsions and is usually contained in a tank or tanks permanently assigned to hold such material.

STRIPPING
— The operation at the conclusion of a discharge whereby the final part of the bulk liquid cargo is removed from a cargo tank.

TOPPING OFF
— That part of a loading operation at which tanks are carefully brought to their final required sounding or ullage. The most critical stage of loading normally conducted at reduced loading rates.

ULLAGE
— The distance from the ullage reference level to the oil surface. The vapour space in a fixed roof tank or sample container. Also known as outage. The depth of free or vapour space left in a cargo tank above the liquid level.

VENTING
— The process of releasing cargo gas or inert gas to atmosphere by way of the vessel's venting system and vent stack.

VESSEL EXPERIENCE FACTOR (VEF)
— The adjusted mean value of the VLRs obtained after several voyages.

VESSEL LOAD RATIO (VLR)
— The ratio of the quantity (TCV) of oil measured on board a vessel immediately after loading less the on-board quantity (OBQ) to the quantity (TCV) measured by the loading terminal, i.e.:

$$\text{VLR} = \frac{\text{Vessel's TCV after loading} - \text{OBQ}}{\text{Shore TCV loaded}}$$

VOID SPACE
— Empty space surrounding cargo tanks. Hold space not occupied by cargo tanks. Sealed off under deck spaces not designated as ballast space.

VOLUME CORRECTION FACTOR
— A factor dependent upon the oil density and temperature which corrects oil volumes to the standard reference temperature. For crude oil such factors shall be obtained from the API-ASTM-IP Petroleum Measurement Tables.

VOLUME FOR DYNAMIC MEASUREMENT CALCULATIONS
— *Indicated volume:* The change in meter reading that occurs during transfer through a meter.

Gross volume: The indicated volume multiplied by the appropriate meter factor for the liquid and flow rate concerned, without correction for temperature and pressure. Note: This includes all water and sediment transferred through the meter.

Gross standard volume: The gross volume corrected to standard conditions—i.e., 15°C and 1.01325 bar.

Net volume: The gross volume minus the volume of water and sediment transferred through the meter.

Net standard volume: The net volume corrected to standard conditions—i.e., 15°C and 1.01325 bar.

VOLUMES FOR STATIC MEASUREMENT CALCULATIONS
— *Total observed volume (TOV):* The volume of oil including total water and total sediment, measured at the oil temperature and pressure prevailing. This may be either the volume in a tank or the difference between the volumes before and after a transfer.

Gross observed volume (GOV): The volume of oil including dissolved water, suspended water and suspended sediment but excluding free water and bottom sediment, measured at the oil temperature and pressure prevailing. This may be either the volume in a tank or the difference between the volumes before and after a transfer.

Gross standard volume (GSV): the volume of oil including dissolved water, suspended water and suspended sediment but excluding free water and bottom sediment, calculated at standard conditions—i.e., 15°C and 1.01325 bar. This may be either the volume in a tank or the difference between the volumes before and after a transfer.

Net observed volume (NOV): The volume of oil excluding total water and total sediment, measured at the oil temperature and pressure prevailing. This may be either the volume in a tank or the difference between the volumes before and after a transfer.

Net standard volume (NSV): The volume of oil excluding total water and total sediment, calculated at standard conditions—i.e., 15°C and 1.01325 bar. This may be either the volume in a tank or the difference between the volumes before and after a transfer.

Total calculated volume (TCV): The gross standard volume plus the free water measured at the temperature and pressure prevailing.

WATER
— *Dissolved water:* the water contained within the oil forming a solution at the prevailing temperature.

Suspended water: The water within the oil which is finely dispersed as small droplets. It may, over a period of time, either collect as free water or become dissolved water, depending on the conditions of temperature and pressure prevailing.

Free water: The water that exists in a separate layer. It typically lies beneath the oil.

Total water: The sum of all the dissolved, suspended and free water in a cargo or parcel or oil.

WATER CUT OR DIP

— The measured depth of free water lying on the bottom of the tank.

WEDGE FORMULA

— An equation relating the volume of liquid material in a ship's tank to the dip, ship's trim, dipping point location and the tank's dimensions when the ship's calibration tables cannot be applied. To derive the equation, assumptions have to be made. the major assumption in the derivation is that the material is free flowing and will accumulate in the after end of a tank when the ship is trimmed by the stern.

WEIGHT CONVERSION FACTOR

— A factor, dependent on the density, for converting volume to weight-in-air. Such factors shall be obtained from the API-ASTM-IP *Petroleum Measurement Tables.*

WEIGHT

— *Gross weight-in-air:* The weight of oil including dissolved water, suspended water and suspended sediment but excluding free water and bottom sediment.

Net weight-in-air: The weight of oil excluding total water and total sediment.□

ELEMENTS OF COMPASS ADJUSTING

AND THE IDENTIFICATION OF ERRORS WHICH SHOULD LEAD A SURVEYOR TO REQUIRE THE COMPASS TO BE ADJUSTED

Captain J. A. Cross, Ex.C, FNI, MCMS, MCIT

Severnside Consultants; Member, The British Nautical Instrument Trade Association

DUE TO THE RAPID ADVANCEMENT of electronic aids in recent years, the magnetic compass has tended to be neglected to an increasing degree by mariners. The drastic reduction in world shipping and the demise of major traditional shipowning companies would appear to have led to a reduction in standards in ship operation.

Increasing confidence, whether or not justified, in satellite navigation and numerous electronic devices has led to this increasing neglect and has necessitated increasing vigilance on the part of surveyors. The purpose of this chapter is to draw attention to this sometimes alarming neglect and hopefully to indicate how surveyors in the various spheres can assist in improving the situation.

Electronic aids

Aids to navigation such as the satellite navigator, gyro-compass, Decca Navigator, radio direction-finder, radar, etc., are, of course, dependent on a power supply, and with (the albeit unlikely occurrence of) a total shutdown could conceivably leave a vessel totally without a course reference.

The one remaining instrument totally independent of power is the magnetic compass. Before the development of electronic aids, vessels were navigated successfully with the efficient use of a magnetic compass associated with a sextant and chronometer. Mariners were fully aware of the capabilities of the magnetic compass and from experience and constant monitoring of errors were quickly aware of any fluctuations in deviation.

With an ever-increasing reliance on electronic aids, what used to be a constant process of periodically checking a magnetic compass has faded, and the instrument can deteriorate to such degree as to be unreliable without detection. When an emergency does arise or when a surveyor or superintendent detects a fault, the cure can sometimes cause delays. A suitable spare bowl has to be found or repairs effected, and adjustment carried out.

Department of Transport requirements

The provision, siting and maintenance of UK ships' compasses are still governed by the requirements of the Merchant Shipping Notice M.616, which provides strict guidelines as to the number of compasses on various classes of ships, their siting and specification, etc. It is clearly stated that the responsibility for ensuring that a ship's compasses are always maintained in good working order rests on the owner and/or master of the vessel. It is further recommended that adjustment should be entrusted to a compass adjuster who holds a Department of Transport certificate of competency.

Until recent years the syllabus for a Master's Foreign-Going Certificate contained sufficient instruction in the elements of compass adjusting to enable the master to carry out reasonable adjustment should the need arise. This part of the syllabus, however, has been reduced drastically and increased the necessity of retaining an experienced certificated adjuster at regular intervals.

Certificates of competency as an adjuster are issued by the Department of Transport following an examination, a set period of workshop experience, and the adjustment of 24 seagoing vessels under the instruction of a certificated adjuster. The requirements for the certificate are clearly set out in UK Merchant Shipping Notice M.664.

Before being accepted for examination, the 24 deviation cards produced are considered by a compass adjusters' panel of the British Nautical Instrument Trade Association and recommendations made to the Department of Transport as to whether the candidate is ready to sit the examination.

The reduction in shipping has also meant a reduction in the associated industry of providing nautical equipment and the number of compass manufacturers has been drastically reduced. This means that the opportunities for training compass adjusters have been similarly reduced, and the inevitable reduction in demand for adjustment is also a discouraging factor.

There are cases constantly reported of uncertificated adjusters practising, sometimes with disastrous results. These are principally restricted to the yachting world and it is to be hoped that owners and masters of commercial vessels ensure the services of a certificated adjuster are retained when adjustment is required. There can be no substitute for experience and this is guaranteed when a certificated adjuster is in attendance.

Prudent practice

Efficient siting of magnetic compasses is vital to accurate operation, and a compass adjuster should be consulted at the design stage. It is unfortunately the case, too frequently, that magnetic compasses are put perhaps in the most convenient position without any thought being given as to whether they are ideally sited when related to the surrounding magnetic field and local disturbance of the vessel. Initial adjustment by an experienced adjuster is vital and, provided no structural alterations are carried out, a good adjustment should be effective for up to two years.

Periodic adjustment, at, say, two-year intervals, should be carried out and the compass suitably monitored for efficient operation so that any irregularities associated with wear and vibration are readily and quickly detected. There is a tendency, particularly with standard compasses sited on the monkey island and viewed from the wheelhouse by a reflector/periscope arrangement, to be out of sight and out of mind. There have certainly been cases when compass adjusters summoned to carry out adjustments just before sailing times have found such compasses totally inoperable.

Surveyors

In recent years, it has been the case that class surveyors and Department of Transport surveyors are increasingly carrying out random examination of compasses and deviation books. It is also a requirement of the principal P&I Clubs to request surveyors to check navigational equipment when carrying out condition surveys. Examination of the vessel's deviation book, provided it is regularly endorsed, should provide a good indication as to the state of adjustment of the magnetic compasses.

Significant changes in deviation with magnetic latitude should give a clear indication that all is not well, and a fluctuation in deviations over a period of time on similar headings should also give cause for concern. The mechanical performance can also be checked by deflecting the compass card from a set heading and ensuring that it returns within half degree after deflection. Any sticking or irregularity must give rise to the suspicion that the pivot/jewel is faulty and requires attention. The period of the compass card can also be easily checked and any sluggishness investigated.

When a surveyor has good reason to suspect that a compass is mechanically faulty, or has excessive deviation, then he will insist that the instrument is attended to at the earliest convenient opportunity.

Legislation

The British Nautical Instrument Trade Association has for some years been active in attempting to persuade the Department of Transport to make compass adjustment for commercial vessels compulsory at, say, two-yearly intervals so as to be in line with other European countries. To substantiate the case, a survey was taken of the work undertaken by 19 compass adjusters over a period of 21 months. The main areas of concern were as follows:

- The length of time that a compass could be expected to remain in close adjustment under normal seagoing conditions.
- The length of time a magnetic compass could be expected to remain in good mechanical condition under normal seagoing conditions.
- To examine increasing numbers of reports of uncertificated persons carrying out compass adjustments on small craft, to ascertain to what extent this was happening, and to what extent their adjustments were of an adequate standard.

On completion of the survey the following conclusions were drawn:

(i) That the standard magnetic compass when properly sited and regularly serviced and adjusted at least every two years, and always after structural alterations, provides an accurate and reliable direction reference.

(ii) A majority of oceangoing vessels do have their magnetic compasses regularly adjusted and serviced.

(iii) A minority of these vessels neglect their magnetic compasses. The navigating officers are unaware of the condition both as regards mechanical condition and deviation.

(iv) In the event of lack of efficiency of other navigational aids such vessels would be in danger.

(v) Some inshore vessels have compasses that are unreliable and some that are useless.

(vi) Some fishing vessels, steel built, have compasses that have been placed in position without any regard to compass performance or the regulations contained in M.616. The requirements with respect to annual overhaul and readjustment have been overlooked.

(vii) Yachts with well placed compasses have an accuracy of direction within 2° and this accuracy is maintained over several years.

(viii) Adjustments of compasses carried out by uncertificated compass adjusters are, on average, less accurate than those carried out by adjusters having taken and passed the DTp certificate of competency.

Whilst the UK Department of Transport has to date decided against periodic compulsory adjustment, it is felt that a strong case has been made to illustrate an increasingly unsatisfactory situation with regard to the magnetic compass.

The observations recorded are not intended as a criticism of all vessels. It is pleasing to note that the majority of vessels are staffed by prudent mariners who are conscientious in their duties. It is the increasing minority which must cause concern. □

USE OF METEOROLOGICAL INFORMATION FOR WARRANTY SURVEYING PURPOSES

Captain A. Blackham, FNI
Noble Denton Weather Services Ltd

SINCE THE middle 1950s, the search for oil on shallow Continental shelves has required platforms for exploratory drilling and, when oil is found, for production facilities. The heavy engineering required to construct these platforms is located onshore so that tow-out and emplacement is always necessary, even for short distances. Of recent years, economic decisions frequently result in the finished product having to be transported for long distances by sea.

Since the object being transported will be subject to stresses, fatigue and accelerations during the voyage which it would not experience when in its final position, and the expense of providing the extra strength is uneconomic, it is necessary to plan such voyages carefully, particularly because damage is not acceptable. It is not acceptable because of the higher cost of replacement, the distribution of the entire schedule and the cash value of lost production (known as consequential loss).

The primary concern of an insurance warranty surveyor is to ensure that all reasonable steps are taken to ensure that the voyage is completed without loss or damage. A study has to be made of the project, firstly to determine its feasibility and secondly, in those cases where it is found not be to feasible, to recommend what has to be done to make it feasible.

A team carries out the study. First, a meteorologist will assess the route and calculate the environmental extremes. Second, a naval architect will use the extremes to calculate accelerations and stresses. Third, a mariner will assess the route, the capabilities of the towing fleet and the actual *modus operandi* of the project to ensure that the normal practices of seamen are followed, and issue of certificate of approval.

Environmental design criteria

Ocean passage route planning is best carried out by using marine climate atlases. The most useful of these are published by the US Navy in the Navair series. These show average wind and wave frequencies, seasonal currents and the incidence of tropical revolving storms. Another useful publication is *Global Wave Statistics,* published by BMT, giving wave heights and periods. The requirement is invariably to avoid loss or damage so great circle routes are not usually recommended.

There are different approaches to selecting a design figure—(a) the value for which there is an X per cent risk of exceedance; (b) the 'ten-year extreme,' which is defined as that value which is likely to be reached or exceeded, on average, once in ten years, seasonally adjusted.

The choice of the ten-year return period is both historical and arbitrary but it is simple and appears to be successful. It is not directly related to risk of exceedance during a passage, since no account is made of exposure time. If a transportation takes one month, and the area through which it passes has a uniform wind and wave climate, then the probability that the ten-year monthly extreme will be exceeded is approximately 10 per cent.

However, if the transportation is shorter than one month, then the probability of exceeding the ten-year monthly extreme is reduced in a simple linear fashion: 30 days' exposure gives 10 per cent risk; 15 days' exposure gives 5 per cent risk; 3 days' exposure gives 1 per cent risk.

It is apparent, therefore, that across-the-board use of the ten-year extremes will lead to variable risk. It is considered that an alternative basis for the choice of environmental criteria for transportations is justified.

Estimating exposure time

For passages taking less than a month it is appropriate to calculate a 'reduced' extreme by estimating the exposure time. A safety margin, usually 25 per cent, is added and consideration must also be given to the fact that the passage will only begin with a fair-weather forecast. Reduced values can then be calculated by multiplying the ten-year extreme by a factor based on the extreme probability distribution.

Longer passages may be regarded as a succession of short passages through differing climatic areas. The seasonal extreme for the worst area is calculated and a 'reduced' extreme is then calculated using the exposure time in the worst area plus the exposure time in all other areas which have an extreme within 10 per cent of the worst.

To calculate the design figure it is necessary to have a frequency table of wind or wave and to carry out an extreme probability analysis. An archive of ship reports is, at present, the best source of data, since it is generally available and is global and the selected area can be adjusted to cover the route under investigation.

Data from ship reports should be quality controlled, estimated wave heights converted to significant heights, and averaging times determined for wind speeds (some are one-hour means, some are one-minute means). It is marine engineering practice to use a one-minute mean wind for design. Not less than 10 years of data are required. A second source of data is measured data but there are no measured data of sufficient quantity to be of use in calculating extremes over the open ocean.

A third source of data is a computer hindcast model. Synoptic data over a 20-year time span are fed into a computer which assigns spot values on a fine-mesh grid with four values per day. The only one at present available covering the whole northern hemisphere is the US Navy's spectral ocean wave model, but this is not suitable for design values as it does not include swell from the southern hemisphere. Output from these models can be very expensive.

Having data in the form of frequency tables, a statistical distribution function is fitted to the cumulative frequency distribution.

TRS approach

In those areas which are affected by tropical revolving storms (TRS) of hurricane force it is necessary to adopt a different approach. All TRS storm tracks which have passed through the area over a period of 25 years are fed into a model and a suitable grid chosen. Each occurrence of a storm affecting a grid point is assumed to be an independent extreme event and the average number of extreme events per year is calculated.

An extreme value distribution function is then fitted to the distribution of ranked extreme values and the threshold below which values are ignored is varied until the best least squares fit is obtained. The extreme value is obtained by extrapolation. Reduced extremes can only be calculated for a typhoon area when an approved procedure for typhoon avoidance will be operated.

Once the environmental design criteria have been established it is then up to engineers to calculate whether motions from those figures are acceptable or not. If acceptable the unit proceeds on the agreed route. If not, it is necessary to establish a weather-routeing concept. This should include:

- Overall procedure.
- Defining the route.
- A series of safe-havens to be agreed. Successive safe-havens should always be chosen so that they can be reached inside the period of the forecast. Although forecasts are available for up to ten days ahead their reliability beyond 72 hours is questionable.
- Nomination of a reliable weather-routeing service. This service should invariably be staffed by master mariners or have the in-house availability of master mariners to ensure that all advices give safe navigation and follow the normal practice of seamen.

Fatigue

For all voyages, the effects of fatigue on the unit being transported should be considered. The required meteorological input into the engineering calculation is usually in the form of a polar diagram indicating the number of waves at specified height levels, with associated periods, likely to be encountered during the whole voyage.

Routine weather forecasts

It is essential to select a weather service which is experienced in marine weather forecasting, is familiar with marine practice and can operate on a global basis. There are probably only three such services—Ocean Routes in Palo Alto, the UK Meteorological Office in Bracknell, and Noble Denton in London. The importance of experience in actual oceanic weather forecasting cannot be over-emphasised. The majority of weather forecast services are primarily engaged in forecasting for either aviation or the general public, which concentrate on quite separate parameters.

For local weather forecasts—e.g., a port or harbour—there is no doubt that local knowledge is most important and the best service will be given by a man on the spot.

There are now a large number of weather forecasting services, both government and private. A government service is not necessarily a better service and not all private services are reliable. The best indicator is track record.

Bakground to forecasts

The first step in the process of forecasting weather is to find out what is happening now. The World Meteorological Organisation (WMO) is located in Geneva. Most countries in the world are members and, by agreement, the WMO lays down all the rules for making observations and for ensuring that they are collected and redistributed to all other countries. This is done by a communication system called global telecommunications system (GTS), with three main centres, Washington, Moscow and Melbourne, and several smaller ones. The GTS is an enormous electronic conveyor belt, linked by landline, radio and satellite, down which large amounts of data can be sent at very high speed (9,600 bauds).

Observations are made simultaneously around the world every six hours (sometimes more frequently) at ground stations and every 12 hours at 'upper air' stations. The most important variables are the pressure, temperature and humidity of the air, either at the surface or in the atmosphere above. Even with high-speed communications, it needs at least a couple of hours to receive enough data to start work.

Most national centres handle all these data by computer, so that they can be stored and so that the right bits can be redistributed or analysed. When enough data have been received, they can then be used for plotting out in graphical form for subjective analysis or they can be fed into another computer for analysis and for projection into the future. This is done by means of a 'model,' which is run twice daily, at 00.00 and 12.00 GMT.

The computer allocates values to a series of grid-points and then, treating the air as a fluid, sees what happens to it in, say, 15 minutes' time, then re-analyses it for another 15 minutes and so on until 24, 48 or 72 hours or longer ahead and at the end of these times prints charts showing air pressure, temperature and humidity. The equations used are complex and non-linear and require very great computing power. One of the most powerful in the world is used by the European centre for medium range weather forecasts at Reading. It is a Cray, which can process 800 million instructions per second and analyses data at 16 levels.

Accuracy and availability

The accuracy of these prognoses is fundamental to the techniques of using them, which means that a forecaster has to use a lot of experience to know which particular model to follow.

The output from these models is internationally available through radio facsimile, with schedules printed in the *Admiralty List of Radio Signals*. They include pressure and temperature patterns at the surface and at various levels of the atmosphere. Having obtained these prognoses, the next stage is to interpret that information into a specific weather forecast. To help in making the forecast, a forecaster must also make full use of other information, including reports of present weather, satellite imagery, temperature and humidity profiles and much experience.

The principal parameters affecting marine operations are wind, wave and swell, although on some occasions where cargo can be affected, lighting, precipitation and atmospheric obscurity (fog) can be important. In a few areas sea-ice, drifting ice and superstructure icing can affect the operation.

Wind speed is determined by the pressure gradient. Most forecast models calculate wind direction and speed automatically in the form of 'spot winds,' or wind speeds can be estimated by the 'geostrophic wind scale' printed on the chart. Further corrections have then to be made for surface friction and cyclonic curvature.

Wave heights

Wave heights are usually given as 'significant' (the average of the highest third), 'maximum' (the highest in a 20-minute period) or 'extreme' (the highest during the period of the forecast). The height of a wave depends on the speed of the wind, the distance over which the wind blows (the fetch) the duration of the wind and water depth. Several nomograms are available to calculate wave height and period from such data. Those most widely used are Jonswap, Bretschneider, Pierson-Moskowitch and Darbyshire-Draper.

Numerical wave models are in widespread use at most main meteorological centres and the output is available on radio facsimile. For oceanic areas, the model is run on a coarse mesh directly from the wind-field model, with fetch and duration being the other variables. For shallow-water areas a finer mesh is used and bathymetry and refraction are considered. The reliability of these forecasts depends mostly on the reliability of the forecast wind.

Should a unit be correctly designed to withstand the 'ten-year storm' the only weather forecast required would be the initial fair-weather forecast for departure.□

References

Blackham A. 1987
'Practical Applications of Meteorology in Heavy-Lift Transportation'
World Meteorological Organisation Seminar, Reading.

Morgan S. K. and Babbedge H.N. 1985
'Transportation Criteria for Offshore and Industrial Equipment'
Royal Institution of Naval Architects Symposium, London.

United States Navy Marine Climatic Atlas of the World 1981 Volume 8 Worldwide Means and Standard Deviations NODC Asheville N.C.

US Department of Commerce National Climatic Center
Marine Surface Observations Data Tapes.

NERC 1975
Flood Studies Report Volume 1.
Hydrological Studies Natural Environmental Research Council, England.

United States Navy Marine Climatic Center.
Consolidated Worldwide Tropical Cyclone Dataset NOAA, Asheville N.C.

Weibull, W. 1951
'A Statistical Distribution of Wide Applicability'
J.Appl Mech. Vol 18.

Bretschneider C. L. 1958
'Revisions in Wave Forecasting; Deep and Shallow Water'
Proceedings of the Sixth Conference on Coastal Engineering ASCE Council on Wave Research.

Darbyshire M and Draper L 1963
'Forecasting Wind-Generated Sea Waves'
Engineering Vol. 195.

Lawes H. D. 1988
Personal Communication.
The Noble Denton Weather Guide, 1988.

Carter D. J. T. 1982
'Prediction of Wave Height and Period for a Constant Wind Velocity Using the Jonswap results.'
Ocean Eng. Oxford 9

CARRIAGE OF STEEL IN BULK CARRIERS —PROBLEMS OF TANKTOP LOADINGS

David Reid, AMNI
Operations Manager, Westport Navigation Inc

WE ARE OPERATORS of time-chartered bulk carriers specialising in the carriage of steel products from steel mills in Brazil and Venezuela. We are responsible for co-ordinating the shipment of about 500,000 tonnes of coil, plate and sheet each year.

Since we charter vessels to perform each of the voyages, we experience many types of vessels/owners and masters. We co-ordinate and plan all the stowage in co-operation with the various mill technical personnel. To a great extent the masters of the ships involved do not have the knowledge required to plan the loading with the care and precision required.

They are advised and asked to verify that the ship's trim and longitudinal strength is within the limits for the voyage, but for all practical purposes they are observers. This system works very well in almost all cases, except when we are confronted with a master who declares himself as an expert in the field of 'tanktop strength,' There appear to be different schools of thinking on what 'tanktop strength' really means and how it applies to the loading of cargo.

Therefore it is the question of 'Tanktop strength' that we wish to address, to set forth what we believe the interpretation should be versus the argument we have been given, and to seek views in this matter.

Interpretation

Tanktop strength or floor strength as given by the shipyard is stated as a weight per unit area that is given primarily to comply with the IMO *Code of Safe Practice for Solid Bulk Cargoes,* section 2/ 2.1.2.1, wherein it is recommended that the master be provided with loading information so as to avoid overstressing the structure. The reason for this is that when loading a high-density bulk cargo, the cargo may be 'peaked' such that there is a greater weight under the peak and thereby a more intense pressure on the tanktop plating. Since bulk cargo is loaded direct on the tanktop plating the load force on any four-sided panel (between longitudinals and solid transverse floors) could result in deflection of the plating. Tanktop strength therefore refers to the plating used in the floor area of the hold as the upper 'skin' of the double-bottom structure.

When loading cargo consisting of individual pieces such as steel finished or semi-finished products, the method of loading is entirely different to that of a solid bulk cargo. Steel is loaded using dunnage to interface between the cargo and the ship's structure. The reason for this is to permit slinging of cargo, passage of condensate, protection of cargo, and to distribute the weight.

Distribution of weight—commonsense dictates that in order to load heavy unit loads in a ship's hold the weight must be transmitted to the strongest part of the double-bottom structure. In practice this is done by placing the dunnage on the tanktop plating directly above the position of the longitudinal framing. Therefore the load is carried in this case not by the floor plating but by the supporting structure underneath.

Therefore the criteria for establishing the maximum to be loaded in the hold is determined by ensuring that (a) the total weight in the cargo hold is within the ship's longitudinal strength parameter and (b) the weight in the hold is distributed over the entire hold floor. One variant to this is in the case of loading steel coils in which case the dunnaging is laid athwartships and not fore and aft as in the case of plates/sheets and slabs. In this case the dunnage spans the area between longitudinals and the weight of the cargo is transmitted through the dunnage to the floor plating and thence to the longitudinal frames underneath.

The coils usually have a diameter of about 1.0-1.25 metres, which means that the spacing of coils across the floor is such that in many cases the coils are located directly over or very close to a longitudinal frame support. The total weight carried on the tanktop plating is therefore minimised, because many coils are supported direct through the structure of the double bottom.

This is we believe consistent with the Iron and Steel Institute of Japan's own theory and diagrams shown in their booklet *A Working Standard for the Stowage and Securing of Export Steel Cargoes*.

Counter interpretation

The following theory has been put forward by a naval architect, and has caused us and the steel mill considerable problems. The theory is as follows:

To determine the quantity of steel to be loaded in a cargo hold the width of the hold is taken and for every metre of floor length the maximum permissible is that given by the following:

Width (m) $\times$ 1 (m) = Total square metres.
Total square metres $m^2 \times$ Tanktop strength T/m^2
= Total permissible cargo in the area (T).

Example: given floor of 21.5 metres and a tanktop strength of 15.0 T/m^2, according to this method of calculation the maximum permissible weight for a 1-metre strip of tanktop is 322.5 tonnes. This is regardless of what the cargo is or how it is dunnaged.

The basic philosophy expressed is that in any square metre there can be no more than 15 tonnes weight. We believe this to be the incorrect use of the ship's data.

Discussion

We believe that tanktop strength is directly referenced to the inner plating and not the ability of the bottom structure to carry a load. For example, if a tanktop strength of $15T/m^2$ is used, this is equivalent to the following in kg/cm^2: $1.5 kg/cm^2$.

Consider a man with normal shoes having a contact area of about 50 cm^2 —if the man's total weight exceeds 75 kilos then he will not be allowed to walk on the tanktop. (This is obviously not the meaning intended and in fact proves how ridiculous the argument is.)

Consider also the following: Steel plates two metres in width are stowed athwartships and bottom dunnage consisting of 7.5cm square dunnage is placed at 1 metre intervals over longitudinals. The weight of the plates in stow is, say, 750mt total. Using our method, the weight on tanktop is effectively nil, because the entire load is transmitted direct to structure. Using the disputed method, the weight on the tanktop allowed is $2 \times 322.50 = 645.0$mt. Therefore the tanktop is over-stressed by 16.27 per cent. The vessel is not seaworthy and charterers/shippers are liable for damage to the vessel, etc.

Consider the actual pressure on the tank plating. The contact area is actual bottom dunnage: 1 piece 2.0m × .075m for every 1m of floor width (21.5), say, 23 pieces; contact area = $2.0 \times .075 \times 23 = 3.45m^2$; total weight = 750mt. Actual pressure on floor at contact point is 217.39 Tm^2 or if one considers at the total weight allowed by the method, that is 645mt, then the actual pressure on the floor is 186.95 T/m^2.

Obviously the actual pressure at point of contact is far different to the so-called 'allowable or permissible' loading of the tanktop. We believe that the reason for this is that it is erroneous to use the tanktop strength factor to calculate what may be loaded on the floor of the hold, except when considering the carriage of solid bulk cargoes which have entirely different characteristics.□

COMMENT

The original publication of this material in *Seaways* gave rise to a considerable amount of comment, including that reproduced here:

Captain P. C. T. Chu, MNI, FICS

THERE HAS never been any problem on stowage calculation concerning tanktop strength for any recently-built ships if average tanktop strength is the only concern. For ships of older construction, say, 20 to 25 years ago, one may not find the designed tanktop strength on the ship's particulars. To find out what is the maximum cargo one can load to each hold, we can apply the following formula:

$w = (c/C) \times W$

where

w is the weight in the hold
W is the total deadweight
c is the capacity of the hold
C is the total capacity.

This is not a formula from any authoritative source but merely a logical deduction.

Another way to estimate the tanktop strength is by employing the formula as below:

$w = A \times 1.4 \times d$

where

w is the weight in the hold
A is the total area of tanktop
d is the draft in load condition.

This assumes the ship is of 100A1 class.

As mentioned previously, ships recently built are all provided with details of designed tanktop strength including which holds are specially strengthened for heavy cargo. The real problem is the 'point load,' that is, the average weight per unit area is within the designed limit, there could be some small areas which are, if we strictly apply the TPM (tons per metre square) definition, overloaded or overstressed.

A traditional way to solve the problem of concentrated load is to use dunnage to spread the load. But the real problem is what should be the sizes of dunnages; would 1 in be too thin or would 6 in be too thick? How close should dunnage be laid?

Before we go further, let us see what will happen to the tanktop if it is overloaded: the answer is buckling if it is slightly overstressed, and a total collapse if it is overloaded too much. It is difficult to describe what is too much or slightly, unless we have calculated and compared the deflection of the tanktop. This is a job beyond a master's competency to do on board ship. That is a naval architect's job. Unfortunately, no such information is readily available to the master, and he often has to rely on experience and rule of thumb, or sometimes watch the deflection of the tanktop while loading, in which case it may be too late to remedy.

In general the following three items determine the stress on the tanktop:

1. Thickness, dimension and disposition of dunnage.
2. Thickness of tanktop plate.
3. Dimension and spacing of floors, longitudinals and stiffeners underneath the tanktop, and of course the load or weight of cargo.

Among the items, shipmasters are able to alter only item 1; there is no way he can change the ship's structure.

The effect of using dunnage to spread the weight is not as good as people have thought except the dunnage is of considerable thickness. The reason behind this is that thin wood can never spread the weight effectively due to its soft nature. The effective area increase by wood is unlikely to be more than the original contact area plus an additional width (equal to the thickness of the dunnage) surrounding the circumstance of the contact area. On the other hand, one must not forget that the thickness of the tanktop plate has the inherent characteristic of spreading weight.

A very rough estimate is that for a non strengthening ship, you can spot load an area with 10 times the designed tpm, provided always the average strength of that area is not overloaded. For example, if the contact area is 0.1 metre sq and the tpm is 15, then in that 0.1 metre sq one can load $0.1 \times 10 \times 15$

which is 15 tons, provided over the 1 metre sq the total load is not over 15 tons, or for a bigger area the average load is not over 15 tpm. The dunnage must not be spaced too wide apart, say for every 15 cm dunnage-covered area there should be not more than 45 cm non-dunnaged area. In other words, if the total tanktop area is 20 m × 28 m, the total contact area required would be about one-fourth of 20 m × 28 m, provided they are evenly spaced. In fact, this can be further reduced if the dunnage is like a square tile instead of a long piece of wood. It is quite obvious that if we cut out some length in the middle part of the dunnage it would not affect the weight spreading.

It is possible to calculate what is the exact relationship between plate thickness, floor and longitudinal spacing and stiffeners versus the area and weight of spot loading. But, unfortunately, so far, no such or similar information is available. Perhaps the 'tyre print load stress curve' of Lloyd's Register or other classification's rules and regulations for construction of ships can provide some idea how the spot weight should be estimated.□

J. H. A. Baker, B.Sc, M.Sc, AMNI

THE SIMPLE proportionality $w = (c/C) \times W$ is misleading since it makes no allowance for trim, alternate hold loading, or tanktop strength, which is the subject under discussion. It is also incorrect since Captain Chu states that W = deadweight. Deadweight includes items such as stores and fuel (generally of the order of 7 per cent of deadweight), and thus Captain Chu's equation would only be correct if W is equal to cargo weight or payload.

Captain Chu's rough estimate for spot loading only looks correct because he uses a spot load area of 0.1 m^2 with a factor of 10 and then multiplies. In fact, 10 times the given TPM of 15, placed upon an area of 0.1 m^2 will give a resultant loading of 1500 tonne/m^2. This could cause serious damage.

To turn to Mr Reid, he is incorrect to asume that the TPM figure applies only to the plating. The 'tanktop' in this case is a matrix of plates, stiffeners, longitudinal girders and intercostal floors (in a longitudinally stiffened hull). The load-bearing mechanism is such that the plate deflections allow loads to be passed outwards to successively stronger sections via the bending moments. Spacings of floors and girders will generally be of the order of 0.6 to 0.8 m.

The conical pile of bulk concentrate is itself a load-spreading mechanism. Each grain is supported by surrounding grains and it is this that causes such a pile to assume a characteristic angle of repose. The tanktop structure distributes and absorbs the resultant load. The finished and semi-finished steel products will also be 'seen' by the tanktop as a distributed load, provided the items have been stowed reasonably uniformly. Dunnage of the dimensions generally used for aiding slinging, ventilation, and protection will add only marginally to the section modulus of the tanktop. However, most items will lie across either floors or girders, and the tanktop can be expected to bear a load which averages to the given TPM, provided there are no obvious concentrations.

I am interested in Mr Reid's shoes which only give him a contact area of 50 cm^2. I measured my own size eight boots and consider that they give an actual contact area of about 400 cm^2. My 68 kg then gives a loading rate of 1.7 tonne/m^2.

If Mr Reid intends to ship a specific item whose point loads are well in excess of the stated TPM (even though overall load area may not be), such as a heavy unit standing on feet, he is recommended to take details of the cargo and plans of the relevant ship to a firm of consultant naval architects. They will be able to calculate the size and positioning of the dunnage required to spread the load. But while he is dealing with cargo such as plates, bars and coils, the stated tonnes per square metre for the tanktop should be his guide.□

B. M. Bell, C.Eng, MRINA

THE QUESTIONS surrounding the cargo loading on a ship's bottom structure are perennial for shipowners' naval architects. The matter hinges on the design criteria used in establishing the scantlings of the double bottom structure, which in the case of general cargo ships will invariably be the standard applied by the classification society concerned. In the case of bulk carriers the future owner or the shipbuilder will specify either the density or the weight of cargo to be carried in each hold which is then applied in the calculations of strength for the entire hull and in the determination of the scantlings of the double bottom structure. The values of cargo loading used for this design work are the same as those which are normally available to the master on a capacity plan or in trim and stability information and are therefore directly related to the ability of the bottom structure to carry load. I fear therefore that Mr Reid's concept of the matter, in that the loading rate is applicable only to panels of plating, is invalid.

But what of the anomaly of the man weighing 75 kilos. It is quite obvious that despite the fact that the area in contact with the man's shoes is subjected to the maximum permitted loading, there is no question of any part of the structure being overloaded. This is because the load is not applied continuously across the panel of plating. This being the case it would appear reasonable to increase the weight of our man wearing size nines to the point where the local load just meets the constraints of structural strength. The point at which this balance occurs is not easy to determine and will depend upon the circumstances of each case.

It can be demonstrated using simple beam theory that a concentrated load of one-third of the permitted loading across a panel of plating, the width of the contact area, can be applied with safety. This, however, is a simplification of the problem since, due to its small depth, the plating does not behave as beam under bending alone and therefore the approach can be used only as another inexact guide. Captain P. T. C. Chu offers the rough guide of ten times the tpm rating for spot loads. Such a high figure must be used with extreme caution even where the average loading is not exceeded.

To calculate the pressure under small areas of contact is quite meaningless, and whilst there is difficulty in using the overall loading rate for a tank top when concentrated loads are present, it is the best guide that the master has.

The use of dunnage is a subject worthy of very close attention on the part of those responsible for loading a vessel. Despite the poor performance of the timber sections normally employed in spreading concentrated loads, dunnage is very important as a means of softening the contact with the tank top and preventing damage from abrasion and local compression on the plating surface.

Mr Reid is to be congratulated on successfully controlling the loading of his ship so as to avoid any lateral loading of the tank top plating by means of the correct positioning of dunnage directly over supporting structure. In the majority of cases masters cannot rely on stevedores to be sufficiently careful in this respect, and since misplaced dunnage can have the very opposite effect, considerable caution is necessary.

Captain Chu gives a formula for calculating the permissible weight of cargo in a hold space; this expression is a useful guide which strictly speaking applies to general cargo ships. I should also point out that an additional term is required, namely the design cargo stowage rate of 1.39 cu m/ton (50 cu m/ton):

$$W = 1.4 \times d \times \frac{1}{1.39} \times A$$

where:

W = permitted weight in the hold
d = loaded draft
A = plan area of hold.

It has been suggested that classification society requirements for wheel loading may throw some light on this subject, and in this vein the following comparison may be of interest.

Taking for example a typical Panamax bulk carrier (ship C in table below), the maximum capacity of fork-lift truck which can be used in the strengthened holds is of the order of 35 tonnes, assuming four wheels at the fork end axle arranged in closely spaced pairs. The tyre print size of a single wheel on such a vehicle might be about 58 × 40 cm, and the wheel loading approximately 19.5 tonnes. Taking two wheels in combination the load calculated on the basis of the permitted loading of 29.0 t/sq m would be: 29 × 0.84 × 1.0, say = 24.36 tonnes, where spacing longitudinals is 840mm.

Therefore ratio of wheel load to distributed load

$$= \frac{2 \times 19.5}{24.36} = 1.60$$

i.e., 60 per cent in excess of the permitted distributed loading.

It should be noted that the wheel load is supported by massive pneumatic tyres; also that the plating in the vicinity of the concentrated loads is not bearing any other cargo load. The table gives some typical figures for acceptable fork lift truck loading on plating.

Ship	Type	DWT (tonnes)	Plating thickness (mm)	Permitted loading (t/m^2)	Maximum capacity forklift (payload tonnes)	Remarks
A	Cargo	15,000	12.5	8.93	3.5	2 wheels at fork end
A	Cargo	15,000	12.5	8.93	5.2	4 wheels at fork end
B	Bulker	26,000	19	15.30	2.3	4 wheels at fork end
C	Bulker	67,000	20	29	35	4 wheels at fork end

The method of calculation used by Lloyd's Register for determining acceptable wheel load is far too cumbersome for general use by ships' masters; however, an adaption specifically for an individual ship is an interesting idea.□

Captain P. Lemarquand, MNI

MR REID did not state what he believes the upper limit to be. Most ships, and especially vessels strengthened for bulk cargoes in alternate holds, have both a maximum tanktop loading factor and a maximum hold tonnage. Often these two figures are the same, as Mr Reid concluded. However, if these figures are not accepted as the upper limit, what arbitrary limit does he suggest?

Our fleet includes a number of geared bulkcarriers carrying coal and ore about the Australian coast and backloading preslung steel cargoes. The actual tanktop loading factors are in many cases very similar to the 15 tonne/metre^2 which Mr Reid used as an example.

Generally the cargo is staggered to cover the entire hold floor, ship side to ship side and hold bulkhead to bulkhead. We use 75 mm square dunnage very liberally, not to facilitate slinging as the chains belong to the steelworks and remain with the cargo, but in addition to protection and passage of condensate it is a very effective securing method. Generally we dunnage under every sling of cargo and apart from tomming off a broken or incomplete tier of, say, tinplate or heavy coils, use no other cargo securing.

By distributing cargo over the entire hold, most ships would be loaded to their loadline with a reasonable margin, prior to achieving the stipulated tanktop limit. We often load heavy coils on edge, along fore and aft bulkheads, winged tightly right across the hold. To prevent a loose stow it is very desirable to complete the tier even if tanktop loading is exceeded. The loading manuals of Japanese-built vessels usually include a vector diagram which shows tanktop loadings may be increased by a factor of 15 to 20 per cent adjacent to a bulkhead.

A perusal of the engineers' boot rack indicates an average sole of 28 cm by 7 cm, or a total bearing surface for the pair of 392 cm^2 and therefore the man can safely walk on Mr Reid's tanktop even if he weighs in at 588 kg!□

Captain E. W. Pretyman, MNI

I AM INVOLVED in loading some 100,000 tonnes of steel and zinc products per annum into single-deck 'Logger'-type vessels of around 18,000 dwt.

I have specifically mentioned the carriage of zinc above, as a formula does exist for calculating tanktop strength, in the Australian Department of Transport Marine Notice 21/1977 *Carriage of Cargoes of Unwrought*

Non-ferrous Metals, which covers zinc loading. Whilst zinc is in slab form, it is bundled in a manner that renders its shape not unlike certain steel cargoes, such as tinplate stillages or the base dimensions of vertical steel coils. It is similar in weight for size. By logical projection this formula could therefore also apply for the loading of steel products.

The formula is given in Section A page 2 of Marine Notice 21/1977 and states:

'Mass of Cargoes

(1) The inner bottom or the decks of a ship should not be loaded above the prescribed limits for that ship. Where this information is not available the following precautions may be followed for guidance only:

a) On the inner bottom the maximum number of tonnes loaded uniformly over the proposed area of the stow should not exceed

$$\frac{db(3L + B)}{4.6} \text{ tonnes}$$

where

d = Summer load draft in metres
b = average breadth of proposed stow in metres
L = length of proposed stow in metres
B = maximum moulded breadth of ship in metres.'

An example of the formula is as follows: 'Calculate the total mass allowable on the inner bottom given d = 9.3m; b = 18.0m; L = 7.0m; and B = 22.0m.' The answer is 1564.82 tonnes. If reduced to the values of Mr Reid's (naval architect's) formula: 'Total square metres × tanktop strength = Total permissible cargo in that area' then tanktop strength = 12.42 tonnes per square metre.

To quote further from Marine Notice 21/1977 section A(2) 'The general distribution of the cargo throughout the ship should be arranged to comply with the allowable maximum still water bending moment and shear force prescribed for the ship . . . ' Based on the information given in the Notice I would conclude that the requirements are quite clear—i.e., for a given area one may only load the mass of cargo as calculated from the formula for that area.

As a rough cross-check on the accuracy of the above formula, it can be seen that by multiplying the area of the tanktop in each hold by the mass per square metre, then the deadweight of the ship is closely approached. There should not in theory be need to exceed the calculated tanktop strength.

A good surveyor is usually the answer to a ship planner's dilemma in this situation. □

ESTIMATION OF SHIP SPEED FROM SHAFT REVOLUTIONS

J. E. Turner, C.Eng, B.Sc, M.Sc, FRINA, AMNEC
Senior Lecturer in Transport Studies, City of London Polytechnic

FREQUENTLY, when investigating the behaviour and track of a ship or ships involved in a collision, it is often necessary to estimate the speed of the ship or ships at various instances before impact occurred. A high level of confidence in the resulting speed estimate is essential in order that the calculations relating to the track of the ship or its kinetic energy can be readily accepted.

The only data available on such occasions are often those of shaft revolutions at various times, with such values of revolutions not corresponding to 'standard' values, such as half speed, for which the appropriate ship speed, in knots, would be known. To determine the ship speed for a given value of shaft revolutions, it is frequently assumed that revolutions and ship speed have a linear relationship and hence if the service speed and revolutions are known then any lesser speed is easily deduced for a particular value of shaft revolutions.

However, inspection of the results of the trials data of most ships would readily show that a linear relationship does not generally exist between the two variables of interest, and the assumption of such a relationship could lead to large errors in estimating ship speed, especially at low values of revolutions. Thus, the prime purpose of this article is to present an alternative simple method by which ship speed may be estimated for any given value of shaft revolutions with a significant level of confidence in the results.

Prediction from shaft revolutions

At constant ship speed the thrust produced by the propeller must equal the total drag of the hull. It may be assumed, without large error, that hull drag varies as the square of the ship's speed (Ref. 1) and thus it follows that propeller thrust varies as ship speed squared.

i.e. $$\frac{R_1}{R_S} = \left(\frac{V_1}{V_S}\right)^2 = \frac{T_1}{T_S} \qquad \ldots\ldots (1)$$

Where R_S and T_S are the hull drag and propeller thrust for the service speed V_S and R_1 and T_1 are the corresponding values at some ship speed V_1.

Now, for the propeller it is known that (Refs. 2, 3):

$$\text{Propeller efficiency, } \eta = \frac{\text{Thrust power}}{\text{Delivered power}}$$

$$= \frac{P_T}{P_D} \qquad \ldots\ldots (2)$$

By appropriate substitution, as detailed in Appendix 1, equation (2) may be re-written as:

$$\text{Propeller efficiency} = \frac{T_S V_S (1 - W_T) \times 0.515}{0.97 \times \text{Service shaft power}} \qquad \ldots\ldots (3)$$

Whence, propeller thrust may be expressed as:

$$T_S = \frac{0.97 \times \text{Propeller efficiency} \times \text{Service shaft power}}{0.515 \times V_S (1 - W_T)} \qquad \ldots\ldots (4)$$

Further, in general, the shaft power may be written as (Refs. 2, 3):

$$\text{Shaft power} = 2\pi NQ \qquad \ldots\ldots (5)$$

where Q is the shaft torque at revolutions N.

By combining equations (1), (4) and (5) it may be shown, as given in Appendix 2, that:

$$\left(\frac{V_1}{V_S}\right)^3 = \frac{\eta_1 N_1 Q_1 (1 - W_{T1})}{\eta_S N_S Q_S (1 - W_{TS})} \qquad \ldots\ldots (6)$$

If it is assumed that the wake fraction, W_T, is reasonably constant for the working range of ship speeds—i.e., speeds for which the ship has steerage way—while both the propeller efficiency and torque vary as functions of shaft revolutions which are not necessarily linear, then equation (6) may be expressed in the form of:

$$\left(\frac{V_1}{V_S}\right)^3 = K_1 \left(\frac{N_1}{N_S}\right)^a K_2 \left(\frac{N_1}{N_S}\right)^b \frac{N_1}{N_S} \qquad \ldots\ldots (7)$$

where $K_1 \left(\frac{N_1}{N_S}\right)^a = \eta_1/\eta_S$, K_1 being a constant

and $K_2 \left(\frac{N_1}{N_S}\right)^b = Q_1/Q_S$, K_2 being a constant

Whence, by combining the terms on the right hand side of equation (7) and taking the cube root, we have:

$$\left(\frac{V_1}{V_S}\right) = K \left(\frac{N_1}{N_S}\right)^d \qquad \ldots\ldots (8)$$

Equation (8) enables the ship speed, V_1 to be deduced for given shaft revolutions, N_1. Obviously, in order to be able to apply equation (8) values of the constant K and the power index d need to be determined. Strictly, the values of such coefficients would vary with ship type, loading condition, type of main engine, number and type of propellers, etc.

However, the aim of this note is to produce a general expression by which speed may be predicted and thus the approach that follows is proposed for the determination of K and d.

Determination of constant K and power index d

Equation (8) represents a curvilinear function which, for further analysis, may be more conveniently written as:

$$\log_{10}(V_1/V_s) = \log_{10}K + d\log_{10}(N_1/N_s) \quad \ldots\ldots (9)$$

Equation (9) may now be taken to be a linear equation for which the constant of the equation, $\log_{10}K$, and the slope of the line, d, may be readily deduced by application of simple linear regression analysis to appropriate data.

For this exercise, data of speed and revolutions were extracted from the results of speed trials for a number of ships and a regression analysis made as per equation (9). The output of the analysis is given in Appendix 3 which, when substituted in equation (8), produces for all practical purposes the expression:

$$V_1 = V_s(N_1/N_s)^{0.854} \quad \ldots\ldots (10)$$

Discussion of the proposed equation

The suitability of equation (10) for predicting ship speed from shaft revolutions may be tested by applying it to known data and comparing the resulting percentage errors with those produced by the conventional linear relationship mentioned at the beginning. Tables I to III show such comparisons for three different types of ship.

Table I Bulk Carrier 34,300-tonne deadweight.
V_s 15.8 knots; N_s 105 rpm

Measured V_1	Measured N_1	Linear V_1	% Error	Equation (10) V_1	% Error
14.1	92	13.84	−1.84	14.10	0
11.8	75	11.29	−4.32	11.85	0.4
9.7	60	9.03	−6.91	9.79	0.93
7.3	45	6.77	−7.24	7.66	4.93

(Source: Ref. 4)

Table II General Cargo Ship 12,400-tonne deadweight.
V_s (loaded) 16.0 knots; N_s 130 rpm

Measured V_1	Measured N_1	Linear V_1	% Error	Equation (10) V_1	% Error
15.6	120	14.77	−5.33	14.94	−4.23
13.0	100	12.31	−5.33	12.79	−1.62
7.0	50	6.15	−12.09	7.06	+1.14

V_s (Ballast) 18.08 knots; N_s 130 rpm

Measured V_1	Measured N_1	Linear N_1	% Error	Equation (10) V_1	% Error
16.5	118	16.41	−0.54	16.63	+0.72
14.3	100	13.91	−2.74	14.45	+1.05
8.16	50	6.95	−14.78	7.99	−2.02

Table III Cargo Liner 10,700-tonne deadweight.
V_s 22 knots; N_s 115 rpm

Measured V_1	Measured N_1	Linear V_1	% Error	Equation (10) V_1	% Error
21	109	20.85	−0.70	21.02	+0.09
20	104	19.90	−0.52	20.19	+0.95
19	96	18.37	−3.34	18.86	−0.74
18	92	17.60	−2.22	18.18	+1.00
17	86	16.45	−3.22	17.17	+1.00

(Source: Ref. 5)

The tables show that, in general, the 'conventional' method underestimates ship speed with, typically, the percentage error ranging from 1 per cent to 15 per cent, with the average being 4.74 per cent. Thus, the behaviour of a ship would also be underestimated with large errors arising in the case of the calculation of kinetic energy which varies as speed squared.

The proposed equation (10) gives much smaller percentage errors which, for the given examples, average only 0.09 per cent too high, with a maximum of plus or minus 4.58 per cent. Thus, the suggested predictor equation appears to be more satisfactory for estimating ship speed from a given value of shaft revolutions and hence enables the calculations relating to ship behaviour, before or during a collision, to be accepted with a greater degree of confidence than before.□

References

1. Dien, R. 'A Proposal for Assessing the Stopping Ability of Ships'—*International Shipbuilding Progress,* Vol. 23, 1976.
2. Rawson, K. J. and Tupper, E. C.—*Basic Ship Theory*—Longmans.
3. Munro-Smith, R. *Applied Naval Architecture*—Longmans.
4. Baseler, R. W. 'Design, Construction and Operation of a 36250-ST Self-Unloading Collier with a Coal-Fired Steam Plant'—*Marine Technology,* Vol. 21, No. 4, Oct 1984, pp 319-333.
5. Meek, M. 'Glenlyon Class—Design and Operation of High-Powered Cargo Liners'—Royal Institution of Naval Architects, 1964.
6. Cass, T. *Statistical Methods in Management 1*—Cassell, 1984.

Appendix 1

Expression for propeller efficiency

$$\text{Propeller efficiency} = \frac{\text{Thrust power}}{\text{Delivered power}}$$

and Thrust power, P_T = Thrust × Speed of advance of propeller

Speed of advance = $V_s(1-W_T) \times 0.515$

where V_s is ship speed in knots

W_T is the wake fraction

Hence, $P_T = T\,V_s(1-W_T) \times 0.515$

Delivered power, P_D = 0.97 × Service shaft power

The service shaft power is about 3 per cent higher than the delivered power when allowance is made for transmission losses.

Whence, propeller efficiency, η, becomes:

$$\eta = \frac{P_T}{P_D} = \frac{T\,V_s(1-W_T) \times 0.515}{0.97 \times \text{Service shaft power}}, \text{ as per equation (3).}$$

Appendix 2

Now, by equation (5) the shaft power for the service condition at revolutions N_s and any other condition at revolutions N_1 would be respectively:

$2\pi N_s Q_s$ and $2\pi N_1 Q_1$

Where Q_s and Q_1 are the respective shaft torques.

It then follows that if η_s and η_1 are the propeller efficiencies for each conditions then, by application of equation (4), we have that:

$$T_1 = \frac{V_s(1-W_{Ts})\,N_1\,Q_1\,\eta_1}{V_1(1-W_{T1})\,N_s Q_s \eta_s}$$

Hence, from equation (1)

$$\frac{T_1}{T_s} = \left(\frac{V_1}{V_s}\right)^2 = \frac{V_s}{V_1}\frac{(1-W_{Ts})}{(1-W_{T1})}\frac{N_1}{N_s}\frac{Q_1}{Q_s}\frac{\eta_1}{\eta_s}$$

$$\text{Hence, } \left(\frac{V_1}{V_s}\right)^3 = \frac{\eta_1\,N_1\,Q_1\,(1-W_{T1})}{\eta_s\,N_s\,Q_s\,(1-W_{Ts})} \quad \text{as per equation (6)}$$

Appendix 3

Application of simple regression analysis

As given in the explanation for equation 10 earlier, data were extracted from published trials results for a small sample of ships of various types. The sample included single screw and twin screw ships with either diesel engines or steam turbine main machinery.

For each ship the quantities of $\log_{10}(V_1/V_s)$ and $\log_{10}(N_1/N_s)$ were calculated with the sample size being 42 pairs of values. With the quantity $\log_{10}(V_1/V_s)$ as the dependent variable and $\log_{10}(N_1/N_s)$ the independent variable, a simple regression was carried out using the model given by equation (9). The computer output was as follows:

Dependent variable: $\log_{10}(V_1/V_s)$

Independent Variable	Coefficient	Standard Error	T-Ratio	Significant
Constant	−0.0027	0.0034	0.785	NS
$\log_{10}(N_1/N_s)$	0.854	0.025	34.583	99%

Coefficient of Determination (R^2) 0.968
Correlation Coefficient (r) 0.984

For this analysis, the degrees of freedom (df) is 40. Thus, at the 0.05 level of significance and df of 40, the significant value of the correlation coefficient is 0.3044 (Ref. 6). As the calculated value of 0.984 is in excess of this value then this demonstrates that the relationship between the variables in question is significant.

The coefficient of determination shows that 96.8 per cent of the variation in the dependent variable is explained by variation in the independent variable. It is possible that with a much larger sample size this level of explanation would be improved further. Similarly, the correlation coefficient value of 0.984 shows that there is a 'high degree of association' between the given variables. Finally, the output shows that the constant is not significantly different to zero while the independent variable is highly significant.

Thus, the regression equation may be written as:

$$\log_{10}(V_1/V_s) = -0.0027 + 0.854 \times \log_{10}(N_1/N_s) \quad \ldots\ldots (11)$$

(s.e.) (.0034) (0.025)

( ) terms are the standard errors

By comparing equation (11) with the model of equation (9) it then follows that:

$\log_{10} K = -0.0027$
whence $K = 0.994$

Hence, the relationship between ship speed and shaft revolutions may be written as:

$$V_1 = V_s \times 0.994\,(N_1/N_s)^{0.854}$$

This, for all practical purposes, may be taken as:

$$V_1 = V_s(N_1/N_s)^{0.854} \text{ as per equation (10).}$$ □

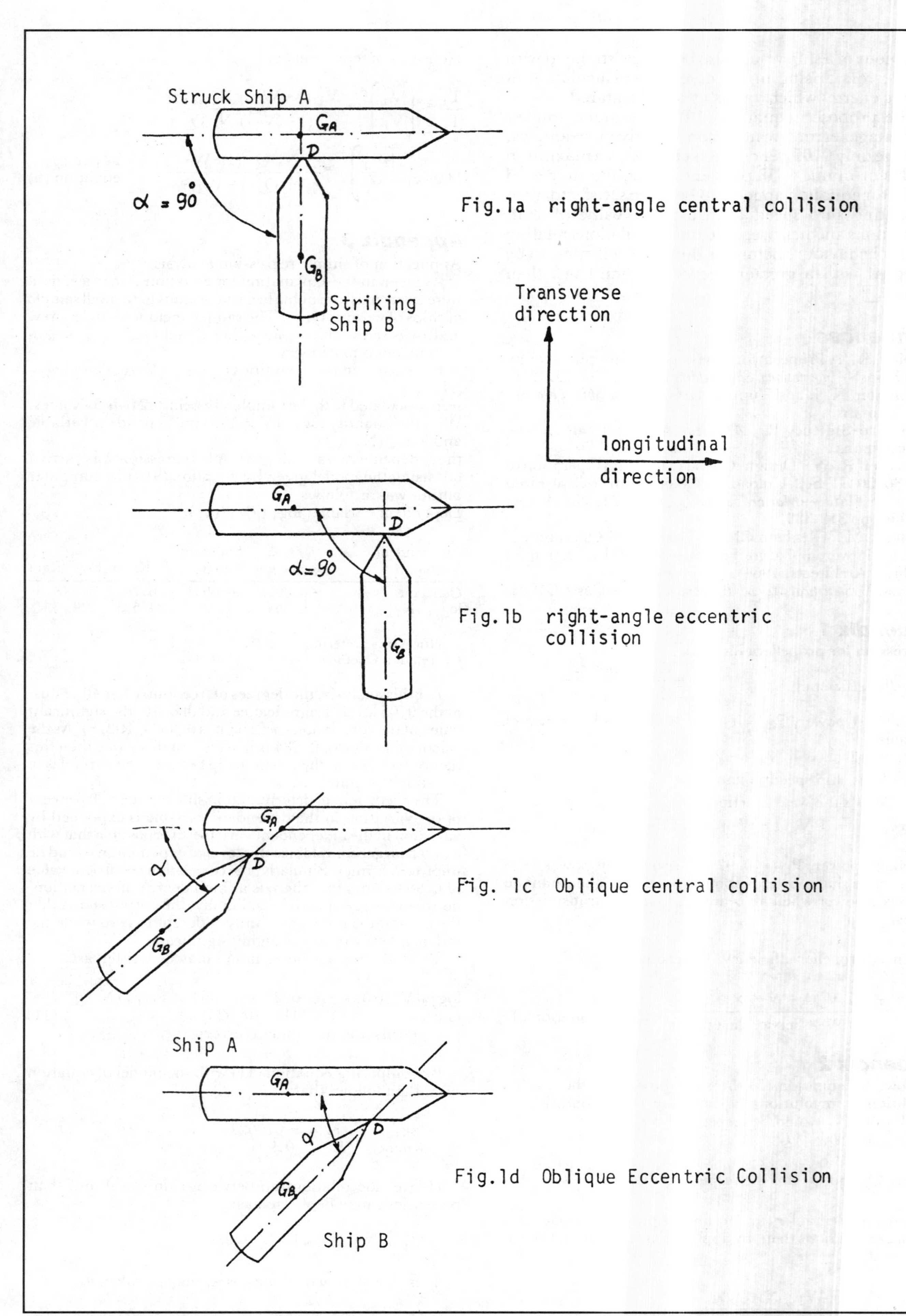

Figure 1: Different situations of collisions.

NOTES ON THE EXTERNAL MECHANICS OF COLLISIONS BETWEEN SHIPS *

M. Pawlowski
Technical University of Gdansk, Poland
E. H. Hegazy
University of Suez Canal, Egypt

IN THE LAST DECADE there has been an increase in interest in all questions related to ship collisions. The reasons for this are, on the one hand, growing public interest in the effect of collisions involving ships with cargoes harmful to the environment—e.g., tankers; on the other hand, the demand for maximum safety in ships powered by nuclear energy makes it essential to examine very closely all aspects connected with collision between ships. In this chapter the authors investigate one of the most important questions related to ship collision, namely, external mechanics of collision.

The mechanics of ships' collision is often divided into 'external' and 'internal' aspects. The external mechanics describe the motion of the ships as a whole, while the internal mechanics deal with the elastic and/or plastic deformations occurring due to collision. The mutual forces of impact betweeen the ships are the links, not only between the colliding vessels, but also between these two parts of the problem of ships' collision.

Many papers have been published on individual aspects of the external mechanics of collisions. Minorsky[1], Woisin [2,3,4], Castagneto[5], Haywood[6], etc., derived expressions for the kinetic energy lost in collision (i.e., energy to be absorbed by the structural damage of both ships). The final expression for the lost kinetic energy depends on the assumption made regarding the direction of the impact force. In most textbooks of engineering mechanics, it is assumed that collision forces act only perpendicular to the plane of contact. This concept has been adopted by Minorsky and Castagneto. Contrary to it, Woisin assumes that the forces act co-linear with the relative speed vector between the two vessels.

The effect of surrounding water has been examined experimentally and theoretically by Motora *et al* [7]. The energy absorbed by transverse vibration of the ship as a beam was estimated in ref.[8]. These investigations were limited to the simplest case of a right-angled collision with the struck ship stationary.

In this paper, based on a more realistic assumption with regard to the acting impact forces, a theoretical analysis of the motion of the colliding vessels is developed. The effects of different factors on the kinetic energy lost during collision—e.g., angle of encounter, location of impact, masses and speeds of colliding vessels, effect of surrounding water, shock effects, etc.—have been investigated and represented in the forms of charts. Also the size of the resulting damage is examined and given in terms of the kinetic energy lost during collision. Calculations are carried out assuming the collision force is either constant or varies linearly with depth of penetration.

Energy balance during a collision

The analysis of the distribution of the energy transferred in a collision is complicated by the large number of variables which are involved. These include (a) displacement of each vessel, (b) speed of each vessel, (c) angle between courses of each vessel (i.e., angle of encounter), (d) point at which struck ship is hit, (e) location of the centre of mass of each vessel, (f) moment of inertia of each vessel about their common centre of rotation after collision, (g) effect of surrounding water, (h) longitudinal bending moment of the struck ship in a horizontal plane, and (i) power being produced by each ship during collision. Some of these factors may be negligible and can be eliminated without introducing significant errors.

The hydrodynamic forces exerted by the surrounding water due to transverse motion of a ship—i.e., both inertia and damping (resistance)—can be represented by an increase in the mass of the ship. This increase in the effective inertia is variously known as equivalent entrained water, equivalent added mass or equivalent hydrodynamic mass. The effect of hydrodynamic forces due to longitudinal motion is neglected. This assumption is substantiated by comparison with the prolate ellipsoid investigated in reference[8]. For a ratio of length/breadth 10 the true mass of the ellipsoid is increased by over 96 per cent when moving 'broadside', and by less than 2 per cent when moving 'end on.' Experimental and theoretical studies[7,9] have shown that the effect of hydrodynamic forces on the struck ship plays an important role during a collision between ships.

Castagneto[5] gave an estimate of the energy which is absorbed during collision as a consequence of the longitudinal bending of the struck ship in a horizontal plane. His calculations showed that this energy is small and can be neglected. Similar calculations have been made by Guide[10], whose results are essentially the same as those given by Castagneto.

According to statistical data about actual collisions, one can say that collisions involving vessels proceeding at full power at the time of impact are very rare. In most cases the colliding ships were usually aware of each other's presence well before the collision, and, therefore, they had sufficient time to reverse their engines[11,12]. On the basis of such observations and the results of the typical example worked out in ref.[12] it is believed that the energy contribution due to

*A paper presented at the 'Fitness for Sea' conference, University of Newcastle upon Tyne, Department of Naval Architecture & Shipbuilding, September 1980.

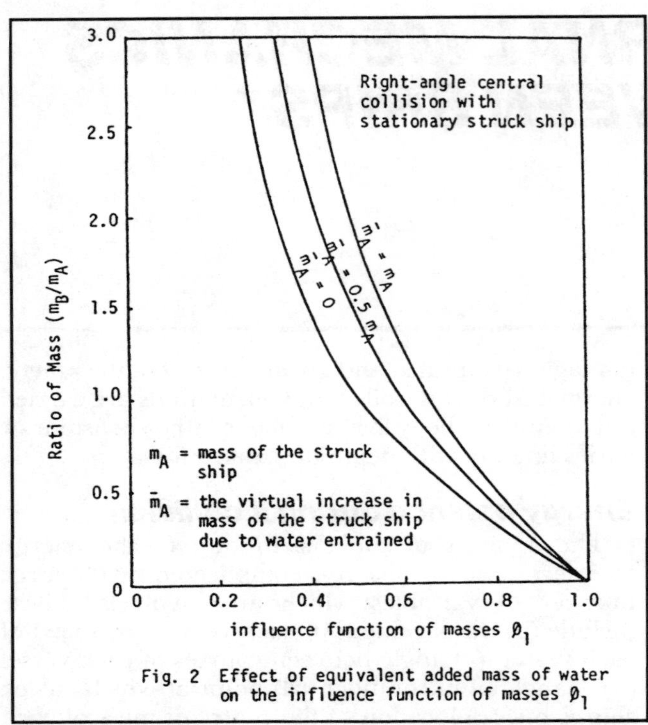

Fig. 2 Effect of equivalent added mass of water on the influence function of masses ϕ_1

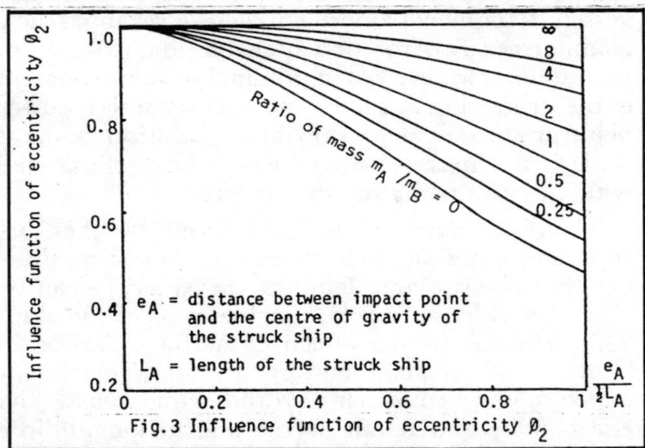

Fig. 3 Influence function of eccentricity ϕ_2

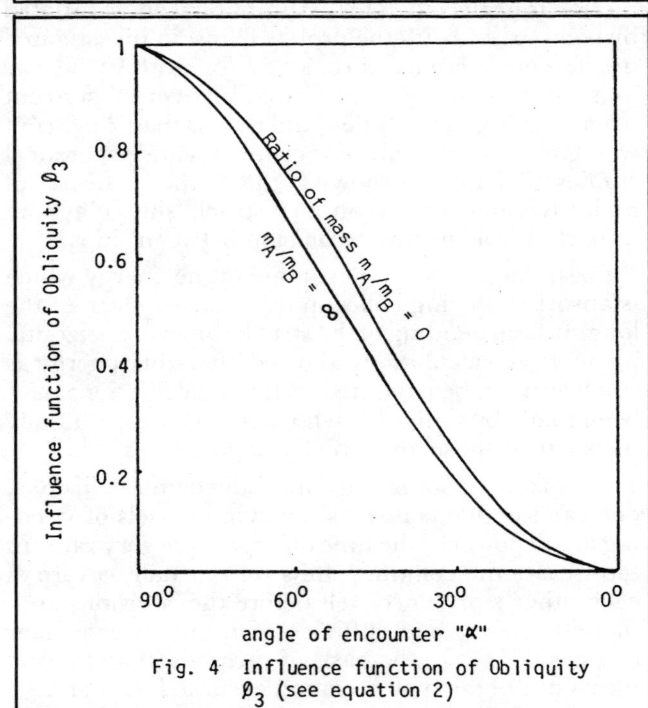

Fig. 4 Influence function of Obliquity ϕ_3 (see equation 2)

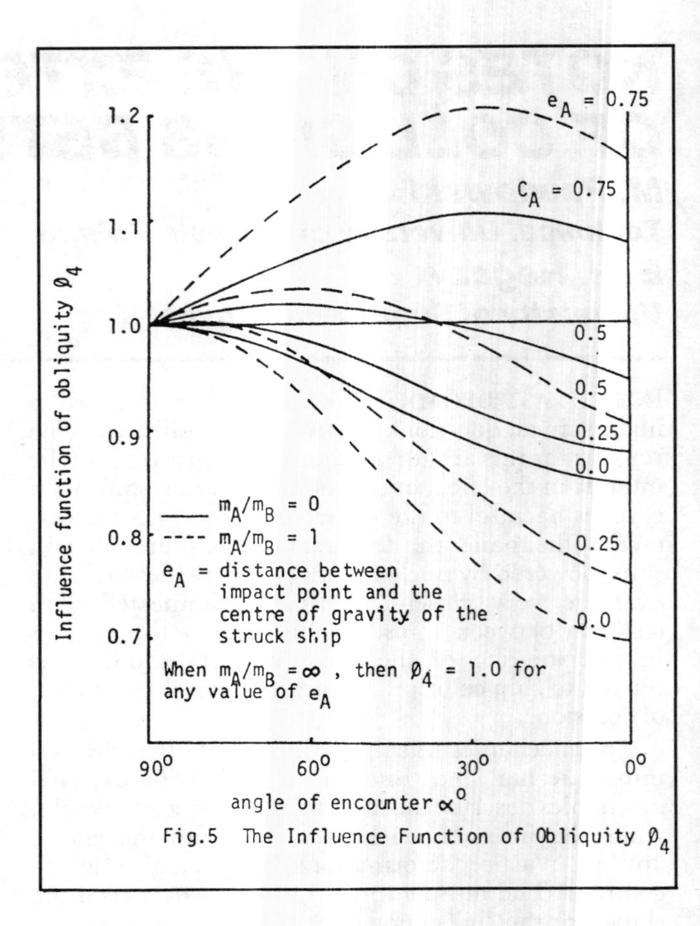

Fig.5 The Influence Function of Obliquity ϕ_4

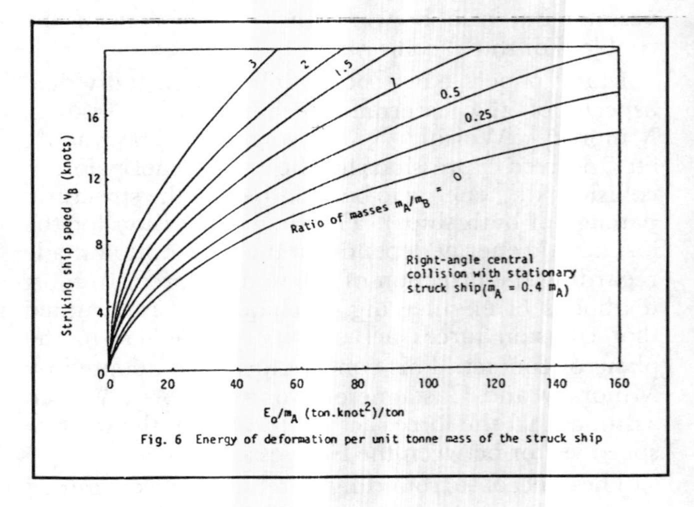

Fig. 6 Energy of deformation per unit tonne mass of the struck ship

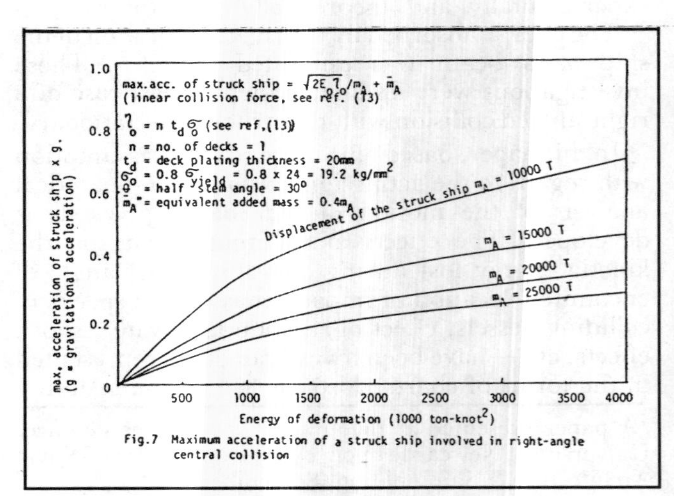

Fig.7 Maximum acceleration of a struck ship involved in right-angle central collision

the shaft horsepower of the vessels need not be considered in the calculation of the energy during collision.

Following the above discussion it can be conjectured that the initial kinetic energy possessed by the colliding vessels before collision is redistributed through the process of deformation of both ships and through the final kinetic energy of the bodily motions of both ships following collision—i.e.:

$$E_T = E_D + E_F \quad (1)$$

where:

E_T = kinetic energy of the colliding vessels before impact,

E_D = energy of deformation (i.e. the kinetic energy lost in collision and to be absorbed by the structural damage of both ships)

E_F = final kinetic energy of the vessel pair following collision.

Calculation of energy of deformation (E_D)

With regard to the struck ship we can differentiate between four different kinds of collisions: (a) right-angle central collision (see figure 1a); (b) right-angle eccentric collision (see figure 1b); (c) oblique central collision (see figure 1c), and (d) oblique eccentric collision (see figure 1c).

As already mentioned in the introduction, the final expression for the energy of deformation depends on the assumption made regarding the direction of the impact force. There is no problem concerning the direction of the impact force for right-angle collisions (cases (a) and (b)), since in such cases it is clear that the impact force is perpendicular to the centre line of the struck ship. In order to deal with oblique collisions we have to make some assumptions with regards to the acting impact force.

In this study two different assumptions are tried out—namely, (i) the impact force is assumed to act perpendicular to the plane of contact, and (ii) the direction of the impact force is assumed to be unknown and to be determined from the analysis of the motion of the colliding vessels. As a result two different expressions for the energy of deformation were obtained corresponding to these two assumptions. The validity and applicability of each of the used assumptions will be discussed later.

Consider the general case of collision in which ship B strikes ship A at any point along her length with any angle of encounter (α) as shown in figure (1d). By equating the initial kinetic energy possessed by both ships before collision to the energy of deformation plus the residual energy of the vessel pair following impact (see equation 1), the following expressions for the energy of deformation E_D have been derived by the authors[13] : by assuming that the impact force acts perpendicular to the plane of contact:

$$E_D = \tfrac{1}{2} m_B v_B{}^2 \phi_1 \phi_2 \phi_3 \quad (2)$$

by assuming that the direction of the impact force is unknown:

$$E_D = \tfrac{1}{2} m_B v^2{}_B \phi_1 \phi_2 \phi_4 + \tfrac{1}{2} m_A v^2{}_A \phi_5 - \tfrac{1}{2} v_B v_A m_A \phi_6 \quad (3)$$

where,

m_B, v_B = the mass and speed at impact of striking vessel,

m_A, V_A = the mass and speed at impact of the struck ship,

ϕ_1, ϕ_2, (ϕ_3 and ϕ_4), ϕ_5 and ϕ_6 are non-dimensional functions representing the effect on the energy of deformation E_D of masses of the colliding vessels, the impact location, the angle of encounter, speed of the struck ship, and the product of both speeds, respectively. The functions ϕ_1, ϕ_2 and (ϕ_3 and ϕ_4) will be called as the influence functions of masses, eccentricity, and obliquity and they are shown graphically in figures 2, 3, 4 and 5 respectively.

It should be pointed out that, in the derivation of equations (2) and (3), all the parameters necessary to describe the motion of the colliding ships before collision as well as the vessel pair following impact were taken into account. The collision process was assumed to be an entirely plastic and two-dimensional problem and any other motions such as heeling of the struck ship, lift of the striking bow, etc., have been neglected. Obviously, the assumption made regarding the direction of the impact force must be chosen in view of the resulting collision damage, since the contact area between the colliding vessels serves as a boundary condition of the problem.

It is believed that the first assumption concerning the orthogonality of the impact force and the plane of contact appears to be valid for collisions in which rupture of the struck ship's hull does not occur (i.e., minor or low-energy collision), since in this case sliding of the contact point is likely to occur along the struck ship. On the other hand, the second assumption (in which the direction of the impact force is unknown and to be determined from the analysis of ships' motion) might be valid for collisions which cause rupture of the hull of the struck ship (major or high-energy collision). Thus expression (2) would be used to calculate the kinetic energy lost during minor collisions and equation (3) during major collisions.

Once the expression for the energy of deformation (E_D) has been developed for the more general case, it is not difficult to find such expression for any other special case of collision (see figure 1).

Consider the right-angle central collision with a stationary struck ship, see figure 1a. Such a case represents the simplest and traditional case of collision, which is usually used for the analysis of collision damage[1, 8, 9]. For such a collision we have: angle of encounter $\alpha = 90°$, distance between the impact point and the centre of gravity of the struck ship, $e_A = 0$ and the speed of the struck ship, $v_A = 0$. As a result we get $\phi_2 = \phi_3 = \phi_4 = 1.0$ and expression (2) and (3) give the same result for E_D (since in this case the direction of the impact force is well defined) as:

$$E_D = E_0 = \tfrac{1}{2} m_B v_B{}^2 \phi_1 = \tfrac{1}{2} m_B v_B{}^2 \frac{u}{1+u}$$

where,

E_0 = energy of deformation for right-angle central collision with a stationary struck ship,

and $u = \dfrac{m_A + \bar{m}_A}{m_B}$, $\bar{m}_A$ = equivalent added mass

This result reveals that in this kind of collision a fraction $\frac{u}{1+u}$ of the initial kinetic energy possessed by the striking vessel before collision is used to do work by the collision force and a fraction $(1/1 + u)$ of the initial kinetic energy remains as a final kinetic energy of the vessel pair after collision.

Equation (4) was obtained, for the first time, by Minorsky in his pioneering paper on ship collisions in 1959[1]. He derived it from the principles of the conservation of momentum for the simple case of collision under consideration. Using equation (4) the energy of deformation per unit mass of the struck ship (i.e., E_0/m_A) is shown in figure 6 against the speed of the striking ship (v_B) for different ratios of the masses of the colliding vessels ($R = m_A/m_B$).

For calculating the influence coefficient of masses ϕ_1 the virtual increase in the mass of the struck ship due to entrained water was taken, according to Minorsky[1], as 40 per cent of the mass of the struck ship (a full discussion of the effect of the entrained water will be given later). Depending on the degree of safety required, this quantity (E_0/m_A) must be kept within a certain limit for any specific struck ship. For example, for nuclear-powered ships this value is limited by the maximum amount of energy that the collision barrier on the side of the ship could absorb before the striking bow would penetrate the reactor compartment longitudinal bulkhead.

Size and distribution of damage

Up to now the only way to obtain the distributions of damages was and still is to gather and investigate statistical data about ships' collisions. But this statistical approach has an important fault—i.e., it does not give in practice the possibility of obtaining the conditional distribution of damages for a given vessel whereas just such a distribution is indispensable in calculating the probability of collision survival. In other words, based upon the statistical data we can take into account at the best, only the effect of a ship's length on the conditional distribution of damages and in practice we are forced to omit other features of a ship and its structure which are so important.

The starting point in the theoretical method is a relationship between the dimensions of the damage in the struck ship and the kinetic energy lost during collision (i.e., energy of deformation) which has been established by the authors[13]. The dimensions of damage include the penetration and the length of damage. By penetration (w) is meant the dimension of the damage in a transverse direction—i.e., normal to the centre line of the struck ship—while the length of damage (L) represents the extent of damage in the longitudinal direction—i.e., along the length of the struck ship. The derivation of the required relationship is based on the following assumptions:

(a) The striking bow has a vertical stem;
(b) The angle of encounter (α) does not change during collision.

The calculations are carried out for infinitely rigid striking bows and soft striking bows assuming the collision force is either constant or varies linearly with depth of penetration. The exact functional form of the collision force is determined from a knowledge of the collision process[9, 14].

The derived formulae for the dimensions of damage give us the possibility of obtaining the damage distributions, i.e., both distribution of penetration and length of damage. This we can do for a given vessel if we know the distribution of all random variables upon which depend the size of damage.

The value of penetration as determined in this section should not exceed a certain critical value for the safety of the struck ship under consideration. For example, for nuclear-powered vessels the critical value is given by the width of the collision barrier on the side of the vessel in way of the reactor compartment.

Duration of collision

The impact duration (τ) is a very important factor in calculating the virtual increase in the masses of the colliding vessels due to entrained water as well as in calculating the acceleration of both ships during collision. The latter should not exceed a certain critical value for the safety of plant or cargoes in ships.

A general expression for calculating the duration of impact (τ) was derived by the authors[13]. It was found that the duration of collision is only related to the initial speed of the striking ship and the amount of penetration (or relative displacement in the case of soft striking bow). It is proportional to the penetration, and inversely proportional to the initial speed of the striking vessel.

Acceleration and shock effect

Now let us examine the values of the maximum acceleration of the struck and striking ships during a collision. As already mentioned, above such values should not exceed a certain critical value for safety of plant and cargoes in ships.

As a result of the theoretical analysis developed in this study[13] expressions for the maximum accelerations of the colliding vessels have been derived for the general case of collision. In figure (7) the maximum acceleration of a struck ship involved in a typical right-angle central collision is shown against the energy of penetration for different struck ship's displacements. The value of maximum acceleration is given as a fraction of the gravitational acceleration (g). It is clear that the maximum acceleration of the struck ship for the range of displacement used (i.e., 10,000T—25,000) is less than 'g'. The value of the maximum acceleration of the struck ship in the typical collision example worked out by Minorsky[1] as calculated by the derived equation was found to be 0.186g.

Obviously, the accelerations of large masses in collision conceivably result in damage to vital engine components, rupture of pipelines, etc. However, the above theoretical calculations, together with the practical example of Minorsky, give calculated values for the acceleration of the struck ship during collision of the order of 0.2g to 0.6g (see figure 7). Moreover, the analysis of 60 major collisions[12] showed that no evidence whatever of shock damage was found in any collision. The testimony of personnel involved in collisions generally revealed that no shock could be felt in any part of the ship.

This testimony indicates that the accelerations are less than those encountered in extreme roll, pitch and heave. Since estimated g values in extreme pitch and roll are of the order of 0.6, it is evident that accelerations to be expected in collision will be low. From the above, it is concluded that the accelerations obtained in ship collisions are low and under no circumstances can be expected to exceed those to which merchant ship's foundations and mountings are normally designed (= 2g).

Effect of surrounding water

As already explained earlier the effect of the hydrodynamic forces resisting the motion can be represented as an increase in the mass of the struck ship. This increase in the effective inertia, variously known as entrained water, added mass or hydrodynamic mass, plays an important role during collision and cannot be neglected. Also it was shown that the effect of hydrodynamic forces due to the longitudinal motion of a ship is insignificant.

Minorsky was the first to examine the effect of the entrained water on the kinetic energy lost during collision in his pioneering paper on collisions[1]. He investigated the effect of the entrained water on the kinetic energy lost in the simplest collision case (i.e., right-angle central collision with a stationary struck ship), where the effect of entrained water is only included in the influence function of masses ϕ_1. Minorsky named the function ϕ_1 the 'coefficient of energy absorption,' since it represents the fraction of the initial kinetic energy possessed by the striking vessel before collision, which is absorbed by the fracture of both colliding vessels.

The theoretical results which are presented in figure (2) (as taken from reference[1]) show that the kinetic energy lost during a collision (i.e., energy of deformation) is relatively insensitive to the actual value of the virtual mass. The largest discrepancy, according to Minorsky's calculations, occurs when the mass of the striking ship is larger than the mass of the struck ship. For example, when the mass of the striking ship is double the mass of the struck ship, then the kinetic energy lost during a collision when the added mass is neglected is two-thirds the value obtained if the added mass is assumed equal to the mass of the struck ship. The theoretical results for the amount of added mass recommended by Minorsky lie about mid-way between those two calculations.

In fact Minorsky assumed that the virtual increase in mass of the struck ship due to entrained water was 0.4 times the mass of the struck ship since previous works on the transverse vibrations of hulls in deep water indicated that the added mass was approximately this value[15,16,17]. More recently, Motora, *et al*[7] conducted some experimental tests on a ship model and obtained good agreement with a theoretical approach which predicts the added mass during a collision. It turns out that the added mass of the model is about 40 per cent of its mass when the duration of impact is short. This agrees with Minorsky's assumption.

However, the actual value of added mass depends on the duration of impact and on the functional form of the external force. The equivalent added mass becomes larger as the duration increases.□

Acknowledgements

This work was carried out whilst the authors were visiting the Department of Naval Architecture and Shipbuilding, University of Newcastle upon Tyne. The authors are most grateful to Professor J. B. Caldwell, the Head of the School of Marine Technology, who has initiated and supervised this study on collisions between ships and has helped to provide literature.

References

1. Minorsky, V.U., 'An Analysis of Ship Collisions with Reference to Protection of Nuclear Power Plants,' *J. Ship Research,* Vol. 3, No. 2, October 1959, pp. 1-4.
2. Woisin, G., 'Der Einflub des Drehimpulses bei einer Shiffskollision besondres in Hinblick auf die Sicherheit Von Atomschiffen,' *Schiff Und Hafen,* 1962, pp. 577-588.
3. Woisin, G. 'Kollisionsprobleme bei Automchiffen,' *Hansa* 1964, pp. 999-1010.
4. Woisin, G., 'Schiffbauliche Forschungsarbeiten fur die sickerheit Kernenergiegetriebener Handelsschiffe,' *Trans. STG 65* (1971).
5. Castagneto, E. 'L'energia Distributtiva nella collisions della navi,' *Technica Navale,* 1962, pp. 731-742.
6. Haywood, F. H. 'Ship Collisions at Varying Angles of Incidence,' *Report No. N.C.R.C./N.163,* Naval Construction Research Establishment, February 1964.
7. Motora, S., Fujino, M., Sugiura, M., and Sugita, M. 'Equivalent Added Mass of Ships in Collisions,' *J. Soc. Naval Arch. Japan,* Vol. 7, 1971, pp. 138-148.
8. Haywood, J. H. 'A Theoretical Note on Ship Collisions,' *Report No. R.445,* Naval Construction Research Establishment, February 1961.
9. Akita, Y., Ando, N., Fijita, Y. and Kitamura, K. 'Studies on Collision-Protective Structures in Nuclear-Powered Ships,' *Nuclear Engineering and Design,* Vol.19, 1972, pp. 365-401.
10. Guide, A. 'Analisi della Similitudine nella prove di Collisione,' *Technica Italiana,* 1964, pp. 213-222.
11. Devanney, J. W., Protopapa, S. and Klock, R, 'Tanker Spills, Collisions and Groundings' MIT, *Report No. MITSG 79-14,* June 1979.
12. 'Collision Study for *Savannah,'* by George G. Sharp, Inc. 1961.
13. Pawlowski, M. and Hegazy, E. H., 'Theoretical Note on External Mechanics of Collisions Between ships,' University of Newcastle upon Tyne, Department of Naval Architecture & Shipbuilding, *Report,* August 1980.
14. Akita, Y. and Kitamura, K. 'A Study on Collision by an Elastic Stem to a Side Structure of Ships,' *J. Soc. Naval Arch. Japan,* Vol. 131, 1972, pp. 307-317.
15. Koch, J. J. 'Experimental Method for Determining Virtual Mass for Oscillations of Ships,' *Ingenieur Archiv.,* Vol. 4, part 2, 1953.
16. Dieudonee[1], J. 'Vibration in Ships,' *Trans. of The Institution of Naval Arch.,* January 1959.
17. Johnson, A. J., 'Vibration Tests on all Welded and all Riveted 10,000 Ton Dry Cargo Ships,' *Trans. N.E.C.I.E.S.* Vol. 67, 1950-1951.

BIBLIOGRAPHY AND REFERENCES

Compiled by the Nautical Surveyors' Working Group

IN A BOOK of this type it is not possible to cover every aspect of surveying and neither is it possible for a surveyor to know everything. A surveyor therefore has to be prepared to read widely and obtain information from reference sources. Most organisations maintain a working library of sorts and most surveyors have a bookshelf of useful titles in their office. A surprising amount of information is published and the over-riding problem is knowing where to find it in a hurry.

Here then are some publications which will provide a useful starting point:

General

There are only a few organisations which have specialised in marine publications and you can start by checking:

1. The International Maritime Organization.
2. Government publications.
3. Classification society publications.
4. British Standards Institute Publications including ISO etc.
5. Lloyd's of London Press.
6. Fairplay Publications.
7. Transactions from The Nautical Institute, the Institute of Marine Engineers and the Royal Institution of Naval Architects.
8. The Institute of Petroleum/American Petroleum Institute.
9. The Chartered Institutes of Arbitration, Shipbrokers and Insurance.
10. The Association of Average Adjusters.
11. Witherby and Co. who publish guides, texts and law books.
12. There are many other organisations with specialist expertise like the Institute of Measurement and Control or the British Chain Testers Association and the International Cargo Handling Co-ordination Association (ICHCA).

For those who need to be kept up to date, there is no substitute for studying conference proceedings and the relevant periodicals. Useful addresses appear at the end of the book section.

BIBLIOGRAPHY

Accident Prevention Manual

National Safety Council of the USA.

This is a major reference work spanning occupational health and safety across all industries and for this reason, it has many valuable lessons for shipping. The sections cover legislation, hazard control, removing hazards from the job, accident record, accident investigation, workers' compensation, safety training, human factors engineering, behaviour, maintaining an interest in safety, publicising safety planning for emergencies, occcupational health and other relevant material.

Arrest of Ships

C. Hill, K. Soehring, T. Hosoi and C. Helmer, Lloyd's of London Press, 1985.

The primary purpose of this book is to acquaint all those who may ever have occasion to take arrest action, because of some grievance. It covers the law and practice and procedures to be adopted. Judicial sales, maritime liens and problems of wrongful arrest are covered and the authors cover the practices in UK, Germany, Japan and the USA.

British Standards Institution Catalogue

Available in all major libraries, this contains the complete list of current BSIs and how to obtain them. Containing 575 pages of small print, it covers all industries. Wire ropes, for example, are not listed under the marine section. It is noted that some 180 standards, notably bridge design, are not yet incorporated: so it is important to pursue the secretariat to check on the latest standards. Tables of IEC and ISO equivalents to BSI are also given.

Bunkers

C. Fisher and S. Hodge, Lloyd's of London Press, 1986.

The foreword puts it simply thus: Bunker quantity and quality disputes have received a considerable amount of publicity and have been a cause of some concern to many people in the shipping industry. This book helps to develop an understanding of marine bunker fuels from refinery to the combustion chamber.

Carriage of Goods by Sea

E. R. Hardy Ivamy, Butterworths.

This book covers the subject in a very clear understandable manner, giving a wealth of case detail to illustrate points. Also included are The Hague Rules, York-Antwerp Rules, Merchant Shipping Acts and specimens of bills of lading, etc.

Claims—*Marine Claims Handbook*

N. G. Hudson and J. C. Allen, Lloyd's of London Press, 4th Ed. 1984.

This handy reference book is in nine parts and covers the nature of claims, steps to be followed by shipowners and/or agents, from casualty to collection of claim; general average and salvage; English and American clauses; particular average claims; shipowners' liabilities; claims presentation and evidence; goods and cargo claims. The appendices cover The York Antwerp Rules and assorted agreements and insurance clauses.

Casualty—*Marine Casualty Investigations*

Proceedings of joint conference organised by The Nautical Institute and the Institute of Marine Engineers, Published by IMarE.

The book covers the legal dimension, the public interest, the technical dimension, international issues, safety and the role of the IMO.

Container and Cargo Securing

Conference proceedings, Safe Ships-Safe Cargo Conference by Cargo Systems. CS Publications Ltd, 1987.

These papers examine losses to ships and cargo; cargo care and lashing security. Specialised papers cover classification rules for container securing, cargo stowage and securing, analysis of lashing systems, cargo securing calculations, new lashing systems, problems with bulldog clips, stevedoring, training and education.

Containers—*Cargo Containers*

Herman D. Tabak. Cornell Maritime Press Inc.

A comprehensive and detailled guide to cargo containers stowage, handling and movement techniques on the international scene—a book which proves to be both interesting to read, and a good source of reference.

Containers—*The Securing of ISO Containers Theory and Practice*

An ICHCA Survey, 1981.

This authoritative guide is in four parts—summary, shipboard securing of ISO containers, going into considerable detail on techniques, tools, wires, lashings, etc., specialist operations and different container types, ship design solutions, regulations and standards. Section 3 examines container terminal practice and section 4 securing for inland transport.

Container Systems

Eric Ratu. John Wiley & Sons.

A book of value for marine surveyors as a reference—contains every possible facet of the intermodal concept.

Dangerous Goods—*The Carriage of Dangerous Goods by Sea*

C. E. Henry. Francis Pinter 1985.

This book has been prepared by the Graduate Institute of International Studies, Geneva, and looks specifically at the role of the IMO in international legislation with respect to dangerous goods. To the operator it is a little formal and academic, but its value lies in the fact that the subject has been exceedingly well researched and the notes and references are of comparable value to the text. In many cases, a surveyor is not reporting just to companies, but to governmental administrations. In this case reports must reflect an awareness of the international approach to safety and this book will help.

Dangerous Properties of Industrial Material

N. Irving Sax. Van Nostrand Reinhold.

Purely a list of chemicals and their technical details and hazardous properties.

Fire Aboard

F. Rushbrook, Brown, Son and Ferguson.

This classic book was first published in 1961 and was updated in 1974, so is limited in its coverage of new detection and extinguishing methods. However, fire prevention remains a question of good operational practices and the valuable list of case histories will provide important pointers for any surveyor investigating fire at sea. Classification and government rules and guidance to surveyors should also be studied.

Fisherman's Manual

4th Edition, World Fishing Publications, IPC.

This softback 120-page A4 publication is a starting point for those who find themselves involved in surveying fishing vessels. The sections cover trawlers and their gear, seine netting, ring netting, fishing with light, set netting, line fishing, shell fishing, fish finding, care of catch, deck machinery, ropes, knots, etc, and a final section in the wheelhouse. Details of the legal requirements for surveying are of course provided by governments.

Government Publications, UK

Many of the publications relate from the time when the UK had the largest Merchant Navy in the world and the publications are both thorough and reflect a high order of quality, which are not really surpassed elsewhere. Typical are the following available from HMSO:

Instructions to surveyors:

- *Aids to navigation on offshore structure* 1980.
- *Crew accommodation* 1979 amended to 1983.
- *Fire appliances* 1981.
- *Fishing vessels* 1975.
- *Lifesaving appliances* 1984.
- *Lights and signalling equipment* 1982.
- *Navigational equipment installations* 1985.
- *Merchant Shipping Radio Installations* 1980.
- *Passenger Ships* 1982.
- *Code of Safe Working Practices.*

Merchant Shipping Notices and Statutory Instruments

M Notices form a detailed compendium of years of experience in the safe operation of ships. They cover everything from examination details to the stowage and carriage of cargoes which are known to be dangerous—e.g., coal slurry. Now published in four volumes, available from HMSO Publications Centre, PO Box 276, London SW8 5DT, UK. (Telephone: 01-622 3316.)

Guide to Government Departments and Other Libraries

Compiled by S. V. King and C. M. Shaw, British Library, Science Reference and Information Service—available in most public libraries.

Inert Gas and Venting Systems

B. W. Oxford. Lloyd's Register, published paper.

This comprehensive paper, which is complete enough to be considered a monograph, covers the history of inert gas, theory, practice, engineering and systems design. It contains a useful glossary of terms as well as information on design, combustion, operation and testing. Part II covers venting systems, flame screens and arresters. The appendix A discusses the scope of survey of various items and B a suggested programme of testing.

Index of Conference Proceedings

British Library Document Supply Centre—available in most public libraries.

International Maritime Organization

Produces: resolutions, technical papers, advisory papers, selected short course training material (see 'useful addresses').

The International Maritime Organization Volumes I & II

Samir Mankabady, Croom Helm.

Volume I gives a good overview of the IMO, its work and conventions, resolutions and protocols. Volume II is a clear and concise appreciation of IMO's rules on collision and the legal aspects of collisions.

International Publications (Others)

Published by Witherby and Co. and cover:

- *Bridge Procedure Guide.*
- *Guides* relating to tanker and terminal operations.
- *Guides* relating to the operation of gas carriers.
- *Helicopter operations.*
- *Measures to Combat Oil Pollution.*
- *Ship-to-ship Transfer Guide.*
- *Tanker Safety Guide* (chemicals).

(See 'useful addresses.')

Lifting Appliances—*Code of Lifting Appliances in a Marine Environment*

Lloyd's Register

This reference work covers derrick systems, cranes and submersible lifting appliances, mechanical lift docks, lifts and ramps, fillings, loose gear and ropes, machinery, electrical and control engineering systems, materials, testing, marking and survey requirements and finally a section on documentation.

Lifting Requirement—*Code of Practice for the Safe Use of Lifting Equipment*

Chain Testers Association of Great Britain.

The reference work is not related specifically to the marine environment but is relevant in many cases. It covers general requirements, chain slings, eyebolts, shackles, lifting beams and spreaders, chain pullies, pulling machines, beam clamps, travelling trolleys, lever hoists and plate clamps. Do's and don'ts are well illustrated and testing methods explained, together with the legal requirements.

Maintenance—*Shipboard Maintenance*

B. E. M. Thomas. Stanford Maritime, 1980.

There are few books dealing with maintenance at sea and this examines the relationship between breakdowns, replacements, preventative measures and maintenance. It is a matter of concern that there are few regulations governing performance standards at sea and this book provides a valuable commentary on the need for systematic maintenance methods. The inventory and recording systems have been overtaken in some companies by computer-based systems, but the principles remain the same. Every surveyor at some time has to ask: How was the equipment maintained?

Nautical Books

Bibliography. Warsash Nautical Book Shop.

A-Z Bibliography is arguably the most comprehensive ever produced by title, author and subject.

Naval Science

The Naval Institute Press, USA.

There is a comprehensive selection of authoritative books from naval architecture to electronic navigation published by the Naval Institute Press, Annapolis, Maryland 21402, USA.

Offshore Legislation—*United Kingdom Offshore Legislation Guide*

Edited by F. Osliff. Benn Technical Books 1983 and updated with supplements.

This is one of the most helpful books available, which has been made possible by a very imaginative approach to legislation by the Department of Energy. The book explains lucidly in Part 1 government requirements and in Part 11 goes into clear language descriptions of relevant Acts of Parliament and Statutory Instruments, definitions of terms used, continental shelf operations, diving safety memoranda, official forms and most comprehensive bibliography embracing relevant information to be found in HMSO, DoE, and DTp libraries, BSI, IMO and the HSE. The whole is well indexed and forms an essential reference work on the legal regime offshore, both fixed and floating.

Oil/Petroleum—*American Petroleum Institute (API)*

Publications and materials index, lists all the publications currently in stock and covers most of the oil and oil-related industries: API Headquarters, 1220 L Street, NW Washington, DC 20005, USA. (Telephone: (202) 682 8375.)

Oil/Petroleum—*Institute of Petroleum, UK*

Catalogue of publications contains valuable information. The series of publications of interest to the surveyor are:

Petroleum Measurement Manual which is produced in parts and covers calculations, calibration, temperature, automatic tank gauging, sampling, density, meters, meter proving, and more specialist measurement techniques, which are dealt with individually.

The *Code of Safe Practice in the Petroleum Industry* is published in parts and each covers electrical safety, marketing, refining, drilling and production, pipelines, airfields, LPG, storage, Bitumen, Pressure vessels, piping, instrumentation. The Institute of Petroleum, 61 New Cavendish Street, London W1M 8AR, UK. (Telephone: 01-636 1004.)

Oil Pollution from Ships

International UK and US Law and Practices.

D. W. Abecassis and R. L. Jarashaw. Stevens and Sons, 1985, 2nd Edition.

This book provides a good overview of both the legal framework and the methods by which compensation is paid and regulated. Any surveyor involved in a pollution incident has not only to be familiar with the technical issues involved, but the legal conditions which apply. Case histories and conventions are covered and the book is designed to outline the legal structure in reasonable detail, highlight points of particular interest or importance and enable the reader to find his starting point for the solution of whatever detailed problem he may have.

Patents—*ABC of Patents*

R. Godfrey. Kenneth Mason, 1972.

It frequently happens that individuals working in a consultancy capacity develop some new idea. By the same token, a number of companies exist on their patent rights alone, employing a unique piece of equipment. This little book enables the reader to understand the patent system. A call to the Patent Office will provide information about the computer record system.

Pollution—*Handbook on Marine Pollution*

E. Gold, Assurance Foreningen Gard, Norway.

This is an exceptionally well laid out book which is divided into nine main sections covering: responsibility; response; environmental code; national responses; national legislation; liability; causes; pollutants and their effects; prevention and reduction. In short, this is an ideal working book from which to establish responsibilities and liabilities and then to take the reader exactly to the appropriate legislation. The appendices provide all the main texts in the international conventions.

Ports—*Design & Construction of Ports and Marine Structures*

Alonzo deF. Zuinn. McGraw Hill.

Basically this is a civil engineer's book but it has a wealth of information from design down to fendering and moorings.

Professional Negligence—*A Guide to the Insurance of Professional Negligence Risks*

D. C. Jess. Butterworth 1982.

Written by a barrister, this book provides a thorough immersion into all the types of risk you might be exposed to and is richly supported by cases (rather frightening). The book covers professional negligence and liability and although not directed specifically to marine surveyors, the message is clear. Insurance, underwriting claims procedures, and duties of the broker are all covered.

Safety Is No Accident

W. H. Tench. Collins 1985.

Written by the Chief Inspector, Aircraft Accident Investigation Branch, the book is a good example of how absolutely impartial accident reporting can be achieved and the level of professionalism which is required.

Safety—*Ship Safety Handbook*

Bureau Veritas compilation published by Lloyd's of London Press, 1986.

This down-to-earth book has been designed to provide a portable ready reference to the latest requirements of Solas and the Colregs. This book with its notes, interpretations, specimen forms and guidance on surveys is intended to lighten the task of owners, officers, surveyors and others connected with ship safety methods—a sentiment which will be warmly shared by all.

Sale and Purchase of Ships Avoiding Snags

H. F. Mansfield. *Marine Engineers' Review*, January 1980, pages 18-20.

This useful article embraces many years' experience of delivery and acceptance of ships. It describes the procedures involved in buying and selling ships. The study of classification records, detailed assessment of the tonnage and the memorandum of agreement embracing some 15 clauses is discussed in detail.

Salvage Law—*Maritime Law of Salvage*

G. Brice. Stevens & Son.

This is one of the definitive works on salvage law. The title may sound heavy but it is actually very readable and interesting.

Salvage Law—*Kennedy on Civil Salvage*

K. C. McGuffie. Stevens & Son.

Coupled with the former is a fund of information on civil salvage law. This book is quite old and may be difficult to come by. Numerous actual cases are given as examples. It is like reading a sea yarn.

Salvage Practice—*Reed's Commercial Salvage Practice*

D. Hancox. Thomas Reed Publications.

The book is probably the most comprehensive guide concerning salvage practice ever to have been written. The 39 chapters cover everything from lifting techniques to administration in two well illustrated volumes. The book has a comprehensive bibliography of its own and is an excellent source of reference.

Survey Handbook—*Lloyd's Survey Handbook (Cargo Damage)*

Edited by K. G. Knight. 4th Ed. 1985, Lloyd's of London Press.

As its name implies, it is a handy first reference when surveying damaged cargo. The first 20 pages contain excellent advice as to how a surveyor should approach the job of surveying damaged goods. Part II covers condensation, infestation, mould, inherent vice, rust, etc., Part III looks at problems with containers and Part IV dangerous and hazardous cargoes (a bit light). Part V commodities—loss and damage. Some useful appendices are provided, together with illustrations.

Surveying—*Marine Surveying Basic Aims and Practice*

J. W. Bull. IMarE, 1967.

This was one of the first monographs to be published dealing with marine surveying and is listed for this reason. Since 1967 the process of surveys have been altered and the traditional role of classification societies, government and independent survey companies is more integrated. The book covers survey organisations, qualifications, for recruits and training, first surveys, re-surveys, damage casualty and condition surveys. The appendices are now out of date but it is worth reading since the principles have not changed.

Stowage—*Thomas' Stowage*

R. E. Thomas, O. O. Thomas and S. Agnew. Brown, Son and Ferguson, Second Ed. 1985.

This book is a completely revised version of the classic *Thomas' Stowage*, known to all deck officers of cargo ships. The new book is metricated and contains a large number of new commodities. The section on damage and claims is particularly relevant. A helpful aspect of this book is its comprehensive index.

Surveys—*Marine Surveys*

C. F. Durham. Fairplay Publications, 1982.

This is a useful book for providing a quick overview of the sort of work that a master mariner might undertake, either independently or as part of a firm. It covers draft surveys, small boat surveys, on and off hire surveys, assessing damage, deciding on the suitability of compartments for cargo and some useful formulae and tables.

Surveying—*Sea Surveying*

Edited by A. Ingham, John Wiley and Sons.

The first book looks at the sea environment, position, sea surveying operations and book two the detailed illustrations. These books are still worth their value for the principles they embrace. More specialist reading can be found in the *Admiralty Hydrographic Surveying Manuals* and the proceedings of the Hydrographic Society in London.

Surveying Small Craft

I. Nicolson. Allard Coles, 2nd Edition, 1983.

There are few books on this subject and so this handy publication is to be valued. In sensible fashion it describes the process of surveying, examination of the hull, wood and fibreglass construction, metal hulls and major component fittings. Boat systems are covered, engines, steering, plumbing etc and associated equipment. A final chapter covers arbitration. The diagrams and photographs are clear and well chosen to illustrate the text.

Survival—*Marine Survival and Rescue Systems*

D. J. House. Cambridge University Press.

Covers evacuation systems and requirements, emergency alarms and on-board training, helicopter operations, boatwork, liferafts, personal survival, rescue boats, medical advice and communications.

S.I. Units

Chiswell and Grigg. John Wiley and Sons, 1971.

This is still a useful book to have at hand when having to make a conversion for which you have forgotten the relevant numbers and relationships!

Terms—*Glossary of Marine Technology Terms*

Published in association with the Institute of Marine Engineers by Heinemann.

This book is one of the best technical glossaries although none ever seems to have the exact acronym that one is looking for.

Valuation—*Ship's Value*

Kaj Pineus. Lloyd's of London Press Ltd, 1986.

A very specialised subject which has been approached by the author in an easily readable manner. Where as the book covers the assessment of ship's values for legal and other purposes, but not technical survey requirements, it is an excellent reference book for the consultant who may require some background before undertaking sale/purchase or salvage surveys.

SOME USEFUL ADDRESSES

The Association of Average Adjusters,
HQS *Wellington,*
Temple Stairs,
Victoria Embankment,
London WC2 2PN, UK.
Tel: 01-240 5516.

Bibliography of Nautical Books,
Warsash Nautical Book Shop,
31 Newtown Road,
Warsash,
Southampton SO3 6FY, UK.
Tel: (04895) 2384.

British Standards Institution,
2 Park Street,
London W1A 2BS, UK.
Tel: 01-629 9000.

Bureau Veritas,
Cedex 44,
92077 Paris La Defense,
France.
Tel: (1) 42 91 52 91.

Chartered Institution of Arbitrators,
75 Cannon Street,
London EC4, UK.
Tel: 01-236 8761.

The Chartered Insurance Institute,
20 Aldermanbury,
London EC2V 7HY, UK.
Tel: 01-606 3835.

Det Norske Veritas,
Veritasveien 1,
PO Box 300,
N1322 Hovik,
Oslo, Norway.
Tel: (010) 472 479 900.

Fairplay Publications Ltd,
PO Box 96,
Coulsdon,
Surrey CR3 2TE, UK.
Tel: 01-660 2811.

A Guide to Marine Transport in Great Britain,
Marine Librarians' Association,
c/o Witherby & Co Ltd,
32-36 Aylesbury Street,
London EC1R 0ET, UK.
Tel: 01-253 5413.

International Cargo Handling Co-ordination Association (ICHCA),
71, Bondway,
London SW18 1SH, UK.
Tel: 01-793 1022.

International Maritime Organization,
4 Albert Embankment,
London SE1 7SR, UK.
Tel: 01-735 7611.

Institute of Chartered Shipbrokers,
24 St. Mary Axe,
London EC3A 8DE, UK.
Tel: 01-283 1361.

Institute of Marine Engineers,
76 Mark Lane,
London EC3R 7JN, UK.
Tel: 01-481 8493.

Institute of International, Container Lessors,
Bedford Consultants Building,
Box 605,
Bedford, NY10506, USA.
Tel: (914) 234 3696

Institute of Petroleum & American Petroleum Institute. (See Oil)

Lloyd's of London Press Ltd,
1 Singer Street,
London EC2A 4LQ, UK.
Tel: 01-250 1500.

Lloyd's Register of Shipping,
71 Fenchurch Street,
London EC3M 4BS, UK.
Tel: 01-488 4796.

The Nautical Institute,
202 Lambeth Road,
London SE1 7LQ, UK.
Tel: 01-928 1351.

Royal Institution of Naval Architects,
10 Upper Belgrave Street,
London SW7X 8BQ, UK.
Tel: 01-235 4622.

Salvage Association,
Bankside House,
107-112 Leadenhall Street,
London EC3A 4AP, UK.
Tel: 01-623 1299

Society of Consulting Marine Engineers & Ship Surveyors,
6 Lloyd's Avenue,
London EC3N 3AX, UK.
Tel: 01-488 3010.

Witherby and Co Ltd,
Marine Publishing,
Book Dept, 2nd Floor,
32-36 Aylesbury Street,
London EC1R 0ET, UK.
Tel: 01-251 5341.

GLOSSARY

SOME OTHER terms used in this document are defined here for the benefit of the reader:

IMO International Maritime Organization. The branch of the United Nations, based in London, dealing with maritime matters.

ILO International Labour Organization. The branch of the United Nations dealing with the well-being of employed persons and all related matters. Based in Geneva.

Inmarsat The International Maritime Satellite Organization, based in London. Deals with all matters related to satellite communications.

New ship Any ship whose keel is laid or construction commenced on or after the date upon which a convention or agreement has entered into force.

Existing ship Any ship constructed before the date upon which a convention or agreement has entered into force and to which the provisions of the convention do not apply.

All ships Both 'new' and 'existing' ships.

Flag State The country of the flag the ship is entitled to fly. The State having jurisdiction over that ship.

Ratification The action of a sovereign country in formally accepting a convention or other international instrument or set of rules.

Adopted The action of IMO or ILO in approving any set of rules to be submitted to governments for possible ratification.

Resolutions Rules created by IMO which represent the member countries' guidance or advice concerning the improvement of safety at sea or prevention of pollution. They are not binding upon governments unless formally adopted, but resolutions are usually incorporated in subsequent amendments to conventions.

Interpretations Agreements reached within IMO as to how the requirements of a particular convention might be applied in practice to achieve uniformity and simplicity.☐

NOTES

NOTES

NOTES